Moses Foster Sweetser

The White Mountains

Vol. 2

Moses Foster Sweetser

The White Mountains
Vol. 2

ISBN/EAN: 9783337292515

Printed in Europe, USA, Canada, Australia, Japan

Cover: Foto ©Andreas Hilbeck / pixelio.de

More available books at **www.hansebooks.com**

THE WHITE MOUNTAINS:

HANDBOOK FOR TRAVELLERS.

A GUIDE TO

THE PEAKS, PASSES, AND RAVINES OF THE WHITE MOUNTAINS
OF NEW HAMPSHIRE, AND TO THE ADJACENT RAILROADS,
HIGHWAYS, AND VILLAGES; WITH THE LAKES AND
MOUNTAINS OF WESTERN MAINE; ALSO, LAKE
WINNEPESAUKEE, AND THE UPPER
CONNECTICUT VALLEY.

With Six Maps and Six Panoramas.

BOSTON:
JAMES R. OSGOOD AND COMPANY,
LATE TICKNOR & FIELDS, AND FIELDS, OSGOOD, & CO.
1876.

PREFACE.

THE main object of this Handbook is to supply the summer-tourist with such information as may render his visit to the White Mountains both pleasant and profitable. The villages and hotels among the mountains have been described with care, and the routes in their vicinity have also been explained. Accounts of the more remote and lofty peaks have been added, to open new fields to adventurous mountaineers. It is hoped that travellers may thus be rendered independent of the services of hotel-keepers, summer correspondents, and guides, and may be enabled to enjoy to the fullest extent the magnificent scenery of the hill-country of New England. Those who wish to study economy of time or money will be enabled to compute the duration or expense of their journey from the data hereinafter given.

The rapid extension of the railroads into this region, and the construction of large and comfortable hotels at favorable points, have given facilities which are now enjoyed by a vast and increasing number of travellers. Summits have recently been visited far in the unbroken wilderness, and others still remain to reward the attack of their first explorer. Mountain exploring clubs have united the more active among the alpestrians of New England, and are making fresh advances over the untrodden heights, combining scientific observations with bold feats of adventure.

This book is not written or modified to aid the interests of any railroad, hotel, or vicinity. So far as our information goes, it is intended to make a fair display of the scenery and advantages of each route, with reference only to its natural capabilities and æsthetic wealth. The Publishers placed large resources at the command of the Editor, with the condition that he should receive no special favors from the common carriers and landlords in the mountain-region. The railroad companies were equally courteous ; and the landlords of the hotels were uniformly obliging and hospitable, as it is their duty, in all cases, to be. And, in the words of Starr King, "We take it for granted, also, that travellers are moved to spend their money and time, not primarily to study the gastronomy of Coös County in New Hampshire, or to criticise the comparative upholstery of the largest houses there ; but to be introduced to the richest feasts of loveliness and grandeur that are spread by the summer around the valleys, and to be refreshed by the draperies of verdure, shadow, cloud, and color, that are hung by the Creator around and above the hills."

The A B C Pathfinder Railway Guide (published by Rand, Avery & Co., Boston) is the best collection of the time-tables of New-England railroads.

The new topographical map embodies the results of the surveys of Bond, Boardman, Hitchcock, Huntington, Vose, Anderson, and others, up to the close of the year 1875, together with the new names given by the mountain clubs of Portland and Boston. The other maps are the result also of much care and research, and will doubtless be of essential aid to the traveller in selecting the best routes and attaining a satisfactory knowledge of the situation of the mountains, lakes, and villages.

The panoramas from Jefferson Hill and Mt. Prospect were made by Prof. J. H. Huntington, who secured perfect accuracy by the use of a camera lucida. The artistic and beautiful views from Mt. Kiarsarge, Mt. Pleasant, and Mt. Washington are careful reductions from drawings which were made by a member of the White-Mountain Club of Portland, for the Geological Survey of New Hampshire. It was intended to have also panoramas from Moosilauke and Mt. Lafayette, but protracted cold and snowy weather settled down when the Guide-book party moved in that direction, and prevented the drawing.

The Editor has thoroughly and carefully explored the greater part of the country described, and has made the ascent of nearly eighty peaks therein, solely with the object of gathering fresh and reliable information. During the summer and autumn of 1875 he secured the assistance and companionship of Prof. J. H. Huntington, the veteran alpestrian and scientist, who also prepared the section of the Introduction which treats of the geology of the White Mountains. Many gentlemen have communicated to the Editor facts of interest about the hill-country; and among these he would gratefully mention Profs. C. H. Hitchcock, C. E. Fay, E. C. Pickering, E. Tuckerman, and G. L. Vose; Messrs. Warren Upham, C. B. Raymond, W. G. Nowell, Lory Odell, and Joel Eastman; and several members of the White-Mountain Club of Portland and the Appalachian Mountain Club of Boston.

But it is impossible to avoid errors in a work of this kind, and it is therefore requested that tourists who find misstatements or grave omissions in these pages will have the kindness to point them out to the Editor, in order that due corrections may be made. The communications of this character already received about the previous volumes of Osgood's American Handbooks have in many instances proved very serviceable.

M. F. SWEETSER,
Editor of Osgood's American Handbooks,
113 Franklin St., Boston.

OSGOOD'S AMERICAN GUIDE-BOOKS.

" There are no better guide-books in the world than Baedeker's, after which these of Osgood's are modelled, and the modelling is intelligent, able, and successful, admirably applying tried methods to new materials." — THE CONGREGATIONALIST, BOSTON.

NEW ENGLAND.

A Guide to the Chief Cities and Popular Resorts of New England, Its Scenery and Historic Attractions; with the Western and Northern Borders, from New York to Quebec. 6 Maps and 11 Plans. Fourth Edition . . . . $ 2.00

THE MIDDLE STATES.

A Guide to the Chief Cities and Popular Resorts of the Middle States, their Scenery and Historic Attractions; with the Northern Border, from Niagara to Montreal. 8 Maps and 15 Plans. Revised Edition 2.00

THE MARITIME PROVINCES.

A Guide to the Chief Cities, Coasts, and Islands of the Maritime Provinces of Canada; with the Gulf and River St. Lawrence; also Newfoundland and the Labrador Coast. 4 Maps and 4 Plans . 2.00

THE WHITE MOUNTAINS.

A Guide to the White Mountains and Franconia Mountains. With full descriptions of each Peak and of the Routes through the "Switzerland of America." 6 Maps and 6 Panoramas . . . 2.00

JAMES R. OSGOOD & CO., Publishers.

CONTENTS.

CONTENTS.

CONTENTS. xiii

MAPS.

PANORAMAS.

ABBREVIATIONS.

M. — Mile, or miles.	hr. — hour.
N. — North.	min. — minutes.
S. — South.	r. — right.
E. — East.	l. — left.
W. — West.	ft. — foot, or feet.

Mt. preceding a name, means *Mount*; but when it comes after the name, it means *Mountain*.

Mt. Clay. — Mount Clay.   Morgan Mt. — Morgan Mountain.

Asterisks denote objects deserving of special attention, or indicate the most comfortable hotels.

THE APPALACHIAN MOUNTAIN CLUB.

EARLY in January, 1876, a meeting of those interested in mountain exploration was called at the Institute of Technology, Boston. This resulted in the formation of a society called *The Appalachian Mountain Club*, having for its objects the exploration of the mountains of New England and the adjacent regions, both for scientific and artistic purposes ; also, in general, to cultivate an interest in geographical studies. Provision is made for five departments of work, each of which is placed under the guidance of a member of the club elected yearly for that purpose. These five, with the other officers of the club, form the council, whose duty it is to superintend the general business of the club.

During the fall, winter, and spring, meetings are held on the afternoon of the second Wednesday of each month ; and throughout the summer field-meetings will be convened at various points among the mountains. The meetings so far held have been well attended, and numerous interesting papers about mountain explorations have been read. Recognizing the importance of a correct system of nomenclature of the mountains, this subject early received the attention of the club, and a committee carefully examined the various names by which many of the peaks are designated. After a full discussion, in which the views of all the available authorities were represented, it was voted to recommend certain names as the most authentic or best established. These names, almost without exception, are those given in this Guide-Book. Of those on which no conclusion was reached should be mentioned Kiarsarge, or Pequawket, and Sandwich Dome, or Black Mountain.

The departments are : Natural History (under Dr. T. Sterry Hunt), embracing all work pertaining to the geology, botany, zoölogy, and entomology of the mountains ; Topography (under Prof. C. H Hitchcock), including the construction of accurate maps of the mountains, and all work necessary for that purpose ; Art (under Prof. C. E. Fay), covering the collection of views of the mountains, improving the prospects by removing obstructions, and aiding in every way possible the true lover of mountain-scenery ; Exploration (under L. F. Pourtales), relating to the formation of parties to visit the more inaccessible points, and otherwise encouraging the opening of new fields of labor ; and Improvements (under W. G. Nowell), in charge of the making of new paths and clearing old ones, building stone, log, and brush camps, erecting correct guide-boards, and other works tending to make the mountains more accessible and attractive. The club will collect books, maps, sketches, photographs, and all available information of interest or advantage to frequenters of the mountains ; and, in order to facilitate comparative study, the library will be extended by the addition of books and maps relating to all parts of the world. The club wishes to connect its special work among the Appalachians with the general results of investigations elsewhere, and will encourage the study of comparative geography in general, opening its meetings to contributions, both scientific and popular, on zoölogical and botanical geography, geology, topography, hydrography, travel, and exploration.

The club has now about 100 members, including both ladies and gentlemen. The annual assessment is $2. In order to carry out the plans of the club and secure uniformity of action, it is desirable that all persons interested in this work shall become members.

INTRODUCTION.

I. Geology of the White Mountains.[1]

By J. H. Huntington.

As the natural scenery of every country, the very feature that attracts the tourist, depends upon its geological structure, every one may well wish to know something about the geology of the country he visits. But one may travel and make extensive tours without map or guide-book, and, at the end of a summer vacation or a year of travel, return to his home and be as profoundly ignorant of the country he has traversed as a man is of astronomy, who has spent his whole life in a mine, and has never seen the light of the sun, or even a geography of the heavens. Such a person of course only cares to travel that he may be able to say, "When I was in Paris," or "When I was at the Yosemite"; "When I ascended Mount Rigi," or "When I went up Mount Washington by rail." Individuals who have such ideas of travel would not care whether a mountain was trap or travertine, granite or gneiss, or whether the noted structures of a famous city were built of adobe or marble, of sandstone or brick. Yet there are tourists who see in every cliff, in every sculptured rock, and in every water-worn pebble, a history they would like to read. The buried cities of the East, before excavations are made, are only mounds of earth, not unlike hundreds of others, but when these mounds are known to hide the secrets of dynasties, that long, long since perished from the earth, they excite an interest that will cause men to undergo the greatest privation and labor, that they may bring to light these treasures of antiquity. Yet the rock-masses whence come the stones of which cities are built have an antiquity compared with which the time from when the most ancient city was founded until to-day is a mere point. In the years that are coming, the tourist who does not know the kind of rocks of the mountain he has ascended, or something of the geological structure of the valley where he has passed his summers, will be regarded very much in the same light as a per-

1 Those who desire a more extended account of the geology of this region are referred to the Report of the Geology of New Hampshire, Vol. I. Physical Geography (C. H. Hitchcock, State Geologist, J. H. Huntington, Principal Assistant, 1874); Vol. II. Stratigraphical Geology; and a Folio of Maps, 1876. Published by the authority of the State of New Hampshire, at Concord.

son now is, who cannot tell the geography of the country in which he has travelled, or something about the scenery of the places he has visited.

In the following pages we shall give (1) a description of the rocks with their mineral constituents, in what appears to be their chronological order, and in this connection we shall note the localities, on the routes of travel, where each variety of rock can be seen in its characteristic form; (2) the present configuration of the land, — the mountains and hills, the gorges and valleys; (3) the minerals and where they are found.

The areas of the different groups of rocks are for the most part exceedingly irregular. This is owing (1) to the wearing away of stratified rocks, and the subsequent uplifting and folding of the strata by forces that acted in different directions; and (2) to the intrusion of granitic rocks that absorbed or partially concealed the stratified rocks.

The Laurentian System.

1. *Porphyritic Gneiss.* — This rock nearly everywhere has the appearance of being granite instead of gneiss, but from its relations to other rocks with which it is associated, and in some instances in the arrangement of its constituents, there is little doubt but that it is a true gneiss. It has a very coarse granular texture. The large crystals of feldspar are from one half to three inches in length, and are generally scattered irregularly through a base of finer material, though sometimes they have their longer axis parallel to the plane of stratification. The quartz is not abundant, and the mica is generally black. Nearly everywhere that the rock is exposed, it is of a rusty brown color from oxidation of the iron that is diffused through it. This rock extends southward beyond the area we are considering. At Lake Village and northward to Centre Harbor, if ledges do not appear, boulders can be seen almost everywhere. At Meredith Village, on account of its being used in building, this rock is very noticeable, and the clear-cut crystals of feldspar appear on every freshly fractured surface. On the road from Centre Harbor to Plymouth, where it begins to descend towards White-Oak Pond, the boulders are remarkably free from iron. N. of Little Squam Lake, where the road turns to the N. beyond the townhouse, the porphyritic gneiss is associated with a ferruginous schist. On Mt. Prospect we have the same association of rocks, but here the crystals of feldspar are much smaller than they are generally in the porphyritic gneiss. On the road from Sandwich Notch towards Sandwich Centre, there are many outcrops, and it appears on the steep southern slope of Whiteface. In Waterville it can be seen on Bald Knob, on Snow's Mountain, and on Cascade Brook, ½ M. above the Cascades. In the N. W. part of Albany, on the extreme northern slope of Passaconaway, where this rock outcrops, the crystals of feldspar are sometimes three inches in length. They are larger here than at any other locality. On the S. E. spur of Mt. Carrigain we find a characteristic variety, and it extends up the

ridge nearly a mile from where it terminates at Sawyer's River. At one point on the river the water falls 25 – 30 ft. over the smooth surface of this rock. This is probably the northern limit of the eastern range of the porphyritic gneiss. Another range of this rock appears in Rumney, on the northern slope of Mt. Stinson, where it is quite free from iron. It outcrops in several places on the road running through the S. part of Ellsworth. It is probably an extension of this range that appears in Thornton, S. of Hatch Hill, and in Woodstock, W. of the Pemigewasset, and farther N. on Bald Mountain. We meet with it on the path up Mt. Lafayette, on the ridge before coming to the Eagle Lakes. Northward it outcrops on the B., C. & M. R. R., between Alder-Brook Station and Wing Road; and it forms the high white bluffs to the W. These localities are the most northern outcrops where the rock has been seen in place.

Bethlehem Gneiss. — The characteristic variety of the Bethlehem gneiss is composed of a light flesh-colored feldspar which largely predominates, chlorite in the place of mica (or if the latter is found it occurs in minute scales), and light-gray quartz. Some of the coarse varieties resemble porphyritic gneiss, but the crystals of feldspar are almost invariably smaller, and are arranged in nodular branches, which are in the midst of the finer micaceous layers of the rock. These layers will usually distinguish this rock from all other varieties of granitic rock found in the White-Mountain area. It has its greatest development in the town of Bethlehem, where it is almost the only rock. The boulders, also, that are so numerous in many places, are of the same kind of rock. On the S. it extends into Franconia; and N., in Whitefield, there are extensive outcrops on the road E. of Kimball Hill. Thence it extends N. E., and outcrops on the road just E. of the Mt.-Adams House, in Jefferson, and also on the hill above the house.

Berlin or Lake Gneiss. — We use this term to designate a gneiss of a coarse foliated texture that is extensively developed in the White-Mountain area. It is composed of light-colored feldspar, dark-gray quartz, and a large proportion of black mica, which occurs in plates of considerable size. On account of the large quantity of mica, the rock is of quite a dark color, though the amount of mica that the rock contains is somewhat variable. It is the rock of the islands of Lake Winnepesaukee and its eastern shores. It forms immense ledges in Wentworth, along the railroad, and extends into Warren; and is found on the road around Mt. Cuba, in Orford. It is the underlying rock of the hills N. of E. Haverhill, and extends through the corner of Benton, and northward through the central part of Landaff into Lisbon. It outcrops again in Littleton, below the village, near the Scythe Factory, on Oak Hill, and on Mann's Hills. It is almost the only rock W. and N. of Whitefield village, and it extends into Lancaster along the western base of Mt. Prospect. It is the rock on the road from Lancaster to Jefferson Hill. N. of the latter place, in the clearing near the

woods, it is cut by the intrusive rocks of the mountain. Northward, in
Berlin and Milan, it is found along the Grand Trunk Railway. There is a
fine outcrop near Milan water-station. On the W. from Milan Corner it is
the country rock, but probably it does not extend northward beyond the
limit of the town. It nowhere rises in high elevations; it is either confined to the valleys, or forms low, rounded hills, of which Bray Hill, in
Whitefield, is a typical example. The gneiss below Berlin station resembles this, but may be the remnant of an older formation. At the station
and above, also N. in Milan, on the railway, there is a gneiss of a different
type: fine-grained, of a dark color, and containing veins of epidote and
calcite, and is evidently a newer rock.

White-Mountain Gneisses. — In the Report of the Geological Survey,
these rocks have been grouped under what is there called the Montalban
series. We find the rocks of this series more extensively developed in our
area than any of those we have already described. The variety of the
rocks is also greater, though the characteristics of each are quite persistent. We find micaceous gneisses verging into mica, and andalusite
schists, genuine gneiss, granitic gneiss, and granite, the latter occurring in
veins. The prevailing rock is a micaceous gneiss. It has generally a
rather coarse granular texture, with well-defined crystalline plates of
mica, and these are often arranged at various angles to the plane of stratification, showing that they were developed subsequent to the sedimenta-
tion of the rock. They also bind the constituents of the rock firmly
together, and being thick-bedded and having few cleavage planes, it is a
remarkably tough rock. In places we find well-defined mica schist interstratified with these gneisses, and elsewhere no less well-defined granitic
gneiss. The granite occurs as coarse granitic vein-stones, and these very
frequently contain tourmaline, and sometimes beryl. Rocks of this series
we find occupying the summit of Copple Crown. Here, as in many places,
it is stained by iron. It extends northward, and is the rock in the
vicinity of Tuftonborough village. It is the rock of Effingham, the N.
part of Ossipee, of Madison, Eaton, and the E. part of Conway. In
Bartlett and the S. part of Chatham, it is cut off by other rocks, but from
the N. part of Chatham it extends W. and N. and forms the summits of all
the peaks of the Presidential Range, of Wild-Cat, the Carter Range, Baldface, Mt. Royce, Imp, and Moriah. It is also the rock of the mountains
in Shelburne, N. of the Androscoggin, and thence it extends N. E. towards
Lake Umbagog. Just S. of the lake it is replaced by an intrusive granite,
but it outcrops in many places on the road from the lake to Bethel.

Western Range. — In Plymouth we have the well-defined micaceous and
granitic gneisses of this series. Above the village, at the quarry, we have
a rare opportunity of seeing the changes that have been effected in the
rock by metamorphism. There is a rounded hill on the side of which the
quarry has been opened in the granitic gneiss, which extends nearly to the

top of the hill. But crowning the summit, instead of the granitic gneiss, in which the lines of stratification are almost entirely obliterated, we have a well-defined micaceous gneiss, the strata of which are set, as it were, into the granitic gneiss. The micaceous gneiss extends northward as far as Waterville. It can be seen at Campton Hollow, on the river opposite W. Campton, and on Mt. Weetamoo, where there is a fine exposure of the characteristic White-Mountain gneiss. To the N. E., the most northern exposure of these rocks is on Sandwich Dome. In Rumney there is a bend of rocks, perhaps converging from that in Plymouth, and it extends N. as far as Benton. The high bluff N. of the river near Rumney station presents a fine exposure of this rock. It differs from the characteristic White-Mountain gneiss, in that some of the strata are quite argillaceous, while others carry fibrolite; and as on Moosilauke and the ridges northward, we have a typical mica schist. It has been suggested that these rocks were formed from material derived from the wearing away of the White-Mountain gneiss. But these rocks, like those in the vicinity of Mt. Washington, contain many coarse granitic veins that carry tourmaline, and very often beryl. Such is a general outline of the rocks of this series. The granitic gneiss can be seen in several places on the road from the Glen House to Gorham. On the Androscoggin River, at Gorham village, it is interstratified with the micaceous gneiss; and there is a fine exposure at Moses Rock, in Shelburne. It is also the rock of Paradise Hill, in Bethel, Me. The mine in Madison, E. of Silver Lake, is in the micaceous gneiss, and this is the rock of the N. part of Jackson; and Grant's Ledge is a characteristic variety. It can be seen at the Glen-Ellis Falls, and at the Emerald and Garnet Pools. At the falls, a short distance above where the water plunges over, there is a wide quartz vein. As this rock will probably be more often observed on the carriage-road and about the summit of Mt. Washington, we shall be a little more specific. In going up the carriage-road, the first rocks we see are more quartzose than the rocks of the mountain generally. At the first mile-post we have gneiss, with granitic veins, while the rock above is almost a quartzite; then the rock approaches mica schist, and the strata present an undulating surface. The hill to the N. E. is coarse granite, that carries tourmaline. After passing a stream we find the rock somewhat ferruginous. A rock that will attract attention farther on is a conglomerate "trap"; then we find coarse intrusive granite, and above the schist has a distinct cleavage, and has beds of granite interstratified. At a sharp turn in the road there are mica schist, granite, and a compact siliceous rock. An examination of the rocks a little beyond reveals a fault. At the second mile-post the rock is thick-bedded, and carries andalusite. At the second turn above the rock is more fissile, the strata are twisted and contorted, and there is a sudden change in the dip. Above, where the road runs northerly, the rock has more the character of a fine-grained sandstone. On the upper side of a long bridge there is a banded granite, then wavy mica schist, followed by coarse granite veins, with tourmaline. Above, where there are many ledges, the undulations appear like waves in a troubled sea. What is remarkable, these wavy strata, after continuing for a short distance, suddenly become vertical. At the third mile-post, where we have again an easterly dip, the smaller waves are transverse to the dip of the strata. Above, for ½ M. we have a series of folds. As we approach the Half-Way House, the ledges are concealed by drift. At the fourth mile-post the strata have a westerly dip. Above, for more than a mile, the curves and folds in the strata are very marked, especially at the fifth mile-post. A spring about eighty rods below this mile-post had a temperature of 37° in August. Towards the summit of the mountain the rocks become thick-bedded, and have both easterly and westerly dips. On the summit there is a small synclinal axis, but it is of little significance as showing the structure of the mountain. In every direction from the summit the rocks have the same general character. Following the ridge over the summits of Mts. Clay, Jefferson, Adams, and Madison, we meet with innumerable folds. On Clay the strata are perhaps more regular. On Jefferson we find that the rocks contain some andalusite, in imperfect crystals, but everywhere the rock is a thick-bedded micaceous gneiss. On the bridle-path to the Crawford House, after we pass Mt. Monroe, the rocks become quite ferruginous, and there appear to be great concretionary masses. These are of a much finer texture than the mass of the rock. The mica-ceous gneiss extends nearly to the Crawford House. The walls of all the deep gorges in the vicinity of Mt. Washington are, for the most part, the gneiss of the mountain. There is also a small area of this rock below Bemis Station, on the P. & O. R. R, and it forms the walls of the flume on Nancy's Brook. This area is entirely surrounded by intrusive granite. The hard granitic gneiss at the Gate of the White-Mountain Notch, and in the two cuts below, the rock of Elephant's Head (at least the top), and that for ¾ M. S. of the Crawford House, probably belong to this series. The rock at Gibbs' Falls is a granitic gneiss, and above is the micaceous gneiss of the mountain.

Labradorite. — It cannot be affirmed with any very great degree of certainty that the labradorite of the White Mountains is a stratified rock, yet it appears to be. It was not suspected as occurring in New Hampshire, until specimens were obtained by me from the great slide in Waterville. Usually, as found in the White-Mountain area, it is a dark granular crystalline rock, and it is so unlike the other rocks that it is easily recognized. The feldspar, when it has not been affected by the weather, is of a dark bluish color, and if examined with a lens, very fine lines (striæ) can be seen on the crystals. With this Labrador feldspar, or labradorite, in some places we find the mineral chrysolite, which can be here recognized by its occurring in grains, and being of an olive-green color. The rock that is composed of these two minerals has been named *ossipyte*. Another variety of the Labrador rock is of a light color, and contains much hornblende and no chrysolite. Every one who visits the slide in Waterville will see the rocks we have here described. Another locality where these rocks occur is on the Mt.-Washington River, some 2 M. from the Saco. If the rock is stratified, it rests on the White-Mountain gneiss. Boulders of labradorite, composed entirely of Labrador feldspar, are found in great abundance on the W. slope of Mill Mountain, in Stark. The beautiful striation of the crystals from this locality is remarkable.

Franconia Breccia. — This rock is made up of fragments of other rocks. In it we find porphyritic gneiss, a dark compact gneiss, hornblende, and other siliceous rocks, and these are cemented together by a light feldspathic paste. In the Franconia region it extends from Mt. Haystack to the Georgianna Falls. In the White-Mountain Notch it outcrops on the side of Mt. Willard. The rock is easily recognized, and it occurs in places of frequent resort. It can be seen just above where the water pours over into the Basin, and it is the rock of the high cliffs at the Profile House. Boulders of it are numerous on the bridle-path on Mt. Lafayette, as it begins to ascend the mountain, and we see the rock in high and overhanging ledges where the path passes through the sharp notch in Eagle Cliff. In the White-Mountain Notch it is cut by the railroad a little more than $\frac{3}{4}$ M. from Crawford Station, and its immense angular fragments can be seen from the cars. It crosses the deep gorge below at Dismal Pool, and appears on the opposite side of the valley in the Silver Cascade, a short distance above the road.

Huronian System.

The rocks of the Huronian system lie altogether to the W. of the Laurentian rock in New Hampshire. They consist of mica, greenish, feldspathic, chloritic, and siliceous schists. The latter pass into hydro-mica schists. These can generally be cut with a knife, and have a greasy feeling.

Mica Schists. — The typical rocks of this class consist of mica and quartz, and can be recognized by the shining surface of the folia. It usually splits readily, on account of the parallel arrangement of the mica. But mica schist verges on the one hand into clay slate, and on the other into gneiss. Rocks nearest to the typical mica schist occur in Errol, on the road along Clear Stream, and on the hills W. of Lake Umbagog. This particular variety seems to be confined to this area, though a similar rock outcrops along the W. side of Kimball Hill, in Whitefield. In the E. part of Lisbon, particularly in the vicinity of Sugar Hill, we find a dark thick-bedded variety that carries large and beautiful crystals of staurolite. On the road S. W. towards Lisbon village, where it crosses Salmon-Hole Brook, the mica schist splits readily into thin plates, and the surfaces of these are thickly studded with garnets. The mica schists that appear in the towns S. along the Connecticut River are exceedingly variable. A high hill in the N. part of Orford has a typical mica schist, and here as on the opposite side of the river below, in Thetford, there is soapstone interstratified with it. Near E. Haverhill, but extending into Piermont, the rock is composed almost altogether of quartz, and it is worked very extensively for scythe-stones. In Northumberland, between Mt. Lyon and the Pilot Range, there is a dark siliceous schist, and this rock is found also on the E. side of the Devil's Slide, in Stark, while a very similar rock forms high and precipitous ledges in the vicinity of N. Stratford

Greenish, Feldspathic, and Siliceous Schists. — These rocks are quite variable. Some are thick-bedded, and have well-defined crystals of feldspar; but chlorite is almost invariably present, and gives the rock a greenish cast. Others are composed largely of quartz, while in places chlorite forms quite a large proportion of the rock. The first two varieties are found interstratified at Groveton, and the last can be seen a mile above Lancaster village. On the S. it forms the high bluff opposite Orford village, whence it extends N. along the river nearly to Bradford, and here it is largely feldspathic. A similar rock is found N. of Lisbon, near the line of Lisbon and Lyman; and in Littleton it is adjacent to the fossiliferous limestone. On the Connecticut River, it outcrops in Vermont near the Portland & Ogdensburg Railroad. In Dalton it is cut by the B., C. & M. Railroad, where it is slightly calcareous. On Mt. Prospect and the range W. in Lancaster, it is highly siliceous. It has the same character near Lancaster, and not far from the spring that supplies the Lancaster House a siliceous limestone is found. The first outcrop in Vermont above the Lancaster bridge is the chloritic variety, but this is limited and is succeeded by the feldspathic variety, which extends along the Connecticut nearly to the line of Stratford. In the N. part of Lancaster hornblende is quite abundant, and on the road that turns N. E. of the Lancaster Fair-Ground massive epidote forms a component part of this rock.

Siliceous and Hydro-mica Schists. — The latter are distinguished by having an unctious or greasy feeling, and generally they are readily cut by a knife. On the one hand they pass into slaty rocks, as at Bradford; and on the other into siliceous schists, as in Stark. Northward from Bradford the cleavage planes become less distinct and the rock is more compact. A cut on the Passumpsic Railroad just below Wells River shows this rock with the strata nearly vertical. There is also a cut through a ledge in the village, on the Montpelier road, where the rock can be seen. The highly tilted strata appear in the Connecticut, at the railroad bridge, and at a narrow gorge $\frac{1}{2}$ M. above, and there are numerous outcrops at Woodsville. It is the principal rock of Bath, and it extends N. and forms the mountain ridge between Monroe and Lyman, including Hunt's and Gardner Mountains. In Dalton, W. of Dalton Mountain, we find a characteristic variety; and northward through Lunenburg and Guildhall, W. of the feldspathic, chloritic schists. From Starkwater station E. along the Grand Trunk Railway, including two extensive cuts near W. Milan, we have what would be called more properly a siliceous schist, though it has some of the properties of hydro-mica schist. The changing but nearly vertical dip of the strata show that in the band, as a whole, there are many folds. A similar rock forms the castellated peaks and spires of Dixville Notch. But here the peculiar cleavage of the strata into longitudinal fragments gives to the Notch its striking features. On the S. wall, half-way through the Notch, there is a profile, sharp in outline and remarkable (as pointed out by Mr. Arthur Fletcher, of Concord) for the change of expression when seen from different points. There is also a siliceous schist of limited area that outcrops in the river in Littleton village and on the railroad above. There is a light-gray siliceous schist in the N. E. part of Lyman that frequently carries gold. This may differ from the schist of Mt. Gardner, already noticed.

Other Rocks. — There is a series of rocks that consists of clay slates, calciferous mica schists, quartzites, and conglomerates, that may belong to the Lower Cambrian rather than the Huronian system. They are described below.

Clay Slates. — Genuine clay slate is a rock quite rare in the White-Mountain area. There is a limited outcrop on the B., C. & M. Railroad, a short distance above E. Haverhill. The quartz veins in the slate of Lyman carry gold, and it is here that the Dodge Mine has been worked for several years. Free gold frequently occurs in the slate adjacent to the quartz. Attempts have been made to work slate for roofing in the N. E. part of Lisbon and in the S. W. part of Littleton. Clay slate passing into argillaceous sandstone is the rock of Dalton Mountain. This may be a newer rock. Below Mt. Willard, on the side of Mt. Willey, the railroad cuts an argillaceous schist. From the summit of Mt. Willard we can trace the line where the granite comes in contact with this schist. The line has a zigzag course and the schist extends W. to Mt. Tom. Boulders of the schist, with crystals of andalusite, can be seen in the stream back of the Crawford House. Boulders containing thin crystals of andalusite, similar to the rock above Mt. Willard, but more indurated, have been found as far S. as Waterville. These, however, probably came from the Pemigewasset region. There is also a small area of slate on the S. side of Mt. Pequawket.

Calciferous Mica Schist. — These rocks consist of bands of argillaceous mica schist, clay slate, hard arenaceous schist, and siliceous limestone. These bands often alternate. There is a small area in the N. E. part of Orford, in the vicinity of Indian Pond. The largest area is in the N. part of Coös County, from Colebrook northward. These rocks underlie the remarkably fertile soil of that region. The absence of boulders, the contour of the hills, and fragments of decaying rock in the drift generally, indicate that this is the country rock.

Quartzite. — This rock when examined by a lens shows that it is made of rounded grains which appear to be imbedded in a glassy-looking cement, while quartz always occurs in veins traversing other rocks, and has none of the granular structure of quartzite. It enters our area from the S., and in Orford it forms almost the entire mass of Mt. Cuba. N. it becomes very narrow, but widens as it approaches Piermont Mountain. The road from Warren to Piermont crosses it, where it passes through the mountain range, W. of the height of land. No outcrops are seen along the B., C. & M. Railroad, but there are enormous masses along the line of Haverhill and Benton. The sharp mountain and precipitous ridge that can be seen from E. Haverhill are formed of it. Between Sugar Loaf and the sharp peak called Black Mountain, it appears as a distinct conglomerate. From Black Mountain it turns suddenly to the E., then to the N., and extends nearly through the town of Landaff, where it disappears.

Conglomerates. — A conglomerate of the most interesting character begins ½ M. S. W. of Young's Pond in Lyman and extends southward into Bath. Nowhere can the foldings, overturns, and dislocations of strata be so well seen as along the line of its outcrop.

Helderberg Rock.

The discovery of fossils in New Hampshire was the most interesting, from a scientific point of view, made by the Geological Survey. The most southern point where the rocks are known to contain fossils is in the Ammonoosuc River, about ½ M. below N. Lisbon. The fossiliferous rocks extend N. E. to an old quarry in the N. part of Littleton. The most interesting locality is on Fitch Hill. It is about fifty rods southerly from the house of E. Fitch, in an open pasture above an orchard. It was here that a fossil was discovered by me that determined the group to which this band of rocks belongs. A band of undoubted Helderberg rock occurs in Lyman, N. from the Dodge Mine, upon a hill near the house of D. Knapp. The rocks that belong to this group are (1) sandstone, (2) coralline limestone, (3) slate, schist, and conglomerate.

Intrusive Rocks.

Whether the intrusive rocks are of later origin than the Helderberg is something that cannot from our observation be determined. But that they are more recent than the Huronian is absolutely certain; for we find them penetrating these rocks in Jefferson, Stark, Stratford, in Granby, Vt., and in various other places. It is an interesting feature of these intrusive granitic rocks that we find them spread out over large areas, and that they follow each other in quite regular succession, as though one overflow of rock had been followed by another of a different kind; and it is worthy of note that the more siliceous or acidic varieties are older than the less siliceous or basic varieties; thus confirming what was long ago pointed out by European geologists in regard to the relative ages of these different kinds of rocks. The names used to designate the different granitic rocks are taken from the names of the places where that variety is the prevailing rock.

Syenite. — This rock, for the most part, is found in the White-Mountain area, and consists of a dark-gray feldspar, hornblende, and mica. The principal locality is on Red Hill, in Moultonborough and Sandwich, and is about $3\frac{1}{2}$ M. in length, and $\frac{1}{2}$ M. wide. It seems to rest on the nearly vertical strata of the granitic gneiss, by which it is surrounded. The syenite appears S. of Lake Winnipesaukee, on Mt. Belknap; and a rock very similar to that of Red Hill is cut by the adit at the tin-mine in Jackson. A syenite differing from the last in the color of its feldspar and in the quantity of mica it carries is found just S. of the village of Colebrook.

Conway Granite. — This granite is a typical variety, — quartz, feldspar, and mica. It is the rock at Diana's Bath, the Cathedral, and the White-Horse Ledge. At Goodrich Falls it decomposes very readily, but on the southerly spurs of Iron Mountain it seems to be firm and compact. Westward, along the valley of the Saco, on the mountains N. of Upper Bartlett, on Mts. Crawford and Resolution, at the Giant's Stairs, and on the side of Mt. Webster, the rock everywhere decomposes. Conway granite is the rock of almost the whole valley of Swift River, W. of the Albany Intervale. At Church's Falls, on Sabba-Day Brook, we see beautifully polished surfaces of this rock. S. of the Intervale, Potash Mountain and to the northward Green's Cliff and Mt. Tremont are made of it. In the great valley of the East Branch of the Pemigewasset, the *terra incognita* of tourists, except on the N. W., this is almost the only rock. But some may penetrate the wilderness as far as the Thoreau Falls or the Mad-River Notch, where it will be found. The range extending S. W. from the Mad-River Notch, including Mts. Osceola, Tecumseh, and Fisher, is composed of this rock, chiefly. It appears in the Franconia Notch, on Mts. Cannon and Profile, at the Basin, the Pool, and the Flume. *The Old Man of the Mountains* is of this Conway granite, here decomposing and friable. The profile is each year crumbling away. In the lower part of the White-Mountain Notch, on Cow, Bemis, and Davis Brooks, there is a reddish granite that resembles the Conway. There is a tabular granite at the Lower Falls of the Ammonoosuc, and a typical variety at the Upper Falls.

Chocorua Granite. — This rock consists of greenish feldspar and quartz. Sometimes it has a little hornblende, sometimes a little mica, and is more like a porphyrite than granite. It forms the mass of Mt. Chocorua, but the character of it is not often apparent, as the feldspar changes by weathering to a dingy white color. In a slide, on the path from Piper's to the summit, the feldspar of the unweathered rock can be seen. It extends westward, on the S. side of Swift River, at least as far as the Champney Falls. It appears on the W. side of Iron Mountain, and on a S. E. spur, and blocks of it can be seen in the abutments of a bridge just above the Iron-Mountain House. Up the valley of the Saco it will be found at Sawyer's Rock, and it is cut by the P. & O. Railroad at Frankenstein Cliff and other places, where the fresh fractures show the greenish feldspar. On Mill Mountain, in Stark, this rock is found, but it passes on the one hand into syenite and on the other into almost a pure feldspathic rock or porphyrite.

Albany Granite. — This rock has everywhere a porphyritic texture, and is composed chiefly of feldspar, both as a paste and as imbedded crystals. It is largely developed in Albany, along the Swift River. It forms the base of Mt. Pequawket and the cliffs of Humphrey's Ledge. It outcrops on the summit of Thorn Mountain, where it comes in contact with a dark siliceous schist, and a very little is seen at Jackson Falls. It appears on the P. & O. Railroad in the Notch, and forms the summit of Mt. Willard. Above the breccia there is a vein cutting the Montalban strata, that resembles the Albany granite. In the Pemigewasset region it is found overlying the Conway granite, and it forms the summits of Mts. Flume, Liberty, and Osceola. On the bridle-path up Mt. Lafayette there is a limited outcrop, just before we come to the porphyritic gneiss and also above the gneiss before we reach the summit, and it appears on the ridge S. This is also the rock of a considerable part of the Ossipee Range.

Other Granite. — A granite, that cannot be classed with either we have described, is found in the upper part of the Waterville Slide. A variety very similar to this will be seen on Mt. Pleasant, Me., but the rock about this mountain is composed chiefly of feldspar.

Porphyrite. — The rocks of this class are always of some dark color, but they often weather so that the surface is quite white. This rock crowns the summit of Lafayette, and is immediately above the Albany granite. It forms the summit of Mt. Lyon (Cape Horn) and the ridge S. of the G. T. Railway in Northumberland, but the specimens from these two localities differ in many respects from similar rocks found elsewhere. The rock of Mt. Starr King, although chiefly feldspar, may belong to the quartz-porphyries.

Porphyries and Breccias, with porphyry as a matrix. — Both these rocks vary greatly in texture, but they are limited to small areas. On the N. E. side of the Ossipee Range, on the ridge most frequently ascended from W. Ossipee, there is a very small area of breccia. In the Pemigewasset region, near the head-waters of the Franconia Branch, on the E. side, there is a considerable area of porphyry, boulders of which can be found along the East Branch of the Pemigewasset. Boulders of porphyry are also found at the slide in Waterville. At Farnum's, in Albany, about eighty rods N. E. of his house, there is a porphyry, with feldspar and quartz; but as we begin to ascend Moat Mountain, from near Farnum's, we find a distinct breccia, made largely of fragments of schist. From this point, until we get within a few rods of the N. peak of the mountain, the rock is almost entirely porphyry, composed of feldspar, with grains of quartz, and very rarely we find fragments of other rocks. On the N. peak the rock is made up chiefly of fragments of schist, the matrix being the porphyry of the mountain. The very distinct breccia is only a few feet in thickness. On the N. E. the porphyry extends nearly to Diana's Bath. On the S. side of Mt. Pequawket there is a band of slate above the Albany granite; and then there are places where the rock is composed entirely of angular fragments of slate, apparently without any cementing material, but the remaining 2,000 feet of the mountain is porphyry, with here and there a fragment of other rock.

Dikes. — At the top of the Crystal Cascade the "trap" forms an interesting breccia. At the head of Tuckerman's Ravine, a "trap" dike of reddish-brown color crosses the path leading from the ravine, just before it reaches the plateau. On the Mt.-Washington River, 1 M. from the Saco, there is a "trap" chiefly of a light-brown color, but on the S. side, where it forms a breccia, it changes to a dark-greenish color. At Berlin Falls, on the railway and in the river, the dikes are numerous and of great width. On the side of the precipitous ledges, N. of the railway, 2 M. above the station, there is an intrusive rock, that in composition is a compact feldspar. The interesting feature of this is, that there is a cave in it which has the appearance of having been artificially excavated. The fact that it is the material that was used by the Indians for arrow-heads lends confirmation to this idea.

Coarse Granitic Veins. — While these are so common there are two or three that deserve special notice. Jockey Cap, in Fryeburg, seems to be one immense vein; and Pine Hill, W. of the village, is largely made of vein-stone. Just above the station at Berlin Falls, there is an immense coarse granitic mass, and extending S. W. it is the rock of Mt. Forist.

Recent Time. — Perhaps the most interesting themes of study, certainly those which will most attract the tourist, are the general features of the country, — the mountain summits, the gorges, the high and overhanging cliffs, the deep-worn channels of the streams, the valleys, the ravines, the travelled boulders, and the general phenomena of the drift. During the cycles of vanished ages, while the immense beds of fossiliferous strata were being formed, a process of slow decay was going on wherever an area of crystalline rocks appeared, so that in some places this decay must have reached the depth of many hundred feet. Then came the great ice age, when the continental glacier was of so vast a thickness that even Mt. Washington was covered by the ice-sheet, and northward the thickness must have been immense. From the molecular pressure in this enormous mass, the great ice-sheet moved southward, or to the point where it thinned out towards its edges. In its onward movement the ice-sheet often carried this decaying rock many miles, sometimes across deep ravines and over mountain-ridges. In this decomposing rock were many hard masses that had resisted decay, and these are the boulders that we find carried so far away from their original beds. How great a thickness of rock was worn away by the ice-sheet we cannot determine with any degree of exactness, as the strata have been pressed into so many folds. That it enlarged and modified many of the ravines and valleys is certain. That local glaciers distributed the boulders in King's Ravine and those in the Carter Notch seems probable. That many of the gorges and valleys must have had their origin in fractures and faults seems also probable, since from the Presidential Range we have valleys radiating in so many directions; and still there is a general resemblance in all those that occur in the same kind of rock, even in those that radiate in opposite directions, and it is only when the rock changes that the main features of the valleys change. Transverse valleys and gorges which break abruptly through the strata, as at Dixville Notch, seem to have had their origin in a fracture, possibly caused by the uplifting of the whole eastern portion of the continent. Gorges like that at Franconia Notch, where they are included between rocks of different geological formations, must have had a primitive origin. There are many valleys that are undoubtedly due to the wearing away of rocks by decay, by water, and by glaciers. Some short, narrow gorges, the so-called flumes, had their origin in the wearing away of trap-dikes. Elsewhere the vertical jointing of the rock allowed the water to penetrate, then freezing and thawing caused it to disintegrate.

Moraines. — Although we have no glaciers, yet the moraines that extend across the valleys show that they once existed, and probably continued long after the continental glacier had retreated far to the N. One of the best marked is in Warren, about a mile above the village, on the road to Piermont. Those about the Twin-Mountain House show a movement different from that of the great ice-sheet, and one that is very noticeable can be seen on the N. side of the river-valley after crossing the bridge at the Fabyan House. On the Connecticut, at Colebrook, and several miles below, the moraines show the retreat of the glacier up the valley.

Minerals. — We notice a few minerals and the places where they can be obtained: *Gold*, Lyman. *Graphite*, Rumney. *Molybdenite*, Whitefield, S. E. of the village. *Galenite*, Rumney (N. E. part), Woodstock, Madison, Lyman (Mt. Gardner), Shelburne. *Spalerite* (zinc blende), Warren, Madison, Shelburne. *Pyrites*, many places. *Chalcopyrite* (copper pyrites), Warren, Lyman (Mt. Gardner), Littleton. *Arsenopyrite* (mispickel), Franconia. *Florite*, Glen, Crawford, and Jackson. *Hematite*, Bartlett (Iron Mountain). *Magnetite* (magnetic iron), Lisbon and Bartlett, and frequently in schist. *Cassiterite* (tin ore), Jackson (rare). *Manganese* (rhodonite) Bartlett. *Quartz* (crystals), Benton, Woodstock, and Greenwood, Me. *Pyroxene*, Warren. *Beryl*, islands of Lake Winnepesaukee, Rumney (N. part), Chatham, in the stream near the path to Baldface. *Amphibole* (tremolite), Warren, Lisbon, and in many rocks. *Chrysolite* (with labradorite), Mt.-Washington River and Waterville. *Garnet*, common in many places. *Cinnamon Garnets*, Warren. *Epidote*, Warren (in boulders). *Mica*, many places. *Labradorite*, Stark (Mill Brook), Mt.-Washington River and Waterville. *Feldspar*, everywhere in the coarse granitic veins. *Tourmaline*, granitic veins about Mt. Washington and Warren, almost massive. *Andalusite*, Mt.-Washington carriage-road, between the first and third mile-posts. *Staurolite*, Lisbon, from Mink Pond to Sugar Hill. Small crystals about Mt. Washington.

The remarkable size of the crystals of some of the minerals and the brilliancy of others have given this and the adjoining region a world-wide distinction.

II. Topography.

The White Mountains cover an area of 1,270 M. of the State of New Hampshire, and are bounded by the lake country and Baker's River on the S., the Connecticut Valley on the W., the Upper Ammonoosuc and Androscoggin Valleys on the N., and the frontier of the State on the E. The contiguous parts of Vermont, and the whole area of New Hampshire N. of the Androscoggin Valley, form mountain districts which are closely connected with this alpine region; and the groups of lofty peaks which run off to the N. E. in Maine give reason for the citizens of that State to claim that "the White Mountains extend to the Kennebec River."

The White Mountains are divided into two nearly equal divisions by the deep and continuous valleys of the Saco and Ammonoosuc rivers, and each of these may be separated into three sections, — the E. division being cut by the Ellis-Peabody and Moose-Israel valleys, and the W. division by the Pemigewasset and E. Branch valleys. The State Geologist considers the mountains as grouped, topographically, in ten subdivisions: 1. The Mt.-Starr-King group. 2. The Mt.-Carter group. 3. The Mt.-Washington range, with a Jackson branch. 4. The Cherry-Mountain district. 5. The Mt.-Willey range. 6. The Carrigain-Osceo'a group. 7. The Mt.-Passaconaway range. 8. The Mts. Twin and Lafayette group. 9. The Moosilauke and Mt.-Profile division. 10. The Mt.-Pequawket area.

The term "White Mountains," which is given to so large an area, and which some would have applied to all the region between the Passumpsic and Kennebec rivers, the lake country, and the sources of the Connecticut, is derived from and more properly applied to the highest range in this district, extending from the N. base of Mt. Madison to the S. base of Mt. Webster, a distance of about 13 M., in the direction of S. S. W. A larger

view of the same range (as entertained by Prof. Hitchcock) considers it as
extending from Pine Mountain, in Gorham, along the line of the Presidential Range, and then crossing the Notch and including the Nancy Range,
Carrigain, Osceola, and Tecumseh, terminating at Welch Mountain, in
Waterville. The length of this chain is about 35 M. Prof. Hitchcock
also groups the western mountains into an irregular chain, about 30 M.
long, which runs from the Sugar Loaves in Carroll, by the Twin Mountains, Lafayette, Profile, and Kinsman, to Moosilauke; and the eastern
mountains form a range nearly 20 M. long, extending from the Carter
Range S. to Iron Mountain, in Jackson.

But geological, political, or geographical divisions are hardly available
for the purposes of a guide-book. Anticlinal lines, township boundaries,
and hill-dividing valleys are naturally of less consequence to tourists than
railways, villages, and hotels. Having, therefore, given these general
ideas of the White Mountains as scientifically considered, the individual
peaks will be treated, in the following pages, in reference to the hotels
and villages, — the objects of attack in relation to the nearest bases of
operations. The popular divisions of the White Mountains, the Franconia
Mountains, and the Sandwich Range will be recognized, and the minor
chains will be considered with relation to the summer-resorts in their
vicinities.

It is claimed that the White Mountains are visible from Chambly, P. Q.,
about 125 M. N. W.; from the hills near Quebec, 165 M. N.; from Mt.
Katahdin, the Camden Mountains, and Mt. Desert, in Maine, the latter
being 155 M. distant; from Po Hill, in Amesbury, Mass., 95 M. S. by E.;
from Mt. Wachuset, 130 M. S. by W.; from Mt. Monadnock, 110 M.;
possibly from Greylock, in Berkshire County, Mass., 155 M. S. S. W.;
from Mount Equinox, in Manchester, Vt., 120 M. S. W.; and from Mt.
Marcy, of the Adirondacks, 140 M. W. by S.

III. Scenery.

Gentlemen who, after visiting the White Mountains, have travelled
among the Swiss Alps, the Sierra Nevada, and the Yosemite, have said
that the scenery of the New-Hampshire highlands never loses its interest
to them, nor is rendered permanently insignificant by contrast with the
other mountain-groups. The foremost charm of the White Mountains is
their almost infinite variety of scenery, inexhaustible in its resources and
unlimited in its manifold combinations. Each of the outer villages and
each of the inner glens commands aspects of the main ranges so distinct
and different as to resemble views in separate lands. Every mile of approach or recession on either of the roads opens a new series of prospects,
each of which has its own peculiar beauty and attractiveness, and reveals
new phases of natural grandeur, new combinations of landscape effects.

The most competent critics say that the proper focal points for the main range are at N. Conway, Bethlehem, Jefferson Hill, and Shelburne, and the effect produced from each of these points is different in almost every regard. From the inner glens of Jackson, the Fabyan House, and the Glen House other and still peculiar arrangements of the main peaks are perceived, — new frameworks, varied groupings, and changed colors. The entering valleys, too, by their wide and sinuous windings, give a series of dissolving views, — as when, ascending the Saco, one turns westward from a course toward the desolate mountains of the Maine border and fronts the needle-like peaks of Chocorua, then passes northward along the ponderous ridges of the Moat Range, toward the graceful pyramid of Kiarsarge, and with Mt. Washington in the far front view, then bends to the W. between the dark peaks of Bartlett and towards Carrigain, the vast watch-tower of the wilderness, and finally ascends the wild gorge to the northward, in many places apparently surrounded by sharp-crested and massive ridges. The feature of *intervales,* the level green meadows which adorn these valleys, is almost peculiar to New England, and adds an element of richness and quiet beauty over which the vast and rugged mountains appear in strongest contrast.

The element of color appears in this region in manifold varieties of combination and brilliance, varying, moreover, not only with the seasons but with the changing hours of the day. The different lithological formations of the mountain-groups also call forth admiration in their scenic aspect, — the brilliant hues of Mt. Webster and Mt. Crawford contrasting sharply with the browns and grays of the higher peaks; and the vivid colors of Moat and Chocorua light up the landscape for leagues. The deadly pallor of the Percy Peaks, the unbroken blackness of Sandwich Dome and Passaconaway, and the blanched crests of Whiteface and Baldface, all enter as elements of variety and interest into the composition of the landscapes. The forests, too, with their intermingling of evergreen and deciduous trees, clothe the slopes of the ridges with changing and restful tints, and fill the remoter valleys with the luxuriant frondage of a primeval wilderness. They give rise, also, to the most magnificent displays of coloring, when the early frosts of autumn and the full ripening of their leaves combine to produce the matchless pageantry of yellow and scarlet, brown and gold, in which the highlands are arrayed. In respect to lakes, also, the New-Hampshire mountains are highly favored above the Colorado and Sierra-Nevada ranges, having two large and navigable sheets of water within sight of their highest crests, and many lesser tarns scattered through the glens and reflecting the storm-beaten peaks above.

Not the least of the charms of this region is the *human* interest that is attached to it, and in which it claims a high precedence over many loftier and more imposing ranges. The valleys which slope away toward the lowlands are occupied by ancient towns whose names have been familiar to many generations of New-Englanders; and the glens are occupied by the descendants of the hardy and heroic ancestry that conquered the wilderness and the savage foe. It is no untamed and unoccupied forest like that from which the Coloradian peaks rise, nor yet a fierce desolation of unfailing ice, such as guards the Alps; — but a land of beauty as well as of majesty, flanked by rich meadow-plains and smiling villages, and whose wildest primeval nooks give birth or sustenance to the great rivers which sustain the life of New England. The traditions of the aborigines and the pioneers have been woven about many of the most interesting of its localities; and the pens of poets and dreamers, scientists and historians, have been busy for over two centuries with these mountains.

The route of the Portland & Ogdensburgh Railroad from N. Conway to the Fabyan House leads through a grandly picturesque region, and affords a comprehensive (though rapid) view of the Saco-River ranges. The Pinkham Notch is less interesting; and the valley of the Ammonoosuc is not as rich in scenery as might be expected from its vicinity to the higher peaks. The stately beauty of the Franconia Notch contrasts strongly with the desolate majesty of the White-Mountain Notch, and deserves close and loving study. The delicate grace and the tender pastoral sweetness of the Pemigewasset Valley is worthy of equal admiration with the bold Tyrolese scenery of the Androscoggin Valley, from Bethel to Berlin; but the idyllic loveliness of the Saco Valley, from Fryeburg to Lower Bartlett, has been sadly marred by the intrusion of unsightly railroad embankments and trestles.

Of all the highways about the mountains that which leads from Gorham or the Glen House to Jefferson Hill is the most renowned for scenic splendor. The drives to the eastward from Gorham and to the northward from N. Conway are full of interest and attractiveness; and the roads from Lancaster to Lunenburg and to Jefferson Hill reveal a series of brilliant panoramas. In the southern ranges the road to Waterville has all the wild surroundings of a mountain-gorge, but is unfortunately a *cul-de-sac*. The road from Centre Harbor to W. Ossipee affords a series of fascinating views of the stately peaks of the Sandwich Range; and the drive from Laconia to W. Alton gives the best views of the lake-country and its environing mountains. The route up the Swift-River Intervale is uninteresting, because high forests overarch the road and conceal the ridges on either side.

The Mountain-shapes. — A gentleman once asked a farmer in one of these glens if he did not enjoy the majestic views around and above him. "Wal, yes," said the rustic; "but if I'd had the sortin' of these hills I'd made 'em a little *peakeder*." This idea will often suggest itself to tourists, who will perhaps weary of the flowing lines and heavy masses of the mountains, and long for a view of the *aiguilles* of Switzerland. It is said that there are but five peaks among the Alps that fall away sharply on all sides; and in New Hampshire the nearest approaches to these are Mts. Adams and Chocorua, Sugar Loaf (in Benton), and the Franconian Haystacks. According to Dr. Jackson, "Deluge after deluge has swept over the surface of the State, rounding the outline of the mountain-masses, smoothing and polishing the rocky strata, and heaping up huge piles of diluvial matter."

The Mountain-views. — After having visited nearly every conspicuous peak in this region, in company, and under favorable atmospheric circumstances, Prof. Huntington and the Editor agreed that each should write down, without mutual consultation, the names of the six mountains from which, in his opinion, the finest views had been obtained. The first-named gentleman wrote the names of Mt. Prospect, Sandwich Dome, Washington, Lafayette, Moosilauke, and Kiarsarge; the Editor preferred Moosilauke, Belknap, Kiarsarge, Washington, Lafayette, and Chocorua (in the order as named). The landscapes presented from the New Hampshire Mountains are of that comprehensible character of which Baron Humboldt says, — "The prospect from minor mountains is far more interesting than that from extreme elevations, where the scenery of the adjacent country is lost and confounded by the remoteness of its situation."

The Lakes afford one of the chief elements of beauty in the highland region, and should be visited by every tourist. The most charming of these is the many-islanded, mountain-walled Squam Lake, which lies between Centre Harbor and Plymouth. It has no peer in all New England. The Ossipee Ponds are less attractive, on account of their dull surroundings and desolate shores. Starr King's wish will be remembered, that the Peabody Glen should be filled with a broad-bosomed lake, in which the noble presidental peaks might be mirrored.

But it is on the outer lakes, Winnepesaukee and Sebago, that the greatest interest dwells. The former is the more beautiful, on account of its multitude of graceful islands and the alpine outlines of the nearer mountains; but Sebago has the unique episode of the passage of the Songo, and views the singularly abrupt peaks of Northern Maine. From either of these lakes Mt. Washington may be seen, surrounded by its brotherhood of far-away blue peaks; and they are equally endowed with pleasant and sequestered villages where the traveller may rest. Comfortable steamboats are found on each, whose tranquil voyages are made through fair and ever-changing scenes.

Visitors to the mountains who demand sensational effects, the close contact of lofty peaks, and the overpowering presence of wild scenery, should stop at Waterville, Upper Bartlett, Jackson, Gorham, or in the glens of the Profile, Crawford, or Glen Houses. Either of these points is surrounded with imposing prospects in narrow horizons, and affords numerous pleasant excursions over rugged peaks or along picturesque falling brooks. A higher artistic pleasure is, however, to be gained from a sojourn at one of the valley villages, — N. Conway, Bethel, Gorham, Jefferson Hill, Lancaster, Bethlehem, or Campton, — at the proper landscape distance from the main ranges, and where beauty and grace are combined with the strength and wildness of the inner glens. Either of these villages would serve as a centre from which many short and interesting excursions might be made, and whence the manifold appearances of the mountains, in storm or sunshine, dawn or moonlight, June or October, could be studied and admired. Let the visitor be provided with a few choice books relative to or suggested by the land in which he is sojourning. Thompson's new edition of the Rev. Benjamin G. Willey's *History of the White Mountains* contains many quaint and interesting stories of the pioneers and their battles with winter and want, storms and floods, the Indians and the wild beasts. *The White Hills*, by the Rev. Thomas Starr King, is perhaps the most fascinating book that has ever been written about these or any other American mountains. The florid beauty of its style is enriched by copious quotations from Ruskin, Wordsworth, Shelley, Goethe, Thoreau, Whittier, and other lovers of nature; and legends of the mountaineers are plentifully strewn through its pages. It is not always exact in its statements of facts (as when, for instance, it makes the Connecticut River empty into *the ocean*, at *New Haven*), and some of the many pictures are more ideal than portrait-like; but there is no other book that will so enable the summer-visitor to enjoy and appreciate the beauty and majesty of the mountains, and to grow richer in æsthetic culture and perception. If the tourist is interested in the natural history of the locality, he can find Hitchcock's noble volumes on *The Geology of New Hampshire* at the house of the town-clerk (or on sale at Eastman's bookstore, in Concord), — telling not only of the rock-formations, but also of the botany, entomology, and meteorology of the State.

'Other countries may possess a richer soil and a gentler sky; but where shall we find the rude magnificence of nature so blent with scenes of entrancing beauty, as among our mountains and lakes? Believe me, it is because our country is yet unexplored, that her scenes of beauty and grandeur, her bright waters and swelling hills, her rich pasturage of living green, mingled with fresh flowers, and skirted with deep and shady forests; her fields teeming with life and vegetation; her mountains rising into the dark blue sky, and blending their summits with the purple clouds; her streams rushing from the hillsides, and hastening to mingle with the sea, or lingering in the solitude of her valleys, and sparkling in the glorious sunshine; — it is because these are unexplored, that they are unsung. The time is not far distant, when the poet will kindle into a rapture, and the painter glow with emotion, in delineating our romantic scenery." (N. A. HAVEN, 1823)

"Compared with the high mountains of the globe, the White Mountains can indeed claim only a very moderate rank, although higher than many of the most famous and venerated summits of the Old World. They are, indeed, scarcely lower than Olympus itself; and their peaks are inhabited by superior names. Though far below the regions of perpetual snow, they are much more elevated than the mountains of England and Scotland. . . . . More than all the other mountains of our country, they have long been an object of interest and curiosity, and every year they are visited by many thousands of our people. Compared with the Alps of Switzerland, they want the immense peaks and ridges covered with perpetual snow, and bathed in all the hues of heaven, and the glaciers, those lakes and rivers of ice invading the warm regions of summer below. But their sides and bases are clothed with one of the most beautiful and varied forests, whose autumnal glories are inferior to those of no part of North America, and are wholly unknown in any country of Europe. The brightest and most varied tints of the American forest are here contrasted and heightened by the dark masses of the spruces and firs, and the bare rocky summits of the mountains.

"The view of the mountains in a clear day, when all their outlines and details are distinctly seen, is only one of their many and various aspects. Their changing garments of clouds of every form, color, and combination, give them their highest beauty and glory. One of the first signals of storm and rain is the gathering of the clouds around their highest peaks, gradually spreading and thickening until the whole mountains are hidden in a gloomy shroud. When the storm is over, and the sun is shining brightly on the country around them, the mountains for a time still remain concealed by the heavy piled-up mass. The dark curtain rises slowly from below, some of the lower peaks are next uncovered, perhaps for a moment the summit of Washington is seen, and at length the veil is broken into fragments, which, growing thinner and thinner, are wreathed about the sides, or settle between the ridges. Sometimes, in calm weather, a broad mantle of white transparent misty cloud, like a thin and finely carded sheet of wool, or like a gauzy veil, is laid over the sides of the mountains. After a thousand combinations of light and shade, sunshine and gloom, the shifting vapor at last dissolves and passes away. The forest below and the mountain-tops are bright and fresh after the rain, the new-born torrents run foaming down the ravines, — the S. W. wind blows dry and soft, and you sit and watch the shadows of clouds sailing above the mountains, or stooping for a moment to kiss their summits as they pass.

"When the rain-storm has been followed by a strong and steady N. W. wind, a little cloudy cap often for several days obstinately adheres to the very summit of Mt. Washington, while all the other summits, and all New England, are under a bright and cloudless sky.

"The rosy light of sunset on the snow, which makes the Alps of Switzerland so glorious, is not very frequent on the White Mountains. But more than once, late in autumn, after the sun had set, and the mountains were becoming dark below, I have seen the whole snowy pyramid of Mt. Washington glowing like a furnace with a bright and intense rose-color, fiery and brilliant, but still soft and most beautiful." (Oakes's *White-Mountain Scenery.*)

"The most romantic Imagination here finds itself surprised and stagnated — every thing which it had formed an idea of as sublime and beautiful is here realized. Stupendous mountains, hanging rocks, crystal streams, verdant woods, the Cascade above, the torrent below, all conspire to amaze, to delight, to soothe, to enrapture, in short to fill ye mind with such ideas as every Lover of nature and every devout worshipper of its author would wish to have." (Dr. Belknap.)

"Now I would ask any of my readers who are candid enough to expose their own ignorance, whether they ever heard, or, at any rate, whether they know anything of the White Mountains. As regards myself, I confess that the name had reached my ears; that I had an indefinite idea that they formed an intermediate stage between the Rocky Mountains and the Alleghanies, and that they were inhabited either by Mormons, Indians, or simply by black bears. That there was a district in New England containing mountain scenery superior to much that is yearly crowded by tourists in Europe, that this is to be reached with ease by railways and stage-coaches, and that it is dotted with huge hotels, almost as thickly as they lie in Switzerland, I had no idea." (Anthony Trollope, 1862.)

IV. History.

The history of the White-Mountain region is hereinafter given in connection with the points on which it is localized. It is enough to say, here, that the vicinity of the main range was avoided by the Indians, from a feeling of superstitious awe.

According to Dr. Belknap, the Indians called these mountains by the name of *Agiochook* (or *Agiocochook*), which the Rev. Edward Ballard interprets "The Place of the Great Spirit of the Forest." President Alden states that the Eastern tribes called the mountains *Waumbekket-methna*, which has been interpreted as "Snowy Mountains." Schoolcraft says that the Algonquins called them *Waumbik*, meaning "White Rocks." Another Eastern tribe called them *Kan Ran Vugarty*, "The Continued Likeness of a Gull."

The Florentine navigator Verrazano was the first European who speaks of having seen the White Mountains. In the year 1524, while cruising along the American coast, and after visiting the present site of Portsmouth, he says: "We departed from thence, keeping our course N. E. along the coast, which we found more pleasant champion and without woods, *with high mountains within the land*." Dr. Kohl, the eminent German antiquarian, says that these were, without doubt, the White Mountains. They were also probably the *Montes S. Johannis* of Michael Lok's map (1582); and are the *montañas* of Ribero's map of the Polus Mundi Arcticus, dated 1529. They also appear on Nicollo del Dolfinato's map, published at Venice in 1560, in the "Navigationi del Mondo Nuovo."

On the great map of Gerard Mercator (Duisburg, 1569), they are firmly drawn, lying to the W. of the splendid city of *Norumbega*.

They are called *Les Montaignes* on the map of the world which was painted on parchment by the Bishop of Viscu, under the orders of Francis I. of France (in 1542). *Montagnas* is found in the place of the White Mountains on Sebastian Cabot's map of the world (1544). The Camden Hills are distinguished, on the early maps, by the title of the Green Mountains (in French or Spanish).

The name of the *Chrystall Hills* was given to the higher peaks by Darby Field (about the year 1642), on account of the gems which he claimed to have seen there. They seem to have received the name of WHITE MOUNTAINS from the sailors off the coast, to whom they were a landmark and a mystery, lifting their crowns of brilliant snow against the blue sky from October until June, and visible from Massachusetts Bay (Mts. Monadnock and Belknap) to the seas beyond Portland. The name of White Mountains first occurs in Josselyn's Voyages, published in 1672. "And there is no ship arrives in New England, either to the W. so far as Cape Cod, or to the E. so far as Monhegan, but they see this Mountain, the first land, if the weather be clear." (CHRISTOPHER LEVITT, 1623.) Dwight says that in his day the mariners of the Eastern coast claimed that they could see Mt. Washington for 30 leagues at sea, or from a point 105 M. distant.

After the exploratory visits of Field and Vines, in 1642, the New-Englanders were too busy in extending their settlements and strengthening their frontiers to pay much attention to these remote wilderness-peaks in a land of enemies. During the French and Indian wars they were occasionally entered by parties of rangers; and the Indians often took refuge within their glens from the merciless forays of the troops of Massachusetts. When the conquest of Canada had been achieved, pioneers began to move into the mountain-region, advancing up the Saco and Pemigewasset Valleys, and facing great perils and trials. Though menaced by the British and Indians on the N., the mountain-towns gave a full quota of soldiers to the Continental army.

" The forests resounded with the woodman's strokes ; the hand of industry rapidly, and as if by enchantment, laid open new fields and erected commodious dwellings ; commerce was extended ; and the means of literary and religious improvement multiplied. . . . . Almost all the roads in which they travelled passed through deep forests and over rough hills and mountains, often over troublesome and dangerous streams and not unfrequently through swamps miry and hazardous ; where wolves, bears, and catamounts obstructed and alarmed their progress. The forests they could not cut down ; the rocks they could not remove ; the swamps they could not causey ; over the streams they could not erect bridges. Yet men, women, and children ventured daily through this combination of evils, penetrated the recesses of the wilderness, climbed the hills, wound their way among the rocks, struggled through the mire, and swam on horseback through deep and rapid rivers." (WHITON.)

In the latter part of the last century the movement of the inhabitants of the cities toward the mountains commenced, the pioneers being the parties of President Dwight and Drs. Cutler and Belknap. Year after year the number of the visitors slowly increased, and the Crawfords, Rosebrooks, and Fabyans found frequent occupation during their short summers in piloting aspiring citizens through the forests to the upper peaks. The completion of better roads and the establishment of comfortable inns aided to swell the tide; and Conway began to see the yearly return of artists. By the year 1850 spacious and well-appointed summer-hotels had been erected at various points, and the stage-lines through the three great notches were well patronized. The Grand Trunk Railway was the first to reach the mountain-land; since which the southern lines have been advancing slowly, until now the region is girded and belted by first-class railroads over which palace-cars traverse the wildest of scenery and whirl around the ridges of the most rugged mountains.

It is impossible to estimate the number of summer-visitors who now enter the White-Mountain region. One railroad alone claims to have carried 160,000 in one season. It is said that over $ 3,000,000 are spent in the State every year by pleasure-travellers. Fogg's *Statistical Gazeteer* says that the annual income from summer-tourists in 17 towns near the White Mountains is $ 636,000; in 16 towns near the Franconia Mountains it is $ 300,000; and in 14 towns in the lake-country it is $ 340,000, — making an aggregate of $ 1,276,000, exclusive of the receipts of several of the great mountain-hotels, the Maine and Vermont border-towns, and the railroads, which would probably swell the sum to above $ 2,500,000.

V. The Indians.

When the first English explorers reached the shores of New England, they found a strong confederacy existing between the various Indian tribes of Maine and New Hampshire, which were then populous and powerful.

The headship of this union was vested in the chief of the Penobscot tribe, who bore the title of *Bashaba*. Soon after the year 1614, however, several war-parties of Tarratine Indians from Acadia advanced stealthily into the Penobscot country, and surprised the royal town at night. The Bashaba and his chief warriors and councillors were slain while fighting, and the power of the Penobscots and the union of the tribes were broken together. According to Sir Ferdinando Gorges's *Description of New England*, a terrible state of anarchy and civil war ensued, the chief sagamores battling with each other for supremacy, while against the divided league foreign enemies made successful campaigns. The valiant Tarratines marched mercilessly throughout the country of the Bashaba, shattering the power of the isolated tribes, and sending their fleets even as far as the Massachusetts coast, where the Indians of Ipswich were harried by a fierce naval foray. " The strong fought for supremacy, the weak for existence. There was no necessity for the war-song or the war-dance. Every brave was compelled to enlist whether he would or not. The signal-fire gleamed on the hill-top. The war-whoop was heard in the valley. New England, before nor since, never saw such carnage within her borders."

The destruction of the villages and their deposits of provisions, and the impossibility of tillage or hunting, caused a wide-spread and desolating famine to fall upon the tribes, already in process of extermination by battle and ambush. In company with the universal war and famine came a mysterious pestilence, which broke out in 1616 on the coast and spread inland in every direction with fatal swiftness. Entire villages were depopulated, and tribes were blotted out. This visitation lasted for three summers, and swept away the strength of all the northern peoples. Morton tells, in his *New English Canaan*, that the bones and skulls that he saw throughout the Massachusetts district made the country seem "a new-found Golgotha."

After the passage of the pestilence and the famine, the remnants of the thirteen tribes of the Connecticut Valley and the White-Mountain region formed a new confederation, designed to resist the Mohawks on the W. and the Tarratines on the E. The noble Passaconaway, formerly a valiant warrior and chieftain, now a venerable and sagacious sagamore of Pennacook, was appointed Bashaba.

The Indians of New Hampshire belonged to the Abenaqui nation, and were called Nipmucks, or fresh-water people, from *Nipe*, "pond," and *auke*, "place." They were divided into 13 tribes, each with its semi-independent chief. The Nashuas lived on the river of that name (meaning "pebbly-bottomed"); the Souhegans occupied the Souhegan Valley (*Souheganash* means "worn-out lands"); the Amoskeags were about Manchester (deriving their name from *namaos*, "fish," and *auke*, "place"); the Pennacooks were at Concord (from *pennaqui*, "crooked," and *auke*, "place"); the Squamscotts were about Exeter (from *asquam*, "water," and *auke*, " place "); the Newichawannocks were on Salmon-Falls River (from *nee*, " my," *week*, " wigwam," and *owannock*, " come "); the Pascataquaukes were toward Dover and Portsmouth (from *pos*, " great," *attuck*, " deer," and *auke*, " place "). " The eighth tribe built a wigwam city at Ossipee Lake (*cooash*, 'pines,' and *sipe*, ' river'), and they were the cultivated Ossipees, with mounds and forts like more civilized nations. A ninth built flourishing villages in the fertile valley of the Pequawket River (now Saco, — from *pequawkis*, 'crooked,' and *auke*, ' place'), and were known as the pious Pequawkets, who worshipped the great Manitou of the cloud-capped Agiochook. A tenth had their home by the clear Lake Winnepesaukee, and were esteemed ' the beautiful Winnepesaukees.' An eleventh set up their lodges of spruce bark by the banks of the wild and turbulent Androscoggin River, and were known as ' the death-dealing Amariscoggins ' (from *namaos*, 'fish,' *kees*, 'high,' and *auke*, ' place '). A twelfth cultivated the Coös intervales on the Connecticut, and were called ' the swift deer-hunting Coosucks ' (from *cooash*, 'pines,' *auke*, ' place ')." The thirteenth were the Pemigewassets.

On Father Ducreux's Latin map of 1660, the Abenaqui nation occupies all the country between the Kennebec and Lake Champlain, including the upper waters of the Androscoggin (*Fluvius Amingocontius*) and Saco (*Choacotius Fluvius*).

" Moſt of theſe Northward *Indians* are between five and ſix Foot high, ſtraight Body'd, ſtrongly compoſ'd, ſmooth Skin'd, merry Countenanc'd, of Complexion more ſwarthy than the *Spaniards*, black *Hair'd*, high Foreheaded, black Ey'd, out-Noſ'd, broad Shoulder'd, brawny Arm'd, long and ſlender Handed, out Breaſted, ſmall Waſted, lank Belly'd, well Thigh'd, flat Kneed, with handſome grown Legs, and ſmall Feet : In a word, take them when the Blood ſkips in their Veins, when the Fleſh is on their Backs, and Marrow in their Bones, when they frolick in their antique Deportments and *Indian* poſtures, they are more amiable to behold (though onely in *Adam's Livery*) than many a trim Gallant in the neweſt Mode; and though their Houſes are but mean, their Lodging as homely, Commons ſcant, their Drink Water, and Nature their beſt Clothing, yet they ſtill are healthful and luſty." (OGILBY's *America*.)

After the abdication of Passaconaway, in 1660, his son Wonnalancet succeeded to the chieftaincy. According to the Puritan fathers, he was " a sober and grave person, of years between 50 and 60. He hath been always loving and friendly to the English." The Apostle Eliot visited him in May, 1674, and preached from the parable of the King's son, after which the Sachem embraced Christianity in a beautiful allegorical address. He lived a pure and noble life, and restrained his warriors from attacking the colonists, even during the deadly heats of King Philip's War. After that struggle, he visited the frontier town of Chelmsford, and asked the minister if it had suffered from attacks. The Puritan answered, " No, thank God." " Me next," rejoined Wonnalancet. At a later day he found it impossible to restrain his people from open hostilities, upon which he gave up the chieftaincy, and retired, with the few families who adhered to him, to St. Francis, on the St. Lawrence River, far away from the crash of war and the undiscriminating fury of the English forays.

He returned to the Merrimac Valley in 1696, but stayed only a short time, finally retiring to St. Francis, where he died.

When Wonnalancet retired, in 1685, Kancamagus, the grandson of Passaconaway, assumed the government. He made several attempts to retain the friendship of the English, as is seen in his letters to Gov. Crandall, but was slighted and ill-treated by them, and finally yielded to the impulses of the martial and patriotic party in the confederation. He organized and headed the destructive attack on Dover in 1686, which was the last terrible death-throe of the Pennacooks ; and was present at the signing of the truce of Sagadahoc, in 1691. He then vanishes from history, and it seems probable that he led the feeble remains of his people to the Abenaqui city of refuge at St. Francis.

"Kancamagus was a brave and politic chief, and in view of what he accomplished at the head of a mere remnant of a once powerful tribe, it may be considered a most fortunate circumstance for the English colonists, that he was not at the head of the tribe at an earlier period, before it had been shorn of its strength, during the old age of Passaconaway, and the peaceful and inactive reign of Wonnalancet. And even could Kancamagus have succeeded to the Sagamonship ten years earlier than he did, so that his acknowledged abilities for counsel and war could have been united with those of Philip, history might have chronicled another story than the inglorious death of the Sagamon of Mount Hope in the swamp of Pokanoket." (Potter's *Hist. of Manchester.*)

The northern tribes of the confederation remained in their ancestral homes for some years longer, under the government of their local chiefs, but were nearly annihilated by military expeditions from the New England towns. (See *Fryeburg, Plymouth,* etc.) They then migrated to Canada, and after their mournful exodus the Saco and Pemigewasset Valleys were opened to the settlers from the lower towns.

"Thus the aboriginal inhabitants, who held the lands of New Hampshire as their own, have been swept away. Long and valiantly did they contend for the inheritance bequeathed to them by their fathers; but fate had decided against them, and it was all in vain. With bitter feelings of unavailing regret, the Indian looked for the last time upon the happy places where for ages his ancestors had lived and loved, rejoiced and wept, and passed away, to be known no more forever."

Concerning Passaconaway, the Great Chief of the Mountain and Merrimac Indians.

The name Passaconaway is derived from two Indian words, *papoeis,* "child," and *kunnaway,* "bear," the Child of the Bear being a fitting chief for the tribes whose ancestral insignia was a mountain-bear. It is estimated that the Merrimac tribes had 3,000 warriors in the year 1600, but the annihilating successions of famine, pestilence, and pitiless invasions of hostile tribes reduced their number, in less than 20 years, to 250 men. There is a tradition that the Mohawks attacked Concord not long before the year 1620, and inflicted terrible damage on the Pennacooks; and a subsequent foray of the western tribes of Passaconaway's league into the land of the Mohawks resulted disastrously.

Passaconaway was probably at the head of the Pennacook confederation before the Pilgrims landed at Plymouth; and Captain Levitt reported having seen him in 1623. In 1629 he and his sub-chiefs granted the coast of New Hampshire to John Wheelwright; and in 1632 he sent in to Boston a culprit Indian who had killed an English trader. In 1642 Massachusetts despatched a strong force to disarm the friendly Pennacooks; but Passaconaway retired to the forest, and there received a just apology from the colonial authorities, after which he voluntarily surrendered his guns. In 1644 he put his "subiects Lands and estates vnder the Gouermt and Jurisdiction of the Massachusetts to be gouerned and protected by them." From this time the forest emperor and mighty necromancer became nominally a sort of Puritan magistrate, administering the laws of the colony upon his astonished liegemen. In 1647 Passaconaway was visited by the Apostle Eliot ("one of the noblest spirits that have walked the earth since the days of the Apostle Paul"), whose preaching deeply impressed the great chief and his sons, and led them to entreat him to dwell with them and become their teacher. He was probably converted to Christianity by Eliot's loving counsels. In 1660, overburdened with years and weary of honors, he abdicated his authority at a solemn senate of the mountain and river tribes holden at Pawtucket Falls. His farewell address to his people was heard by two or three English guests, and was reported by them to have been a splendid piece of oratory. The following sentences are extracted from it:—

"Hearken to the words of your father. I am an old oak, that has withstood the storms of more than a hundred winters Leaves and branches have been stripped from me by the winds and frosts,—my eyes are dim,—my limbs totter,—I must soon fall! But when young and sturdy, when my bow no young man of the Pennacooks could bend,—when my arrows would pierce a deer at a hundred yards, and I could bury my hatchet in a sapling to the eye,—no wigwam had so many furs, no pole so many scalp-locks, as Passaconaway's. Then I delighted in war. The whoop of the Pennacooks was heard on the Mohawk,—and no voice so loud as Passaconaway's. The scalps upon the pole of my wigwam told the story of Mohawk suffering. . . . . The oak will soon break before the whirlwind,—it shivers and shakes even now; soon its trunk will be prostrate,—the ant and the worm will sport upon it. Then think, my children, of what I say. I commune with the Great Spirit. He whispers me now: 'Tell your people, Peace—peace is the only hope of your race. I have given fire and thunder to the pale-faces for weapons.—I have made them plentier than the leaves of the forest; and still shall they increase. These meadows they shall turn with the plough,—these forests shall fall by the axe,—the pale-faces shall live upon your hunting-grounds, and make their villages upon your fishing-places.' The Great Spirit says this, and it must be so! We are few and powerless before them! We must bend before the storm! The wind blows hard! The old oak trembles, its branches are gone, its sap is frozen, it bends, it falls! Peace, peace, with the white man'—is the command of the Great Spirit; and the wish,—the last wish of Passaconaway."

"In reflecting upon the character of the Merrimack Sagamon, the conviction forces itself upon one, that at the head of a powerful confederacy of Indian tribes, honored and feared by his subjects, and capable of moulding their fierce passions to his will, the history of New England would have told another story, than the triumph of our Pilgrim Fathers, had Passaconaway taken a different view of his own destiny and that of his tribe,—and exerted his well-known and acknowledged power against the enemies of his race." (POTTER's *Hist. of Manchester*)

"It is a notorious fact that the English trespassed on his hunting-grounds and stole his lands. Yet he never stole anything from them. They killed his warriors, — yet he never killed a white man, woman, or child. They captured and imprisoned his sons and daughters, — yet he never led a captive into the wilderness. Once the proudest and most noble Bashaba of New England, he passed his extreme old age poor, forsaken, and robbed of all that was dear to him, by those to whom he had been a firm friend for nearly half a century." (LITTLE'S *Hist. of Warren*.)

Goethe or Sir Walter Scott could not construct wilder or more fascinating stories than those that were narrated and believed by the Indians and colonists about Passaconaway. In early life he seems to have been a brave and skilful warrior, and in middle age a *powow*, one of a semi-sacerdotal class among the aborigines whose members were "part philosophers, part magicians, and part wizards." One of the Puritan fathers called him the Indian Balaam, and the parallelism between the two characters is certainly remarkable. When the English settlers reached the Massachusetts coast he put all his necromantic arts in operation against them, but failed so signally that he became convinced that they were under the protection of the Great Spirit, and so avoided a conflict with the more carnal weapons of bow and spear.

"Hee can make the water burne, the rocks move, the trees dance, metamorphise himself into a flaming man. Hee will do more : for in winter, when there are no green leaves to be got, he will burne an old one to ashes, and putting those into the water, produce a new green leaf, which you shall not only see, but substantially handle and carrie away ; and make of a dead snake's skin a living snake, both to be seen, felt, and heard. This I write but upon the report of the Indians, who confidently affirm stranger things." (WOOD'S *New-England Prospect*.)

The legend of the apotheosis of Passaconaway on Mt. Washington suggests the mysterious story of St. Aspinquid, who, according to the tradition, was an Indian sage, born in 1588, converted to Christianity in 1628, and died in 1682. His funeral was on Mt. Agamenticus, and was attended by many sachems, who had a great hunting-feast and brought to his grave 6,711 slain animals, including 99 bears, 66 moose, 25 bucks, 67 does, 240 wolves, 82 wild-cats, 3 catamounts, 482 foxes, 32 buffaloes, 400 otter, 620 beaver, 1500 mink, 110 ferrets, 520 raccoons, 900 musquashes, 501 fishers, 3 ermines, 38 porcupines, 832 martens, 59 woodchucks, and 112 rattlesnakes. On the mountain-tomb was carved the inscription : —

"Present useful ; absent wanted ;
Lived desired ; died lamented."

St. Aspinquid is said to have preached the Gospel for 40 years, and among 66 nations, "from the Atlantic Ocean to the Californian Sea." Mr. Thatcher thinks that Passaconaway and St. Aspinquid were the same, since their age and reputation so nearly agree; and advances a theory that Passaconaway retired to Mt. Agamenticus during King Philip's War, received the name of Aspinquid from the sea-shore Indians, and died a few years later.

The Apostle Eliot and Gen. Gookin saw Passaconaway when he was in the white winter of his 120th year. After his abdication of the Pennacook sovereignty he was granted a narrow tract of land in Litchfield by the Province of Massachusetts, where he lived for a short time. The time and manner of his death are unknown, but the traditions of the Penna-cooks relate that he was carried from them, in the winter season, by a weird, wolf-drawn sleigh, and borne to the summit of Mt. Washington, whence he was received into heaven.

> *Strange man was he!* 'T was said he oft pursued
> The sable bear and slew him in h,s den,
> That oft he howled through many a pathless wood,
> And many a tangled wild and poisonous fen,
> That ne'er was trod by other mortal men
> The craggy ledge for rattlesnakes he sought,
> And choaked them one by one, and then
> O'ertook the tall gray moose, as quick as thought,
> And then the mountain cat he chased, and chasing caught.
>
> *A wondrous wight!* For o'er 'Siogee's [1] ice,
> With brindled wolves, all harnessed three and three,
> High seated in a sledge, made in a trice,
> On Mount Agiocochook, of hickory,
> He lashed and reeled, and sung right jollily
> And once upon a car of flaming fire,
> The dreadful Indian shook with fear to see
> The king of Pennacook, his chief, his sire,
> Ride flaming up to heaven, than any mountain higher."

VI. The Nomenclature of the Mountains.

Men of culture have mourned, for many years, the absurd and mean-ingless origin and associations of the names of the White Mountains. Be-ginning with a misnomer in the title of the whole range, they descend through various grades of infelicity and awkwardness to the last names imposed in the summers of 1874–75. The confused jumble of the titles of the main peaks suggests the society of the Federal City and the red-tape and manœuvring of politics and diplomacy, rather than the majesty of the natural altars of New England; and the Franconian summits are not more fortunate. The minor mountains are for the most part named after the farmers who lived near them, or the hunters who frequented their forests. The names in themselves are usually ignoble, and it may be questioned whether the avocations of a mountain-farmer or a beaver-trap-per are sufficiently noble or so tend to produce high characters as to call for such honors as these. Other peaks commemorate in their names cer-tain marked physical productions or resemblances, and this is certainly a desirable mode of bestowing titles. But the farmers who christened them were men of narrow horizons and starved imaginations, scarce knowing of the world's existence beyond their obscure valleys, and so we find scores of mountains bearing similar names, and often within sight of each other. Others were christened in memory of puerile incidents in the lives of unknown and little men, or of dull legends of recent origin. Some were named after popular landlords and railroad men; some after famous foreign peaks; and some have the titles of the towns in which

[1] Lake Winnepesaukee, or Winnepiseogee.

they stand. Others bear resonant Indian names, the only natural out-
growth of the soil and the only fitting appellations for the higher peaks.

After a brief and superficial study of maps, the Editor has selected the
following series of names now applied to some of the mountains in and
near this region, to show at once their poverty and the confusion resultant
upon their frequent duplication.

The names of early hunters and settlers are preserved on Mts. Stinson,
Carr, Webster's Slide, Glines, Tom (Crawford), Russell, Hatch, Hix,
Bickford, Lyman, Eastman, Snow's, Royce, Carter, Hight, Morse, Orne,
Ingalls, Crawford, Smart's, Kinsman, Big and Little Coolidge, Cushman,
Fisher, Morgan, Willey, Parker, Pickering, Sawyer, Gardner, Hunt.
Probably Welch, Israel, Green's Cliff, and hundreds of names in Western
Maine have a similar origin. There are also summits named for Bill
Smith, Bill Merrill, and Molly Ockett, and Western Maine has an Aunt
Hepsy Brown Mountain. Farther N., where the lumbermen abound,
there are mountains whose popular names are so vile as to be omitted
from the maps.

Among the Indian names now attached to the mountains are Ossipee,
Squam, Weetamoo, Tecumseh, Osceola, Passaconaway, Wanosha, Chocorua, Waternomee, Kinneo, Moosilauke, Pequawket (or Kiarsarge), Aziscoös, Chickwolnepy, Sabattos, and Pemigewasset.

Eminent American statesmen are honored in Mts. Washington, Adams,
Jefferson, Madison, Clay, Monroe, Franklin, Clinton, Jackson, Webster,
Hancock, and Lincoln.

The following mountains bear the names of the townships in which they
are located: Campton, Plymouth, Stratford, Dixville, Randolph, East
Haven, Westmore, Burke. The Percy Peaks, Mt. Cardigan, and Mt.
Dartmouth preserve the ancient names of the towns in which they stand.
Some early legend or simple incident connected with them gave rise to the
names of Resolution, Pilot, Deception, Mitten, Cuba, Sunday, Nancy.

Other groups of names are Cow, Horse, Sheep, Bull, Wild Cat, Caribou, Moose, Deer, Rattlesnake, Sable, Bear, Eagle; Iron, Tin, Ore; Pine,
Spruce, Beech, Oak, Cedar, Blueberry, Cherry.

Popular landlords are commemorated by Hayes, Corning, and Forist;
railroad officials by Anderson and Lyon; mountain-explorers by Agassiz,
Hale, Starr King, Willard, Lowell, Belknap, Carrigain, and Field. Lafay-
ette and Pulaski have their peaks, and so have Seneca and Pliny. Car-
mel, Pisgah, Moriah, and Hor are duplicated in New Hampshire and
Vermont; and Teneriffe and Cape Horn are here hidden from the ocean.
The following names are inexplicable: Puzzle, Silver Spring, Umpire,
Goose Eye, Patience, Sloop (or Slope), Thorn, Young.

Among the mountains which have been named after some physical
peculiarity are the Haystacks, Bald, Table, Giant's Stairs, Double Head,
Gemini, Prospect, Pleasant, Sandwich Dome, Tripyramid, Flat, Cannon

(or Profile), Flume, Potash, Sugar Loaf, Owl's Head, Mist, Sentinel, Cone, Avalanche, Baldcap, Baldface, Copple Crown, the Diamond Peaks, Bowback, Long, Crescent, Cherry, Imp, Surprise, Streaked, Speckled, Twins, Green, Black, Blue, the White Caps, Whiteface, Red Hill, Red Rock.

The last nomenclatural degradation is found in the various Hog-back Mts., and in the villainous names given to the fine peaks of the Ossipee Range, which are called the Black Snouts by the neighboring rustics.

A fruitful source of confusion is the frequent duplication of names on neighboring mountains. Bald Mt. is a common name in New Hampshire and Maine, and is bestowed on dozens of peaks, so that it loses all its distinctive value and expression. Speckled Mt. is a favorite term in Western Maine, where it has several localities. Camel's Hump (or Rump) is found in four places; Tumble Down Dick occurs in Gilead, Peru, and Brookfield; Haystack, in Albany, Franconia, Bartlett, and Westmore; Owl's Head at Lake Memphremagog, in Benton, Carroll, Pemigewasset, Stark, and Jerusalem; Sugar Loaf in Benton, Stratford, and the Twin-Mountain district; the pleasantly suggestive Rattlesnake is in Rumney, Conway, Bartlett, Porter, and Casco. Hog's-Back appears in Benton and Stratford, and on the Maine frontier. Bear Mt. is in Waterford, Albany, Wentworth's Location, and Stoneham. Prospect is at Lancaster, Freedom, Holderness, and Carlisle's Grant. Black Mt. is in Sandwich, Jackson, Lincoln, Newry, Milan, Peru, and Sweden; and there are two of that name in Benton. Pleasant is in Lancaster, the Presidential Range, and Denmark. There are also duplicates of Mts. Glines, Monadnock, Kearsarge, Saddleback, White Cap, Green, Ragged, and Tom; and the favorite prefix " Bald," besides the many cases in which it appears alone, is compounded in Baldface, Baldcap, Bald Ledge, Bald Knob, Bald Peak, and so on.

Still further confusion results from the fact that one mountain sometimes has several names, according to the different villages from which it is viewed, or from other considerations. Thus a certain peak lies between Monroe and Lyman, and on one side it is called Monroe Mt., on the other, Lyman Mt. It is also called Mt. Gardner, because connected with that range; Hunt's Mt., by the U. S. Coast Survey; and Bill Smith's Mt., after a farmer who lives near it. Another peak is called Middle Mt., because it lies between Chocorua and Passaconaway; Bald, because of its bare ledges; Hunchback, for its low and massive form; Deer, by the Albanians; and Paugus because Lucy Larcom so designated it. Still another was named Mad-River Peak by Prof. Guyot, Tecumseh by a village photographer, and Kingsley's Peak by a gentleman by the name of Kingsley, who recently ascended it and fancied that he was its discoverer.

Out of this blind maze of homely and hackneyed names must arise the significant nomenclature of the future. Why should our mountains not bear such noble names as those of Switzerland, the Allaleinhorn, the

Jungfrau, the Mischabel, the Wetterhorn, Monte Rosa, — names which resound like the roll of the avalanche ? Why can we not equal the Helvellyn, Skiddaw, and Catchedicam of English Cumberland on our more stately mountains? It must necessarily be a slow process, but it has already commenced well, and by the second centennial the entire nomenclature of our New-England highlands may be reformed. The Alps of Western America are being badly treated in this regard, and surveyors and geologists are allowing their names to be attached to peaks that rival Mont Blanc. Arizona emulates Maine in its Bill Williams Mt., and the depths have been reached in Mt. Jeff. Davis.

VII. Season.

The most favorable season for a visit to the higher mountains is in July and August, for then the cool air of the highlands affords the most grateful relief to the burning heats of the cities. The hotels and boarding-houses are then filled with guests, and parties are frequently formed to visit the interesting points in the vicinity of each. Metropolitan society transfers its headquarters and its modified ceremonials to the shadows of the mountains, and the villages are filled with busy and exotic life. On account of the clemency of the temperature, camping-parties can then attack the higher mountains and explore the great ravines.

But, for the comparatively few persons who can choose their own time, who have vigorous physical powers, and who love Nature with an ardent and undivided love, the months of September and October will be found more favorable for the visit. Then is the season of the harvests, of the magnificent coloring of the autumnal forests, and of clear and bland air. Accommodations are more easily obtained at the hotels; and whereas in August the transient tourist is often obliged to sleep on sofas or floors in overcrowded houses, in the later months he is sure of comfortable quarters and quiet rest. One of the best times to enjoy the scenery is in late September and early October, after the sky has been cleansed by the equinoctial storm and before the higher peaks have been covered with snow. But at that time there are very few tourists remaining in this region, since the approach of the fall trade and the opening of the schools draw most of the visitors back to their city-homes. The month of June is also more favorable to the lover of nature than the later summer months, because then the air is usually clear and balmy, and the fresh colors of the northern forests appear in their best estate. "From the middle of June to the middle of July foliage is more fresh; the cloud scenery is nobler; the meadow grass has a more golden color; the streams are usually more full and musical; and there is a larger proportion of the 'long light' of the afternoon, which kindles the landscape into the richest loveliness. . . . . In August there are fewer clear skies; there is more fog; the meadows are apparelled in more sober green; the highest rocky crests may be wrapped in mists for days in succession; and a traveller has fewer chances of making acquaintance with a bracing mountain breeze. The latter half of June is the blossom season of beauty in the mountain districts; the first half of October is the time of its full-hued fruitage."

The higher peaks usually retain their snow until the latter part of May, and are cold and damp, forming unfavorable ground for excursions. Sometimes in September, and invariably in early October, the snow revisits them, usually to remain for the next eight months.

Pedestrian tours should be made in June or the autumn months, because the heats of July and August are too intense, even in the higher regions, to render long-continued physical exercise either comfortable or healthy. The crisp and electric air of October is far more invigorating and inspiring than the sultry languors of August, and then the tourist is regaled with the gorgeous richness of the reddening forests, contrasted, oftentimes, with the bright crests of virgin snow upon the loftier peaks. In early June the air is often sparkling and exhilarating; but visitors to the remoter glens and to the forest are liable then to suffer from the attacks of mosquitoes and black flies.

The lake-country of New Hampshire should be visited in the autumn, for the heats of summer usually lie heavily over this section, and render physical exertion unpleasant. It should be explored after the return from the mountains, rather than before, in order to avoid the sultriness of the summer days.

The completion of the railroads through the mountains has rendered easy the task of visiting them during those portions of the year which are not in the "season" proper. Already "autumn-leaf parties" have been formed in the cities for brief excursions through the Notch, and the custom will probably gain in popularity and interest. There is also a peculiar interest in traversing this region in the winter or in March, when the scenery is more than ever alpine, and is brilliant in spotless white. Then the valleys are filled with vast masses of snow, the peaks stand out like white crystals against the lucid blue sky, and the higher ravines are filled with incipient glaciers. Starr King says that at this time the White Mountains become "a mimic Switzerland." The ascent of the ridges and the exploration of the ravines can then be conducted on snow-shoes, the traveller gliding thus over dwarf forests and ragged rocks on a carpet-like covering of snow.

VIII. Pedestrian Tours.

It affords ground for rejoicing to lovers of American physical manhood that walking-parties are beginning to traverse some of the more picturesque districts of the Eastern States, in search of scenic beauty and vigorous health. Foremost among the regions thus visited is that beautiful mountain-land which lies between Lakes Sebago and Winnepesaukee and the Connecticut Valley, including the most majestic and diversified scenery in New England, if not in the Atlantic States. This district affords peculiar advantages to the pedestrian on account of good roads, short distances between villages, an honest and kindly rural population, and vicinity to the great Eastern cities. The scenery is of a most varied and interesting character, passing from shadowy woodlands to fertile intervales and mountain rimmed valleys, and from the breezy distances of the open lake-country to the imposing ravines of the higher ranges of peaks. There is also every variety of walking-ground, whereby the unskilled pedestrian can take short and easy saunterings along the plains of Sandwich and the Pemigewasset Valley, the Cherry-Mountain Road, or the Saco Valley; the practised woodsman can encamp among the unexplored fastness of the great wilderness of Pemigewasset; the alpestrian finds the noble presidential peaks and the far-viewing Sandwich Range, with many a deep, unvisited ravine, offering high rewards to his prowess and endurance; and the fisherman meets the quietest pools, homes of myriads of trout, among the gorges of Albany or through the dolphin-colored Pilot Hills. In the following pages the Editor trusts that every traveller may find something to suit his taste, whether it be for the nobly majestic, the wildly primitive, or the tranquilly beautiful.

When the busy citizen has grown weary under the pressure of business or study, and loses his ability to eat or sleep, or to take pleasure either in present or anticipated comforts, let him visit the mountains and inhale their electric air, forgetting for the month his home-cares, and adapting his thoughts to the ennobling surroundings. The sojourn in a summer-hotel is well and beneficial, but the journey on foot is better, since it gives incessant variety and ever-changing themes of diversion. After a few days of marching, he will cease to complain of sleepless nights or zestless meals, and will find the leathery steaks of the village-inns more delicious than the choicest triumphs of the Parisian *chefs*. The pedestrian tour is of high value to men of sedentary habits, giving them a valuable and needed change of habit, expanding their shrunken lungs, and teaching their limbs pliancy and strength. It is pleasing to see so many of the undergraduates of the New England colleges taking up this form of exercise and visiting the mountains in small squads on active service. In the course of time it may be that the White Mountains shall be as favorite walking-ground as the Scottish Highlands or Swiss Alps now are, and that the nervous American energy may acquire a legitimate strengthening of solid Anglo-Saxon endurance.

An objectionable feature connected with most of the pedestrian tours in this region is the absurd rate of speed at which they are carried on. Several parties of students who have passed through the mountains of late years have made forced marches which would astonish even defeated raiders, retreating through a hostile country. No just perception of the scenery can be obtained by any such foot-cavalry exploits, and the excessive and unusual strain on the physical system is likely to do more harm than good. The average daily march should not exceed 15 M., and should be much less if mountains are ascended. The inexperienced walker should commence moderately; though there are but few healthy men of the cities who cannot march 12-15 M. a day without exhaustion. At the easy rate of 3 M. an hour, this would make 4-5 hours of walking, daily, and this should be done in the cooler hours, perhaps at early morning and towards evening. Ten hours a day are left for other purposes than walking, and for enjoying and comprehending the noble scenery through which the route is laid.

In these days of the advocacy of female suffrage and woman's rights, it needs hardly to be stated that American ladies can accomplish nearly everything which is possible to their sturdier brethren. Among these equalities is the ability for a light pedestrian tour, and already our dauntless sisters have threaded the upward intricacies of Tuckerman's Ravine, ascended the rugged Crawford bridle-path, and visited even the remote peak of Mt. Madison. The Pemigewasset Valley has seen bridal couples sauntering slowly on foot along its Arcadian meadows; and beach-costumed maidens, ranging the woods from their camps half-way down Mt. Washington, have played the part of Oreads among the fair New-England hills.

The custom which once prevailed of tourists walking through the mountains and putting up at farm-houses is no longer in vogue. The alarming development of the "tramp" scourge during the last few years, and the ferocious brutality of many of the tramps, have caused the people of the New-England rural districts to look with suspicion on all unknown footmen, and to make their houses as castles against them. Even the stringency of the laws of New Hampshire against this class of unfortunates has failed to deliver her borders from terrible outrages at their hands. The people are therefore careful to exclude wayfaring strangers from their homes, and, since the rural inhabitants are not adepts at physiognomy or personal analysis, the hungry and foot-sore pilgrim in search of the grand and beautiful in Nature may often be turned away from the farm-house doors, on suspicion of being a ruffian or a burglar. If he enters at all, it will only be after a long and humiliating inquisition, and then rather as a matter of humanity than as a source of interest and pleasure. The Editor has in mind the ludicrous consternation caused on several occasions by the advent of his party among the scanty populations of secluded glens, resulting not only in grave inconveniences to the rustics, but also in compelling the Guide-Book people to march hour after hour by closed-and-barred houses to the more hospitable taverns in the open country.

The following inexpensive inns are within easy marches of each other:
N. Conway, the Washington House, N. Conway House; Jackson (9 M.),
Trickey's Jackson Falls House; Copp's farm-house (14 M.), 3 M. be-
yond the Glen House; Gorham, Eagle House; Mt.-Adams House (11 M.
from Gorham, 10 M. from Copp's); Jefferson Hill (6 M.), Jefferson Hill
House; Lancaster (7 M.), Williams House; Lunenburg Heights (7 M.),
Chandler House; Littleton, Union Hotel; Bethlehem (5 M.), several board-
ing-houses; Franconia (6 M.), Lafayette House; Lincoln (12 M.), Tuttle's;
W. Campton (15 M.), Sanborn's; Plymouth (9 M.), Plymouth House.

The Upper-Bartlett House is inexpensive; and so is the White-Mountain House. Other inns of this grade are named in the subsequent pages.

A fortnight's mixed tour, including the sail across Lake Winnepesaukee
and visits to the summits of nine first-class mountains, could be arranged
as follows: —

First day — By railroad to Weirs, Alton Bay, or Wolfsborough, and thence by
steamboat on Lake Winnepesaukee to Centre Harbor. Second day — Centre Harbor
to Plymouth, by Squam Lake (12 M.). Third day — Plymouth to Mt. Prospect and
Campton Village (12 M.). Fourth day — Campton Village to the Flume House (18 M.).
Fifth day — Flume House to Mt. Lafayette and the Profile House. Sixth day —
Profile House to Sugar Hill and Bethlehem. Seventh day — Bethlehem to Kimball
Hill and Jefferson Hill. Eighth day — Jefferson Hill to Gorham (17 M.). Ninth
day — Mt. Hayes, and Gorham to the Glen House. Tenth day — Glen House to the
top of Mt. Washington. Eleventh day — Descent by the bridle-path to the Crawford
House. Twelfth day — Railroad through the White-Mt. Notch to N. Conway;
ascent of Mt. Kiarsarge. Another route might lead from N. Conway to Jackson
Falls (12 M.), and the ascent of Thorn Mt. (3 M.); Jackson Falls to the Glen House
(13 M.) or Copp's (16 M.), visiting Glen-Ellis Falls and the Crystal Cascade; Glen
House to Gorham (8 M.) and the ascent of Mt. Hayes; Gorham to Jefferson Hill
(17 M.); Jefferson to Lancaster (7 M.); Lancaster to Bethlehem (16 M.), visiting
Kimball Hill; Bethlehem to Lafayette House (6 M.) and Profile House (10 M.);
Profile House to N. Woodstock (9 M.); N. Woodstock to Campton Village (15 M.);
Campton Village to Plymouth (8 M.)

Mt. Washington may be ascended from this route either from the Glen House (by
the highway or Tuckerman's Ravine) or from Bethlehem (by railway); and the White-
Mt. Notch and the Crawford House may be visited during the detour.

The Lake-Country. — Wolfeborough to Alton Bay (12 M.); Alton Bay to Mt.
Belknap and Laconia (20 M.); Laconia to Plymouth by rail, and Plymouth to Centre Harbor by Squam Lake (12 M); Centre Harbor to Red Hill and Centre Sandwich (12 M.); Centre Sandwich to Moultonborough and Melvin Village; Melvin to
Wolfeborough (12 M.) Six days. Meredith Village to Centre Harbor, Red Hill,
and Centre Sandwich; Centre Sandwich to Chocorua Lake; Mt. Chocorua, and
down to W. Ossipee; W. Ossipee to Moultonborough, visiting the Whittier Peak;
Moultonborough to Centre Harbor, and by steamboat to Alton Bay; Alton Bay to
Laconia, by Mt. Belknap.

In the Woods. — The average walking-time in the woods is about a mile
an hour. On level ground among well-developed trees of the second
growth $1\frac{1}{2}$ – 2 M. an hour may be made; but the progress is much slower
among older trees or along ridges. "Fallen timber" is one of the worst
obstacles in this mode of travelling. It is due to fires and strong winds,
and leaves the ground encumbered with piles of tree-trunks and bristling
with spiky boughs. The Editor has struggled without stopping for over
two hours to cross a belt of fallen timber less than $\frac{1}{4}$ M. wide. But the most
effectual barrier which Nature puts before her mountain-shrines is the

dwarf spruce-tree, which is found on some of the higher ridges in compact groves. These trees are 4–7 ft. high, and are armed with sharp and inflexible boughs. They grow very close together, and the spiky limbs are interlaced with each other in an inextricable manner. It is so difficult to force a way through these jungles that travellers generally walk over them when they are low, and creep under when they are high. The Twin-Mountain Range is prolific in these thickets, and other peaks have patches of them, one of which extends across the ridge of Mt. Lafayette, about ½ M. S. of the peak.

"A young forest looks poetic in the distance, especially if it is a birch one, and steeps itself every evening in yellow sunset light. But attempt to go through one, where no path has been bushed out, and your admiration will be cut down, as Carlyle would say, 'some stages.' What with dead trunks that promise foothold, and in which you slump to the knees; *chevaux de frise* of great charred logs that bristle with sharp black spikes; openings of tall purple fireweed, hiding snags that pierce through your boots; snaky underbrush that trips you; intertangled young limbs that fly back and switch your eyes; rocks half covered with moss that wrench the ankles; slanting sticks that lie in wait for your pantaloons . . . ;— the poetry of wild forest-clambering turns out pretty serious prose. It is about like fighting a phalanx of porcupines." (STARR KING.)

In selecting the ground for a forest-camp attention must be paid to the two main requisites, — wood and water. As to the latter, the mountain-streams are always clear and pure, and cold springs may often be found. In very dry weather, when the brooks are exhausted, water may sometimes be secured by following down some old torrent-bed, and looking under the large rocks in its course. Having found water, the next thing to be sought is wood, and the stream may be followed until the desired kinds are met. The first requisite is a "stub," or trunk of a dead fir-tree, many of which are found in these woods, standing upright in white decay, and easily pushed over, but retaining tough dry hearts, suitable for kindlings. During the latter part of the day's march care should be taken to secure broad sheets of birch-bark, with which to start the fire. Then several hard-wood trees (maple, birch, or beech) should be felled and cut up into logs, for the backlogs of the fire. It is desirable to pitch the camp as near as possible to these trees, in order to save the labor of carrying the heavy logs for some distance. Having secured enough fire-wood to last all night, one or two fir-trees are cut down and stripped of their small boughs, which are carried to the tent and laid carefully on its bottom, very much as shingles are fastened on a roof, — the tier nearest the fire having their broken ends toward the fire, but the others having their soft parts toward that side. The weary voyagers thus secure a bed which is soft, springy, and aromatic.

Tourists should be careful about the use of fire in the woods or on the mountains, during dry weather. The camp-fires should be well guarded and kept from spreading among the leaves, and they should be extinguished before being left. Serious and destructive forest-fires have resulted from carelessness in this regard; and, since the woodlands in New Hampshire are in no sense public or State property, the burning of a piece of forest involves a direct loss to its owners. Large bonfires which have been kindled and left on the mountains have spread to the valuable timber below and caused heavy losses. A portion of the Green Hills of Conway was burned over in this manner, and the unlucky tourists who kindled the fire were mulcted in heavy damages.

An arduous but highly interesting excursion, and one practicable only for skilful woodsmen, could be arranged on the Presidential Range, traversing the whole ridge in three days. On the first day an early start should be secured from Copp's, and the party could cross Mts. Madison and Adams, encamping in the ravine beyond. On the second day, cross Mts. Jefferson and Washington, and encamp in Tuckerman's Ravine. On the third day ascend from the ravine to the bridle-path at the base of Washington's cone, and go down over Monroe, Franklin, Pleasant, and Clinton to the Crawford House. This route is feasible only in pleasant weather, since the advent of a storm (even in August) would render an encampment in the higher ravines almost untenable.

IX. Equipments.

The clothing should be strong and well made, with buttons riveted on, and with capacious pockets. If the pedestrian wishes to stop at the best hotels along his route, and cares to be presentable there, he should carry appropriate clothing to don at these places. In case of a prolonged tour he would need a valise for superfluous clothing and books, which could be sent by express from point to point ahead. A reserve of this kind would be found of great aid and value under either of several conceivable contingencies. A flannel shirt, with a rolling collar of the same material, is about all the chest-covering which is comfortable in warm-weather walking. Linen collars and cuffs are quickly melted by perspiration; the waistcoat is quite superfluous; and the coat (a light English shooting-jacket, buttoning across the breast, is the best) should be rolled up and carried on the haversack.

Shoes should be selected with great care, and should fit neatly. A very tight shoe pinches the foot when the blood settles downward, and a loose one is apt to chafe. The bottoms of the soles and heels should be garnished with rows of soft iron hob-nails, to prevent wearying slips while ascending steep grassy hills, and to avert the dangerous falls which sometimes result from climbing inclined ledges with smooth soles. If the nail-heads are of hard iron, there will be a possibility of their acting like skates and causing the very disasters which they were meant to avert. The Congress gaiter is perhaps the best shoe to walk in, since the elastic sides afford support to the ankles and room for the slight expansion of the foot. It does not annoy one by getting untied or by losing off buttons;

it does not harden into a stove-pipe rigidity, as boots do after a rainy walk; and it keeps the ankles dry as low shoes cannot. Woollen socks protect the feet from being chafed better than any other kind. A pair of slippers should be carried to relieve the weary feet at night and when they have been wet. Besides being very comfortable, it is beneficial and refreshing to bathe the feet after every day's walk, and some luxurious pedestrians carry a small bottle of bay-rum with which to cool their ankles at night.

If it is designed to attack remote wilderness-peaks or to penetrate the pathless woods, the traveller should get a pair of duck overalls, for there is no material known among tailors that can resist the attacks of the sharp boughs and rugged crags in these regions.

The Haversack. — The best way in which to carry a limited amount of personal luggage is in a haversack, which may, at times, be used as a knapsack, or carried in the hand. Strong and capacious haversacks of white canvas, neatly ornamented, may be bought at the trunk-stores in Boston for $ 2 50 – 3.00 each. They may be worn as a knapsack by having a small loop inserted near the bottom of the inner side, through which passes the middle of the long strap. Two stout thongs should be attached by their middles to the rings by which the long strap is fastened to the haversack, and upon these the rolled-up coat may be secured during walking-time.

Parties who design to encamp among the mountains should be provided with axes, with which to cut fire-wood and tent-poles. It is expedient to have the axe furnished with a closely fitting cover of thick leather, which may be bound upon the metal by thongs. Dangerous accidents may thus be avoided, and the axe is less likely to be injured by rust.

A knapsack is indispensable to the forest-traveller, on account of the quantity of provisions which he must carry. This equipment should be furnished with a stout frame, in order to prevent the breakage of its contents.

The Editor has encamped in shelter, "A," and Sibley tents, and in bark and bough camps, but for ordinary forest-marching he prefers a light shed-tent, which may be made of nine yards of white cotton drilling, cut into three lengths, which are sewed together at the sides, firmly bound at the ends, and provided with five rope-holes at each end. Canvas tents require heavy and well-adjusted poles, and are burdensome to carry; while bark or bough camps are difficult to make, and the latter are leaky in wet weather; but the shed-tent may be set up in a few minutes with sticks from the woods, — it is light, and it throws off all the rain that can fall on it when in the forest. In front of the open side of the tent the camp-fire is placed, and serves both for cooking purposes and for warming the night air, besides smoking out the black flies and mosquitoes. The weight of a shed-tent of this character is about 2½ pounds. It is set up by running a short pole through the rope-loops on one end, and putting the pole on two forked sticks about 5 feet high, which are set in the ground or supported by two other forked sticks from behind. The tent is then drawn back tightly and fastened to the ground by extemporized tent-pegs through the rope-loops on the lower end. A good shelter is thus formed on the back and overhead, which is generally sufficient in the windless woods, though fir-boughs may also be piled at the sides in case of storm.

A *field-glass* will, of course, be a valuable companion for travellers in this region. It is preferable, for landscape purposes, to a spy-glass or tele-scope, since it opens a broader field, and is more easily manipulated. The small but costly Tolles telescopes are of remarkable power, and may be used by visitors who desire the most distant views. For mountain-work the field-glass should be covered with canvas, or otherwise ad-equately protected against the rocks and trees.

A *rubber overcoat* will be found of service to tourists who make long excursions in the woods and thickets during or after rain-storms. An umbrella, besides being very awkward to carry, is of but little use in these regions, since it cannot protect the clothing from the drenching splashes of wet bushes. The rubber coat averts all trouble from this source, and sheds the rains from above with equal facility, protecting also the field-glass, haversack, and other equipments. It is also comfortable to sleep in during night-encampments whereto no blankets can be carried. The Editor wore an 18-ounce rubber coat (costing $ 10) during all the storms that the Guide-Book survey encountered, traversing many leagues of tangled forests and jungles, wherein woollen clothing was reduced to tat-ters, but so firm and strong was the light water-proof fabric that it escaped without a rent.

Tourists who do not enter the primitive forests, and pedestrians gener-ally, will find a valuable companion in a light, strong-sticked *umbrella*, which can be used as a walking-staff, a shelter against the sun, or a protec-tion against showers. It will also be useful as a support in climbing mountains, though if it is to be devoted to this purpose it should be pro-vided with a stout canvas cover, to protect it against the abrasion of rocks and bushes.

Every one who enters the pathless woods or advances along the remoter ridges should carry a serviceable *pocket compass*, by whose aid he may recover his reckoning in case he gets lost. The cheap compasses that are sold in many of the shops are usually unreliable.

Alpenstocks and mountain-staffs are useful only on mountains where good roads or paths have been made, or above the line of large vegetation. Most of the mountains are covered with woods and thickets, and both hands are needed constantly to keep the boughs and bushes from sweep-ing the face. It is well, therefore, to avoid encumbering one hand with a nearly useless stick.

X. Guides.

There are but a very few guides left among the White Mountains, since most of the popular routes are now so plain and easy as to render that profession unprofitable. Their services command from $ 1.50 to $ 3 a day. The farmers who live near the mountains are usually familiar with the ridges and ravines in their vicinity and will impart information freely, if approached in a proper manner. Oftentimes these men can be engaged as guides over ground where they are acquainted, and they are generally intelligent and reliable.

No one should attempt *alone* the passage of the great ravines or to visit out-of-the-way mountains, since in case of any accident resulting, it might be impossible either to get out or to summon assistance.

XI. Routes.

The rapid extension of the railroads into the mountain-district has sub- stituted for the formerly arduous task of travelling from point to point a luxurious and rapid transit, while by lifting the tourist on higher grades it affords better opportunities for outlooks. The sybaritic traveller now traverses the savage defiles and ascends the rugged valleys while reclining among the cushions of a palace-car, passing thus over ground that was formerly visited only by weary days of horseback-riding on miry and rocky roads. Either of the mountain-resorts may be reached by an easy day's journey from Bostom; and one may breakfast on Beacon Hill and sup on the summit of Mt. Washington.

There are several long *stage-lines* in the mountain-district, including those from N. Conway to the Glen, the Glen to Gorham, Lancaster to Jefferson Hill, Littleton and Bethlehem to the Profile House, Plymouth to the Profile House, and Centre Harbor to W. Ossipee. These are served by large Concord stages, drawn by four or six horses each, and afford to the summer-loiterer the variety of a journey in the olden style. The out- side seats should be secured, if the weather is not threatening, in order to get the views on all sides. It is to be regretted that a stage-line has not been established between Gorham and Jefferson Hill, and over the Cherry- Mountain Road.

Many combinations may be made among the existing railroad, steam- boat, and stage lines, in order to form routes to the mountains. A few of these are set forth below, and full descriptions of their stations and points of interest may be found in the companion to this volume, Os- good's *New England*.

Routes from New York.

1. By the Hudson River and Lake Champlain. This route may be taken by either the N. Y. Central & H. R. R. or the river steamers to Albany; thence northward by the Rensselaer & Saratoga R. R.; up Lake Champlain by the steamboats (or including Lake George by a short detour); and then across Vermont by the Central-Vermont and Wells-River route, or by the Portland & Ogdensburg R. R.

The most direct route from New York to the mountains is that by way of New Haven, Hartford, Springfield, Bellows Falls, White-River Junction, Wells River, and Wing Road. The distances are: New York to New Haven, 74 M.; Hartford, 110; Springfield, 136; Brattleborough, 196; Bellows Falls, 222; White-River Junction, 262; Wells River, 302; Fabyan House, 343; (Lancaster, 344).

Either of the routes from Boston to the White Mountains is available by passengers from New York and the West, and Boston may be reached by the N. Y. & Boston Express Line (via Springfield) or the Shore Line of railroads, or by the steamboat and rail routes via New London, Stonington, or Fall River.

The routes from Portland may also be used by reaching Boston on either of the above-mentioned lines, and passing thence to Portland; or by taking a sea-voyage on one of the vessels of the Maine Steamship Company, which run between New York and Portland via Long-Island Sound and Vineyard Sound (fare $5, meals not included). Portland may also be reached by the new through route from Worcester through Nashua and Rochester, Worcester being on the N. Y. & Boston Express Line, or by the Norwich Line (on the Sound).

From Boston.

The routes from Boston to the White Mountains are:—

1. By through trains on the Boston, Lowell & Nashua R. R. and Concord R. R. to Concord, and thence by the Boston, Concord & Montreal and White-Mountains, N. H., R. R. to the lake-country, the Pemigewasset and Upper Connecticut Valleys, Littleton, Lancaster, Bethlehem, the Twin-Mountain and Fabyan Houses, and the top of Mt. Washington (see Route 2).

2. By the Eastern R. R., along the sea-coast to Portsmouth, and thence to Wolfeborough and N. Conway, connecting at the latter point with trains on the P. & O. R. R. for the Notch and the top of Mt. Washington (see Route 1).

The Boston & Maine R. R. does not approach the mountains, but it has a branch to Alton Bay, on Lake Winnepesaukee, whence tourists can cross the lake by steamer to Wolfeborough (on the Eastern R. R.) or Weirs (on the B., C & M. R. R.).

The routes from Portland are also available to travellers mountainward from Boston, Portland being easily reached either by the Eastern R. R., the Boston & Maine R. R., or the comfortable vessels of the Portland Steam Packet Company.

From Portland.

1. The Portland & Ogdensburg R. R., passing Sebago Lake and N. Conway, and traversing the Notch (see Route 5).

2. The Grand Trunk Railway, following the upper Androscoggin Valley, and passing Bethel, Gorham, and Groveton (see Route 6)

Passengers from points E. of Portland can either take one of the above-mentioned routes or go to Boston by steamer, and take one of the routes from that place.

Passengers from Quebec visit the mountains by the Grand Trunk Railway. Passengers from Montreal can take either the Grand Trunk Railway (Route 7) or the Southeastern-Counties Railway (Route 8), in the latter case changing cars at Wells River.

Passengers from Niagara Falls and the West can use either of the routes from Montreal or Albany, or pass to the mountains by way of Boston.

XII. Round Trips.

The railroads that pass near the mountains make series of excursion-routes every summer, with reduced rates on the whole circuit, and coupon-tickets which cover the entire trip. Some of them publish little pamphlets containing lists of these routes and details of their combinations, with the prices in each case, — and these lists are given or sent *gratis* on application at the offices of their general passenger agents. Among these are the Central Vermont R. R. (office at 322 Washington St., Boston), the Boston, Concord & Montreal R. R. (5 State St., Boston), the Portland & Ogdensburg R. R. (offices on Exchange St. Portland), and the Grand Trunk Railway (at 280 Washington St., Boston).

The following rates are according to the *Eastern R. R.* tariff of 1874, and are here given in order that the tourist may gain some idea of the expense of the journey. (The completion of the P. & O. R. R. from Bemis Station to the Fabyan House will reduce the expense in that section somewhat, by substituting railroad-travelling for staging.)

1. Eastern R. R , Boston to Wolfeborough and return, $5 00.
2. Eastern and Grand Trunk Railways to Portland and Gorham, and stage to Glen House, $ 7.00.
3 Eastern R. R. to N. Conway and stage to Glen House, $ 8.00.
4. Eastern R R. to N. Conway, P. & O. R. R. to Portland, Eastern R. R. to Boston, $ 10.00.
5. Eastern R. R. to Wolfeborough, steamer to Centre Harbor and return, Eastern R. R. to Boston, $ 5.50.
6. Eastern R. R. to Wolfeborough, steamer to Weirs, B., C. & M. and Lowell R. Rs. to Concord and Boston, $ 6.80
7. Eastern R. R. to N. Conway and return to Wolfeborough, steamer to Weirs, B., C. & M. and Lowell R. Rs. to Concord and Boston, $ 10 75.
8. Eastern R R. to N. Conway, stages to Glen House and Gorham, Grand Trunk Ry. to Portland, Eastern R. R. to Boston, $ 15.00.

The "Summer Excursionist of the *Central Vermont Railroad* " is the best and most copious list of routes in New England, being printed on tinted paper, and embellished with pictures of many of the summer hotels. It is given or sent free on application to the office, and contains the details and prices of over 350 excursion-routes, from which the following five are taken (edition of 1875).

1. By rail from Boston to Concord and Plymouth ; stage to the Profile House and Littleton ; rail to Wells River, White-River Junction, Bellows Falls, Fitchburg, and Boston. Fare for the round trip, $ 16.65.
2. By rail from Boston to Concord and Weirs ; steamer to Centre Harbor : stage to W. Ossipee ; rail to N. Conway, Fabyan House, and Bethlehem : stage to Profile House and Littleton ; rail to Wells River, Bellows Falls, Fitchburg, and Boston. Fare, $ 22.25.
3. By rail from Boston to Fitchburg, Bellows Falls, Wells River, and Littleton ; stage to Profile House and Bethlehem ; rail to Ammonoosuc station and up Mt. Washington ; carriage to Glen House ; stage to N. Conway ; rail to W. Ossipee ; stage to Centre Harbor ; steamer to Weirs ; rail to Concord and Boston. Fare, $ 29.15.
4. Rail from Boston to Fitchburg, Bellows Falls, Wells River, and Littleton : stage to Profile House and Bethlehem ; rail to Fabyan House and back to Boston by Fitchburg. Fare, $ 19.00.
5. Sound steamer to New York ; Hudson-River steamer to Albany ; rail to Niagara Falls ; rail to Lewiston ; steamer to Toronto : rail or steamer to Montreal ; steamer to Quebec and back ; rail to Montpelier, Wells River, and Littleton ; stage to Profile House and Bethlehem : rail to Fabyan House, Crawford House, N. Conway, and Wolfeborough ; steamer to Centre Harbor and Weirs ; rail to Concord and Boston. Fare, $ 55.00.

XIII. Hotels and Boarding-houses.

The White-Mountain region is provided with numerous large and first-class hotels, in which nearly every luxury of the cities may be found, together with cuisines of far-famed excellence. The chief of these are the Kiarsarge House, at N. Conway; the Glen House, in the Peabody Glen; the Crawford House, at the White-Mountain Notch; the Fabyan House, on the Ammonoosuc plain; the Twin-Mountain House, in Carroll; the Profile House, in the Franconia Notch; and the Pemigewasset House, at Plymouth. The rates of board at these great caravansaries are $4 – 4.50 a day, with varying reductions for tourists who remain for some weeks, — dependent on the location of rooms and the number occupying them, and on the duration of the sojourn. Large reductions in the price of permanent board are made in June and September, before the opening and after the close of the popular season of travel. During the height of the season the hotels are usually crowded, and transient guests are sometimes obliged to sleep in the offices or parlors. In several of these houses the waiters are students of Dartmouth and Bates Colleges, or young ladies from the New-Hampshire academies and seminaries.

Smaller than either of the aforementioned houses, yet in many respects of a high grade of excellence, are the following hotels: The Intervale, McMillan, and Sunset Pavilion, in and near N. Conway; the Conway House, at Conway Corner; the Plaisted and the Waumbek, at Jefferson Hill; the Lancaster, at Lancaster; the Sinclair and the Maplewood, at Bethlehem; and the Sumner, at Dalton. These houses charge about $3 a day, and make large reduction for permanent boarders. They are well-conducted and comfortable, and, with perhaps two exceptions, have richly provided tables.

There are many second-class inns among the outer valleys, whose rates for transients vary from $1.50 to $2.50 a day, and their accommodations are subject to a like variation.

The villages on the outer borders of the mountain-district have several comfortable hotels, among which may be named the Chandler House, at Fryeburg; the Oxford and "The Elms," at Bethel; the Oak-Hill and Thayer's, at Littleton; the Spring House, at Newbury; and the Bear-camp-River House, at W. Ossipee.

Boarding-houses among the mountains are numbered by hundreds, and are found in all the villages and out on the roads on every side. They vary from the large and commodious buildings which have been erected near the favorite resorts down to the story-and-a-half farm-houses, in whose antiquated chambers small families can take shelter. The usual rates in the better houses are $7 – 10 a week, and in the farm-houses $5 – 7 a week. Lists of these houses will be found in the subsequent pages. Among the best of them are Willey's and Seavey's, at N. Conway; the groups at Kiarsarge Village and Lower Bartlett; the Thorn-Mountain, at Jackson; Lary's, at Gorham; Gates's, in Shelburne; the Hillside, at Lancaster; the Starr-King and Jefferson-Hill Houses, at Jefferson Hill; the Mt.-Adams House, in Jefferson; Dodge's and Fiske's, near Whitefield; Goodenow's, at Sugar Hill; the Franconia and Lafayette Houses, in Franconia township; the Bellevue, Strawberry-Hill, Mt.-Agassiz, Prospect, and others, at Bethlehem; Parker's and Fox's, in Woodstock; Greeley's, at Waterville; Merrill's and Foss's, in Thornton; Foss's, Chase's, Mitchell's, and Sanborn's, in Campton township; Blair's, near Plymouth. All the Connecticut-Valley towns have similar houses. The number of farm-houses is large where people can get plain fare and old-time surroundings in return for a very moderate compensation.

The Mountain-top Hotels. The Mt.-Washington Summit House has probably as many of the comforts of a first class hotel as its situation and climate allow. It is a large and firmly built house, well warmed, but indifferently ventilated; and sets a good table. The hotel on Mt. Pleasant (Maine) is in many respects the most comfortable of the summit-houses, having broad verandas and neat environs. The house on Mt. Kiarsarge is smaller and less comfortable, but meals and shelter can be obtained there. The Mt.-Hayes House is a ruin; and that on Mt. Moriah has disappeared. The Prospect House, on Moosilauke, was deserted and in the last stages of dilapidation at the time of the Editor's visit, but its proprietors intend to restore it to comparative habitability.

The Lake-Country. — The chief hotels near Lake Winnepesankee are the Senter House, at Centre Harbor, and the Pavilion, at Wolfeborough, both of which are large and first-class hotels, pleasantly situated with regard to views over the water. The Moulton House, at Centre Harbor, and the Glendon House, at Wolfeborough, are smaller summer houses, comfortable and pleasant, and charging $3 a day. The Laconia Hotel and the Willard House are at Laconia, near Lake Winnesquam. The Bearcamp-River House (formerly Banks's) is pleasantly situated at W. Ossipee station, not far from the Ossipee and Silver Lakes.

Boarding-houses are scattered plentifully through the lake-country, most of them being at Wolfeborough, Centre Harbor, and Sandwich. Long Island has two large houses; Meredith has two; Moultonborough and Tamworth have several; and the Bay-View House, near Laconia, is one of the best. On the shores of Squam Lake there is but one small cottage for the reception of boarders.

New hotels should be erected at Gorham, Upper Bartlett, and Warren, three of the best centres of mountain-excursions. Large and first-class boarding-houses should be opened on Sugar Hill, whence one of the grandest of all the mountain-views is gained; on Shepard Hill, whence is obtained the most fascinating prospect of Squam Lake; in Shelburne, near the Lead-Mine Bridge; and in Chatham.

XIV. The Villages.

N. Conway is, and will probably continue to be, the summer-capital of the mountain-region. Its inhabitants have done and allowed many things which have deteriorated from its attractiveness, and have neglected important improvements; but the Saco intervales, though mutilated, are still beautiful, and there is no place in or near the mountains whence so many pleasant excursions may be made. The hotel-accommodations are sufficient and good, and the railroad communications with Boston, Portland, and the Crawford and Fabyan Houses are easy and frequent. Bethlehem also has large accommodations for summer-boarders, but its environs are comparatively uninteresting. Being at a greater altitude than N. Conway, it is much cooler, a marked advantage in the month of August. Jefferson Hill has good hotels, but is several miles from the railroad. Its environs are of some attractiveness, and the view of the loftiest of the White Mountains, as obtained from this point, is one of the best and most comprehensive in the State. Gorham is the nearest village to Mt. Washington, and there are many interesting places in the vicinity; but it has no large hotel, and its peculiar charm as a quiet mountain-hamlet is injured by the presence there of the repair-shops of the Grand Trunk Railway. Bethel and Fryeburg are rich and beautiful villages near the mountains, in Western Maine, — the former being on the Androscoggin and the latter on the Saco. Lancaster and Littleton are on the W. slope, and are prosperous and pleasant towns, having good hotel-accommodations and commanding broad panoramic views. Newbury, in Vermont, is a lovely village, tranquil and neat, with a prospect of the Franconia and Benton Ranges; and Warren, if it had better accommodations and a less straggling settlement, would be a good centre for many profitable excursions. Campton Village and Plymouth are S. of the Franconia Range, and in close contiguity with the Waterville mountains. They have spacious quarters for summer-visitors, and are rich in their surroundings. Meredith and Alton Bay being shut off on long arms of Lake Winnepesaukee, Wolfeborough and Centre Harbor are the best points for a sojourn in the lake-country, and they each have good hotels and boarding-houses. Wolfeborough is on the railroad, but it has the disadvantage (for a summer-loiterer) of being a brisk manufacturing village; while Centre Harbor is a small rural hamlet. The former looks out on Wolfeborough Bay; the latter commands a view down the whole lake.

XV. Expense.

The expense of a sojourn in the mountain-region varies widely according to the manner in which it is undertaken. Tourists who avail themselves of the first-class hotels and frequently use carriages and guides will find $6-9 a day none too much — and can enjoy nearly every metropolitan luxury for that price. Those who establish themselves in the summer boarding-houses in the outer valleys may reduce their expenses to $6 – 10 a week, and be sure of comfortable and substantial living. Pedestrians who wish to reduce their daily outlay still farther must carry their own provisions and encamp at night. A pedestrian tour in which the smaller public-houses are depended on for food and lodging will cost not less than $15 a week.

XVI. Miscellaneous Notes.

Black Flies and Mosquitoes. — The traveller among the deep forests and uninhabited glens is apt to meet terrible and pitiless enemies in the form of black flies and mosquitoes, especially during May, June, and July. They come in such vast numbers, and with such unappeasable hunger, that it is almost impossible to keep them away for a moment, and their stings are so sharp and empoisoned as to wellnigh madden their unfortunate victims. Various preparations of tar and oil, and other ingredients, are used to anoint the hands, face, and neck, to keep off these ferocious insects; but their feeling and odor are unpleasant, and it is the height of discomfort to march through a warm morning, perspiring freely, and with the face smeared with these abominable compounds. In case of short halts, protection is obtained by making a "smudge," — a small fire on which fresh bark or green boughs are placed. The copious smoke which arises scatters the insects and keeps them away. The night-camps are guarded in the same manner. During breezy days no inconvenience is experienced from this source; but when the air is still, the flies rise in voracious swarms, sometimes attacking people even on the mountain-tops and driving them down.

Water. — The difficulty of obtaining water is one of the worst trials in mountain-climbing in this region. Heated by the exercise of walking, freely perspiring, and craving frequent draughts of cool water, yet, from the nature of the ground one is frequently obliged to pass many hours without such refreshment. As hereinafter shown, there are several mountains which have springs near their summits; and water may be found in the hollows of the flat rocks almost anywhere, soon after a rain-storm. When tourists are about to ascend a mountain where no water can be found, each of them should carry a bottle of cold tea, to be drank sparingly and at wide intervals. Strong liquors are weakening in their effects when such work as mountaineering is on hand.

Clothing. — Visitors to the summit of Mt. Washington and the other high peaks should be prepared with suitable overcoats and shawls, in order to meet the low temperature which often prevails there. Warm clothing is frequently needed as a change by visitors in the early part of the season at Bethlehem, Jefferson, and the Profile and Glen Houses.

THE APPROACHES TO THE MOUNTAINS.

FULL descriptions of the cities and interesting localities along the lower portions of these routes are given in Osgood's *New England:* the Eastern R. R. in Routes 37 and 31; the Boston, Concord, and Montreal R. R. in Routes 29 and 30; the Boston and Maine R. R. in Route 38; the Portland and Ogdensburg R. R. in Route 39; the Grand Trunk Ry. in Route 40; the line from New York, by Springfield and the Connecticut Valley, in Routes 21 and 24; by the Hudson River and Lake Champlain in Routes 52 and 53.

The following itineraries are intended only to show the sequence of the stations and the views from the trains. In other parts of the book are descriptions of the villages and localities on and near the several lines where they approach the mountains. These may be found by reference to the index. They cover the stations on the Eastern R. R. from Ossipee to the N., the B., C. and M. R. R. from Laconia, the Portland and Ogdensburg R. R. from Portland, the Grand Trunk Ry. from Paris Hill to N. Stratford, the Passumpsic R. R. from Orford to Wells River.

Minor Routes.

There are also two routes which are sometimes chosen, leading from Boston to the mountains by the Fitchburg R. R. The first diverges from the latter line at Concord Junction, and runs to Nashua by the Nashua, Acton & Boston R. R., and thence to the N. by Route 2. The second follows the Fitchburg line to Fitchburg, the Cheshire R. R. (by Mt. Monadnock) to Bellows Falls, the Central Vermont R. R. to White-River Junction, the Conn. & Passumpsic Rivers R. R. to Wells River, and the B., C. & M. R. R. to the mountains.

Another route is by the lower Montreal route to Concord and White-River Junction, and thence by the Passumpsic R. R.

1. Boston to the White Mountains.

The Eastern Railroad.

Stations. — Boston to Somerville, 2 M.; Chelsea, 4; Revere, 5; Lynn, 11; Swampscott, 12; Salem, 16; Beverly, 18; N. Beverly, 20; Wenham and Hamilton, 22; Ipswich, 27; Rowley, 31; Newburyport, 36; Salisbury, 38; Seabrook, 42; Hampton, 46; N. Hampton, 49; Greenland, 51; Portsmouth, 56; Kittery, 57½; Elliot, 63; Conway Junction, 67; S. Berwick, 69; Salmon Falls, 70; Great Falls, 73; Rochester, 79; Hayes, 84; S. Milton, 85; Milton, 87; Union, 93; Wolfeborough Junction, 97; (Wolfeborough, 108;) Wakefield, 99; E. Wakefield, 103; N. Wakefield, 106; Ossipee, 111; Centre Ossipee, 115; W. Ossipee, 121; Madison, 125; Conway, 132; N. Conway. 137. (Crawford House, 154; Fabyan House, 168.)

This is the quickest route from Boston to the mountains, and its trains reach N. Conway in less than 6 hrs. Pullman palace-cars are attached to the trains, in which travellers can enjoy the luxury of a parlor for a small additional outlay. Throughout the first section of the route, seats on the r. side of the cars are preferable, since they give frequent views of the ocean, and the best prospects of the coast-cities. When approaching the mountains, the l. side of the cars commands panoramic views of the Ossipee and Sandwich Ranges, and of Mt. Chocorua and Moat Mountain. The section lying between Kittery and Ossipee is the least interesting part of the road.

On arriving at N. Conway, passengers have ample time to dine before the departure of the P. & O. train for the Crawford and Fabyan Houses, or the stages for the Glen House.

On leaving the Eastern station in Boston, the train runs out across the Charles River, with the populous heights of Charlestown on the r., and the factories of E. Cambridge on the l. When it reaches the Somerville meadows, the McLean Asylum for the Insane is seen on the l.; and soon afterward the Mystic River is crossed. Chelsea is then traversed, with its churches and the soldiers' monument on the r.; and as the long marshes of Lynn are crossed the ocean is seen on the r., with the high promontory of Nahant and the nearer hotels on Revere Beach. The tourist next sees the busiest part of the great shoe-manufacturing city of Lynn, and then speeds away by the summer-villas of Swampscott, viewing the ocean on the r., to the ancient maritime city of Salem, older than Boston, and famous for the witchcraft horrors of 1692, and other interesting episodes of the earlier centuries.

The train then crosses the North River, with an island-studded arm of the sea on the r., and passes the shoe-factories of Beverly and the great ice-houses of Wenham Lake, near the camp-meeting grounds. The quaint and quiet old Puritan hamlet of Ipswich is next seen, on the r., with its high church-towers; and then the line traverses several leagues of moorlands and salt meadows, with the ocean-fronting sand-hills of Plum Island cutting the horizon on the r., and stops again at the ancient sea-city of Newburyport, famous for its pleasant environs and antiquated houses, as well as for the deposits of silver in its vicinity. The train here crosses the broad Merrimac River on a massive and costly bridge, whence the city is finely displayed in retrospective views, and the ocean is seen on the r., beyond the Plum Island lighthouses. On the l. is the swelling eminence of Po Hill, over the village of Amesbury, where resides John

G. Whittier, the poet of New England. Beyond the Seabrook forests the train traverses the broad Hampton marshes, with occasional glimpses of the sea, and a complete picture of the summer-hotels on Hampton Beach and Boar's Head. Stages run from Hampton station to Hampton Beach, and from N. Hampton to Rye Beach. Then comes Portsmouth, another venerable coast-city, filled with quaint traditions and antique houses, and the centre from which many interesting excursions may be made, — to the Isles of Shoals (by steamboat), to York Beach, New Castle, Straw's Point, and the U. S. Navy-Yard at Kittery. A fine view of the city and the navy-yard is obtained on the r., as the train crosses the long bridge over the Piscataqua River.

11 M. beyond Portsmouth Conway Junction is reached, where the mountain division diverges from the main line of the Eastern Railroad. The train ascends the valley of the Salmon-Falls River, stopping at the prosperous manufacturing villages of Salmon Falls and Great Falls, and at Rochester meets the Portland & Rochester, Nashua & Rochester, and Dover & Winnepesaukee Railroads. **Rochester** (*Dodge's Hotel; Mansion House*) is a pleasant village of about 5,000 inhabitants, situated on the Norway Plains, near the Cocheco River. It manufactures large quantities of woollen goods and shoes ; and has 5 churches and several large schools. It was granted by Massachusetts in 1722, and incorporated in 1792 ; but until the Conquest of Canada it suffered much from Indian forays. The streets are wide and pleasant, and the village attracts numerous summer-visitors.

Milton (Ben Franklin House) has several small boarding-houses, and is near the picturesque Milton Ponds, on which summer-visitors find good boats. Teneriffe Mt. is 2 M. distant, and rises 600 ft. above the ponds, affording a broad and beautiful view. The train now runs N. to *Union* (Union Hotel), and *Wolfeborough Junction* (Sanborn House), whence a short branch-road runs to the shores of Lake Winnepesaukee. Occasional views of Copple-Crown Mt. are gained on the l.; and the train runs N. to Wakefield (small inn), a pleasant village near the highland-guarded and historic Lovell's Pond, — and E. Wakefield, which is near the Balch and Pine-River Ponds and the beautiful Lake Newichawannock. There are several boarding-houses in this town, and 200 – 300 summer visitors stay here every season, attracted by the number and beauty of the ponds. The line now bends to the N W., and passes N. Wakefield, approaches the high hills of N. Wolfeborough on the l., and traverses the uninteresting drift-plains of Ossipee. *Ossipee Corner* (two inns) is the capital of Carroll County; and from *Ossipee Centre* the Green and Ossipee Ranges may be visited. Frequent views of the near Ossipee Range are now gained on the l.; Green Mountain appears on the r.; and there is a transient glimpse of Ossipee Lake.

Beyond Centre Ossipee the train traverses a belt of second-growth woods as far as the rural station of Bearcamp, beyond which it crosses the Bear-camp River. The wooded hills of the Ossipee Range now roll back on the l. and open a fine view of the Sandwich Range on the N. W. At *W. Ossipee* station the Bearcamp-River House is seen on the l., above the track, and the great Tamworth valley opens beyond. As the train passes on, varying views are given of the Sandwich Range on the l., the nearest mountain being the superb white peak of Chocorua, next to which is the low and ledgy Paugus, then the dark and pointed Passaconaway, then the high blanched cliffs of Whiteface, the lofty flat top of Sandwich Dome, and the crest of Mt. Israel, isolated on the plain. The pyramidal Whittier Peak is seen at the end of the Ossipee Range, and the alpine heights to the N. gradually fall behind each other as the train advances.

On the r., the Green Mountain in Effingham is frequently seen while passing N. from Ossipee Corner, and a glimpse of Kiarsarge is also obtained far in advance. Soon after leaving W. Ossipee, *Silver Lake* is seen close at hand, and its white sandy beaches are followed for a long distance with Gline Mountain beyond and Madison Village at the head of the lake. On the l. are the great E. outworks and craggy buttresses of Chocorua, and the view soon includes the sharp S. peak of Moat Mountain, in advance. The country about the railway is covered with dense thickets and second-growth trees, and is cheerless in character. Just before reaching Conway Corner, Frost Mountain and the Burnt-Meadow Mountains in Brownfield are visible; and as the train slows up at Conway, a glimpse of Mt. Pleasant (in Maine) is gained on the r. front over the village. From *Conway* the view includes Chocorua on the l., separated by the Swift-River Valley from the long flanks of Moat Mountain, near at hand, on whose r. is the Mt.-Washington range, the next blue peaks being Wild-Cat and Carter Dome, between which is the cleft of the Carter Notch. R. of these is the lower swell of Thorn Mountain, with the truncated pyramid of Double-Head and the graceful cone of Kiarsarge. The train soon swings around to the N. W., and advances between the Moat Range on the l. and the Green Hills on the r., crossing the Saco River and traversing the renowned intervales of that stream. After running for several miles over these rich and verdant meadows, the boarding-houses of N. Conway are seen on the r., and the White-Horse and Cathedral Ledges draw near on the l. The tower of the Kiarsarge House rises towards Mt. Kiarsarge, and the train soon runs up to the station in N. Conway.

North Conway, see Route 11.

2. Boston to the Franconia and White Mountains.

The Boston, Lowell and Nashua R. R.; the Concord R. R.; the Boston, Concord and Montreal, and White Mountains, N. H, R. R.; and the Mt.-Washington Branch R. R.

Stations. — Boston to E. Cambridge; Milk Row; Winter Hill, 2 M.; Somerville Centre; Willow Bridge; College Hill, 4; Medford Steps; W. Medford; Mystic; Winchester, 8; E. Woburn, 10; N. Woburn, 12; Wilmington, 15; Billerica, 19; N. Billerica, 22; Lowell, 26; N. Chelmsford, 29; Tyngsborough and Dunstable, 32; Concord Junction, 36; Nashua, 40; Thornton's Ferry, 46; Merrimac, 49; Goff's Falls, 53; Manchester, 57; Martin's Ferry, 62; Hookset, 66; Suncook, 67; Robinson's, 70; Concord, 74.

E. Concord, 76; N. Concord, 78; Canterbury, 82; Northfield, 85; Tilton, 90; E. Tilton, 94; Laconia, 99; Lake Village, 101; Weirs, 105; Meredith, 109; New Hampton, 113; Ashland, 117; Bridgewater, 120; Plymouth, 123; Quincy, 129; Rumney, 131; W. Rumney, 134; Wentworth, 139; Warren, 141; Warren Summit, 144; E. Haverhill, 151; Haverhill and Newbury, 156; N. Haverhill, 161; Woodsville, 166; Wells River; Bath, 170; Lisbon, 175; N. Lisbon, 180; Littleton, 185; Wing Road, 192; Fabyan House, 206; Whitefield, 196; Dalton, 200; S. Lancaster, 203; Lancaster, 207; Northumberland Falls, 213; Groveton Junction, 219.

Passengers leaving Boston at 8 A. M., reach Concord at 10.35, Weirs at 11.50, Plymouth at 12.35 P. M. (30 minutes for dinner), Wells River at 2.40; Lancaster at 4 35, Bethlehem at 3 50, Fabyan House at 4.25; Profile House at 6. Passengers leaving Boston at noon reach Plymouth at 5 45; and there are other trains along this route. A train also leaves the Boston & Maine station, in Boston, at 7.30 A M., and connects with this line.

This line follows the Merrimac Valley for a great distance, passing through the rich manufacturing cities along its course. It then gives fine views over the waters of Lake Winnepesaukee, including the Ossipee and Belknap Mountains and the Sandwich Range. The meadows of Newbury, the glens of the Ammonoosuc, and the inner valleys of the Presidential Range are then traversed in succession. Palace-cars run on this line.

The train runs out from the stately building of the Boston terminus, and crosses the Charles River, with Charlestown on the r., then traverses the long district of Somerville, by several suburban stations, and at College Hill runs by *Tufts College*, on the l. Beyond Mystic Pond it reaches Winchester, and passes on through the rural stations in Woburn and Billerica to **Lowell**, the City of Spindles. This city is only about 50 years old, but it has over 70 factories, employing 15,000 operatives, and producing over 120,000,000 yards of cotton cloth annually, besides immense quantities of woollen goods, carpeting, shawls, hosiery, and prints. The main water-power is derived from the Pawtucket Falls, near which the Indians formerly had a populous village.

Seats on the r. are now preferable, as the train ascends the r. bank of the Merrimac River, with pleasant views over its broad waters. Beyond Tyngsborough it enters the State of New Hampshire, and soon reaches **Nashua**, a busy manufacturing city of 10,543 inhabitants, engaged in the manufacture of cotton goods, locomotives and iron goods, locks, and many other articles, its water-power being derived from falls on the Nashua River. It occupies the site of an ancient Indian village, and was in the warlike border-town of Dunstable. The city was founded in 1823, on a sandy pine-plain. A branch railroad runs N. W. to the hill-villages of Amherst, Mt. Vernon, and Wilton, which are frequented by summer-visitors.

The line now follows the W. bank of the Merrimac River by several rural hamlets, crosses the river near Goff's Falls, and reaches the city of **Manchester** (23,536 inhabitants), an important railroad centre and manufacturing city, near the Amoskeag Falls. One company alone employs 8,000 operatives; and the chief products of the city are cotton cloths, prints, hosiery, paper, castings, and iron-wares, locomotives, and steam fire-engines. Manchester has 14 churches, 2 daily papers, 8 banks, 8 hotels, and a large public library. It was founded about the year 1831, on the site of the most famous Indian fishing-ground on the river; and it is now the largest city in New Hampshire.

Running N. by the Amoskeag Falls, the twin Uncanoonuc Mountains are seen on the l.; and the train reaches *Hookset,* famous for its great brick-yards and granite-quarries. The Pinnacle is a far-viewing hill on the W. side of the river. 7 M. beyond this point the train reaches **Concord** (*Eagle Hotel; Phenix House*), the capital of New Hampshire, and a handsome city of about 14,000 inhabitants, with 11 churches, 2 daily papers, and 7 banks. The famous Concord coaches and stages are made here, the works employing 250 men; and the granite-quarries, 1–2 M. from the city, are worked by 500 men, and turn out nearly $ 800,000 worth of stone annually.

Visitors will be interested to visit the State House, an imposing granite building fronted by colonnades and surmounted by a dome from which may be seen Mt. Belknap, the Uncanoonucs, Crotched Mountain, Mt. Kearsarge, and (on a clear day) Mt. Moosilauke. The Doric Hall contains the battle-flags of the State regiments in the Secession War, and several trophy-cannon. The Council, Senate, and Representatives' Halls contain many portraits of eminent New-Hampshire men. Tourists can also visit the library and museum of the N. H. Historical Society, on Main Street, the Birchdale Springs, and the granite-quarries. Concord occupies nearly the same site as the Indian town of *Pennacook,* the capital of the great confederation over which Passaconaway bore sway (see page 26).

The Boston, Concord & Montreal, & White Mts. N. H., Railroad.

The mountain-train runs out to the N., and soon crosses the Merrimac, and follows near its E. bank. *Canterbury* station is 4 M. from a large Shaker village; and soon after passing it Mt. Kearsarge is seen on the l. The town of Northfield is then traversed, and the train crosses the Winne-pesaukee River to *Tilton* (Dexter House), a prosperous manufacturing village, and the seat of the N. H. Conference Seminary. As E. Tilton is approached the first view of the Sandwich Range is gained over a pond on the l., — the peaks running from l. to r. in the following order, — Sandwich Dome, Tripyramid, Whiteface, Passaconaway, Chocorua, and Ossi-pee. The Winnepesaukee River is crossed and recrossed several times, and its general course is followed closely by Union Bridge and along the expansion of Sanbornton Bay. The train then comes out on the S. shore of Lake Winnesquam, which is skirted for miles, passing the summer-resort stations of the Winnesquam House and the Bay-View House. Near the latter point one of the finest views on the route is gained from the l.

side of the train, looking over Lake Winnesquam, and including Mts. Moosilauke, Kinneo, Cushman, the distant serrated ridge of Mts. Lafayette and Liberty, Mt. Tecumseh, looming over the Squam Range, and the black mass of Sandwich Dome, on the r. of which Whiteface, Passaconaway, and Paugus are seen.

The train now runs through the centre of the busy town of **Laconia** (*Willard House*), with its churches and factories visible on either side. The round summits of the Belknap range are seen near at hand on the r.; and as the line swings around the shore of Round Bay, the first glimpse of Mt. Washington is obtained, low down on the horizon, on the far front.

The road now passes through *Lake Village* (Mt.-Belknap House), and crosses the outlet of Lake Winnepesaukee for the last time. While on the bridge the Ossipee Mountains are seen on the r., with Chocorua on their N. slope. The W. shore of Long Bay is now followed for 4-5 M., passing Sheep and Goat Islands, beyond which are the round domes of Mt. Belknap. From various points the r. side of the train gives views of Mts. Whiteface, Passaconaway, the superb Chocorua, and the black Ossipee Range. Soon after crossing Pickerel Cove a belt of woods is traversed, and the train reaches **Weirs**, where connection is made with the Lake-Winnepesaukee steamboats, at the station. From this point a broad reach of the lake is seen, with several islands; over which is the dark and prolonged Ossipee Range, flanked on the l. by the noble alpine peak of Chocorua. The lower mass of Red Hill is also visible, with the crests of Paugus, Passaconaway, Tripyramid, Whiteface, and Sandwich Dome extending to the l. These peaks are seen for a long time, as the train advances; and Copple Crown soon comes into view on the r. rear, and Tecumseh on the r. front, over the narrow waters of Northwest Bay. The pasture-crowned Rollins Hill is just across the bay; and for a moment Lafayette is visible on the r. front.

Meredith (*Elm House*) is now reached, the village being at the head of the bay on the r. Then Lake Waukawan appears on the r., picturesquely placed in the forest, over which are seen the peaks of Moosilauke, Plymouth Mountain, Mt. Prospect, Sandwich Dome, the sharp apex of Tripyramid, with Whiteface and Passaconaway on the r., the nearer ridge of Red Hill, and the low peaks of the Ossipee Range. Another island-strewn forest-pond is soon passed, over which appear the Squam Range, Sandwich Dome, Tripyramid, Whiteface, Passaconaway, and Red Hill. The Squam River is crossed, and *Ashland* village is seen on the r., over which are Mt. Prospect and the peak of Osceola, with the dark mass of Sandwich Dome. The line soon enters the fair Pemigewasset intervales, near Plymouth, and views the flat topped Mt. Prospect on the r., Tecumseh farther N., and on the r. front the high sharp peaks of Mts. Lafayette and Liberty, with Mt. Cannon on the l. and the pyramidal apex of Mt. Flume on the r. The train now reaches **Plymouth** (**Pemigewasset House*), the

headquarters of summer-tourists in this region and the point of departure
for the Campton, Waterville, and Pemigewasset-Valley stages.

Beyond Plymouth the line crosses Baker's River near its confluence
with the Pemigewasset, and runs nearly W. by Quincy Station, beyond
which it recrosses the river and passes under the low black cliffs of Hawk
Mountain, with Mt. Stinson on the r. The Glendahlia School is on the r.,
and the hamlet of *Rumney* is soon seen beyond, across the valley and at
the outlet of the broad Stinson Valley, which has Mt. Stinson on the r.,
the Mt.-Carr Range on the l., and the bold mountains in Ellsworth at the
head. Passing the hamlet at Rumney station, the high cliffs of Rattlesnake Mountain are seen on the r., across the valley. Running N. W.
by the station at W. Rumney, with the Groton hills on the l., the train
soon crosses the river and passes along its E. bank to *Wentworth*, a pretty
hamlet on the l., with a great white church. The course is now nearly N.,
around the shaggy foot-hills of Mt. Carr (on the r.), and follows the river
to **Warren** (*Moosilauke House*), a long and straggling village among the
highlands. The road thence follows the Mikaseota (Black-Brook) Valley
to the N., with views of Mts. Carr, Waternomec, Kinneo, and the towering mass of Moosilauke on the r. Mts. Mist and Webster's Slide are on the
l. as the train ascends to *Warren Summit*, the water-shed between the
Merrimac and Connecticut Valleys, traversing an almost unbroken forest
over which the S. peak of Moosilauke is often seen.

Near the Summit the train passes through a rock-cutting ¾ M. long, and
in some places 60 ft. deep. This cut cost $ 150,000, requiring the labor of
150 men for a year and a half (1851 – 2). On the r. are seen the S. peak
and central ridges of Moosilauke, but slightly foreshortened by the angle
of vision. As the descent toward the Connecticut Valley is commenced
the fine cliffs of Owl's Head are seen on the r., nearly perpendicular, and
of a dark purplish color. From *E. Haverhill* a good view is gained of
Black Mountain and Sugar Loaf, on the r., with the upper ridge and hotel
on Moosilauke visible over Blueberry Mountain, in retrospect. These
peaks remain in sight for several miles as the train descends to the N. W.,
along the alder-fringed valley of the Oliverian Brook.

At *Haverhill* the Connecticut Valley is reached, and the line turns to
the N., giving frequent views of the exquisite meadows of *Newbury*, with
the hamlet of that name beyond the river and under Mt. Pulaski. This
is one of the most fascinating sections of the route, the grace and loveliness of the broad intervales and winding river being contrasted with the
savage aspect of the adjacent mountains. Passing through N. Haverhill,
the new village of *Woodsville* is soon reached, whence the Connecticut is
crossed to **Wells River**, the junction of the Connecticut & Passumpsic
Rivers Railroad, the Montpelier & Wells-River Railroad, and the present
route.

The train recrosses the river, with Mt. Gardner on the l., and commences

the ascent of the Ammonoosuc Valley, crossing to the N. bank near the
inflowing of the Wild Ammonoosuc. The nearer hills and forests shut
out the mountain-views for many miles. *Bath* is a quiet old hamlet on
the S. bank, beyond which the river is again crossed, and the village of
Lisbon is soon seen on the l., in the valley below. The course is now to
the N. E., and **Littleton** (*Oak-Hill House; Thayer's*) is soon reached,
with the high school and the Oak-Hill House conspicuous on the ridge
beyond, and several small factories by the river-side. From this point
stages run to the Franconia Notch. Beyond Littleton the line passes
Scytheville and Alder-Brook station, and soon reaches the junction at
Wing Road. The peak of Mt. Lafayette is visible from this point, and
one of the best distant views of the Twin Mountains is obtained thence.

The *Mt. Washington Branch*, see Route 3.

The main line continues to the N. E., through a more rugged region, and
soon reaches *Whitefield*, whence the peaks of Haystack and other mountains are seen on the r. In the valley below are the immense lumber-mills
of Brown Brothers. The course of John's River is now followed as far as
Dalton station, which is a short distance N. of the Sumner House, a
secluded summer-resort. Rounding the Martin-Meadow Hills, and running over the fair intervales of the Connecticut River, with the Lunenburg
Heights on the l., the train soon reaches the beautiful village of **Lancaster**
(*Lancaster House*), the capitol of Coös County, and the station whence
stages run to Jefferson Hill. The dark ridges of the Pilot Range are seen
on the r., as the train runs N. E. over the Connecticut meadows, through
the village of *Northumberland*, with Cape Horn on the r. and the Guildhall hills on the l. Glimpses are gained of the Percy Peaks, in front,
and the end of the line is soon reached, at **Groveton Junction,** where
it connects with the Grand Trunk Railway (Route 7).

3. Boston and the West to the White Mountains.

The Mt.-Washington Branch Railroad.

Travellers who approach the mountains on the W., either by the B., C. & M. &
W. Mts., N. H., R. R., or by connecting lines, will enter the remoter defiles by this
branch, which diverges from the main line at Wing Road. The scenery is not all
that might be wished or supposed, from the proximity of the mountains, because
the route lies in the deep valley of the Ammonoosuc, and is often enclosed in dense
forests.

Stations. — Wells River to Wing Road, 27 M.; Bethlehem, 31; Twin-Mountain
House, 36; White-Mountain House, 40; Fabyan House, 41; Base of Mt. Washington, 47.

The train runs S. W. from *Wing Road*, closely following the course of
the Ammonoosuc River, into a more primitive region. Although Bethlehem is but 1–2 M. to the S., it is not seen, although a glimpse of Mt.
Agassiz is obtained on the r., as Bethlehem Hollow is approached; and
soon afterwards there is a fine view of Mts. Lafayette and Haystack, the

3*

Twin Mountains and Mt. Hale. *Bethlehem* station is about 2½ M. from the summer-hotels at Bethlehem Street, and has several large lumber-mills in its vicinity. The town of Carroll is now entered, and a prospect of Lafayette, Haystack, and the immense N. Twin Mountain is gained on the r. The *Twin-Mountain House* is next seen, on the l., across the river, standing out in bright relief against the dark hillside. Then a portion of Mt. Hale comes into sight on the r., with the lower eminences of the Sugar Loaves.

The winding and rapid river is still followed, and the desecrated Ammonoosuc Falls are seen on the r., with much of their natural beauty marred by the intrusion of a saw-mill. Wider plains are now reached, with the dull Mt. Deception on the l. and the lower foot-hills of the Rosebrook Range on the r. Near the White-Mountain House (on the l.), sudden views of the Presidential Range break upon the sight; and the tourist is soon before the **Fabyan House,** at the intersection of the present route with the Portland and Ogdensburg R. R. (Route 5) Beyond this point the train advances directly toward Mt. Washington, up the Ammonoosuc Valley, and with occasional views of the great mountains in advance. About ⅓ M. from Marshfield it connects with the celebrated Mt.-Washington Railway. (See Route 92.)

4. Boston to the Lake-Country of New Hampshire.

The Boston & Maine Railroad.

Stations. — Boston to Somerville, 2 M.; Edgeworth, 3; Malden, 5; Oak Grove, 6; Wyoming, 7; Stoneham, 8; Greenwood, 9; Wakefield, 10; Reading, 12; Reading Highlands, 13; Wilmington, 16; Wilmington Junction, 18; Ballardvale, 21; Andover, 23; S. Lawrence, 26; N. Andover, 28; Bradford, 33; Haverhill, 33½; Atkinson, 37; Plaistow, 38; Newton Junction, 41; E. Kingston, 45; Exeter, 50; S. Newmarket, 54; Newmarket Junction, 55; Newmarket, 57; Durham, 62; Madbury, 64; Dover, 68; Pickering's, 74; Gonic, 76; Rochester, 78; Place's, 82; Farmington, 86; Davis, 90; New Durham, 92; Alton, 95; Alton Bay, 96.

The train runs out from the Boston station across the Charles River, between Charlestown and E. Cambridge, and then passes through Somerville, near the track of the Eastern Railroad. After crossing the broad Mystic River, it traverses Medford and approaches the curving hills of Malden. The suburban stations of Wyoming, Melrose, and Stoneham are passed rapidly, and beyond Crystal Lake (l side) the large and prosperous village of Wakefield is entered. Lake Quanapowitt is next seen (on the r.), and the line traverses the towns of Reading and Wilmington, where it intersects the Salem & Lowell Railroad. *Andover* is a pleasant old Puritan village, amid pretty rural scenery, and is famous as the seat of the Congregational Theological Seminary, which was founded in 1808, and has educated over 2,600 ministers. There are several other educational institutions at this place. The train then reaches the Merrimac River, and passes **Lawrence,** one of the most beautiful of the manufacturing cities of New England. It has nearly 30,000 inhabitants, with 18 churches, 2 daily papers, a fine city hall, and a large and imposing Catholic church. Lawrence was founded in 1844, and is celebrated for its manufactures of cotton and woollen cloths and other goods.

The train now follows the r. bank of the Merrimac to Bradford, where it crosses to *Haverhill,* a pleasant city of Essex North, 18 M. from the sea. It has about 14,000 inhabitants, and is largely engaged in the manufacture of shoes. Beyond this point the line enters New Hampshire, passing through the rural towns of Rockingham County to *Exeter,* an ancient village in which is located Phillips Academy, the Eton of New England. Traversing agricultural Newmarket (where the Concord & Portsmouth Railroad is crossed), hay-producing Durham, and level Madbury, the train reaches the busy little manufacturing city of **Dover,** on the Cocheco River. This is the oldest settlement in New Hampshire, dating from 1623; and was nearly destroyed by a fierce Indian attack in 1689.

The branch line to the lake runs N. W. from Dover up the Cocheco Valley, intersecting several other railroads at Rochester (see Route 1). It then traverses the town of Farmington, in which Henry Wilson was born in 1812; and on the l. the Strafford Blue Hills are visible. It crosses New Durham, with glimpses of Copple Crown on the r., and running by the hamlet of Alton, soon reaches the terminus at **Alton Bay,** where the steamboat *Mt. Washington* is in waiting to carry passengers out over Lake Winnepesaukee.

5. Portland to the White Mountains.

The Portland & Ogdensburg Railroad.

Stations. — Portland to Westbrook, 5 M.; S. Windham, 11; White Rock, 13½; Sebago Lake, 16¾; Steep Falls, 24½; Baldwin, 32; W. Baldwin, 33½; Hiram Bridge, 36½; Brownfield, 43; Fryeburg, 49; Conway Centre, 55; N. Conway, 60; Intervale, 62; Glen Station, 66; Upper Bartlett, 72; Bemis, 78; Crawford House, 87; Fabyan House, 91.

The Portland & Ogdensburg Railroad is now nearly completed throughout its whole extent, and seems destined to become one of the chief thoroughfares to the Atlantic seaboard. The credit of the city of Portland has been largely pledged to aid in its construction, and it is hoped that it will form a great freight-line for Western grain, which will be shipped at Portland. The broad lumber-districts of Maine and New Hampshire are also to be opened by this route, and the rich farming districts of Northern Vermont. The western termini will be at Montreal and Ogdensburg. The section between Fryeburg and the Fabyan House is regarded as a triumph of engineering, and leads through some of the most imposing mountain-scenery in the Atlantic States. No other railroad in this region traverses such wild gorges, or looks out on such majestic peaks, close at hand The mountain-section may also be visited by passengers from the connecting trains on the Eastern and B , C. & M. Railroads. After passing N. Conway, seats on the r. of the cars should be secured, as the finest views are afforded on that side. Observation-cars are run on this section. They are open all along the sides, and afford unobstructed prospects; but are sometimes rendered uncomfortable by the cinders blown back from the locomotive.

Sebago Lake, the Mt.-Pleasant region, Fryeburg, and the mountain-views from Portland and Gorham are described in the division of this book relating to Western Maine.

On leaving Portland the train winds around under Bramhall Hill, and enters *Westbrook*, a populous and diversified town of 6,780 inhabitants, included in several large villages where there are manufactories of cotton cloth, paper, twine, wire, and iron goods. Great quantities of canned goods are prepared here; and the total value of the manufactures of the town amounts to $ 3,500,000 a year. The line thence ascends the Presumpscot Valley to *S. Windham*, in an ancient border-town, and near the Mallison Falls. The Oriental Powder Works are located at this point. From White-Rock station stages connect for the hamlet of N. Windham. The train runs across the upper part of the picturesque town of *Gorham*, and soon afterwards emerges from a deep cutting on the shore of Sebago Lake, which is followed for a considerable distance, with the outspread waters stretching away on the r.

After leaving the lake the train runs N. W. through a thinly settled region, and reaches *Steep Falls*, where it meets the Saco River, henceforward its constant companion for many leagues. On the l. are the hills of Limington; and the forests of the town of Baldwin are now traversed to the N. W., by the stations of Baldwin (whence stages run to Cornish, Porter, and Freedom) and W. Baldwin. Between the latter point and Hiram Bridge is a picturesque portion of the line, lying along the Saco River, and giving views, on the l., of the old Wadsworth homestead, and the white-foaming Great Falls of the Saco. Passing *Hiram Bridge* (Mt.-Cutler House), the village of that name is seen on the r., and the train sweeps around the base of Mt. Cutler. Occasional glimpses of Mt. Pleasant are gained on the r., far away, as the old pine-plains of Hiram are traversed and the line enters Brownfield. The village of *Brownfield* is about 1½ M. from its station, with which it is connected by stage ; and stages also run N. to Denmark and Bridgton. Near this station the Burnt-Meadow and Frost Mountains are seen on the l., not far from the track. The town of *Fryeburg* is now entered, and a portion of the beautiful village of that name is seen on the r. from the station. Stages run thence to Lovell, Stow, and Chatham.

Beyond Fryeburg the train enters the State of New Hampshire, and frequent views over the Saco meadows are afforded. The noblest forms of the eastern mountains are now visible, as the long curves turn the view-line in various directions. On the r. front is the graceful cone of Mt. Kiarsarge, with its crowning hotel, flanked on the r. by Mt. Gemini, and on the l. by the Green Hills of Conway. On the l. is the lofty blanched crest of Chocorua ; and the ledge-lined slopes of Moat Mountain are approached rapidly. The rural station of *Conway Centre* is soon passed, and the train swings around the bases of the Green Hills, traverses a belt of woods, crosses the ravine of Artists' Brook on a high trestle, and stops at the station in N. Conway.

N. Conway, see Route 11.

On leaving the station at N. Conway a good view of Mt. Washington and the connected ranges is obtained on the l. front, and beyond the village the famous Ledges are seen. The train traverses a high-arched pine forest; meets the Eastern-Railroad extension on the l., and reaches the Intervale station, where the great Intervale House is seen on the l. The White-Horse and Cathedral Ledges are across the meadows on the l., N. of and near which is the ravine of Diana's Bath, bounded on the N. by Mt. Attitash and the long spur of Humphrey's Ledge. On the r. is Mt. Kiarsarge, nearly hidden by the heights of Mt. Bartlett. The line now runs along the edge of the narrowing but still beautiful meadows, with the Saco River on the l., bordered by lines of trees. The lonely church in Lower Bartlett is seen on the r., beyond which is the high truncated cone of the S. peak of Double-Head, flanked on the r. by a small white segment of Baldface. After passing Humphrey's Ledge, Iron Mountain is on the l. front, supported by the range on the N. of the river. Soon after crossing the East Branch, the Saco River is seen on the l., and Mts. Stanton and Langdon come into view, while a momentary glimpse is given of Mts. Haystack and Tremont, near the head of the valley.

At **Glen Station**, carriages are in waiting to convey passengers to Jackson, and the Glen-House stages meet some of the trains. At this point the view includes Iron Mountain on the r. front, hiding Mt. Washington, and the blue heights of Mt. Wild-Cat and Carter Dome on the r., separated by the remarkable cleft of the Carter Notch. Still farther to the r. are the wooded knolls on the Thorn-Mountain range. Just beyond Glen Station, the line crosses the Ellis River, with a noble view of the Carter Notch from the bridge. Pleasant open glens are now traversed, with fine retrospects of Kiarsarge on the l. rear, and the high slopes of Iron Mountain ascend on the r. The Saco River is next crossed, with Mt. Haystack ahead on the r.; and then, on the W. of Iron Mountain, up the Rocky-Branch ravine to the r., the long and lofty plateau of Mt. Resolution is seen, with the Giant's Stairs falling upon its r. verge.

The Saco is near at hand on the r., over and beyond which are the imposing dark cliffs of Mts. Stanton and Pickering, close at hand and long visible. Fields and farms are passed on the r., but the valley narrows perceptibly, and the mountains encroach more and more on the lowlands. Towards the front views are given of the conical peak of Haystack, the wavy crests of Tremont, and the majestic dark summit of Mt. Carrigain. Bear Mountain is approached on the l., — a long and chaotic ridge covered with woods, on whose l. is the level plateau-top of Table Mountain. Glimpses of the high overhanging peak of Mt. Crawford are obtained on the r. front, beyond the ravines of Razor Brook. As the train draws up at *Upper-Bartlett* station, the ledgy flanks of Mt. Langdon are seen on the r., beyond which is the conical crest of Mt. Parker, over George's hotel.   On the l. are Table, Bear, and the forest-bound Haystack; and Mt. Carrigain may be seen from the outer platform of the station, looming over the great Pemigewasset Forest.   Partial views of Tremont are gained on the l., and a glimpse of the wilderness-peaks on the W. is obtained as the train crosses Sawyer's River.

As the line bends from W. to N. and advances toward the narrowing Notch, it slowly ascends the ridges on the W., keeping so near their sides that the view in that direction is limited.   It is not so, however, with the prospect on the r., which affords an ever-changing panorama of stately peaks, rising from the narrow forests below in thronging lines.   At first the positive cliffs of Hart's Ledge loom up on the r. and are slowly rounded in a great outer curve.   Then the symmetrical cone of Mt. Hope is seen, and next come the reddish ledges of Mt. Crawford.   As the train crosses Nancy's Brook, the deep flume which the water has cut in the obdurate rock should be noticed, and the foaming falls of the stream.   While slowing in to *Bemis Station*, the old Mt.-Crawford House is seen on the r., over which is the majestic elevation of Mt. Crawford, robed in forests and cut into by a deep ravine.

The train is now on an upward grade of 116 ft. to the mile, and advances along the faces of rugged cliffs.   Above Bemis Station the bed of Davis Brook is crossed; and across the valley, on the r., the bold terraces of the Giant's Stairs are seen, up the ravine of Sleeper's Brook.   Glimpses of Frankenstein Cliff are obtained in advance, and the line crosses Bemis Brook, which comes down foaming from the forest-hidden Arethusa Falls. Fine views of the plateau-summit of Mt. Resolution are obtained on the r., below the Giant's Stairs; and the line is now on a high grade, far above the forests in the Saco Valley.   Mt. Crawford shows finely on the r.   The train next traverses rock-cuttings and emerges in front of the imposing walls of Frankenstein Cliff, one of the loftiest precipices in the mountain-district.   The deep ravine S. of the Cliff is crossed on an iron trestle-bridge, 80 ft. high and 500 ft. long.   Passengers should lean over the sides of the cars and mark the apparent slenderness and the rare gracefulness of the lofty iron piers which support the trestle.   From near this point a good view of Mt. Washington is obtained on the r., up the long ravine of the Mt.-Washington River.   Crossing the gorge of the brook that flows from the Ripley Falls, the line winds around the mountain at a high elevation with rock-walls on the l., and on the r., far below, the unbroken forest, overlooked by the dark Montalban Ridge.

A noble retrospect of Mt. Crawford is next enjoyed, and on the r. front is the ledgy crest of Mt. Jackson, on the l. of the Mt.-Washington-River Ravine, and separated from Mt. Webster on the l. by a short, deep gorge. Swinging around the upper slope of a long rocky ridge, the train comes in sight of the upper part of the valley, with the white *Willey House* 350 ft. below on the r., while in front is the superb alpine peak of Mt. Willey, rising sharply from the great plateau of Pemigewasset and covered with light-colored ledges.

In precise language, the White-Mountain Notch is the chasm extending from the Willey House to the Gate, a distance of about 3 M. As this section is entered, the immense purple cliffs of Mt. Willard are seen in front, striped with lines of fracture and dotted here and there with clinging trees. On the r. is the sharp slope of Mt. Webster, denuded of trees by the procession of avalanches, and banded by the long lines of slides, wherein red and yellow are the chief colors. These vivid stripes start from the very crest of the long summit-ridge, and extend down into the forests that enclose the Saco. As the Brook Kedron is crossed, a pretty cascade is seen on the l.; and soon afterwards a glimpse of Mt. Deception is gained in advance, through the Gate of the Notch, while Mt. Crawford rises in retrospect. The black forests of Mt. Clinton now emerge from behind Mt. Webster; and on the upper slope of the latter are seen the long lines of white light that mark the courses of the Silver Cascade and the Flume Cascade, sweeping down through the trees and over a mile of highly inclined ledges. These bright columns of falling water remain long in sight, and are especially beautiful soon after a heavy rain-storm.

As the train bends to the r. towards and around Mt. Willard, the whole extent of the upper Saco Valley is opened to view, stretching away to the S. for leagues and finally closed by the dark mountains of Albany. After crossing the gorge of the Willey Brook (on a trestle 80 ft. high), the retrospective view continues to open on the r., including miles of the track, the southerly ranges, the edge of the broad Pemigewasset plateau, and the E. peaks of the Nancy Range. The most impressive feature of this prospect is the vast concavity of the Saco Valley below, with its carpet of tree-tops and the narrow stripes of road and river banding its centre in sinuous lines. Just beyond a section-house on the Mt.-Willard slope, a glimpse is obtained of the Hitchcock Flume, far up on the mountain, on the l. The Silver and Flume Cascades are approached on the r., and far below the dark waters of Dismal Pool are seen. The train now swings rapidly around Mt. Willard, and soon passes through the new *Gate of the Notch* which the railroad has made for itself, leaving the old and naturally formed Gate of the carriage-road and river on the r. The two Gates are separated by a massive pier of rock that has been allowed to remain in place.

The train passes the pond which is the source of the Saco River, and soon stops at the station in front of the **Crawford House** (see Route 50), the famous old hotel being visible on the r. front. From this point the train runs on a down grade of 80 ft. to the mile to the Fabyan House, 4 M. distant, through a region that is now in the rough stages of new settlement. Much of the way is in the woods, down the bed of a prehistoric lake. Occasional views of the higher mountains are gained during the latter portion of the route, from the r. side of the train, Mts. Pleasant and Washington being the most conspicuous. The line lies near the course of a branch of the Ammonoosuc River; and the Deception and Dartmouth Ranges are seen in advance.

The **Fabyan House**, see Route 54.

6. Portland to the White Mountains.

The Grand Trunk Railway.

Stations. — Portland to Falmouth, 5 M ; Cumberland, 9 ; Yarmouth, 11 ; Yarmouth Junction, 12 ; Pownal, 18 ; New Gloucester, 22 ; Danville Junction, 27 ; Mechanic Falls, 36 ; Oxford, 41 ; S. Paris, 47 ; W. Paris, 55 ; Locke's Mills, 65 ; Bethel, 70 ; W. Bethel ; Gilead, 80 ; Shelburne, 86 ; Gorham, 91.
This route runs near the bases of the main peaks of the White Mountains, following the trend of the valleys on the N. Its trains are provided with parlor-cars.

The train leaves the station in Portland, runs around the base of Munjoy Hill, and crosses Back Cove on a long bridge, with the islands of Casco Bay on the r. The town of Westbrook is then traversed, and at 3 M. from Portland the train crosses the Presumpscot River on a bridge 300 ft. long, and then passes in succession through the maritime towns of Falmouth, Cumberland, and Yarmouth, with occasional glimpses of Casco Bay on the r. Leaving the coast, the line turns to the N. and intersects the Maine Central Railroad at Yarmouth Junction, entering the agricultural towns of N. Yarmouth and New Gloucester. The village of *New Gloucester* may be seen on the hill about 1 M. W. of the station, in a beautifully diversified and carefully cultivated country. Six M. beyond is *Danville Junction*, where the Lewiston Division of the Maine Central Railroad diverges to the N. E. The Grand Trunk line turns toward the N. W. and passes the obscure stations of Hotel Road and Empire Road. *Mechanic Falls* is a busy little manufacturing village, whence a short local railroad runs N. to Buckfield, 13 M. distant. The train now ascends the pretty valley of the Little Androscoggin River, by the lumber-station at Oxford, and enters Paris, the village of *S. Paris* being just N. of the station. The long levels of the Casco-Bay towns have now been succeeded by a rolling and ridgy country, dotted with bold hills and premonitory of the approaching mountain-scenery. Soon after leaving the station the beautiful highland hamlet of *Paris Hill* is seen, off on the hills to the r., and the train passes the station of Snow's Falls. Three M. beyond N. Paris the line strikes an ascending grade of 60 ft. to the mile, on which it climbs for 4 M., through the shaggy wilds of Greenwood. Near the summit of the grade, front oblique views are gained of the noble peaks of Sunday-River White Cap and Goose Eye, and Mts. Adams and Jefferson. At the station of *Bryant's Pond* the train is 700 ft. above the sea. The line runs along the E. side of the pond, viewing its fine highlands and the long flanks of Mt. Christopher, and soon afterwards enters the glens of the

Alder Stream, passing along the N. E shore of South Pond, near Locke's Mills. After a long descent through a wild and mountainous region the train emerges in the rich Androscoggin Valley, at Bethel. The idyllic village of **Bethel** is on the l., and on the r., several miles distant, are the peaks of Speckled Mountain and the Sunday-River White Cap.

From Bethel the course is nearly W., and for many leagues traverses a region of remarkable beauty and picturesqueness, where the rich meadows and the gracefully winding stream of the Androscoggin make constant contrast with the wild grandeur of the close-bordering mountains. As the fertile intervales of Bethel are traversed, Locke Mountain is seen on the r., across the sinuous river, and on the l. are the ledges of Sparrow-Hawk Mountain. Just beyond the station of *W. Bethel* the line crosses Pleasant River, and soon afterwards a fine view of Mt. Moriah is afforded, up the valley, with the peaks of Adams and Jefferson peering over its long ridge. Similar glimpses are often gained now, on the front oblique, as the train rushes through the Arcadian glens of Gilead. Five M. beyond W. Bethel the train passes the cliffs of Tumble Down Dick and crosses the impetuous and changeable Wild River on a bridge 250 ft. long.

" **The ride in the cars from Bethel to Gorham is very charming. If the railroad approached no nearer to Gorham than this point, a stage-ride along the same route could hardly be rivalled in New Hampshire. What a delightful avenue to the great range it would be! The brilliant meadows, proud of their arching elms; the full, broad Androscoggin, whose charming islands on a still day rise from it like emeralds from liquid silver; the grand, Scotch-looking hills that guard it; the firm lines of the White-Mountain ridge that shoot, now and then, across the N., when the road makes a sudden turn; and at last, when we leave Shelburne, the splendid symmetry that bursts upon us when the whole mass of Madison is seen throned over the valley, itself overtopped by the ragged pinnacle of Adams.**" (KING.)

The narrow valley of Gilead is hemmed in on either side by lofty and rugged mountain-walls, and the railroad is driven near to the river, which winds through fair meadows in short and graceful curves. About 1 M. beyond the Gilead station the frontier between Maine and New Hampshire is crossed, and the line enters the long valley of Shelburne. Frequent and inspiring views of Mts. Washington, Madison, Adams, and Jefferson are now gained on the l. oblique, over the ridges of Mt. Moriah; and on the r. side are Mts. Ingalls and Baldcap. To the l. of the *Shelburne* station is seen the Winthrop House, with Mt. Winthrop towering over it; and a little way beyond the station the highly inclined cliff of Granny Starbird's Ledge is seen on the l. Sweeping around the far-projecting base of Mt. Moriah, the train soon crosses the Peabody River and enters the village of Gorham.

Gorham, see Route 28.

7. Montreal to the White Mountains.

The Grand-Trunk Railway.

Stations. — Montreal to St. Lambert, 5 M.; St Hubert, 10; St. Bruno, 15; St. Hilaire, 22; Soixante, 28; St. Hyacinthe, 35; Britannia Mills, 42; Upton, 48; Acton, 54; New Durham, 66; Richmond, 76; Windsor, 96; Sherbrooke, 101; Lennoxville, 104; Compton, 114; Richby, 118; Coaticooke, 121; Norton Mills, 131; Island Pond, 148; Wenlock, 155; N. Stratford, 163; Groveton, 175; Stark; Starkwater; W. Milan; Milan, 194; Berlin Falls, 199; Gorham, 206.

Crossing the great Victoria Bridge, at Montreal, the train runs near the broad St. Lawrence River, with fair views of the rich and stately city. Passing the Boucherville Mountain on the l., it descends to Belœil, where it crosses the Richelieu River on a costly bridge 1200 ft. long, under the shadow of the Belœil Mountain. Farther distant are the high ridges of Rougemont, and the train runs N. N. E. over a rich and level country, inhabited by an industrious French peasantry, through Soixante, St. Charles, and St. Rosalie, and with continuous views of Yamaska Mountain. The populous French town of *St. Hyacinthe*, with its mediæval appearance and great Roman-Catholic college, is next passed, and a long bridge carries the line across the Yamaska River. Then a great expanse of open and level country is traversed, with quaint little hamlets seen now and then on either side; and then thinly populated and uninteresting forests are traversed for over an hour. Crossing the St. Francis River on a bridge 320 ft. long, the pleasant village of *Richmond* is entered.

The beautiful stream of St. Francis is followed from Richmond for 27 M., to the prosperous town of *Sherbrooke*, passing the romantic island-strewn rapids of the Big Brompton Falls, and traversing many costly cuttings and embankments. Sherbrooke is a busy town at the confluence of the Magog and St. Francis Rivers, with 4 churches and a beautiful surrounding country. It has a large and lucrative trade with the Eastern Townships.

Lennoxville is beautifully situated at the confluence of the Massawippi and St. Francis Rivers, and is the seat of Bishop's College (burnt in January, 1876). A little way beyond, the line enters the valley of the Conticooke, which is ascended to the United-States frontier, traversing the pleasant agricultural district of the Eastern Townships and passing several village-stations. The foot-hills of the Green Mountains are seen on the r., and the Norton and Middle Ponds are passed on the W. Ascending the pretty valley of the Phering River, the train soon reaches **Island Pond**, the frontier station, with its summer-hotel and railway dining-room. Island Pond itself is 2 M. long, and is surrounded by a hard beach of white quartz sand, while the views from Bonnybeag and other adjacent hills are of much interest and beauty.

The line now runs S. E. along a natural terrace, past the Spectacle Pond, and down the long Nulhegan Valley, through a vast forest where lonely ridges rise in rapid succession. The Connecticut River is crossed by a bridge 320 ft. long, near N. Stratford, with the long and massive

Bowback Mountain on the l., and on the r. the N. and S. Notch Mountains, about the Smuggler's Notch, in Brunswick. The line now follows the Connecticut Valley for 12 M., over meadows prolific in corn, hay, and oats, and bordered by mountains on either side. The scenery is a combination of the beautiful and the frowning, and is of high interest to the traveller. The blanched summits of the Percy Peaks are soon seen on the l., and remain in sight for nearly 8 M., being slowly rounded by the railroad.

Beyond Stratford Hollow the line deflects to the S. E., and soon leaves the Connecticut River, passing over to the Ammonoosuc, with Cape Horn and the Pilot Mountains on the r., and the rich plain of Lancaster beyond. At **Groveton Junction** the Grand Trunk line meets the Boston, Concord & Montreal Railroad (Route 2), and passengers for Lancaster, the Franconia Mountains, etc., change cars.

Soon after starting from Groveton Junction the Grand Trunk train stops at Groveton village, and then crosses the Ammonoosuc River, with Cape Horn and Mt. Bellamy on the r., and the Percy Peaks and Bowback Mountain on the l. front and l. Portions also of the Stratford and Sugar-Loaf Mountains are seen to the N.; and on the other side the Pilot Mountains soon swing into view. As the train speeds to the E., the S. peak of the Percies advances over the higher N. peak, and finally eclipses it. The line leaves the river for about 4 M., and runs under the Pilot Mountains, then crosses the river and stops at *Stark*, with the precipice of the Devil's Slide on the l., and Mill Mountain close at hand on the r. The former is a sheer cliff 5,600 feet high, and bears evidence of ancient natural convulsions. Mill Mountain is 2,000 ft. high, and is sometimes ascended from Stark by a walk of 1½ M. through the woods. Beyond Starkwater station fine views are given on the r. and in retrospect, including the Pilot and Crescent Ranges, the Percy Peaks, and Green's Ledge (sharply cut off on the S.). These summits are seen to good advantage across wide and apparently level plains, and present a specially fine prospect to travellers on the late afternoon trains. Just before and after leaving the station at *W. Milan* the traveller who looks forward from the r. side of the train gains a beautiful distant view of the Presidential Range, which is arranged in stately order. The line now leaves the banks of the rapid Ammonoosuc, and follows the course of Dead River through a dull and uninteresting country. At the lonely water-station of Milan the track is 1080 ft. above the sea. Head Pond is soon passed, on the r., and the traveller gains frequent glimpses of the White Mountains on the r. The train soon crosses to the course of another Dead River, passes a small pond, and approaches *Berlin Falls*. On the l., over the diverging track of the Berlin Lumber Company, the far-away blue peak of Goose Eye is seen; and the train soon passes the fine cliffs of Mt. Forist, and stops at Berlin Falls.

Between Berlin Falls and Gorham there is a high descending grade, the track falling at the rate of 50 ft. to the mile. Occasional glimpses of the Androscoggin River are gained, and on either side are mountain-ranges.

8. Montreal to the White Mountains.

The South-Eastern Railway.

This route is over the upper portion of the Montreal & Boston Air Line, passing through an interesting section of Canada and Vermont, and near Jay Peak, Lake Memphremagog, and Willoughby Lake. The cars are changed at Wells River, where the mountain-trains are met, on the B., C. & M. R. R.

Stations.— Montreal to St. Johns, 27 M.; S., S. & C. Junction, 29½; Versailles, 31; W. Farnham, 41; Farndon, 46; Brigham, 48; E. Farnham, 51; Cowansville, 54; Sweetsburg, 56; W. Brome, 59; Sutton Junction, 63; Sutton Flat, 66; Abercorn, 72; Richford, 75; E. Richford, 80; Mansonville, 89; N. Troy, 92; Newport Centre, 98; Newport, 106.

Coventry, 111; Barton Landing, 115; Barton, 121; S. Barton, 125; Summit, 128; W. Burke, 134; Lyndonville, 142; Lyndon, 113; St.-Johnsbury Centre, 147½; St. Johnsbury, 150; Passumpsic, 153; Norrisville, 157; Barnet, 160; McIndoes, 163; Ryegate, 167; Wells River, 171.

Lancaster, 213; Fabyan House, 212; Plymouth, 213; Weirs, 231.

The trains of this line follow the Grand Trunk Railway as far as St. Johns, where they diverge to the E. and cross the Canadian county of Missisquoi, stopping at the stations of W. Farnham and W. Brome. Beyond Sutton the massive highlands which culminate in Jay Peak are approached, and the line soon enters Vermont, crosses the town of Newport, and reaches the village of that name, on Lake Memphremagog (see Osgood's *New England*). A train now passes on to the rails of the Passumpsic R. R., and runs S. by Crystal Lake and Barton village, crossing the St.-Lawrence water-shed near S.-Barton station, whence Jay Peak is seen in the N. W. From W. Burke the charming excursion to Willoughby Lake (6 M. N.) may be made. The bold Burke Mt. is then passed, and at Lyndonville the headquarters of the Passumpsic R. R. are seen. **St. Johnsbury** (two good hotels) is a rich and prosperous town of 5,000 inhabitants, with an athenæum, an art-gallery, a large academy, and a soldiers' monument ("Crowning the Fallen Heroes"). The immense manufactories of Fairbanks' scales are established here, employing 500 – 600 men. At St. Johnsbury the present route crosses the Portland & Ogdensburg R. R. The train runs S. along the Passumpsic River, by McLeran's Falls and Barnet (famed for butter), to the mills at McIndoes Falls, on the Connecticut. The long ridge of Mt. Gardner now appears on the E., and is followed down to **Wells River.**

Wells River to the White Mts., see Routes 2 and 3.

9. Saratoga and Lake Champlain to the White Mountains.

The travel in this direction has hitherto passed over the Central Vermont R. R to White-River Junction, or, by changing at Montpelier, to the Montpelier & Wells-River R. R. Several changes of cars have to be made on either of these routes.

Stations.—Burlington to Winooski, 3 M.; Essex Junction, 8; Williston, 12; Richmond, 17; Jonesville, 20; Bolton, 23; Waterbury, 31; Middlesex, 36; Montpelier, 41. *M. & W.-R. R R.*— Coffee House, 43; E. Montpelier, 47; Plainfield, 50;

Nesmith Brook, 53 ; Marshfield, 56½ ; Kinney's Mill, 57½ ; Summit, 60½ ; Peabody's Station, 62½ ; Ricker's Mill, 66 ; Groton, 69 ; S. Ryegate, 72½ ; Boltonville, 75 ; Wells River, 73 , (Fabyan House, 114.)

The Central-Vermont section of this route leads through rich and picturesque regions; but the Wells-River section is dull, traversing long reaches of forest.

The Portland & Ogdensburg R. R.

promises to complete its Vermont divisions by the summer of 1876. This will then be the best route from the lake-country of New York and Vermont to the White Mountains, as cars will run through without change. Passengers from Burlington will take the Burlington & Lamoille-Valley R. R., which connects with the main line at Cambridge (no change of cars); and passengers from the N. can take the trains on the main line at Swanton. This line traverses the Green Mountains, just N. of Mt. Mansfield (which is reached by stages from Morrisville station), and opens a new region of interesting scenery. It then crosses the rich farming towns which have been called "the Garden of New England." The Connecticut River is bridged at Lunenburg and Dalton (see Routes 62 and 60), and the train passes S. E. into the White Mountains, and through the Notch and N. Conway to Sebago Lake and Portland. The Montreal branch of this route will leave the main line at Sheldon, affording still another way from Montreal to the mountains.

Stations. — Those between Johnson and the W. termini are not now located (Feb., 1876), but daily trains are running between Johnson and Portland. Johnson to Hyde Park, 7 M. ; Morrisville, 10 ; Wolcott, 18 ; Hardwick, 24 ; E. Hardwick, 28 ; Greensboro', 31 ; Walden, 39 ; W. Danville, 44 ; Danville, 47 ; St. Johnsbury, 59 ; E. St. Johnsbury, 63 ; W. Concord, 67 ; Lunenburg ; Dalton ; etc.

10. White-River Junction to the White Mountains.

The Connecticut & Passumpsic Rivers Railroad.

Stations. — White-River Junction to Norwich, 5 M. ; Pompanoosuc, 10 ; Thetford, 15 ; N. Thetford, 17 ; Fairlee, 22 ; Bradford, 29 ; S. Newbury, 33 ; Newbury, 36 ; Wells River, 40 ; Lancaster, 88.

Many tourists from the S. and W. parts of New England, and from the States beyond the Hudson River, would naturally approach the mountains in this direction. It is also on the direct line from New-York City by the Connecticut Valley. Parlor-cars run on this route, and sleeping-cars on the night trains. At White-River Junction the Passumpsic R. R. meets the Northern (N. H.) R. R., the Connecticut-Valley division of the Central Vermont R. R., and the main line of the latter route.

Soon after leaving the Junction the train crosses the White River and runs N. to *Norwich*, whence stages carry passengers across the Connecticut River to **Hanover** (*Dartmouth Hotel*), ¾ M. distant. This beautiful village is the seat of Dartmouth College, which has a large museum of the minerals and rocks of the mountain-district. Beyond this point the line crosses the Ompompanoosuc River and reaches the copperas-exporting station of *Pompanoosuc*. Sweeping around under Oak Hill, distant views are gained on the r., of Smart's Mountain and Moosilauke, over the thinly populated town of Lyme. From *Thetford* station stages run to Thetford village, W. Fairlee, Vershire, Chelsea, and Lyme. Following closely the course of the picturesque Connecticut, N. Thetford is reached, whence much copper ore is sent to Baltimore. The next station is *Fairlee and Orford*, with the Yosemite cliff on Mt. Fairlee on the l., and the embowered hamlet of Orford across the river on the r. Passing through the wide gap between Sawyer's Mountain and Soapstone Hill, Wait's River is crossed, and the train halts at the large village of *Bradford* (l. of the track), with Piermont Mountain across the river on the r.

The valley now opens on the r., and beyond S. Newbury the high-placed hamlet of Haverhill is seen on the bluff to the r. Towards and beyond the beautiful village of *Newbury* (where the Spring House is seen on the l.), the train runs near the famous Ox-Bow Meadows, one of the most fertile and valuable tracts of land in the two States. Interesting views of the Benton mountains are gained on the r., including Black Mountain, Sugar Loaf, Blueberry, and the lofty plateau of Moosilauke. Woodsville and Mt. Gardner are seen on the r. front; and the train soon enters Wells-River station, where passengers for the mountains change cars.

Wells River to the mountains, see Routes 2 and 3.

THE MOUNTAIN VILLAGES AND PASSES AND THE ADJACENT PEAKS.

11. North Conway.

Hotels. — The * Kiarsarge House is a spacious and comfortable hotel, accommodating 300 guests, and charging $4.00 a day and $14-28 a week. It is close to the railway station and the busiest part of the village, and thus loses the attributes of quietude and seclusion while it gains in convenience of access. The views of the mountains and meadows from its verandas are rich and extensive. The building is lighted with gas; and a portion of it is fitted with a steam-heating apparatus, for guests who remain through September and October. The house has a billiard-room, a barber-shop, reading-rooms, spacious parlors, a central rotunda which is the rendezvous of the summer population, and a high tower which commands a fine view of the valley and the White Mountains.

The * Intervale House is about 1½ M. from the village, and occupies a beautiful situation overlooking the meadows of the Saco. It accommodates nearly 150 guests, and its rates are $3.00 a day, and $10-18 a week. The Intervale station of the P. & O. Railroad is near this hotel. The Sunset Pavilion is at the N. end of the village, opposite Christ Church, and back from the street. Nearly 100 guests can be accommodated here, the rates being $3.00 a day, and $10-20 a week. This hotel is on Sunset Bank, a bluff which overlooks the Saco intervales and Moat Mt., with Kiarsarge and Washington on the r. The McMillan House is about 1 M. S. of the Eastern-Railroad station, opposite Sunset Hill, and near the Congregational church. It commands one of the best views of the White Mts. Nearly 100 guests are accommodated here, at $3.00 a day, $10-15 a week. The Artist's-Falls House is about 1 M. from the village, in a sequestered glen near the Green Hills. It has room for 75 guests, at $8-14 a week. The Washington House is near the P. & O. station; the N. Conway House and the Randall House are near the Sunset Pavilion; the Eastman House and Mason's Hotel are in the centre of the village. These houses charge $7-12 a week.

Boarding-houses. — Among the chief of these are J. C. Willey's and J. M. Seavey's, opposite the Kiarsarge House; the Dinsmore Cottage, near the post-office; T. C. Eastman's; and Whittaker's Echo House, near the post-office. There are also two or three boarding-houses to the S., on the Conway road. Near the Intervale House are Barnes's, Mrs. Pendexter's, Tasker's, and the Pendexter Mansion.

At *Kiarsarge Village* are the following summer boarding-houses: — The Merrill House and Summer House (75 guests each), Russell Cottage, Hillside House, Wheeler's Orient House, and Barnes's (40 guests each).

The rates at the boarding-houses are $7-14 a week, and their guests enjoy a greater degree of restful quiet than do the visitors at the hotels. The accommodations are usually comfortable, though simple; and the cuisine is neat and substantial, rather than of wide variety.

Railroads. — The Eastern Railroad runs two express-trains each way daily, between Boston, Lynn, Salem, Newburyport, Portsmouth, and N. Conway. Distance, 137 M.; time, 5-6 hrs.; fare, $5. (See Route 1.) The Portland & Ogdensburg Railroad runs three trains each way daily, between Portland, Sebago Lake, and N. Conway. Distance, 60 M.; time, 2½-3 hrs. The same line runs trains from N. Conway to Bartlett, Bemis Station, the Notch, and the Crawford and Fabyan Houses (see Route 5).

Stages leave N. Conway for the Glen House every morning and afternoon. Distance, 20 M.; time, 5 hrs.; fare, $3.

Post-Office and telegraphic facilities may be obtained in the village; and there are shops for the sale of photographs, clothing, drugs, confectionery, and other articles.

Churches. — The Congregational church is in the S. part of the village, toward McMillan's; the Baptists have the brown church nearly opposite the Eastern-Railroad station; and Christ Church (Episcopal) is in the N. part of the village.

Distances. — The following list is copied from the tariff of the Kiarsarge House, and shows the usual drives in the vicinity of N. Conway, their distances, and the price to be paid by each occupant of a seat in the carriages. (The Editor copies the statement of distances without indorsing it as correct.)   To the base of Mt. Kiarsarge, 5 M., $ 1.25 ; to Diana's Bath, the Cathedral, and Echo Lake, 3 M., $ 1.25 ; to the Artist's Falls, 1½ M., 75c. ; Around the Square, 5 M., 75c.; to the Jackson Falls, 9 M., $ 1.50 ; to Thompson's Falls, 4 M., $ 1.00 ; to the Bartlett Boulder, 7 M., $ 1.50 ; to the Washington Boulder, 6 M., $ 1.25 ; to Conway Corner and return via Conway Centre, 7 M., $ 1.50 ; to Thorn Hill, 9 M., $ 1.50 : to the Champney Falls, 16 M., $ 3.00 ; to the Carter Notch, 14 M., $ 2.50 ; to the Crystal Cascade and the Glen-Ellis Falls, 16 M., $ 3.00 ; to Humphrey's Ledge, 14 M., $ 1.50 ; to Fryeburg, 10 M., $ 2.00 : to Chocorua Lake, 15 M., $ 3.00 ; to Jockey Cap and Lovewell's Pond, 13 M., $2.00 ; the Ridge Ride, 5 M., $ 1.50 ; the Dundee Road, 12 M., $ 2.00 ; to Buttermilk Hollow, 16 M., $ 2.00.

The village of N. Conway is on a long terrace about 30 ft. above the intervales of the Saco and ¾ M. from the river.   The Green Hills guard it on the E., forming a double line of low shaggy summits near the street ; and on the W., across the Saco Valley, is the long and massive Moat Mountain, noble and imposing in its colors and outlines, and the most conspicuous object seen from the village.   A little E. of N., and about 4 M. distant. is the crest of the graceful pyramid of Mt. Kiarsarge, whose long slopes approach within 2 M. of the street.   To the N. N. W., about 16 M. distant, is the peak of Mt. Washington, about which several of the other main mountains are clustered.   In an opposite direction the valley of the Saco opens to the S., over long stretches of fertile lowlands, banded by the groves that enclose the river.   The village is 521 ft. above the sea, or 32 ft. lower than Centre Harbor.

N. Conway is the chief summer-resort among the White Mts., and is occupied by city-people from early May until late October.   The height of the season is in August, when over 3,000 tourists are sojourning here. During the heated term it is warmer than Bethlehem, but cooler than the villages of the lake-country.   Evening gayeties are much patronized, and there are hops, concerts, and readings in the halls of the chief hotels.   The adjacent roads are visited, every pleasant day, by riding parties ; and rambling pedestrians explore the neighboring forests and hills, or fish for trout along the falling brooks.   It is the beauty and variety of its environs that gives N. Conway the foremost rank among the mountain-villages, added to the fact that it is at the proper focal distance from Mt. Washington and its compeers.

The population of Conway was 1,607 in the year 1870, as against 1,765 in 1850.   It has four villages, N. Conway. Kiarsarge, Centre Conway, and Chatauque (or Conway Corner).   At N. Conway is the N.-Conway Academy, occupying the ugly little building near the Kiarsarge House; also the Kiarsarge Boarding-School, a semi-military Episcopal institution.   The upper story of the Academy has been used as the studio of George Inness, the famous landscape-painter.   In the S. part of the village is the pretty cottage and studio of Benjamin Champney.

" We struck across the valley, which is intersected by the Saco River. Never did valley look more delicious; shut in all round by mountains, green as emerald, flat as water, and clumped and fringed with trees tinted with the softest autumnal hues." (HARRIET MARTINEAU.) The Hon. Miss Murray likened this valley to those of Braemar and Invercauld, though giving Conway credit for greater sublimity.

" In Conway you see the curves of the hills on their long swell, rising slowly from valley to summit ; and, on the northern slope, the mountain-wave seems to have broken and rushed abruptly to the plain. Such is the general aspect of the landscape, and one can easily picture to himself a beauty of the scenery that is almost feminine, as it appears at Conway. Not only the hills, but the village itself, and the gentle meadows of the Saco, add to the soft charm of this very Arcadia of the White Hills. Here Nature seems for once to have thrown aside her harsh and severe character in this granite heart of New England, and to have abandoned herself to a genial and happy repose."

The most beautiful features of the scenery of N. Conway are the broad intervales of the Saco River, which spread a level floor of the richest verdure from the foot of the village-terrace nearly to the base of Moat Mt., over 1 M. distant, and extend for several miles to the N. and S. This scene imparts an Arcadian air to the quiet village, and quiets even the electric American mind by its sweet pastoral beauty. The lower portion of the intervales has been ruined by the erection thereon of the ragged embankments and unsightly trestles of a railroad; but rich views may still be enjoyed from the upper parts of the village and from the Intervale House. Closer views of the meadows may be obtained by descending the Artist's Brook, or by the road to the Ledges. It is said that the intervales were originally covered with forests of white-pine and maple, except in the vicinity of the Indian villages. " Game was nowhere so plenty; fish and fowl and animals were almost as thick as in the jungles of Africa." The earlier settlers built their rude houses on the meadows, but these constructions were swept off in the flood of 1800, and after that the people lived on the terraces above. The intervales throughout the town are from 50 to 220 rods wide, and are richer than those of the Merrimac Valley, though the season here is two weeks shorter.

The name of *Saco* is derived from three Indian words, *sawa*, "burnt," *coö*, "pine," and *auke*, "place." The river is here from 8 to 12 rods wide, and from 2 to 7 ft. deep. Its course is rapid, over a rough and stony bed; and it has been known to rise 30 ft. within 24 hours, flooding all the meadows and sweeping up against the flanking terraces. The base of the river-plain is sandy; and it is fertilized every spring by the disintegrated rocks of the Notch mountains, which are brought down by the raging waters and spread out over the valley.

" Now the sun sends mingled light and lengthened shadows over the picturesque labors of the haymakers, in the broad, green, beautiful meadows that spread, a mile wide, waving with grass and grain and patches of glistening corn, clear to the mountains' feet, to the hieroglyphic rocky faces of the curious ledges, that form its outposts in front, and to the winding Saco River, whose course is marked with gracefully overhanging elms and oaks and maples, that also stud the plain in scattered groups, and shade the brooks that ramble, musically gurgling, to the river. A lovelier plain was never spread before a poet's feet, to woo the willing thoughts abroad. A scene of plenty, purity, and peace. On our r., in the N., loom the White Mts., blue and misty, and yet boldly outlined. There is Mt. Washington, rearing his broad Jove-like throne amid his great brothers and supporters; these, with innumerable lesser mountains (each Olympian enough when clouds cap and conceal the grander ones behind them), gaze solemnly and serenely down our broad valley, and look new meanings in the ceaseless changes of the air and light."

Pleasant forest-rambles are found back of Christ Church, across the track, and also in the *Cathedral Woods*, N. of the village. Beyond the latter are the fine buildings of the Bigelow estate, though the mansion was recently destroyed by fire. A short distance beyond is the summer village near the Intervale House, far quieter than N. Conway, and with richer views over the Saco meadows and the great Moat range. This locality is esteemed by many as the most beautiful in the mountain-district.

The **Artist's Falls** are in the E. environs of the village, over 1 M. from the Kiarsarge House, and are visited by a road which turns to the E. from the highway just below the bridge over Artist's Brook. They are a short distance beyond the Artist's-Falls House, in a pleasant woodland region, with picturesque surroundings. The descent of water is small, and occurs on a rivulet which flows from the heart of the Green Hills. The falls are to be considered rather as a centre of pretty forest-scenery and the objective point of a short ramble than as intrinsically remarkable. The routes to Peaked Mt. and Middle Mt pass near this point.

The lower reaches of Artist's Brook are more beautiful, where, below the old mill, it broadens out into a quiet and meandering course over the rich intervales, among clusters of graceful elm-trees, and with noble views of the mountains on either side. This fair scene of summer-rambles may be visited from near the Washington House or McMillan's, though no convenient way of access has been prepared.

The view of the White-Mt. range from McMillan's is one of the best in the village, and opposite this point is the *Sunset Hill*, whence a broad and satisfactory prospect is afforded over the valley and the ranges beyond.

About 1 M. from the village, towards Artist's Falls, are *Eastman's Trout-Ponds*, where about 2,500 trout are kept. A small fee is charged for admission.

Kiarsarge Village is about 1¼ M. from N. Conway, on the road to Mt. Kiarsarge, in the upper valley of Kiarsarge Brook. It is a collection of summer boarding-houses, with a small Episcopal chapel; and is favorably situated with regard to Kiarsarge and the Green Hills.

The *Thorn-Hill Drive* is a favorite excursion from N. Conway. It passes the Intervale House and Lower Bartlett, and ascends one of the spurs of Thorn Mountain, whence is obtained a noble view of the mountains, — including Mts. Pleasant, Franklin, Monroe, and Washington, over the Eagle Mts. Boott's Spur is seen running S. E. from Washington, and portions of Tuckerman's and Huntington's Ravines are beyond. Mt. Adams is visible through the Pinkham Notch, and Mt. Wild-Cat is on its r. Several pleasant retrospects are obtained from this road.

Around the Square is a favorite drive, 5 M. long, which passes through Kiarsarge Village and along the base of Mt. Kiarsarge. The *Dundee Road* is a 10-M. drive, leading through a succession of sequestered glens, and affording interesting mountain-views. It diverges to the r. from the Glen

Road at Lower Bartlett, and ascends the narrow valley between Thorn Mt. (l. side) and Mt. Kiarsarge (r. side), then bears to the l. between Tin Mt. and the stately peaks of Double-Head, approaches Black Mt., and affords striking views of the Presidential Range. The return-drive may be made by way of Jackson and the Glen Road, Jackson being 9 M. from N. Conway. The Washington and Bartlett Boulders are two remarkable rocks which are much visited, the one being 5 M. S. (near Pine Hill), the other 6 M. N. Among the longer drives from this point are those to the Glen-Ellis Falls, 16 M.; Sligo, 13 M.; the Carter Notch, 12 M.; Fryeburg, 10 M.; Buttermilk Hollow, 12 M.; Swift-River Falls, 18 M.; and Chocorua Lake, 18 M. (See the Index, for localities.)

" One always finds, we think, on a return to N. Conway, that his recollections of its loveliness were inadequate to the reality. Such profuse and calm beauty sometimes reigns over the whole village, that it seems to be a little quotation from Arcadia, or a suburb of Paradise. . . . . Certainly, we have seen no other region of New England that is so swathed in dreamy charm. A few years ago the Moat Mountains were ravaged with fire; and yet their lines give such delight that few mountains look so attractive in verdure as they in desolation. The atmosphere and the outlines of the hills seem to lull rather than stimulate. There are no crags, no pinnacles, no ramparts of rock, no mountain frown, or savageness brought into contrast, at any point, with the general serene beauty. Kiarsarge is a rough and scraggy mountain, when you attempt to climb it, but its lines ripple off softly to the plain. Mt. Washington does not seem so much to stand up, as to lie out at ease across the north. The leonine grandeur is there, but it is the lion not erect but couchant, a little sleepy, stretching out his paws and enjoying the sun. And tired Chocorua appears as if looking wistfully down into

A land
In which it seemed always afternoon.

" And then the sunsets of N. Conway! Coleridge asked Mont Blanc if he had ' a charm to stay the morning star in his steep course.' It is time for some poet to put the question to those bewitching elm-sprinkled acres that border the Saco, by what sorcery they evoke, evening after evening, upon the heavens that watch them, such lavish and Italian bloom. . . . . For pomp of bright, clear, contrasted flames on a deep and transparent sky, the visitors of N. Conway, on the Sunset Bank that overlooks the meadows, enjoy the frequent privilege of a spectacle which the sun sinking behind the Notch conjures for them, such as he rarely displays to the dwellers by the Arno or the inhabitants of Naples.

" It would require more space than our volume will allow, to do justice to the various charms into which this wide circle of beauty is broken by walks and excursions and drives. One of the prominent pleasures of a clear and cool day is to find different points for studying Mt. Washington. In what novelties of shape, dignity, and effect he may be thrown by the rambles of a morning! We may see his steep, torn walls rising far off beyond a hill which we are ascending, and which hides from us most of the foreground . . . . ; or may catch a glimpse of him through a couple of trees that stand sentinel to keep other mountains of the range from an intrusion that will reduce his majesty; or may seek a position near a grove whose breezy plumes afford the most cheerful contrast of motion and color to set off his gray grandour and majestic rest; or from different points near the Saco may relate him, by changing angles, into fresh combinations with the level verdure of the meadows, or with some curve of its brooks, or some graceful thicket of its maples." (STARR KING.)

Conway was granted to Daniel Foster in 1765, and was soon occupied by the pioneers of a rude rural civilization, whose cabins supplanted the old Indian town. In 1772 a road was granted from Conway to the Connecticut River, by way of Gorham, Northumberland, and Lancaster. The early settlers of Conway were from Durham and Lee, towns near tide-water in the S. E. part of New Hampshire.

In June, 1775, Andrew McMillan sent down from Conway, praying that troops might be sent to guard the town, as 15 of her citizens were in the American army besieging Boston. The State sent a supply of powder to Conway, and during the following summer Capt. Joshua Heath was ordered to enlist ten rangers and scout through the Conway woods for three months. At this time (census of 1775) Conway had 273 inhabitants and 2 negro slaves. The Congregational church was organized in 1778. In 1781 a postal service was inaugurated here by the State, consisting of a mounted mail-carrier, who rode fortnightly from Portsmouth to Conway, Plymouth, Haverhill, Charlestown, Keene, and Portsmouth, his compensation each three months being "70 hard Dollars, or paper money Equivalent."

When the party of Drs. Belknap and Cutler passed through Conway, in 1784, "The good Women understands there were 3 Ministers in ye Compy were in hope we should *lay the Spirits* wh have been suppofed to hover about ye White Mountains — an opinion very probably derived from ye Indians who tho't thefe Mo ye habitation of some invifible beings —— & never attempted to afcend them."

In the early part of the present century, when all the freighting between the Coös country and the sea-shore passed this way, Conway was a busy place, its street being frequently traversed by wagon-trains a half-mile long. In 1797, President Dwight visited the town, and speaks favorably of the intervales and the view of Mt. Washington ; saying also that "The inhabitants appeared to be in comfortable circumstances, and the houses were decent."

Conway is underlaid by the so-called Conway granite. The geological theory is that near the close of the Labrador period a tremendous earthquake occurred in the White-Mt. region, followed by an eruption of igneous granite, which flooded the Franconia and southern ranges with a sea of fire. "Were there ships of steel, they might have floated on this liquid lake, for the surface was as level as the ocean." Its outlet was by the Saco Valley, and here "remove the overlying rock, and the top of the granite will appear as flat as a western prairie."

12. The Green Hills.

This cluster of highlands lies to the E. of N. Conway, and covers an area of about 16 square miles, having 8 well-marked summits in a double line, and facing to the E. on the plains of Fryeburg. The State Geological Survey gives their height as 2,390 ft., or more than 1,860 ft. above N. Conway, — but as to which of the peaks this measurement was made upon nothing is said. The altitude seems to be overstated. The lower parts of the range are partially covered with forests, and the interior glens are heavily timbered. Some portions have been burnt over by forest-fires, leaving bare peaks on the W., from which pleasant views are gained. Rude logging-roads ramify through the ravines, and ascend some of the slopes.

Artist's Ledge may be reached by a woodland path which crosses Artist's Brook near the falls and leads to the foot, whence the ascent is easily made over the smooth rocks. * *The view* from the Ledge includes Chocorua on the l., with a broad sweep of the Saco Valley to the N., bounded by the rugged ridge of Moat Mt. N. Conway is outspread on the terrace below, and over the Kiarsarge House is the White-Horse Ledge, over the Sunset Pavilion is the Cathedral Ledge, and over the Intervale House is Humphrey's Ledge. Farther to the r. are the bare rocks of Iron Mt., and Thorn Hill is at the head of the valley, flanked on the r. by Thorn Mt., and nearly over Kiarsarge Village. Farther to the r. is Kiarsarge, rising proudly over the Green Hills. Mt. Washington and the adjacent peaks are about N. N. W., nearly over Thorn Hill.

" As a composition, the view from Artist's Hill is very symmetrically proportioned, and is superior to any other in the variety and graduations of its forms. Mt. Washington, which is always the leading object of interest, occupies the central position. The inferior hills rise from the level meadows on either hand, step by step towards his summit, which dominates over the whole scene. In this noble symmetry of multitudinous details it differs from most of the other general views of the White Hills. . . . . In views like that from Campton, or the Artist's Hill in N. Conway, color is displayed, not in simplicity or sombre breadth, but in variety and splendor, and in the intermingling of several contiguous and contrasting scales. . . . . The blue-brown of Shelburne, the yellow-purple of Milan, and the violet-citrine of Campton, each showing an ascent of tone or hue, lead up to the orange-russet or purple, the most delicate, rich, and subtle of all, that dominates and typifies the unsurpassed magnificence of the color-harmonies of Conway." (STARR KING)

Peaked Mountain is easily ascended from Artist's Ledge, which is not far from the summit. The crest consists of a thin rocky ridge, with a few small trees upon it; and looks across a shallow but steep-sided ravine to Middle Mt., on the S. The Editor was overtaken by a heavy rain-storm while on this peak, and is therefore unable to describe the view, which, however, is not unlike those from Artist's Ledge and Middle Mt.

Blackcap, or *Blackhead*, is the highest of the Green Hills, and lies in the E. line, though it is visible from N. Conway. It is covered with trees, and is not valuable as a view-point. It may be reached by passing up the ravine N. of Peaked Mt., and ascending by the outer ridges. S. of this, and partially connected with Peaked Mt. by a low ridge, is the long crest of *Green Mt.*, which is overspread with light green foliage. *Rattlesnake Mt.* is S. of and lower than Middle Mt., and has no good views. Another long ridge is N. of Peaked Mt., running nearly to Kiarsarge, and covered with ledges and low, dense thickets. The first peak N. of Peaked Mt. is called Lookout Point; and the higher point to the N. E. (formerly called Green Mt.) is now known as Hurricane Mt.

The more northerly of the Green Hills are partially bare, and may easily be ascended from Kiarsarge Village and the inhabited glen to the N. E. They are not so valuable, for view-points, as Middle and Peaked Mts., because of the propinquity of Mt. Kiarsarge.

Middle Mountain is the most desirable point of the Green Hills on which to make an ascent. It is about an hour's walk from N. Conway to the summit, and the hill is over 1,500 ft. above the sea. The best route is by the road which diverges from the village-street alongside the Congregational church and runs across the P. & O railroad.

In the valley below it crosses a small bridge; and about 50 rods beyond, the road diverging to the l. should be taken. Following the main path, and disregarding the logging-roads which diverge to the l., r., and l., the tourist enters a clearing, on each side of which are the ruins of shanties. Bearing around and just beyond the old camp on the r., the hill-path turns off to the r., soon crosses a rude little bridge, and then ascends the slope. This path is clear and well-outlined, and can easily be followed. Another path diverges from the road before the shanties are reached, and ascends the ridge in the track of the old lumber-slide. The road which leads from the village to these points affords a pleasant forest-ramble, curving gracefully through fine groups of pine-trees, and being out of sight of clearings, fences, or houses.

The paths come out on the broad bare ledges which front the swell of the ridge, and it is difficult to follow them beyond. The remainder of the ascent is over the ledges, alternating with luxuriant jungles of sweet-fern, and striped with old burnt trees. Fascinating retrospects may be obtained during the resting-times along the slope. The crest consists of a long and narrow ridge, carpeted with grass, edged with ledges, and adorned with a few clean-stemmed trees.

One of the pleasantest features of this trip is the passage over grass-carpeted roads through tall forests. Much of this wood has already been cut away, but enough remains to form a forest where one can wander adventurously for many hours in a labyrinth of old and unused logging-roads, secure from intrusion, and free to follow fancy down the winding woodland aisles.

* *The View.* — The most conspicuous object to the W. is the long frowning ridge of Moat Mt., with its definite craggy crests boldly outlined against the sky. The S. peak rises from the Swift-River Valley by a long even slope, and is succeeded on the N. by rugged central ledges, which are separated by a strongly marked ravine from the high N. peak. Projecting from the base of Moat are the White-Horse and Cathedral Ledges, with their black-and-white fronts, below which is the village of N. Conway, on the edge of the broad and delightful Saco intervales. Over the r. flank of Moat, and to the l. of the line of vision extending to the White-Horse Ledge, is the lofty and slightly notched summit of Mt. Carrigain, its r. flank cutting the N. buttress of Moat over the White-Horse Ledge. The distant peaks to the r., above the Cathedral Ledge, are Mts. Anderson and Nancy, on whose l. is the crest of Lowell, just peering over the foot-hill of Moat, above White Horse. To the r. of Moat and over the r. of the Cathedral Ledge is the red peak of Mt. Crawford, with the table-land of Mt. Resolution on the r. and the Giant's Stairs still farther to the r. Above these are the crests of the Field-Willey Range, and Mt. Bond is seen far away, to the l. of Crawford. The long sloping ridge which terminates at Humphrey's Ledge runs out from Mt. Attitash, at the r. of Moat Mt., and over it in the N. E. is the ledgy and unpointed Iron Mt. in Jackson. When the Editor ascended Middle Mt., in September, 1875, the Presidential Range was hidden by a black and impenetrable embankment of massive clouds. It is, however, safe to say that Mt. Washington is visible, with Monroe, Franklin, and Pleasant on its l. and Adams on the r. Clinton is over the l. of Iron, and Pleasant over the r. of Iron. Over the l. of Mt. Bartlett is the sharp gorge of the Carter Notch, with Mt. Wild-Cat on the l. and the Carter Dome on the r. The narrow clearings of Jackson extend up toward the wooded Eagle Mts.

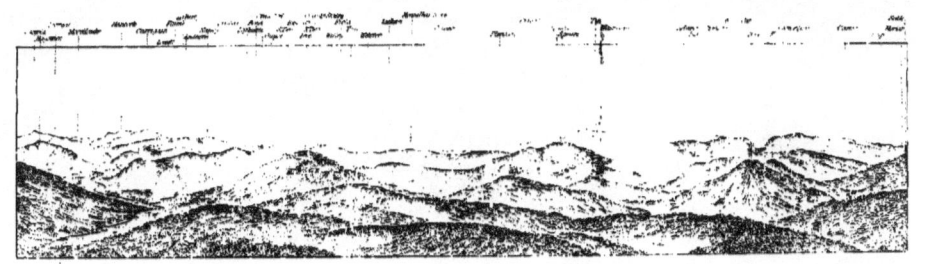

t
r
l:
fi

ii
a;
e\
a1
je
wi
w1
fla
Ho
its
Th·
anc
hill
Cat
of A
Abo
seen
nate:
Mt.,
son.
Presi
of in:
visibl
r. C
the l.
Wild-·
Jacks·

Nearer at hand is the rounded Thorn Hill, on whose r. rises Mt. Bartlett, prolonged into the higher crest of Kiarsarge, which is crowned by a hotel. A large area of Western Maine is shut out by Blackhead, one of the tall and wooded Green Hills which extends for a long distance to the N. E. Over its r. foot rises Mt. Pleasant, with its hotel, and Upper Moose Pond is seen on the l. Nearer, and in line with Pleasant, is Mt. Tom, a bold rounded knoll, with Pleasant Pond on the r. Under the latter is Lovewell's Pond, with its island, near which is Fryeburg village, on the green intervales and near the yellow coils of the sinuous Saco beaches. Still to the r. and farther away is Sebago Lake, which is nearly hidden by the Saddleback and Peaked Mts. The view now dwells on the fertile Saco meadows in the foreground, with the lower hamlets of Conway and the island-strewn surface of Walker's Pond, which is nearly due S., over the lowermost of the Green Hills. Several miles away, over Conway Corner, is a part of Silver Lake, with the Green Mt. of Effingham beyond. To the l. of this line are the hills along the Maine border, — Frost Mt. and the Burnt-Meadow Mts. in Brownfield, Mt. Cutler and Tear Cap in Hiram, and Trafton Mt. in Cornish. Farther to the r. is a part of the Ossipee Range, and then, with its r. flank resting on Moat Mt., comes the superb crest of Chocorua, its red rocks rising sharply into the N. and S. peaks and its long S. flank running out to hide Chocorua Lake. Below, in the plain, is the round knob which is called the Haystack. On the r. of Chocorua is Mt. Paugus, beyond which rises the round-crested Passaconaway. The narrow space between the latter and Moat Mt. is cut by the cone of Tripyramid.

The question as to whether Mt. Lafayette is visible from Middle Mt. has been settled in the affirmative. At a time in the early autumn, when the snow-line was at about 5,000 ft. elevation, a white crest is visible far away over the gap between Moat and Iron Mts. This can be none other than Lafayette.

13. Mt. Kiarsarge, or Pequawket.

Pequawket-Kiarsarge lifts its symmetrical cone into the blue sky over N. Conway, the most alpine of all the heights about that village, and it always preserves the same sharp outline, whether seen from Sebago Lake, Portland, the high peaks of Osceola and Moosilauke, or the houses on Mt. Washington. The bridle-path which traverses its rugged flanks is yearly climbed by thousands of tourists from the adjacent villages, some on horseback but most on foot. The Editor has passed over 50 persons ascending in scattered groups, within less than two hours. A carriage-road is now being built to a point near the summit, and when this is completed the number of visitors here will hardly be exceeded by those to Mt. Washington. The route is near that of the bridle-path. Although it is but little more than half as high as Mt. Washington, it commands a view which is but little inferior, while for the commingled charms of mountain, meadow, and village scenery, it has scarcely a rival. The height above the sea is 3,251 feet, as determined by the trigonometrical measurements of the U. S. Coast Survey. The early morning is the best time for the visit, for though the light is garish and white, the great ravines of the Presidential Range are then best seen. In the late afternoon, the ocean is clearly visible, and the level light falls brightly on the sails of the shipping.

Starr King called this "the queenly mountain," and desired that it should be named "Martha Washington." Theodore Parker spoke enthusiastically of its view of "the chief mountains *en famille.*"

The *Mt.-Kiarsarge House* is a small hotel on the summit, where visitors can get meals and lodgings. It can accommodate 15 – 20 guests, and is often patronized by people who wish to get the sunset and sunrise effects. The rates are $ 4 a day. The best general view is gained from its cupola, for here the observer is elevated high above the intercepting objects in the foreground.

"Do not some of our readers recall the fascination of the diorama exhibited to those whom Pequawket allows to pass above its elegant shoulders? Do they not call to mind the mob of mountains that first storms the sight from the N. and W., as though Mt. Washington had given a party, and all the hills were hurrying up to answer the invitation? Can they not see again with the mind's eye the different effects of color and shadow upon the hues of hills, according to their distance, height, and the position of the sun, and how they soon group themselves in relation to the two great centres, — the notched summit of Lafayette and the noble dome of Washington? Do they not recall the soothing contrast to these shaggy surges of the land in the far-stretching open country of the S., gemmed with lakes and ponds, brilliant with cultivation, sweeping out like a vast and many-colored sea . . . . over which, far in the S. W., the filmy outline of Monadnock gleams like a sail just fading out upon a vaster sea?"

The lower parts of Kiarsarge are composed of common and trachytic granites, with occasional limited areas of slate. The upper 2,000 feet consist of an igneous felsite, full of rounded pebbles and angular fragments of slate. This fiery flood of molten rock was thrown out before the Helderberg period, in the same eruption that formed the ridge of Moat Mt. The geographical position of Kiarsarge is in the towns of Chatham and Bartlett, about 4 M. from the Maine boundary.

The *Bridle-Path* is plain and well-marked, and affords as easy travelling as is possible with the heavy grade. It starts from a farm-house about 2 M. from N. Conway and ½ M. from Kiarsarge Village, where several well-trained saddle-horses are kept for the use of ladies or others who wish to ride up. $ 2 is charged for each horse, and $ 2 for a guide, who can take care of several horses. No extra charge is made if people keep the horses all night on the mountain. Parties of 2 – 3 persons are carried from N. Conway to the base of the mountain for 50c. each. The distance to the top is nearly 3 M., the following being the pedometric measurements: From the house to Prospect Ledge, 1 M. (less 1-20), thence to the spring, ½ M.; thence to the Bartlett path, ⅔ M.; thence to the summit, ⅓ M.

Crossing the fields behind the house, the ascent of the S. W. shoulder is soon entered upon, and is continued up to **Prospect Ledge**, whose vicinity is marked by a guide-board, near which a succession of remarkable outcroppings of breccia is crossed. There is a wooden seat at the Ledge, where one can rest and look over the valley.

The view from Prospect Ledge includes the Green Hills, Mt. Pleasant, the villages of N. Conway, Conway Centre, and Conway Corner, and a broad, rich area of the Saco Valley, extending far away into Maine. At the foot of Moat Mt. are the White-Horse and Cathedral Ledges, and to the r. is Tremont, with a part of Mt. Carrigain.

The path ascends through the forest to the spring, which is on the r., in a stone coping. About ¼ M. beyond, the woods are left behind, and the path comes out on broad ledges, which are followed around toward the E. Branch, with Mt. Bartlett close at hand on the l. Beautiful views are afforded on the W. and N.; and the path finally takes a turn directly up the mountain, soon reaching the summit. The time necessary to ascend from the base is 1½ - 2 hrs.

The old paths from Fryeburg and Lower Bartlett are still occasionally used. The former passed through the shallow depression between the Green Hills and Mt. Gemini, and ascended the E. S. E. slope. The Bartlett path led up from near the East-Branch House, along the slopes of Mt. Bartlett and over the connecting ridge, striking the Conway path far up on the mountain.

** *The View.* — The broad Saco Valley opens toward the W., banded horizontally by a narrow belt of clearings, along which stretches the sinuous river, blue at morning and silvery at evening, ruled off by the straight white line of the P. & O. Railroad track. On the r. the Rocky-Branch valley enters the Saco obliquely from the N. W., broadening out near the point of confluence. At the apparent head of the Saco Valley, and through the directly continuing depression of the tributary Sawyer's River, is the long blue ridge of Moosilauke, nearly 50 M. distant, with its crowning hotel visible by aid of the telescope. Nearer at hand, in the great angle between the Saco and the Rocky Branch, are the comparatively low and rocky summits of Mts. Stanton and Pickering, Willoughby Ledge and Hart's Ledge. Above, and to the r. of Moosilauke, is Mt. Hancock, on whose r. is the immense mass of Mt. Carrigain, rising from the S. The pyramid of Mt. Lowell is against the side of Carrigain, in poor relief. At the apparent head of the Rocky-Branch glen is Mt. Langdon, above which is Mt. Anderson (r. of Carrigain and Lowell), with the slide-striped Mt. Flume to the l., far away, and Mt. Liberty (of the Franconias) directly above. The chain of Lowell and Anderson is terminated on the r. by Mt. Nancy, over whose r. falling flank is Mt. Lincoln, continuous with which is the thin serrated crest of Lafayette, about 30 M. distant. To the r. of the Rocky-Branch glen is the well-defined mass of Iron Mt., banded with yellow ledges and cut by a broad, shallow ravine, and presenting a pointless summit. Over Iron, to the l., are Mt. Parker and the knob-like crest of Mt. Crawford; and over the highest part of Iron is the broad and high-lifted plateau of Mt. Resolution, with its fringe of red granite ledges. To the r., and falling on the r. end of Resolution, are the upper terraces of the Giant's Stairs. Above Iron Mt. and again above the Crawford range is the Twin Range, with Mt. Bond to the l. of Crawford and r. of Lafayette, partly rounded on the top. The South Twin Mt. is directly over the plateau of Resolution, and the North Twin, marked by a long white slide, is over the Giant's Stairs. From the uppermost of the sharply cut steps of Giant's Stairs, Mt. Willey rises precipitously to the r., on whose r. and in a continuous line are Mts. Field and

Tom, beyond the Crawford House. Under the latter is the descendi.ᵧ l. flank of Mt. Webster, with Jackson (on the r.) cutting the sky line with its small sharp peak.

Returning to the near foreground, Thorn Mt. is next seen, within 3 M. across the East-Branch valley, its sides and crest covered with trees. Over its highest point is the flat top of Mt. Clinton, in the Presidential Range; and over its right flank, up the glen of Miles Brook, is the hemispherical top of Mt. Pleasant. Over Tin Mt., the low knoll r. of and on the same ridge with Thorn, are the wooded lines of the Eagle Mts., above Jackson village. Farther toward the horizon is the flat terrace of Mt. Franklin, r. of Pleasant, to the r. of which are the two sharp and craggy humps of Mt. Monroe. The eye next rests on the vast cone of Mt. Washington, and if the day is clear the chief features of its nobly formed E. and S. flanks may easily be discerned. L. of and below the cone is the long embattled projection of Boott's Spur, on whose S. E. flank are unexplored ravines. Over the r. flank of the Spur is a portion of the upper reach of Tuckerman's Ravine. More to the r. are the Lion's-Head crags, the gorge in which Raymond's Cataract is situated, the broad, stony plain of the Alpine Garden, and the jutting rocks of the Nelson Crag, which rises at the head of the tremendous cliffs of Huntington's Ravine.

Entering the foreground again to the r. of Thorn Mt., the eye follows the pleasant Dundee Valley up to the pastured sides of the lowly Black Mt., over which is a heavy spur of the Wild-Cat range. Above and beyond the latter is the stately pinnacle of Mt. Adams, with the crag on the r. which looks down into King's Ravine. E. of Adams, and beyond the lake-infolding ridge between, is the blue, rounded summit of Mt. Madison. Close at hand, within 4 M. of Kiarsarge, are the nobly defined and singular twin peaks of Double-Head, apparently wooded, the S. peak cutting off the W. slope of the N. peak. Directly over Double-Head and Black Mt. is the sharp gorge of the Carter Notch, with Mt. Wild-Cat on the l., and the ponderous ridges of the Carter Dome on the r. To the r. of Double-Head is the low and shapeless Mt. Sable, above whose crest, and r. of Carter, is the high head of Mt. Moriah, nearly due N. On the r. of Sable succeed the dull rounding slopes of Mt. Eastman and the remarkable white crests of Baldface, extending in a long ridge to the S. Immediately to the r. of Baldface are the rock-ribbed round summits of Mt. Royce, between which runs the boundary of the State of Maine. Royce is remarkably picturesque towards evening, when its ravines are filled with black and opaque shadows, and its ledges are thrown out in vivid contrasting colors. Through the notch between Baldface and Royce are the remote blue peaks of Mt. Ingalls and Goose Eye, with other summits near the boundary; and above and between the peaks of Royce is the Bear-River White Cap, which is in the vicinity of the Grafton Notch, towards Lake Umbagog. To the r. of Royce, and separated from it by the

Brickett Notch, is Speckled Mt., a vast and shapeless mass, over which are the shapely crests of Sunday-River White Cap and Puzzle Mt., in Newry.

Mt. Slope is nearly in line with Royce, about 6 M. distant; and still nearer, a long and broken ridge runs N. N. E. from Kiarsarge, bearing the twin peaks which are known as *Mt. Gemini* on the old maps. (See Carrigain's map, of the year 1816.) Back of the line of Speckled are the lower ridges of Red-Rock Mt. and Mt. Calabo. About N. E. is the fine pyramid of Mt. Blue, in Avon, on either side of which, in a very clear day, may be seen still more remote peaks of the great Maine wilderness. The most important of these are Bigelow (over the r. of Speckled), Saddleback, and Abraham (on the l. of Blue), and Spruce and Averill (on the r.), with perhaps Katahdin.

The adjacent ponds to the E. next attract the attention, the first one on the l. being the long-drawn-out Upper Kezar Pond (8 – 10 M. in length), of which several detached sections are seen. Far out over the head of the pond are the bold hills of Albany and Woodstock; and near the central section is the hamlet of Lovell Centre, backed by the low and abrupt Mt. Sabattos. Nearer, and a little to the l., is the Charles Pond, r. of which is the hamlet of N. Fryeburg, with the larger Lovell Village over it. Still nearer are the Kimball Ponds, separated by a belt of woods, and situated on the State line ; while the dark and forest-environed Mirror Lake (or Shingle Pond) is seen below, almost on the flank of the mountain. Over the l. Kimball Pond is Kezar Pond, on the confines of Fryeburg and Bridgton.

Mt. Pleasant is seen toward the E., lifting its long ridge out of the apparent plains of Maine, with its central peak crowned by a large hotel. To the l. of Pleasant is Upper Moose Pond, over which is a portion of Long Pond, with Bridgton village on the W., and Bear and Hawk Mts. (in Waterford) on the N. Over Pleasant, and on the extreme horizon, the ocean may be seen, extending to the r. for many leagues, with Portland to the r. of Pleasant. This portion of the view is best seen at early morning, when the sun is reflected from the water, or about an hour before sunset, when the light falls nearly level on the sails of the shipping. 400 vessels have been seen off this shore at one time. The low but well-marked summit of Mt. Tom is seen under the r. flank of Pleasant, and just r. of the foot of Pleasant is Pleasant Pond, with a part of Moose Pond. To the r., and nearer, is the square Lovewell's Pond, bearing a single island, near which is the gray ledge of Jockey Cap. Nearer the beholder is the white village of Fryeburg, peeping out from its luxuriant groves on the exquisite intervales below. The Saco River is seen for many miles, winding through a well-cultivated and populous farming country, studded with small hamlets. To the r. of Mt. Pleasant, and over Pleasant Pond, is the beautiful expanse of Sebago Lake, broken by Raymond Cape and several islands. On the nearer side of the lake are the heights of Saddle-

back Mt. (in Baldwin) and Peaked Mt. (in Sebago). The village of Den-
mark is to the l. of and beyond Lovewell's Pond. To the S. of Kiarsarge,
and close by, are the Green Hills of Conway, the chief of which (on the
E) is Blackhead; and Peaked Mt. and Middle Mt. are farther S. A little
S. of S. E., and more distant, is a chain of hills in Maine, bounding the
Saco Valley, of which the chief are Frost and Burnt-Meadow Mts. in
Brownfield, Mt. Cutler and Tear Cap near Hiram Bridge, Bill-Merrill Hill
in central Hiram, and Trafton Mt. and its connected heights in Cornish.
Over the Green Hills is the island-studded Walker's Pond, near Conway
Corner, over which is a part of Ossipee Lake. In the foreground are
Silver Lake (in Madison), Knowles Pond, and Pequawket Pond, with the
long plains of Madison and Eaton. The Gline Mt. in Eaton lies nearly S.,
over which is the low and broad-based ridge of Green Mt. in Effingham.

The view next includes the blue crest-line of the Ossipee Range, on
whose r. summit the Melvin Peak is conspicuous. Far down the plain, to
the r. of this range, a semicircular dot on the southern sky, is the summit
of Monadnock, 100 M. distant, near the Massachusetts line. A short distance to the r., and resting on the l. flank of Chocorua, is the sharp peak
of the true Kearsarge, nearly 60 M. distant. The field of vision now includes N. Conway, close at hand below, with the outspread intervales of
the Saco on the W., over which towers the imposing ridge of Moat Mt., all
its peaks being visible. On the farther edge of the interva'e are the White-
Horse and Cathedral Ledges, with Echo Lake below the former. The main
peaks of Chocorua are uplifted with powerful effect over the S. end of
Moat, cutting the sky-line with their sharp serrated and craggy edges.
Over the middle of Moat is Paugus, and over the N. peak is the swelling
dome of Passaconaway, with Whiteface on the l., while on the r. are
the bristling points of Tripyramid, over Bear Mt. Looking more to
the W., over that high spur of Kiarsarge which is called Mt. Bartlett,
Humphrey's Ledge is seen making out from Mt. Attitash, with Table
Mt. scarcely distinguishable over it, and the bold but formless pile of
Bear Mt. still higher up. Over the r. shoulder of the latter is Tecumseh, with a curious knob on its r. Up the Saco Valley, on the l., is Tremont, with its symmetrically rounded peaks, below and against which is
the conical Haystack. To the l., and more distant, is the long apparent
ridge near the Greeley Ponds; and over the S. peak of Tremont is the
stair-like ridge of Osceola.

Concerning the Name of "Kiarsarge."

The old theory of the origin of the name *Kearsarge* (of which *Kiarsarge* is a
modification) was, that the mountain in Warner which bears it rightfully was
frequented, about the middle of the last century, by a hunter named Hezekiah
Sargent, and that the name was derived from a popular abreviation, the mountain
being called 'Kiah Sarge's Mountain Recent investigations have thrown doubt on
the hunter's existence, and have proven that it was so named before 1725, full 25
years before he is reported to have seen it. The name is therefore probably derived

from the Indians, perhaps from a corruption and clipping of *Cowischewaschook*, a title which they are known to have applied to it, and which means "Notch-Pointed Mountain of Pines." A simple and more natural derivation is that given by certain older antiquarians, who said that it was a euphonization of *kees*, "high," and *auke*, "place."

It has long been regretted that two of the most famous peaks of the New-Hampshire mountains, visible from each other, and of nearly equal repute as summer-resorts, bear names which are almost exactly similar. One of these is Mt. Kearsarge in Warner, the other is Mt. Kiarsarge near N. Conway, and although the orthography is slightly different the pronunciation of the two words is usually the same.

The presence of two important mountains bearing similar names in the same State has given rise to so much confusion that many persistent but unavailing efforts have been made to restore to the Conway peak its supposed ancient name of Pequawket. In the sumptuously prepared report of the Geological Survey it is everywhere called by that name, and the U. S. Coast Survey uses the same.

The earlier name of this mountain was Pequawket, derived from the Indian name for all this portion of the Saco Valley (see *Fryeburg*). On Holland's map (of 1760) it has no name, but on Carrigain's map (of 1816) it is called "Pequawket, formerly Kearsarge," indicating that the latter title was applied between 1760 and 1816, and that attempts were already being made to give the name of Pequawket. Within the time indicated occurred a migration of settlers from the vicinity of the true Kearsarge, who probably reasoned, "Lo, here is a goodly mountain, like unto that whence we came. Let us therefore have a new Kearsarge in this place."

The Editor has adopted the name Kiarsarge in view of its strong local tenure, believing that it cannot now be displaced, and that it will be less confusing to tourists. Judge Joel Eastman, of Conway, says: "All, from the oldest to the youngest, still call the mountain *Kiarsarge*, throughout this section of country, and any attempt to change its name will be futile. It will still go by the name of Kiarsarge until the Day of Judgment, and afterwards, if the memory of the things of this world remains after that day. . . . . The people of Boston, New York, and all the cities out of New Hampshire may call the mountain *Pequawket*, and that won't alter its name, so long as its godfathers and the generations after them call it *Kiarsarge*." Judge Lory Odell says: "The new name given to this mountain by Carrigain more than half a century ago has never been adopted or used by the Pigwackets (people of Conway) at all. By them it is now and always will be called Kearsarge. They know it by no other name. . . . . But in addition to the fact that the members of my tribe do still and always have stuck to the name of Kearsarge, I have another reason, in a fact which I would have perpetuated, for not adopting Carrigain's new name of *Pigwacket*. This reason is founded upon the fact stated in a recent letter to me from the Hon. G. V. Fox, formerly Assistant Secretary of the Navy, in which he says: 'When we selected the name for the *Kearsarge* sloop-of-war, in which the gallant Winslow sunk the *Alabama*, *we had no thought whatever of the Kearsarge in Merrimac County*. Some one interested in the Bartlett Mountain, the highest Kearsarge and commanding certainly the finest view, embracing as it does the beautiful Conway intervale, ought to write its history and correct, as I have often done, the error concerning the naming of the sloop-of-war *Kearsarge*.'" This statement is however strongly controverted by Dr. Bouton, the State historian, and other gentlemen. The Editor has seen letters from the Hon. Gideon Welles which lead him to believe that the frigate was named for the Mt. Kearsarge in the town of Warner.

14. The White-Horse and Cathedral Ledges.

These remarkable cliffs are on the W. side of the Saco, and attract many visitors. Hitherto they have been inaccessible to pedestrians, since the road leads down to a ford, where the water is too shallow to swim and too deep to wade, with a bottom of rolling stones. In and just after rainy weather, the swiftness of the river has prevented carriages from fording; and the only route to the Ledges was by the Conway bridge, several miles below. Preparations are being made to build a new bridge opposite N. Conway, at a cost of $20,000, in the year 1876.

Perhaps the most interesting feature of this excursion is the long drive across the intervale, in the midst of a rich agricultural plain and with picturesque mountains on all sides. Fine views of Moat, Kiarsarge, Double-Head, and the Green Hills are enjoyed from the meadow-road. (See also page 73.)

The base of Moat Mt. is surrounded with singular semi-detached bluffs, ending in bold rocky faces, and covered with trees. The most notable of these are the Eagle Ledge and the Haystack, on the S., near the Swift River; the White-Horse and Cathedral Ledges, W. of N. Conway; and Humphrey's Ledge, on the N. The **White-Horse Ledge** is about 700 ft. high, and derives its name from the fancied resemblance of a light-colored spot on its front to a white horse dashing up the cliff. New-comers at N. Conway are seldom allowed to rest until they have seen, acknowledged, and complimented the equine form of this amorphous spot. The lines streaming down the cliff are the result of rain-water, changing the color of the granite and renovating the vegetation.

Echo Lake is about 3 M. from N. Conway, across the Saco, and is visited by turning to the l. from the bridge-road and descending the first divergent road on the r., which reaches the shore in a few rods. There is a restaurant here, with boats on the water, and a small cannon which is fired to arouse the echoes, on the payment of a small fee. The lake covers but a few acres, and has a bright sandy shore. The White-Horse Ledge hangs over it on the W., and is reflected as in a mirror, showing its buff, purple, and gray tints in rich duplication. The most favorable time to visit this point is at late afternoon, towards sunset.

Thompson's Falls are about 4½ M. from N. Conway, and are not often visited. They are behind the great bluff of the White-Horse Ledge, and are formed by a small stream that afterwards empties into Echo Lake. The falls are graceful, rather than otherwise notable, and are surrounded by great masses of shattered rocks.

The Cathedral is 3 M. from N. Conway, and is reached by crossing the Saco and passing over the W.-side road through a gate between stone blocks, whence a short and pleasant road leads through the woods to the base of the ledge. A path leads thence up over the *débris* into the great arch under the black overhanging cliffs. The Cathedral Ledge is just N. of the White-Horse Ledge, and rises to a height of 960 ft. above the Saco meadows. There is a small cavern among the boulders at the foot of the Ledge, which is called the *Devil's Den.*

" An easy climb of 100 ft. carries one to a singular cavity in this Ledge, which visitors have named ' The Cathedral.' And truly the waters, frosts, and storms that scooped and grooved its curves and niches seemed to have combined in frolic mimicry of Gothic art. The cave is 40 ft. in depth, and about 60 in height, and the outermost rock of the roofing spans the entrance with an arch, which, half of the way, is as symmetrical as if an architect had planned it. . . . . The whole front of the recess is shaded with trees, which kindly stand apart just enough to frame off [Kiarsarge] in lovely symmetry, — so that a more romantic resting-place for an hour or two in a warm afternoon can hardly be imagined."

Diana's Bath is a little over 3 M. from N. Conway, on the W. side of the Saco, and is reached by turning to the N. on the W.-side road and passing the Cathedral Ledge, then turning to the l. on a well-travelled side-road, which soon conducts to the little refreshment-booth near the Bath. A short forest path leads thence to the ledges above. A small and limpid stream which flows out from the ravine under the N. peak of Moat Mt. and S. of Mt. Attitash here glides over a nearly level floor of granite, and makes a plunge of 10 ft. into a deep and rounded basin, which is filled to the brim with crystalline water. Along the stream, in this vicinity, are numerous potholes, some of which are ten ft. in diameter and ten ft. deep, still containing the rounded stones which have cut them out. Logging-roads penetrate the forest in various directions, giving opportunity for pleasant rambles among the quiet woods. The Bath is on Cedar Brook.

Humphrey's Ledge is 6 - 7 M. from N. Conway, and is reached by keeping to the N. on the W.-side road. It is a bold and rocky bluff, which is connected with Mt. Attitash, the N. ridge of Moat Mt., by a long ascending spur. It may be climbed, without difficulty; and the view from its summit is beautiful, including a wide area of the intervales, with the environing mountains.

15. Moat Mountain

is the fine ridge which borders the Saco Valley on the W., opposite N. Conway, and is a very conspicuous feature in the views from that village. It has been burnt over several times, and even the soil has thus been destroyed, so that it presents a remarkably bare and rugged appearance. The crest-line of the ridge is about 3 M. long, consisting of the N. and S. peaks and several intervening rocky hummocks, separated by shallow ravines, in which are burnt and fallen trees. The N. peak is 3,200 ft. high, and the S. peak 2,700 ft. The walk along this noble ridge is one of the most fascinating among the mountains, and gives views of scores of famous peaks, with the villages and fair meadows of Conway on the E. The people of N. Conway should have a path cleared and definitely marked to the summit of the N. peak, and thence S. through the tree-strewn hollows on the ridge.

" Less interesting in shape than many of the other ranges of hills in this neighborhood, Moat Mt. has remarkable beauty and variety of color, when the great masses of rock that largely compose it expose their red and yellow and purple surfaces over great areas, made desolate by the burning of the woods along its sides. Here are seen the last red clouds of sunset, and above its ragged summit lingers the last glow of the evening sky. . . . . Frequently by day the farms and orchards that cover its base are bathed in bright sunshine, while the upper regions of the mountain are hidden by dense and dark thunder-clouds, which roll about in round masses dun as smoke." (*Appletons' Journal.*)

The usual route up Moat Mt. is thus marked out : Follow the cart-road up stream from Diana's Bath (see above), crossing two fences and a corduroy bridge, and for the next mile avoid swinging to the l. Take the foot-path at the end of the road, and after a long stretch of good walking cross the Cedar Brook at a ruined bridge, and recross it about 100 yards beyond. Then ascend the S. bank through light underbrush to the lower ledges near a tall tree, whence a knoll is seen and aimed for ahead. Thence a bare ledge conducts up on the shoulder, the way leading between two large boulders. From the crest of the shoulder the top is visible, but an area of burnt and fallen trees must be crossed, and also a swamp, in which water can be found, either in the brook (which should be followed) or in the pitcher-plants. A gentleman of Salem, who has ascended by this route eight times, thinks that it is about 4 M. from Diana's Bath to the summit. He has made the ascent and return (to the Intervale House) in 5½ hrs.

The times and pedometer-measurements on this ascent of Moat Mt. are as follows : From Diana's Bath to the point where the stream is left, 1 hr., 1.65 M. ; thence to bare ledge, 25 min., ⅛ M ; thence to bare pole, ¼ M., 18 minutes ; thence to two boulders, ⅛ M., 25 min. ; thence to the brook on the shoulder of the ridge, ¼ M., 20 min. ; thence to the N. peak, ½ M., ½ hr. ; thence to middle peak, 1¼ M., ¾ hr. ; thence to the "Bear Peak," ¾ M , ½ hr. ; thence to the S peak, 9 M , 30 min. ; back to middle peak, 1.65 M., 50 min. ; to school-house on main road, 5 M., 2¼ hrs.

The ascent of Moat has also been made by passing around Echo Lake and entering the forest beyond the White-Horse Ledge, clambering directly to the N. peak. The chief obstacle on this route is a short but tangled cedar-swamp.

There is a good path leading from B. Farnum's 3 – 4 M. from Conway Corner, on the Swift-River road, which reaches the top of the S. peak of Moat in about 1½ M. It was made by the people who ascend to pick blueberries, vast numbers of which are found along the lower and middle slopes. Care must be taken not to lose the path where it crosses the bare ledges, as such wandering would necessitate a toilsome ascent through tangled thickets.

The following extract from a letter written by a gentleman who lived at Conway for many years, will explain the origin of the name of this mountain, and establish its proper orthography : " I was born in this town (Conway) more than 70 years ago. All I can say about the origin of the name of the mountain is that for as long as I can remember, people in other parts of the town, when they went into the district on the W. side of the river, always called it going up to the Moats or over to the Moats, and they called the mountain Moat Mountain. When I was younger than I am now, I asked old people why this district was called the Moats, and the mountain Moat Mountain. I was told that beaver-dams were found along the foot of the mountain, which were called 'moats,' and from that cause the mountain was called Moat Mountain " It may also be worthy of note that this name, "Moat," first appears on Dr. Belknap's map of the State, dated 1791.
Geologically, the Moat Range is the newest of the White Mts.

The N. peak of Moat Mt. is about 3,200 ft. high, or only 50 ft. lower than Kiarsarge. It is composed of high ledges, and is not encumbered by trees or bushes. In its N. side is the ravine of Diana's Bath, on whose farther side, and lower than Moat, is the ridge of Mt. Attitash.

* *The View.* — The long village of N. Conway is looked down into across the rich and variegated meadows of the Saco, and its great hotels are seen in all their external details, with the quaint Muscovite domes of the rail-road-station on the edge of the terrace.   Over the ledge at the foot of the mountain is Kiarsarge Village, and the Intervale hamlet is farther up the valley, towards the pyramidal peak of Kiarsarge.   Back of N. Conway are the Green Hills, with Peaked and Middle Mts. laid in against the higher hills on the E.   The Green Hills do not reach the sky-line, and over them is seen a vast area of Western Maine, in which N. Fryeburg, Lovell, and other hamlets appear, with Mt. Sabattos, and the highlands of Water-ford and Norway.   Over the r. slope of the Green Hills is the long bulwark of Mt. Pleasant, with a hotel on its middle peak, below which are Pleasant Pond and Lovewell's Pond, the rocky mound of Jockey Cap, and the white village of Fryeburg.   The long sweep of the Saco Valley, from Lower Bart-lett to Fryeburg, is the most fascinating element in the picture, and the ser-pentine river winds widely through its rich meadows, by clusters of graceful elms and before sequestered farm-houses.   To the r. of Mt. Pleasant is the silvery gleam of Sebago Lake, with Saddleback Mt. plainly in sight, and the Frost and Burnt-Meadow Mts. of Brownfield nearer at hand.   About S. E., over the Burnt-Meadow Mts., is the city of Portland, on the edge of the hori-zon, with Casco Bay beyond.   The high hills of Hiram, Mt. Cutler and Tear Cap, are in this direction.   Conway Centre shows its white houses farther to the S., near the long sheet of Walker's Pond, with the highlands of Cornish and Limington far away beyond.   Then come the nearer ranges of Eaton and Freedom, — Cragged and Bickford Mts., on the State line, and Gline and Lyman Mts. on the W., with Mt. Prospect farther S.   Conway Corner is then seen in the foreground, near the S. peak of Moat, which appears at the end of the long rocky ridge.   On its r. the Green Mt. in Effingham cuts the sky-line, on whose r., and nearer, are the broad, bright sheets of Ossipee Lake and Silver Lake, with Whitton Pond still nearer, over which is W. Ossipee.   The Ossipee Range next meets the horizon with its long blue line; and between Eagle Ledge and the massive buttresses of Chocorua the valley of the Swift River enters, running to the W. for many miles, across which are the vast ridges and blanched peaks of Chocorua, cut into by the profound ravine of the Champney Falls.   To this succeeds the Sandwich Range, — the low and craggy Paugus, the pointed crest of Whiteface, and the dark swell of Passaconaway, with the white mound of Potash at its foot, over which is the sierra of Tri-pyramid.   A little S. of E., and to the r. of Tripyramid, is the distant ridge of Moosilauke, near the Connecticut River.

Table Mt. is about 2 M. distant, and is marked by a white spot, over which rises the long and uneven crest-line of Bear Mt., with the lofty peaks of Tecumseh and Osceola beyond.   Farther to the N. is the bold prominence of Green's Cliff, flanked by the well-marked white knobs of Tremont, with Haystack nearer and less lofty.   Then Hancock comes into view, with the formidable notched summit of Carrigain on its r., and the peaks of Lowell, Anderson, and Nancy extending towards the E. Farther away, just over the r. flank of Carrigain, is a portion of Mt. Lafayette. Mt. Bond is over the r. of Anderson; and the Twin Mts. appear over Mt. Nancy, under which is Hart's Ledge. Farther to the r. is the lofty double crest of Mt. Willey, and the red overhanging peak of Mt. Crawford is directly in the White-Mt. Notch, with the rounded disk of Mt. Willard on the r. Adjoining Crawford is the flat top of Mt. Resolution, with the Giant's Stairs beyond, their characteristic shape not discernible. Somewhat nearer appear the cone of Parker and the ledgy sides of Langdon, over which is the Montalban Ridge, hiding all but the peaks of Webster, Jackson, Clinton, Pleasant, Franklin, and Monroe. Washington then appears, W. of N., with the lofty plain of Boott's Spur wheeling out from its S. flank, indented by the Gulf of the Slides. Close at hand, across the Saco Valley, are the crags of Mts. Stanton and Pickering, over whose r. are the reddish ledges of Iron Mt., far beyond which, and over the high cliffs of Huntington's Ravine, on the r. of Mt. Washington, are the noble peaks of Adams and Madison. The view now passes over the Jackson glen, with its little hamlet, and the partially cleared sides of the low Eagle and Black Mts., to Mt. Wild-Cat, on the r. of the Pinkham Notch, which is separated from the Carter Dome, on the r., by the wedge-shaped Carter Notch. Farther to the r. is the massive ridge of Moriah, over Black Mt., with Mt. Ingalls beyond, in upper Shelburne; and Goose-Eye is still more distant.

Nearly N. N. E., across the Saco Valley, is Thorn Hill, back of which is Thorn Mt., and Double-Head appears beyond, over which, and r. of Mt. Sable, are the white peaks of Baldface. Nearly over the indented summit of Eastman is the rounded swell of Mt. Royce, on whose r. is the long, bold ridge of Speckled Mt. As the view extends over Humphrey's Ledge and up the E.-Branch Valley, l. of Kiarsarge, several conspicuous peaks are seen in Western Maine. The thin ridge which runs to the N. from Moat, around the head of the Diana's-Bath Ravine, heads into the lower summit of Mt. Attitash, which is practically a spur of Moat Mt.

The ridge extends S. from the N. peak and soon breaks down in a series of great steps. Passing thence over a minor hummock, a curving hollow is traversed, and the central peak is crossed. On its S. side is a deeper hollow in which the traveller will be embarrassed by the fallen trunks of burnt trees, over which he must clamber. This is the only obstacle in the walk along the ridge, and extends but a few rods. The higher ground towards the S. peak is now ascended, and that peak is soon reached.

The S. peak is 2,700 ft. high, and is surmounted by high bare ledges. It falls off on the S. into the valley of the Swift River, whence it is easily ascended.

* *The View.* — Toward the N. E. is the village of N. Conway, so near that most of its buildings can be distinguished readily; and Kiarsarge Village is a little to the l. In the foreground, and extending for several miles to the N., are the lovely intervales of the Saco, whose predominant color is a rich green, alternated occasionally with the lighter colors of grain-fields or the deep browns of ploughed ground. Through this fair plain the Saco River runs in a series of graceful meanderings; while picturesque groves are scattered here and there along its sinuous banks. Near the foot of the mountain Echo Lake is seen, over which is the Intervale House; and the stately pyramid of Mt. Kiarsarge is still farther to the N. E., capped by a square-sided house. To the l. of the lake is the White-Horse Ledge, beyond which the valley is seen, running N. into the mountains, above the houses of Lower Bartlett. Over N. Conway are the graceful rolling lines of the Green Hills, through whose N. depression is a part of Upper Kezar Pond, in Maine. The Peaked and Middle Mts. are near the lower part of the Green Hills.

Nearly due E. is the long ridge of Mt. Pleasant, with a hotel near its centre; and farther to the r. is Lovewell's Pond, near the village of Fryeburg and the rock of Jockey Cap. In this direction the view extends for leagues down the Saco Valley, passing over Conway Centre. To the r. of Mt. Pleasant a portion of Sebago Lake is seen, flanked on the r. by Saddleback Mt., in Baldwin. Over Walker's Pond, which is nearly S. E., and 7 M. distant, the Burnt-Meadow Mts. of Brownfield appear; far beyond which the city of Portland and the waters of Casco Bay may be seen, if the day is clear. The hamlet of Conway Corner is 5 M. S. E., beyond which are Cragged, Legion, and other mountains near the State line, with the Cornish hills still farther away, over the lower part of Walker's Pond. The yellow and sinuous band of the Swift River runs from Conway Corner towards Moat, and is bordered by occasional clearings. Farther to the r., and more distant, are Gline and Lyman Mts., in Madison, over which is Mt. Prospect, in Freedom; and still farther to the r. is the broad-based bulk of the Green Mt. in Effingham. To the r. of the latter is the blue mirror of Ossipee Lake, flanked by the nearer sheets of Silver Lake and Whitton Pond.

The Ossipee Range is due S., many leagues distant, and runs to the r. until it meets Chocorua. The E. peak of Chocorua is first seen, next to which stands the majestic higher crest, about 6 M. distant, across the Swift-River Valley. To the r. of the W. peak of Chocorua is the lower top of Mt. Paugus; and then comes Whiteface, towering over the long ridge of Passaconaway, whose peak lies to the r., and is marked by a slide. At its foot is the white and skull-shaped Potash, over which appears the high three-headed ridge of Tripyramid. The course of the Swift-River Valley is then followed to the W., ending at the white crest of Tecumseh, over the r. flank of Tripyramid, and near the Pemigewasset Valley ; on whose right is the tall and remote peak of Osceola. Still further to the W. is the dim blue ridge of Moosilauke.

In the nearer foreground are Bear Mt., with white spots on its side, and the flat-topped Table Mt.; over which are the white caps of Tremont and the pointed cone of Haystack. On the l. of and nearly over Tremont is the lofty Mt. Hancock, falling away rapidly on the l.; and the still higher peak of Carrigain is on the r., over Haystack. The sharp, black spire of Mt. Lowell appears on the r., with Mts. Anderson and Nancy farther to the E. Through the gaps between these the blue ridges of the Twin Mts. are visible, over the W. spurs of Moat; and farther to the right is the high crest of Mt. Willey.

The view next rests on the N. peak of Moat, to which a long and nearly straight ridge leads up. On its r., and far away, is the high hemisphere of Mt. Pleasant, with the plateau of Franklin beyond, on the r., and the crags of Monroe still higher. More to the N. E. is the crest of Washington, towering over and behind the white ledges of Iron Mt., with the alpine plain of Boott's Spur extending to the r. Nearly N., through the Pinkham Notch, the clear-cut pyramid of Adams is seen, flanked by the lower crest of Madison. On the E. of the Notch is Mt. Wild-Cat, which is separated from the Carter Dome by the profound ravine of the Carter Notch. The pleasant glen of Jackson is below and toward Mt. Carter, and contains the pasture-dotted hill-ranges of the Eagle and Black Mts. Thorn Mt. is on the r. of the valley and l. of Kiarsarge, and beyond it are the mamelons of Double-Head, the high pallid crests of Baldface, the wooded summits of Eastman and Slope, and the rounded ledges of Royce.

16. Conway.

Hotels. — The Conway House is a large and well-built hotel, about ¼ M. from the station ($ 3 a day, $ 8 – 10 a week). The Grove House and Pequawket House are also in the village, and charge $ 6 – 9 a week.

Railroads. — The Eastern R. R. has a station at this point, 132 M. from Boston; and there is a station on the Portland & Ogdensburg R. R. within 3 M.

Distances. — To the Washington Boulder, 1 M ; Allard's Hill, 2 ; N. Conway, 5 ; Echo Lake, 6 ; Buttermilk Hollow, 5 ; Potter's Farm, 7 ; J. Piper's (path up Mt. Chocorua), 6 ; Chocorua Lake, 9 ; Fryeburg, 8 ; Chatham, 16 ; Swift-River Falls, 8 ; Champney Falls, 10 ; Shackford's (Swift-River Intervale), 14. The Conway House has a good livery-stable.

The hamlet which is usually called Conway has also the names of Conway Corner and Chatauque (the latter being an ancient and popular local title). It is pleasantly situated on the rich level land at the confluence of the Pequawket, Swift, and Saco Rivers, and is surrounded by picturesque rural scenery, without much sensation or variety. Chatauque is usually visited by people who wish to enjoy quiet rest and pure mountain-air, and who dislike the brilliant display of N. Conway. It is 5 – 6 M. from the latter place, by an uneventful road. There are a few small mills here, and a Congregational church.

The high peak of Mt. Washington is seen from this point, far away up the Saco Valley, with the ponderous mass of Moat Mt. near at hand on the l., and the graceful pyramid of Kiarsarge on the r. About 1½ M. N. E. is the shaggy top of Pine Hill; and the foot-hills of Mt. Chocorua are near by, on the W. The road on the W. bank of the Saco may be followed to the N. for 6 M. to Echo Lake and the Ledges (see page 85), whence the return may be effected by crossing the river and descending through N. Conway. The *Washington Boulder,* which some have called the largest in the world, lies 1 M. to the N. E. of the village, and is well worthy of a visit. *Allard's Hill* is a low spur of Mt. Chocorua, about 2 M. from Conway. It commands a beautiful view over the Saco Valley, and around the environing mountains. This is one of the pleasantest short walks in the vicinity.

Conway Centre is a small hamlet near the Saco River, about 3 M. N. E. of Conway, containing the mansion of Judge Joel Eastman. *Conway Street* is a line of farm-houses in the E. part of the town, N. of the Saco, and fronting on the pleasant intervales. It lies between Green Mt. and Fryeburg, and is on the W. road to Chatham.

Buttermilk Hollow is 5 M. from Conway, and is reached by passing down over the old county road. It is a beautiful little glen, containing a picturesque lakelet which is often called Buttermilk Pond. The view from the lower side of the glen up the Saco Valley to Mt. Washington is one of the finest in this region, and is in its best estate during the latter part of the afternoon. This drive was one of the favorite excursions during the earlier times of summer-touring among the mountains.

Potter's Farm is a locality on the E. of Walker's Pond (12 M. from N. Conway), whence a peculiarly rich * view is obtained, including the chief Conway and White Mts., seen to fine effect across the pretty, island-studded lake. Starr King first brought this point into notice, saying: "Fortunate will the tourist be who can find any other view, along this whole favorite avenue to the mountains, that he can call more fascinating." The lake (which deserves a better name) is 3 M. long, with an area of 1½ M., and contains four islets. The view includes the Green Hills of Conway, a little W. of N., over which is the graceful cone of Kiarsarge. About N. N. W., between the near Pine Hill and the Green Hills, the visitor looks up the valleys of the Saco and the Ellis to the main range of the White Mts., which is overtopped by Mt. Washington. Farther to the l. are the ledges opposite N. Conway, with the immense bulk of Moat Mt. heaving into the sky above them. Then the great depression of the Swift-River Valley is seen, with Table Mt. on its r. side; while the peaks of Chocorua appear, 6 – 7 M. distant, a little N. of W. Farther to the S. are the long ranges of Ossipee, and the Eaton hills close in on the S., near at hand. In the E. are the high hills of Brownfield and Fryeburg.

A few miles S. of Conway the old stage-road crosses a hill-top whence is obtained a noble panoramic view of the distant mountains. The view from *Eaton Hill*, in the earlier days of stage-travel, was esteemed as the culminating pleasure of the long ride from Centre Harbor to Conway.

Echo Lake and the Ledges, see Route 14; Moat Mt., Route 15; Mt. Chocorua, Route 140; Fryeburg, Route 171; Chatham, Route 174; the Swift-River Valley and the Champney Falls, Route 141.

" Until about the beginning of the War of 1812, the settlement in the S. W. corner of Conway consisted of scarcely anything more than a saw and grist mill, Abbott's one-story tavern, and perhaps one or two houses besides. About that time it began to become a village, and the jokers among the Pigwackets gave first one and then another nickname, most of them not very complimentary. *Goose Creek*, I remember, was one of them. While this was going on, the newspapers of the day reported that Gen. Dearborn, who, if I remember rightly, was not much of a favorite with the people at that time, had crossed the line into a district in Canada, at the N. W. extremity of Lake Champlain, whose name is printed in Bouchette's *British Dominions in North America*, indifferently *Chateaugay* or *Chateauguay*, evidently an Indian name with French orthography, and had captured and burnt a hamlet, with the house of Squire Odell, together with the Squire, no doubt a worthy magistrate of His Majesty, in it. For some notion or other about this military exploit, which at this distant date I am unable to explain, the Pigwackets determined that Chateaugay was the proper name for and that it should be thenceforward through all time the name of the then just blooming village. Like myself, however, the other Pigwackets did not know how to pronounce it, and they at once corrupted and changed it into a word whose pronunciation may be written *Shat'-i-gee*, and with that pronunciation it has remained among the Pigwackets the name of that rather sightly and pretty village until the present day.

"*Me judice*, the platform of the Shatigee railroad-station, affords the best view which can be had anywhere of the many near and distant mountain-summits which surround and overlook the beautiful valley of the Upper Saco." (*Judge Lory Odell's letter to the Editor.*)

17. N. Conway to the Glen House.

The distance is about 20 M. and the fare is $ 3. Stages leave N. Conway morning and afternoon, the former being available by passengers on the previous afternoon train from Boston. Tourists who come up on the morning train have time to dine at N. Conway before the afternoon stage starts out. Pedestrians can save about 6 M. of walking by taking the train to Glen Station, where also the stages meet some of the down trains.

Distances. — N. Conway to Glen Station, 6 M. ; Jackson Falls, 9 M. ; the Rogers Farm, 12 M. ; the Glen-Ellis Falls, 15¼ M. ; the Crystal Cascade, 17 M. ; the Emerald Pool, 19 M. ; the Glen House, 19¾ M.

Elevations of various points on the road. N. Conway, 521 ft. ; Lower Bartlett, about 520; Goodrich Falls, 708 ; Jackson, 759 ; the height-of-land in the Pinkham Notch, 2,018 ; Glen House, 1,632.

The stage runs out from N. Conway to the N. W., through the Cathedral Woods, and soon passes the Intervale House, with Mt. Kiarsarge on the r. and Moat Mt. on the l., opening pleasant views over the Saco intervales. A continuous line of detached houses extends from the Intervale House to Lower Bartlett, where the E. Branch is crossed. Beautiful prospects are afforded on this section of the route. The road bends around to the W., with Thorn Mt. on the r. and Moat Mt. on the l., and the Dundee and Thorn-Hill Roads are seen diverging to the r. At Glen Station the Ellis River is reached, and the stage turns to the N. up its long valley, passing through alternate strips of forest and clearing, and soon crossing the stream on a covered bridge at the former site of the Goodrich Falls. The hamlet of Jackson is then traversed, with Jackson Falls on the r., Thorn and Tin Mts. on the E., and Iron Mt. on the S. W. The road then ascends the upper Ellis Valley, with the Eagle Mts. on the r., obtaining several brilliant views of the White Mts. from open ridges among the forests. The prospect from the Rogers Farm is especially notable. Beyond the old Cook Farm the Ellis River is crossed, and the forests grow thicker, the road following closely the course of the river, and passing between Mt. Wild-Cat and the great ridges below Boott's Spur. When near the height-of-land the entrances to the Glen-Ellis and Crystal Falls are passed, and impressive but momentary views open to the l. After ascending the long *Spruce Hill* the road begins to drop down into the Peabody Glen, and soon reaches the **Glen House** (see Route 24).

18. Lower Bartlett.

Hotels. — East-Branch House and Pequawket House, 75 - 100 guests each, § 7 - 12 a week. *Boarding-houses* of C. A. Tasker, Cornelius Stilphen (near Glen Station), and others.

Distances. — East-Branch House to Jackson, 4 M. ; to Thorn Hill, 5 ; Dundee, 10 ; Iron Mt., 4 ; Bartlett Boulder, 3 ; Glen House, 16 ; Around the Square, 7 ; Artist's Falls, 5 ; the Ledges and Echo Lake, 7.

Lower Bartlett, or Bartlett Corner, is 3½ - 4 M. from N. Conway, near the great bend of the Saco Valley, and is nearly surrounded by Mts. Kiarsarge, Moat, and Thorn. The road to the S. as far as the Intervale House is lined with frequent farm-houses, and there are several pleasant drives in the vicinity.

Bartlett is a rugged town which includes two mountain-ranges, and is rich in picturesque scenery. The narrow meadows of the Saco are about the only lands fit for cultivation. The population is 629, included in and about the hamlets of Lower Bartlett and Upper Bartlett (see Route 37). The hills are famous for an abundance of berries in their season. A path leaves the road a short distance below the East-Branch House, and ascends Mt. Kiarsarge, passing over the ridge of Mt. Bartlett. This trail has recently been re-opened. The favorite drive in this vicinity is by the Dundee Road to Jackson and back over the Thorn-Hill Road (10 M.). Noble mountain-views are afforded from many points on this drive. One of the most fascinating prospects of the Saco intervales is obtained from the little church near Lower Bartlett, about 3 M. from N. Conway. This view is best enjoyed towards evening, when the valley is flooded with sunset light, and then "one might believe that he was looking through an air that had never enwrapped any sin, upon a floor of some nook of the primitive Eden."

At Lower Bartlett is the confluence of the Saco River and the East Branch, a rapid stream which rises in the Wild-River Forest and forms the deep valley that separates the Carter Range from Mt. Baldface. Within ½

M. of the mouth of this stream is the mouth of the Ellis River, which flows down from Tuckerman's Ravine and the outer spurs of Mt. Washington. Each of these streams is about 12 M. long. A cart-track diverges from the Dundee Road, 3 – 4 M. N. of Lower Bartlett, and runs into the East-Branch valley, whence a trail leads in 1½ M. to *Mountain Pond*, a small sheet of water which was formerly much frequented by moose. The pond is about 3 M. from the farms in Chatham, and is near Mts. Sable, Eastman, Slope, and Baldface, forest-clad summits which rise out of a dense wilderness and are rarely visited by tourists.

The *Rocky Branch* empties into the Saco about 3 M. W. of Lower Bartlett, after a rapidly descending course of 12 M. over a rocky bed. It rises in the gorge between New River and the Montalban Ridge (not in Oakes's Gulf, as some say), and descends on the E. side of the latter, through tangled and nearly impassable glens. A rude road ascends this valley for 3 – 4 M., between Mts. Pickering and Stanton on the l. and Iron Mt. on the r., passing through the farming community known as *Jericho*, or *Sligo*. Pedestrians can follow the cart-track across Iron Mt. from the end of this road, or pass along the S. slopes, near the iron-mines, to Jackson. The summit of Iron Mt. may be reached on this side (see Route 23).

The township of Bartlett was originally granted to William Stark, Capt. Vere Royce, and other veterans of the French and Indian wars, and was settled about the year 1770. In 1777 a new group of pioneers came up from the town of Lee, but their horses, dissatisfied with the new country, started off for Lee, got lost in the mountains, and all perished. The nearest market for the settlers was at Dover, whither they went in winter, on snow-shoes, dragging hand-sleds. In summer they freighted their meagre produce down the Saco in dug-out boats. A large portion of the sustenance of the early settlers was derived from the flesh of the deer and bears which they shot or caught in cunningly devised traps. The first farms were opened in the glen of Upper Bartlett. The town was named in honor of Josiah Bartlett, first American governor of New Hampshire, and a signer of the Declaration of Independence.

One of the pioneers of Lower Bartlett was Hon. John Pendexter, who came up from Portsmouth, over 80 M., on foot, dragging his household furniture on a hand-sled, and accompanied by his wife, who rode on an old horse, seated on a feather-bed and holding a child in her arms. Out of the wilderness they created comfort, and became a highly honored family. The lands of Bartlett were much injured by the slides of 1826, and wolves made sad havoc among the flocks. Judge Pendexter had a rich estate in Lower Bartlett, and about 10 M. above was the broad estate of the Hon. Obed Hall, formerly an officer in the battles on the Delaware, and then a Congressman.

19. Mount Attitash

is a new provisional name given by the Editor to the lofty N. spur of Moat Mt., whence Humphrey's Ledge projects into the Saco Valley. It is joined to the N. peak by a long bending ridge which forms the head of the ravine of Cedar Brook.

The view was first made public by some scientific gentlemen who ascended it in 1875, mistaking it for Table Mt. They found its sides teeming with blueberries, whence the above title, *Attitash* being the Indian name for blueberries. (See also Whittier's poem, "The Maids of Attitash.") It is ascended from the road 1½ M. W. of the Glen Station, 1 hr.'s march leading to the first bare spot and another to the summit. The road should be left near a barn on the l., and some aid may be obtained by the logging roads on the slopes of the mountain.

* The View. — W. of N., 2 M. distant across the Saco Valley, is the ledgy rampart of Mt. Stanton, over which, about 2 M. farther, are the reddish-yellow rocks on Iron Mt., culminating in a rounded crest. Over the l. flank of Iron is the noble peak of Washington, with its houses in cloudland. Over its long r. flank peers a tiny part of the crest of Adams; and farther to the r. is the curving head of Madison, over the r. flank of Iron, falling away into the Pinkham Notch. On the E. of the Notch is the long and monotonous ridge of Mt. Wild-Cat, descending rapidly on the r. into the Carter Notch, on whose r. is the fine peak of the Carter Dome. Nearer at hand on this line are the partially cleared Eagle Mts., above Jackson. To the r. of and beyond Carter are parts of the Moriah range; and the two white crests of Baldface are next seen shining from dark rounded tops, nearly over Sable. On the r. of Baldface are the peaks of Double-Head, the nearer one having a flat top and the farther one being more pointed. Tin and Thorn Mts. are nearer, and in a line towards Double-Head, and Thorn Hill is still nearer. Then come the twin crests of Mt. Gemini, running to the r. into Kiarsarge, whose noble cone rises above the nearer spur of Mt. Bartlett, and is marked by a hotel. Farther to the r. are the clustering Green Hills of Conway, over and to the r. of which the view passes far into Western Maine. Beautiful vistas are there given over the Saco Valley, with its many summer homes. Then the high near N. peak of Moat Mt. closes in on the S., filling the horizon in that direction.

Most of the view to the S. and S. W. is shut out by the near ridges of Moat, which ascend to a great height. About W. S. W. parts of Tripyramid and Sandwich Dome are visible, with Tecumseh beyond. The view then passes over the adjacent Bear Mt., with its forest-clad double ridges, on whose l. is the low mound of Green's Cliff, with the uneven crest-line of Osceola over it. On the r. of Bear Mt. is the high summit of Tremont, with Haystack near it. The shapeless top of Hancock is then seen, crossed on the r. by the high mass of Carrigain, on whose r. and adjoining are the peaks of Mt. Lowell and the Nancy Range. Far away over the r. of Carrigain are Mts. Lincoln and Lafayette, of the Franconia Range; and parts of the Twin Mts. and Mts. Bond and Guyot appear over the Nancy Range.

Looking more to the N., across the Saco Valley, the noble alpine peak of Willey is seen, falling off sharply on the l., and with the clear-cut but lowly head of Crawford below. Next to the r. is the cone of Mt. Parker, flanked by the rounded top of Resolution, between which is the peak of Webster. Over the r. of Resolution is the crest of Giant's Stairs, whence the dark Montalban Ridge runs N. to Mt. Washington, with parts of the Presidential Range beyond.

5 *

20. Jackson.

Hotels. — The Jackson-Falls House accommodates 60 guests, charging $2.50 a day, or $9 - 10.50 a week. It is an old and well-known house, near the Falls. The Thorn-Mt. House ($9.00 a week) is a modern first-class boarding-house, just across the river from the former. N. T. Stillings has just completed another large boarding-house, at the W. end of the village. J. M. Meserve's Iron-Mt. House is on the Glen Road and under *Copp's Hill*, about ¼ M. S. of the Falls, toward the Glen Station.

Stages. — The Glen-House stages pass through the village twice daily each way, between the Glen and N. Conway. *Railroad.* — The hotels send carriages to the Glen Station, 3½ M. distant, to take passengers to and from the P & O. trains.

Distances. — The following table was prepared by the hotel-keeper: Jackson to Winniweta Falls, 3½ M.; to the Fernald Farm, 4 - 5 M.; to the Glen-Ellis Falls, 9 M.; to the Crystal Cascade, 10½ M.; to the Glen House, 12 M.; to Grant's Ledge, 5 M.; to the Carter Notch, 10 M; to Thorn Mt., 3 M.; to Iron Mt., 4 M.; to Double-Head, 4½ M.; the Hillside Circuit, 5 M.; the Thorn-Hill Road, 8 M.; the Dundee Road, 10 M.; N. Conway, 9 M.; the Cathedral Ledge, etc., 10 - 12 M.; Mt. Kiarsarge, 9 M.

The town of Jackson has 474 inhabitants, and occupies 31,968 acres, of which over 26,000 are unimproved, covering the Tin, Double-Head, Sable, Black, Eagle, Wild-Cat, Carter Dome, and Iron Mts., and their spurs. The hamlet of *Jackson City* is in the S. part of the town, at the confluence of the Wild-Cat Brook and Ellis River, and is a favorite resort for summer-visitors, being situated in a pretty glen 759 ft. above the sea, at the centre of several interesting excursion-routes. It has a small Baptist church, where, during the summer, services are conducted by Episcopal and other divines. Over 500 tourists sojourn at this hamlet during parts of every summer, resulting in an annual profit to the inhabitants of $25,000. Pleasant views of Tin, Thorn, Moat, and Iron Mts. are obtained from this point. The *Jackson Falls* are in the village, and are visible from the high-way bridge over the Wild-Cat Brook. The stream is precipitated over a dark ledge in white and glistening bands, and falls into quiet pools below. These falls are very attractive in seasons of high water, and are easily approached on either side. Along the upper course of the brook are favorite resorts of the artists and trout-fishers who visit Jackson in summer.

Good trouting is found in the streams near Jackson, and towards the Carter Notch. Bears were unusually numerous and audacious in this vicinity in the year 1875, and the inhabitants believed that they came mostly from the White-Mt. Notch, whence they had been frightened by the explosions of the blasts on the new railroad. During the same season a deer weighing 300 pounds was killed in the town.

This town was named *New Madbury* by its first settlers, because most of them came from Madbury. In the year 1800 it was incorporated by the name of *Adams;* but in 1829, during the political contest between Mr. Adams and Mr. Jackson, all the voters in the town (except one) voted for the latter, — and the town soon assumed his name. The first settler was Benjamin Copp, who moved here in 1778, and remained alone until 1790, when 5 Madbury families joined him. The highland regions of the town are now much used as grazing ground for cattle and sheep, among whom the bears sometimes make havoc. The Jackson people became discontented during the late civil war, on account of the crushing burdens of taxation, and, after some acts of violence on their part, it was found necessary to occupy the place with national troops, who were quartered in the church.

The *Fernald Farm* is 4 – 5 M. from Jackson, and is reached by an old deserted road, very hilly and rough. It is near the top of the Eagle Mts., on a level with the head of Tuckerman's Ravine, and not more than 4 – 5 M. from Mt. Washington. This point affords the best view of the great mountain and its ravines that can be obtained on this side, being favored both by its altitude and nearness. The most favorable position for the observer is on the ledge about 40 rods from the old house. As seen from this side, Washington assumes its proper pre-eminence among the other peaks of the main range; and the shape and profundity of its ravines are well comprehended. The best time for a visit is at morning, when the light and shadow effects in the gorges aid in studying their forms.

The *Eagle Mts.* form the low range which runs nearly N. N. W. from Jackson, between the Ellis River and Wild-Cat Brook, forming a spur of Mt. Wild-Cat. The summit of this range which rises E. of the Rogers Farm is called *Spruce Mt.* This range has been cleared at several points, and deserted farms are found high up on its flanks.

This range derived its name from the fact that one of its upper crags was formerly the abode of bold and rapacious eagles. To it is attached the legend of the White-Mt. hermit, who is reported to have dwelt in a cavern near its S. slope. According to the tradition, Thomas Crager was a man of Massachusetts in the 17th century, whose wife was executed as a witch, and soon afterwards a marauding party of Indians carried away his only child, a well-beloved daughter. Weary of life, he left the settlements and plunged into the wilderness, vainly seeking his child among the villages of Pequawket, and finally retiring to a cave among these mountains, where he dwelt for many years, living on game, and unmolested by the awe-struck and superstitious Indians. Long afterwards, by the aid of an old Androscoggin Indian, he found his daughter among the Indians on the St. Lawrence River, living as a squaw.

The road to *Black Mt.* affords a fine series of views of Double-Head, Thorn Mt., and Mt. Washington. It runs out to the N., and in a little over 1 M. passes the divergence of the Dundee Road. At the angle of these roads is an old and deserted Free-Will Baptist church, which is used as the studio of G. S. Merrill, the landscape-painter. The walls are covered with sketches and paintings of mountain-scenery. The Black-Mt. road keeps to the l. and passes along the flanks of the ridge, which is a long, low spur of the Carter Dome. Descending into the valley on the W., it crosses the Wild-Cat Brook about 2 M. beyond the church, and the return may be effected down the valley of this brook. A cart-track which diverges to the l. from the Dundee Road about 1 M. from the church leads into the wilderness to a point on the East Branch of the Saco, within 2 M. of the summit of *Mt. Sable.*

The *Winniweta Falls* are on Miles Brook, which flows into the Ellis River near the Rogers Farm, about 3 M. N. W. of Jackson. They are reached by a rugged path ¾ M. long, through the woods, and fall about 25 ft. over broken ledges. 1 M. farther up in the forest is a long rapid, on the same stream.   Nearly 3 M. beyond Miles Brook (on the Glen Road) is the old *Cook Farm,* just before reaching which the tourist obtains a noble and comprehensive view of Mt. Washington.

The *Goodrich Falls* were about 1½ M. below Jackson, but they were nearly ruined by the erection of a mill, in 1875.   The great ledges are still in position, and are worthy of notice, that on the S. being 60 ft. high. In seasons of high water the falls still present a fine appearance, being among the heaviest in the mountains.   The best view of the rocks and the deep basin below is obtained by descending the steep bank on the N. side, and passing around the shore.   One of the noblest prospects of the White Mts. is enjoyed from the road between the falls and Jackson.

21. Thorn Mountain.

Thorn Mt. is S. E. of Jackson, in the town of Bartlett, and is a high and forest-lined rocky knoll rising at the S. end of the ridge on which the lower eminence of Tin Mt. is founded.   There are but few mountains in this region where the labor of ascent is so slight and the view thus gained is so beautiful.   The road to the N. E. is taken at the village, and its first r. branching road leads to the farm-house (2½ M. from Jackson), whence a path about ½ M. long conducts to the summit.   This road commands a series of very noble views, including the Washington and Carter ranges and the great peaks in Pemigewasset and towards Waterville.   Not more than a mile from the village, within reach of an easy after-supper ramble, is one of the best view-points on the S. side of the mountains.   The path leads up from the end of the fields back of the farm-house, and, although its outlet is not well-marked, it should be found, in order to save a perplexing struggle with the thickets.

* *The View.* — The S. W. peaks of the Presidential Range are seen over the long dark crest-line of the Montalban Ridge, beginning with Mt. Webster, which falls off to the l. on the r. of and beyond Giant's Stairs.   Next to the r. is the sharp little peak of Jackson, and the bubble-like crest of Pleasant is seen over Jackson village, and up the valley of Miles Brook. The ridge sweeps up to the r., by Franklin, to the bulging crags of Monroe.   In the foreground is the pretty alpine hamlet of Jackson, with its large boarding-houses, near the edge of verdant meadows; and more to the r. are the long, low ridges of the Eagle Mts., dotted with the lighter green of pastures.   Above and beyond the latter are the narrow banded slides that stripe the head of a deep ravine which falls from the frowning and craggy Boott's Spur above.   Farther to the r. are the thin, straight lips of Tuckerman's Ravine, over which towers the crest of Mt. Washington.   Still farther to the r., the white gleam of Raymond's Cataract is seen, flashing down the high slopes of Washington at a sharp angle, with the tremendous cliffs of Huntington's Ravine on the r.   Through the Pink-

ham Notch is seen the sharply cut pyramid of Adams, falling to the r. on the Wild-Cat range, which looks across the profound chasm of the Carter Notch upon the Carter Dome. Extending towards Wild-Cat is the long, low, and many-headed ridge of Black Mt. A part of Mt. Madison is over Wild-Cat.

Double-Head is next seen, about 8 M. distant, N. by E., its nearer summit being a narrow level plateau, the farther one more pointed. Over the l. of this dual eminence is the high and monotonous ridge of Mt. Moriah, toward the Androscoggin Valley; and more to the r. are the summits of Slope, Sable, and Eastman. Nearer and more conspicuous is the lofty pyramid of Kiarsarge, with its summit-hotel; whence Mt. Bartlett runs out to the r. and the double peak of Mt. Gemini extends to the N. N. E., hiding the mountains of Maine. On the S. is a broad and beautiful section of the Saco Valley, extending as far as Gline Mt. in Eaton and the Green Mt. in Effingham, with verdant meadows traversed by the blue and yellow band of the river and flanked by white hamlets. On the l., and below Kiarsarge, are the broken ridges of the Green Hills of Conway; and on the r. is Moat Mt., with its long and ledgy ridges and low white peaks, over which is Chocorua.

Toward the S. W. the foreground is occupied by a part of the Lower-Bartlett glen, whose dark woods and light-green intervales make a pretty mosaic, across which the P. & O. Railroad stretches. The range S. of the Saco runs W. from Moat, and includes the long, black, and rolling plateau of Bear Mt., next to which is the graceful cone of Haystack, and on the r. is the dark and hummocky ridge of Tremont, with burnt ground on top. Over the S. E. end of Bear Mt. is the high rounded crest of Passaconaway, hiding Whiteface; and the saw-like ridge of Tripyramid is over the r. flank of Bear Mt. On the l. of Tremont, at the end of the Sawyer's-River valley, are the upper lines of Osceola.

The view-line now falls on the range between the Saco and the Rocky Branch, with Mts. Stanton and Pickering in the foreground, surrounded with ledges, and Langdon beyond, ascending to a graceful lowly crest. Through the notch between Langdon and Iron Mt. appears the mighty peak of Carrigain, whence step-like terraces fall away to the S., while over its l. flank parts of Hancock are seen. The high point of Mt. Parker is next seen, with the gently curving plateau of Resolution on its r., and with the tree-dotted ledges of Iron Mt. below, and not more than 4 M. distant, across the Ellis Valley. Over Parker is Mt. Lowell, upon which the r. flank of Carrigain falls, and on whose r. are Mts. Anderson and Nancy. On the r. of Resolution are the two sharply-cut upper steps of the Giant's Stairs, from which the ponderous Montalban Ridge runs to the r. up to Mt. Washington. Over the r. falling flank of Giant's Stairs appears the crest of Mt. Willey, cut off sharply on the l. and adjoined on the r. by Mt. Field.

Tin Mt. is one of the N. crests of the Thorn Mt. ridge, and is sometimes visited by tourists of a geological turn, being easily accessible from the village. It is about 1¾ M. along the ridge from Tin to Thorn.

The first discovery of tin in the United States was made at Jackson, where it occurs on Tin Mt. It is in fine small veins traversing mica slate and granite rocks, and accompanied by fluor spar and arsenical ore. The town also contains magnetic iron ore, phosphate of iron, tungstate of manganese and iron, fluor spar, mispickel, copper, and pyrites.

22. Double-Head.

Double Head is a picturesque mountain in the E. part of Jackson, consisting of two flat-topped peaks, whereof that to the N. is 100 ft. the higher, and that to the S. is divided into three minor crests. It has been considered topographically as a spur of the Carter Range. According to Dr. Jackson's measurements (of doubtful accuracy), the height of the mountain is 3,120 ft. The name "Double-Head" is found on maps of the last century. The peculiar architecture of and the fine views from this mountain render it an interesting point of ascent. Ladies have visited the summit, but only after most fatiguing work. The N. peak commands the best general view and prospect of the Presidential Range; the S. peak looks out over the fair Saco Valley.

The best route for attacking Double-Head is to pass out from Jackson to J. H. Dearborn's, 2½ – 3 M. distant on the Dundee Road, and at the foot of the mountain. The road thither reveals some fine panoramic views over the Presidential Range, and toward the S. W. The distance from Dearborn's to the summit is about 1½ M., and there is a vague and easily lost path, passing along the course of a small brook. No serious difficulties are experienced on the ascent. The objective point is the saddle between the two peaks, whence either may be ascended, or each in succession.

* *The View.* — The entire ridge of Moat Mt. is seen in the S. W., 8–10 M. distant, its three peaks being well marked and distinct, with the white spire of Chocorua just to the l. of the N. peak, and a dim blue segment of the Ossipee Range on the l. of the middle peak. Thorn Mt. is between Double-Head and Moat, with Tin Mt. on its N. ridge. To the r. of Moat Mt. and apparently continuous with it are the Table and Bear Mts., S. of the Bartlett valley; over which appear the low crest of Paugus, the ridgy mass of Whiteface, and the rounded swell of Passaconaway, — the latter being nearly S. W. Much nearer, across the Ellis and Rocky-Branch valleys, are the craggy steeps of Mts. Stanton and Pickering, with Tripyramid cutting the sky over them with its line of serrated summits. To the r. of all these, and below, is the graceful cone of Tremont, with Haystack adjoining and nearly equidistant. The view now rests on the multiplied ledges of Iron Mt., broad-based and massive, about 5 M. distant across the Ellis valley. Over its l. flank is Mt. Langdon, with Osceola far beyond, and over the r. is the vast pile of Mt. Carrigain, reaching far into the sky. Close to the latter is the pointed top of Mt. Lowell, with a section of Mt. Hancock through the gap on the l., and Mts. Anderson and Nancy on the r. The second ridge below Carrigain culminates in Mt. Parker, and rises on the r. to the high plateau of Mt. Resolution, on the r. of Nancy, with the peak of Crawford peeping over its l. shoulder. Far away in this direction is the lofty plateau of Moosilauke. The well-defined upper terraces of Giant's Stairs are just to the r. of Resolution, and over them is a remote blue peak toward the W., while Mt. Bond rises to the r. far in the Pemigewasset Forest. On the r. of the latter is the serrrated summit of Lafayette, cutting the sky-line; and on its r., continuous with Bond, are the ponderous heights of the S. and N. Twin Mts.

The long and undiversified rising slope of the Montalban Ridge stretches to the r. from Giant's Stairs, thickly clad with woods, and reaching the upper swell of Mt. Washington. Over this, and to the r. of the Twins, is the alpine peak of Mt. Willey, descending sharply on the S., and marked by slides and broad bare ledges. Then, in the same ridge, appear the crests of Mts. Field and Tom, under which are Mts. Webster and Jackson. Farther to the r. appear the well-marked peaks of Clinton, Pleasant, Franklin, and Monroe, sweeping to the N. N. E. to the base of the cone of Mt. Washington.

About N. W. is the stately peak of Mt. Washington, 10 – 12 M. distant, yet so favorably situated with relation to Double-Head that nearly all the great eastern ravines are visible. The apparent peak which breaks up before the l. flank of the mountain is Boott's Spur, in which is the broad chasm of the Gulf of the Slides, whose walls are striped with light-colored bands. To the r. are the steep inner slopes of the N. wall of Tuckerman's Ravine, curving around at the head towards Bigelow's Lawn. The shallow oval depression next adjoining contains the falling waters of Raymond's Cataract, and is separated by a narrow ridge from the dark depths, surrounded by majestic cliffs, of Huntington's Ravine. To the r. of Mt. Washington is the peak of Mt. Adams, and over the r. slope of Wild-Cat, which rises on the E. of the Pinkham Notch, is the rounded summit of Mt. Madison. The deep cleft of the Carter Notch is on the r. of Mt. Wild-Cat, and is walled on the W. by the lofty and imposing Carter Dome. Farther to the r. are the embattled ridges of Carter and Moriah, stretching their firm lines across the sky. Nearly N. up the East-Branch valley are the wilderness-peaks of the Wild-River Forest, — Mt. Sable, with its unmarked crest; Baldface, whose upper slopes are white and brilliant; Royce, a double-knolled mountain on the Maine border; and Eastman, nearer and round-topped. To the l. of the N. peak of Baldface and far away is Goose-Eye, or some other of the mountains near the Grafton Notch, and another cluster of Maine peaks is seen between Baldface and Royce, including those in Newry and Andover. N. of E., about 3 M. distant, is the secluded basin of Mountain Pond, with Mt. Slope about 1½ M. beyond. To the r., and more distant, is the long sheet of Upper Kezar Pond, with the highlands of Waterford and Albany beyond. Farther to the S., and in a line over each other, are the Kezar and Upper Moose Ponds, and Highland Lake, with the long and hotel-surmounted sierra of Mt. Pleasant farther to the r. Nearer is the heavy ridge which runs W. from Kiarsarge, upon which are the graceful and similar peaks of Mt. Gemini. To the S. E., over the dark mound of Mt. Tom, in the plains of Fryeburg, is the distant glimmer of Sebago Lake. Since Double-Head is visible from Portland, in this direction, conversely, Portland should be seen from Double-Head, on a clear day and with a strong glass, — and beyond Portland is the sea. Farther to the r., over Lovewell's Pond, are the two peaks of Saddleback Mt., in Baldwin. The view in this direction is now limited by the high pyramid of Mt. Kiarsarge, whose summit is disfigured by a hotel.

In the view from the N. peak, the S. peak of Double-Head, with its triple head, comes in on the r. of Kiarsarge and intercepts the view of the Saco Valley, though Walker's Pond, Ossipee Lake, and the Green Mt. in Effingham are seen over this ridge. The view from the S. peak, otherwise inferior to that from the N. peak, surpasses it in the beautiful prospect over the Saco Valley and the Moat Range; and this alone is sufficient to repay for the labor of climbing up its easy slope.

23. Iron Mountain.

Iron Mt. is in the towns of Jackson and Bartlett, between the Rocky Branch and the Ellis River. It is a massive, low, and broad-based eminence, with extensive ledges on its sides, and is famous for its vast deposits of iron ore. The position of the mountain with reference to the Presidential Range makes it an interesting view-point. Blueberries and raspberries are plentiful along the upper slopes. The true peak is near the S. end of the crest-line, and is covered with low bushes. Until the opening of the mines this mountain was called *Baldface.*

There is a rough road which crosses from Jackson to Jericho, in the Rocky-Branch valley, and passes within 1 M. of the top, but it is very hard for horses on account of its steep grade. This road runs W. from Jackson, crosses the Ellis at the first bridge, then turns sharply to the r., and soon encounters the ascent. The main path enters the woods near a barn on the crest of the ridge, but is hard to find, so it is as easy to go on to the red farm-house on the l. and clamber thence about ¾ M. over the ledges and through the bushes. It is nearly ½ M from Jackson to the divergence of the Iron-Mt. road, beyond the bridge; and 2 M. from that point to the red house. During the latter portion of the route, the road ascends 1,150 ft. The best route from Lower Bartlett would be to go up by the old road to the iron-mines, and thence clamber to the crest. There is a cart-road from the ore-house to Jericho, in the Rocky-Branch glen.

The Geological Survey states the height of Iron Mt. as 2,000 ft. (by estimate), but the barometric measurements of our Guide-Book survey made it about 800 ft. higher.

Above this valley, some 1,404 ft., and 1 M. from the Rocky Branch are the rich deposits of iron ore which have been and are to be worked. One of the veins measures 37 ft. from N. to S., and 16 ft. from E. to W., and farther down the slope are other shafts and adits, reaching the ore-veins. It is probable that the entire mountain is intersected with these veins, containing inexhaustible quantities of iron which makes the finest quality of steel.

The View. — A little W. of N., over the great wilderness between the Rocky Branch and the Ellis River, the high crags of Boott's Spur are seen, with the Gulf of the Slides. Over these is the crest of Mt. Washington, with massive outworks on the E., above which, and through the Pinkham Notch, Mt. Madison is visible, with a rounded summit and a long slope to the r. Then comes the double peak of Mt. Wild-Cat, on whose r. is the Carter Dome, with a line of lofty cliffs fronting in on the Carter Notch. The Eagle Mts. extend their low and partially cleared flanks to the N. from Jackson. About N. W. is the lower end of Black Mt., inconspicuous in the valley; and farther to the r. are Mts. Sable and Eastman, the white caps of Baldface, and Mt. Slope. Over the village of Jackson are the peaks of Double-Head, with Tin Mt. nearly in line and nearer, and Thorn Mt. on the same ridge to the r., while farther back, on the sky line, are the twin peaks of Mt. Gemini. The hotel-surmounted cone of Kiarsarge is next seen, 6 M. distant, with the subordinate Mt. Bartlett on the r., a bold ripple on the descending flank. Then come the Green Hills of Conway, falling away to the S. in graceful undulations, beyond the plain of Lower Bartlett. N. Conway is in the S. E., and the lovely valley of the Saco stretches away beyond the borders of Maine, past Conway Corner and Walker's Pond, and enclosed on the r. by the highlands of Eaton, Brownfield, and Cornish, on the l. by the Denmark and Baldwin hills.

To the S., close at hand across the Rocky-Branch glen, is Mt. Stanton, over which is the N. peak of Moat Mt., whose S. peak is seen more to the l. Farther W., and on a line with Moat, is Table Mt., over which is the superb peak of Chocorua. The long and massive Bear Mt. adjoins Table, on the W., and its r. flank is overlooked by the dome of Passaconaway, with Whiteface on its r. flank. Near the base of Passaconaway is the round white crest of Potash; and Mt. Pickering is in the nearer foreground. Mt. Langdon is 2 – 3 M. distant, across the Rocky Branch, beyond which are the crests of Tremont and Haystack; and farther to the S. W. is the sierra of Tripyramid. Still more distant is the entire ridge of Osceola, with a segment of Tecumseh on its flank, and Kancamagus under it. The stately peaks of Carrigain next appear, over the long and sloping pyramid of Parker; and over their l. flank peers the round head of Hancock. Lowell lies against Carrigain, and on its r. are Anderson and Nancy. Then there appears the remote cluster of the Franconia Mts., on whose r. Mt. Resolution runs up to the immense steps of the Giant's Stairs, which are seen in their best estate from this point. Farther N., over the wooded Montalban Ridge, appear the crests of Webster and Jackson, with the flat top of Clinton, and, farther still, up the length of the Rocky-Branch glen, is the round curve of Mt. Pleasant. Then comes the almost indistinguishable plateau of Franklin, flanked by the beetling rocks of Monroe.

5 *

24. The Glen.

The **Glen House** (W. & C. R. Milliken, proprietors) accommodates 500 guests, at $ 4.50 a day, with reductions for long sojourns, and liberal rates in June and September. The parlors are large and luxurious (100 × 60 ft. in area) ; and the cuisine is among the best in the White Mts. The hotel is provided with a store, under Mr. Eastman's care, where books, stationery, pictures, and small articles of clothing are sold. There is a bowling-alley near the house, and a barber-shop inside. During the season a band is kept here, which discourses music from the stand in front of the house, or plays for dancing-parties in the parlors. There is a telegraph office at the hotel. The house is opened on the 17th of June and closed Oct 1st.

Stages run to Gorham to connect with all trains on the Grand Trunk Railway (time 1½ hrs. ; fare, $ 1.50) There are two stages daily each way between the Glen House and N. Conway, via Jackson and Glen Station (time, 4 hrs. ; fare, $ 3). Stages leave for the summit of Mt. Washington at 8 A. M. and 2½ P. M., connecting with the Mt. W. Railway, the return stages connecting with the lines to Gorham and N. Conway. The fare up and down (including tolls) is $5 ; the fare down is $ 3. It is claimed that the system of stages and horses connected with the Glen House is the best in the Atlantic States, both in equipments and drivers.

Routes.—From New York or Boston to Portland, and thence to Gorham, by the Grand Trunk Railway (Route 6), thence by stage ; or by the Eastern Railroad or the Portland & Ogdensburg Railroad to N. Conway or Glen Station, and thence by stage (Routes 1, 5, and 17) ; or by either of the routes over Mt Washington. Passengers who leave Boston by the Eastern Railroad at 8 A. M., dine at N. Conway, and reach the Glen House at 6.30 P. M.

Distances (from the hotel-card). — Glen House to the Garnet Pool, 1 M. ; the Imp, 2 M. ; Gorham, 8 M. ; the Emerald Pool, 1 M. ; Thompson's Falls, 2 M. ; the Crystal Cascade, 3½ M. ; Tuckerman's Ravine, 3½ M. ; the Glen-Ellis Falls, 4 M. ; the summit of Mt. Washington, 8 M. ; Jackson, 12 M. ; Glen Station, 15½ M. ; N. Conway, 20 M. ; Jefferson Hill, 26 M. ; Lancaster, 33 M.

The Glen House is built on a terrace above the Peabody River, on the lower slopes of the Carter Range, and looks out directly on the five highest mountains in New England, which are but 3 – 4 M. distant, and are not masked by any intervening objects. They form a crescent-shaped line, with the concavity towards the Glen, their order from r. to l. being indicated by the formula MAJ. Clay Washington. Mt. Madison (5,365 ft.) is on the r., l. of which is the sharp and symmetrical pyramid of Mt. Adams (5,794 ft.), then the massive crest of Jefferson (5,714 ft.), the low humps of Clay (5,553 ft.), and the hotel-crowned peak of Washington (6,293 ft.) is on the l., peering over lofty spurs and secondary peaks. The high crags of the Lion's Head are seen on the l., near the opening of Tuckerman's Ravine. The deep gorge of the Great Gulf opens into the range towards Mts. Adams and Jefferson, containing the dense forests of the W. Branch. This noble view is presented from the piazzas and front rooms of the hotel. On the E. is the high and massive Carter Range, which is rarely visited, on account of its tangled thickets ; and more to the S. are the slopes of Carter Dome and Wild-Cat. Below the hotel, on the W., is the pleasant valley of the Peabody River ; and above, on the W., is a far-viewing clearing, in one corner of which is a reservoir.

The Glen House is 1,632 ft. above the sea, 820 ft. above Gorham, and 1,111 ft. above N. Conway. The dry, pure, and fragrant air of this locality affords relief and exemption from the annoyances of rose-cold or hay-fever, many of whose victims escape its attacks by sojourning here.

There is good trout-fishing in this vicinity, especially on Nineteen-Mile Brook and the W. Branch of the Peabody. The forest-scenery in these glens is wild and luxuriant, the Pinkham Woods sweeping down on the S., and the Carter Range being draped with long reaches of heavy forests. In early June large patches of snow are visible from the piazza, striping the sides of the mountains and embanked in their ravines. In October the forests assume their gorgeous autumnal coloring, and form a most brilliant scene, often contrasted richly with the early snows on the higher peaks. The varying and weird effects of clouds and mist along the mountains, alternately closing and breaking away, afford one of the most remarkable spectacles from the Glen House. The long white front of the hotel presents a bright appearance as seen against the dark background of the adjacent forests.

" The Glen House is at the very base of the monarch ; and Adams, Jefferson, Clay, and Madison bend around towards the E., with no lower hills to obstruct the impression of their height, — so that from the piazza and front chamber windows of the hotel, the forest clothing of the five highest mountains of New England is distinctly seen, with all the clefts and chasms and the channelling of the rains, up to the bare ridge from which the desolate cones or splintered peaks ascend. . . . . The best time for the effects of light on the peaks is early in the morning, when the rocky portions of the ridge are often burnished with surpassing beauty, or from four to six in the afternoon of midsummer, when the lights and shadows are most powerfully contrasted

" Mt. Adams looks the highest, at all times, from the Glen House ; and, in fact, although it is nearly 500 ft. lower than Mt. Washington, a greater elevation, on one steady slope, is seen in looking at it than Mt. Washington reveals. The summit of Mt. Washington lies back of the shoulder seen from the Glen, so that the effect of a thousand feet of height is really lost. And yet, after the first surprise has passed, there is no comparison between the two mountains for grandeur, as they tower above the hotel. Washington is more massive. The lines that run off to the S. E. from the summit, and especially those that sweep around the Great Gulf and Tuckerman's Ravine, are far more grand and fascinating, to the eye of an artist, than the symmetry of the slim pyramid of Adams. One can never tire of looking at their sharp, curving edges, into whose steely hardness the torrents and rock-slides have torn deep dikes, that in the afternoon are delicate engravings of graceful shadow. And seen through a southerly air, or a light shower, that shows, much more plainly than clear air does, the number and the graceful flow as well as vigor of these lines, we learn that the great privilege of the Glen is the opportunity it gives for studying from below the granite braces of the cone of Mt. Washington." (STARR KING)

Prof. Vose has advanced the theory that " the main chain of the White Mts. was formed by a fragment of the western slope of an immense anticlinal wave, of which the crest would have been over the Peabody Valley, and of which perhaps a fragment of the eastern slope may be found in the opposite and parallel range of the Carter Mts. ; in which case the Peabody Valley would be a valley of denudation." *Amer. Assoc. Advance. Sci., Proc. XVI.*

There is a legend to the effect that a Mr. Peabody, of Andover, Mass., was present at the birth of the river which has for many years borne his name. He was passing the night in an Indian cabin near the height-of-land, when his entertainers were aroused by a loud roaring, and they had barely escaped from the hut when an impetuous torrent sprang out of the hillside and swept it away. This incident is recorded in White's *History of New England.* In 1784 the old Shelburne road was followed down this valley by Dr. Belknap's party, " one going before with an axe " to clear away the fallen trees. With their utmost exertions they made but 1½ M. an hour.

The **Garnet Pools** are less than 1 M. N. of the Glen House, and are reached by a path running to the l. from the Gorham road, at the foot of a low hill. They are a series of basins in the bed of the Peabody River, carved out and polished by the action of the water and the rolling stones on the ledges.

The Imp is "a grotesque colossal sphinx" which appears on one of the peaks of the Carter Range, the profile being formed by the upper crags of Mt. Imp, and having a weird resemblance to a distorted human face. This appearance is best observed at late afternoon, and from Copp's farm, 1½ – 2 M. N. of the Glen House, on the old road to Randolph, W. of the Peabody River.

Thompson's Falls are about 2 M. from the Glen House, and are reached by taking the N. Conway road to the S. and diverging for ¼ M. to the l., at the guide-board. They are on a small tributary of the Peabody River, and form a chain of cascades ½ M. long, sweeping down through pretty forest scenery, and furnishing rich ground for pleasant rambles. The * view of Mt. Washington and its E. ravines, from the head of the main fall, is one of the best in the mountains, and has been reproduced in *Picturesque America.* "It is a wild forest scene, and the flow of the glistening cascade under the wide expanse of leaf and bough is exquisitely musical and charming in the solitude of the wilderness."

The Emerald Pool is a quiet basin in the Peabody River, where the water rests for a brief space in its rapid downward career. It is just N. of Thompson's Falls, and on the other side of the road. It is a favorite resort for artists and lovers of nature, on account of its quiet beauty and sylvan richness.

The * **Glen-Ellis Falls** are about 4 M. S. of the Glen House, near the N.-Conway road, and are approached by a plank walk ¼ M. long, diverging to the l. from the road at a guide-board. They are on the Ellis River, and at the base of Mt. Wild-Cat, whose formidable ridges tower above to a great height. They were formerly known as the Pitcher Falls, in allusion to their shape, but received the present and less significant name (Glen Ellis, or Elise) from a Portland party, in 1852. The stream slides down about 20 ft. over inclined ledges, and then springs downward for 70 ft. through a deep groove which it has worn in the rocky cliff, being twisted to the l. by a bulge in the ledges, and making almost a complete turn. After viewing the fall from the edge of the rocks above, the visitor should descend the long stairways on the r. to the spray-moistened ledges below, where he can form a better idea of the power of the fall by looking up along its line, across the deep green pool into which it heavily plunges. A variety of rich views may be obtained from points near the bottom of the ravine; and beautiful combinations of the white column of water, with the bearded woods on either side, are gained thence. This fall is probably the finest in the White Mts.

" If we wished to take a person into a scene that would seem to be the very heart of mountain wildness, wi-hout wishing to make him climb into any of the ravines, we should invite him to visit this fall of the Ellis River. The best view of the fall is obtained by leaning against a tree that overhangs a sheer precipice, and looking down upon the slide and foam of the narrow and concentrated cataract to where it splashes into the dark green pool, 100 ft. below. And then as we look off from this point above the fall, we see the steep side of Mt. Wild-Cat crowded to the ridge with the forest. It is not the sense of age, but of grim, almost fierce wildness, that is breathed from the scenery, amid which this cataract takes a leap of 80 ft. to carry its contribution to the Saco."

About ¼ M. N. of the Glen-Ellis Fall, on the W. side of the road, is a lofty cliff over which falls a small rivulet, making a long, white, vertical plunge which is a very beautiful object in seasons of full water. It is near the road, and the silvery gleam of the fall is visible through and above the trees. It is said that this cascade is on the tiny stream which occupies the old bed of the New River, whose new course is crossed a few rods farther N., nearer the Crystal River (see Route 71).

The **Crystal Cascade** is on the Crystal, Cutler's, or Ellis River, below the outlet of Tuckerman's Ravine, and on the W. side of the Pinkham Notch. It is reached by following the N.-Conway road for about 3 M. from the Glen House, and diverging to the r. at a guide-board, whence a good path leads in about ½ hr. to the cascade. The best view-point is on the opposite side, on a little cliff which fronts the water. The vista includes about 80 ft. of fall, over successive step-like terraces of slaty rocks, crossed by igneous dikes. In high water it affords a brilliant sight, but at other seasons the stream dwindles away into white threads of water.

" Visitors should not forget that the proper point from which to see it is not the foot of the fall itself, but the top of the little cliff directly opposite. No contrast more striking can be found among the mountains than that of age and youth, which is furnished from that point. The cliff is richly carpeted with mosses that have been nourished and thickened by centuries. . . . . The rocks of the neighboring precipices look old. They are cracked and seamed as though the forces of decay had wound their coils fairly around them, and were crumbling them at leisure. The lichens upon them looked bleached and feeble. Those protruding portions of its anatomy indicate that Mt. Washington has passed the meridian of his years. But the waterfall gives the impression of graceful and perpetual youth. Down it comes, leaping, sliding, tripping, widening its pure tide, and then gathering its thin sheet to gush through a narrowing pass in the rocks, — all the way thus, from under the sheer walls of Tuckerman's Ravine, some miles above, till it reaches the curve opposite the point where we stand, and, winding around it, sweeps down the bending stairway, shattering its substance into exquisite crystal, but sending off enough water to the right side of its path to slip and trickle over the lovely, dark green mosses that cling to the gray and purple rocks." (STARR KING.)

The road from the Glen House to Jefferson Hill diverges to the l. from the Gorham road near the old saw-mill, about 1½ M. from the hotel, crosses the Peabody River and its W. Branch, and passes the Copp farm, whence the Imp is seen. It then ascends through the forest to the height-of-land in the notch between Mt. Madison and the Pine Mt. of Gorham, the mountains being hidden, for the most part, by the large trees on either side. From the height-of-land it descends rapidly to the Moose River, after crossing which it enters the road from Gorham to Jefferson Hill, near Randolph Hill (see Route 70).

The road from the Glen House to Gorham descends the Peabody Valley all the way, and gives several good views of the peaks on the W., though much of its course is through thick woods. The Imp and Moriah ridges are also seen on the r.

For Tuckerman's Ravine, see Route 72; the Great Gulf, Route 74; the Mt.-Washington road, Route 91; Mt. Washington, Route 93; Mts Jefferson, Adams, and Madison, Routes 86, 85, 84; the Carter Notch, Route 25; Mt. Wild-Cat, Route 26.

25. The Carter Notch.

There are two routes (but no paths) to this natural curiosity, one from the Glen House and the other from Jackson. The former is the shorter and easier, and is $3\frac{1}{2}$ – 4 M. long, leading up the valley of Nineteen-Mile Brook (19 M. from Conway line) between Mts. Carter and Carter Dome and Mt. Wild-Cat. The brook is back of the Glen House, and is followed for a considerable distance by the aqueduct, beyond which ensues a long and wearisome scramble up the bed of the brook and along the adjacent ridges. 1 M. an hour is the best time that can be made, and the ascent is quite arduous. The height-of-land is about $3\frac{1}{2}$ M. from the hotel and 1,750 ft. above it; and beyond this point the visitor descends quickly to the lakes, passes on the l. of the first, by the camp, on the r. of the second lake, and up on the boulders below.

The route from Jackson leads up the Wild-Cat Valley for 5 M., and then follows logging-roads and the course of a brook into the Notch, about 4 M. farther.

The highest point in the Notch is 3,320 ft. above the sea, and the lakes are 3,150 ft. high. It lies between Carter Dome on the E. and Mt. Wild-Cat on the W , and contains some grand rock-scenery. On the W. are fine cliffs, 300 – 400 ft. high, which give back strong echoes to a rifle-shot. On the E., high up on Carter Dome, is a singular cluster of projecting crags, forming rugged profiles against the sky. The lakes are in the central part of the chasm, that to the N. being 200 × 300 ft. in area and the other 150 × 75 ft. Their water is very clear, and abounds in fish. On the terrace between them, to the W., is a snug hunter's camp, where Bishop Jaggar and other gentlemen have sojourned. At the S. entrance of the Notch is a lofty line of immense boulders, piled on each other in inextricable confusion, and affording some of the most remarkable rock-scenery in the mountains. They are rugged and deeply pitted, like the rocks on the cone of Mt. Washington, and probably remained for ages on the crests of the adjacent peaks. One or two of these boulders are over 70 ft. long each. The best point from which to view the ravine is the top of the highest of these boulders, but the route thither is very difficult to find, leading sometimes under arches of huge rocks and sometimes through narrow gates between high walls. From this point the Notch opens on the N., with the cliffs and crags of Mt. Wild-Cat and Carter Dome on the l. and r.; and on the S. is a portion of the valley towards Jackson, with its enwalling mountains. The confused piling together of these huge black boulders forms a weird and singular scene, and is the chief object of interest in the Notch.

The Jackson and Glen-House people have agreed with each other to make a path through this pass, in order to open its fine scenery to the public. At present the journey is too arduous to be repaid by the short time which one can spend here. Persons who encamp at the lakes, and ascend the trout-brooks, have opportunity to study and enjoy the scenery at leisure.

26. Mount Wild-Cat.

Mt. Wild-Cat is S of the Glen House, between the Carter Notch and the Pinkham Notch, with spurs running S. into Jackson. It was named *East Mt.* on Belknap's map of 1791, in allusion to its position with reference to the Pinkham Notch. There is a legend that two hunters named Hight and Carter were once passing through the notch E. of this mountain, when they separated and ascended the ridges on each side, whereof one was named Hight and the other Carter. It is uncertain which of the mountains, E. or W., assumed either of these names, but that on the W. of the Carter Notch has usually been called Mt. Hight. Prof. Guyot named it Mt. Wild-Cat, and latterly the people who wished to retain the name of Hight, being unwilling to derange Guyot's nomenclature, have applied it to the S peak of Mt. Carter. The Appalachian Mountain Club has finally applied the name of the *Carter Dome* to the S. peak of Carter. The true Mt. Hight (Guyot's *Wild-Cat*) is 4,350 ft. high, and is covered with woods. A clearing has been made high up on the W. side, which commands perhaps the best view attainable of Mt. Washington and the great ravines on the E. It is reached by a good path from the Glen House, which passes close to the lower corner of the reservoir and enters the woods beyond. The ascent is easy, even for ladies; and requires less than an hour, the distance being about 1¼ M. The trees along the path are profusely " spotted."

* *The View.* — On the l., over the Pinkham Woods, are the heights which hem in and hide the New-River Ravine, N. of which is the vast outer circle of Tuckerman's Ravine, with nearly all its S. wall strewn with slides, and a small portion of the heading cliffs. The ravine itself is seen opening into the heart of the mountain, its floor forming an apparent terrace above the hollow in which lies Hermit Lake. Above the S. wall are the imposing crags on Boott's Spur; and above the N. wall are the frowning rocks of the Lion's Head. On the r. of the latter is the sharp white fall of Raymond's Cataract, glistening down the cliff-side for hundreds of feet, beyond which are the lofty gray precipices of Huntington's Ravine, whose depth is partly hidden by an intervening ridge. Above these points is the high terrace of the Alpine Garden, with the Chandler Ridge still higher on the r., and the summit-hotel over all. On the r. flank of the mountain are seen various sections of the carriage-road, with the Half-way House; over which is the massive peak of Mt. Jefferson, beyond the Great Gulf. Then the splendid pyramid of Adams is seen, with long slides on its sides and craggy peaks on N. and S. Looking up the deep ravine between Adams and Madison, one views the low saddle in which Star Lake lies, on whose r. is Mt. Madison, a lofty and stately peak over the

Glen House, flanked by the long and stony ridges which run down to Copp's and the height-of-land.

On the r. of Madison, and beyond, is the Crescent Range, over which are the more distant forms of the Pilot Mts. and Green's Ledge, in Berlin. Due N., down the Peabody Glen, is the low and broad-based Pine Mt., on whose r. is the ledgy mass of Mt. Forist, at Berlin Falls, with Mt. Hayes much nearer, on the r. Between Pine and Hayes the long Androscoggin Valley is seen, trending away to the N. through Berlin and Milan, with the high farms on Berlin Heights on the l. of Berlin-Falls village. The view is then closed by Mt. Imp, close at hand down the Glen.

27. The Carter Range.

Properly speaking, this chain of peaks extends from the Androscoggin Valley to the Carter Notch, including Mts. Moriah, Imp, Carter, and Hight. In a narrower sense it includes that portion of the ridge between the Imp and the Notch; and the name of Mt. Carter is applied only to the peaks nearly behind the Glen House. This group has been but partially explored, and the State Geologist says that "it is the least known of all the mountain districts. I do not find any explorer of it anxious to continue his investigations therein." It falls on the W. to the Peabody Glen; and more gradually on the E. to the Cold-River valley and the trackless forest. The N. peak of Carter is 4,702 ft. high, and the S. peak is 4,830 ft. high. It is to the latter that the name of *Carter Dome* has recently been given, in allusion to its remarkable convexity of outline, and to its situation on the Carter Range.

The peaks of Mt. Carter are very rarely visited, so great is the labor of the attack. From people who have been over the ridge the Editor learns that there is a large area of dwarf spruce on its upper parts, through which it is very difficult to pass. The summits of Carter are covered with forests, and are therefore ineligible as view-points.

28. Gorham.

Hotels. — Gorham House, a large country-inn, at the centre of the village ($ 3 a day) ; Eagle House ; Lary's large summer boarding-house, about 1 M. N. of the village, in a populous neighborhood. The Alpine House was formerly located at Gorham, and was one of the largest hotels in the mountains, though unfortunately located as regards views. It was destroyed by fire in 1872. The Mt.-Washington House also was burnt in 1875. A new hotel should be erected in the environs of Gorham, away from the machine-shops and in view of the mountains, in order that the beautiful scenery in this vicinity might be once more available to tourists.

Railway. — The Grand Trunk line (Route 6) runs to Portland (91 M.) in 4½ – 5 hrs. By taking the train to Groveton Junction (31 M. N. W.), a connection is made with the B., C. & M. R. R. (see Routes 7 and 2), running to Lancaster and Littleton. Beyond Groveton the Grand Trunk line passes on to Montreal and Quebec (Route 7).

Stages run to the Glen House semi-daily, on arrival of trains. Mountain-wagons are despatched frequently but irregularly (when parties are made up) to Mt. Washington and Jefferson Hill. The Gorham House has a large livery-stable.

Guides. — The company of skilful guides that used to frequent the Alpine House has now passed away, and visitors will find it difficult to get competent pilots for the Moriah and Madison trips.

Distances. — Gorham to the summit of Mt. Hayes (Route 29), 2 M. ; of Mt. Surprise (Route 30), 2¼ M. ; of Mt. Moriah (Route 31), 4½ - 5 M. ; to the Glen House (Route 24), 8 M. ; Mt. Washington, 16 M. ; Jefferson Hill (Route 66), 17 M. ; Lead-mine Bridge (Route 32), 4½ M. ; Lead-mine, 6¼ M. ; Shelburne station (Route 32), 6 M. ; Alpine Cascade, 5 M. ; Berlin Falls (Route 34), 6 M. ; Milan (Route 36), 13¼ M. ; Randolph Hill, 5 M.

The town of Gorham contains 1,161 inhabitants, and covers 18,146 acres, of which over 16,000 are unimproved, lying on the slopes of rugged ridges. The village is an important station on the Grand Trunk Railway, and is the seat of the machine-shops for the Eastern Division of that line, employing 155 men and producing $250,000 worth of work annually. There are about 900 inhabitants in the village, with 3 churches and 12 stores. It is situated on the Androscoggin River (formerly called the *Amariscoggin*), a wide and rapid stream, near its junction with the Moose and Peabody Rivers.

Gorham is the nearest village to the great peaks N. of Mt. Washington, and although they are hidden from the streets by the low-lying mass of Pine Mt., some of the best possible views are gained from points close by. Mts. Moriah, Carter, and Hayes, and the Pilot Mts. are visible from the village. In the earlier days of White-Mt. travel the noble scenery and adventurous excursions in this vicinity made Gorham one of the favorite centres for tourists. The Rev. Thomas Starr King spent several seasons here, writing the greater part of his charming book, *The White Hills.* The village is 812 ft. above the sea; and the dry and bracing air of the broad valley is invigorating and healthful. The close proximity of the great peaks of Madison, Jefferson, and Adams gives rare grandeur to the views from the environs; and this is said to be one of the proper focal points for the view of Mt. Washington. On the S. the Peabody Glen opens away into the highlands, flanked on the E. by the lofty crests of Mts. Moriah and Carter; and on the N. W. are the long serrated lines of the Pilot Mts., the scene of brilliant displays of color towards evening. The rugged hills N. of the Androscoggin River tower closely on the N. of the village, throwing out their rocky cliffs to the verge of the stream.

"No point in the mountains offers views to be gained by walks of a mile or two that are more noble and memorable. . . . . For river scenery, in connection with impressive mountain forms, the immediate vicinity of Gorham surpasses all the other districts from which the highest peaks are visible. The Androscoggin sweeps through the village with a broader bed, and in larger volume, than the Connecticut shows at Lancaster or Littleton. As a general thing, Gorham is the place to see the more rugged sculpturing and the Titanic brawn of the hills. Turning from N. Conway to the Androscoggin Valley is somewhat like turning from a volume of Tennyson to the pages of Carlyle; from the melodies of Don Giovanni to the surges of the Ninth Symphony; from the art of Raffaelle to that of Michel Angelo." (STARR KING.)

Soldier's Hill is a small eminence in the village, on the r. of the street which runs to the suspension-bridge. It commands a beautiful view of the valley, extending on r. and l. for miles, and affords a good observatory from which to reconnoitre the adjacent mountains.

The best view from Gorham is obtained from the vicinity of Lary's, about 1 M. from the station, and the best time to enjoy it is at late afternoon, towards sunset. Its chief features are the noble prospects of Mts. Moriah, Madison, and Adams.

Mt. Adams, as seen from a point on the road about 1½ M. above Gorham, is "the highest elevation which we can look at in New England from any point within a few miles of the base. Indeed, it is the highest point of land overlooking a station near the base, that can be seen E. of the Mississippi." The peak of Adams (5,794 ft. high) is about 7 M. from the point before mentioned (868 ft. high), above which it towers to the height of 4,926 ft., while Mt. Washington rises but 4,722 ft. above the Fabyan House, and 4,661 ft. above the Glen House.

From the same vicinity is enjoyed the best view of Mt. Moriah, which rises 3,785 ft. above the valley, or over 500 ft. higher than Mt. Lafayette stands over the Profile House. "With the exception perhaps of the Moat Mt. in N. Conway, the long lines of its declivity, towards the E., flow more softly than any others we can recall. They wave from the summit to the valley in curves as fluent and graceful as the fluttering of a long pennant from a masthead. The whole mass of the mountain, moreover, is clothed with the richest foliage, unscarred by any land-slide, unbroken by any ravages of storm and frost, even in its ravines. . . . . But nothing can be more graceful and seductive than the flow of these lines of Mt. Moriah seen through such a veil [of shower]. They do not suggest any violent internal forces. It would seem that they rose to melody, as when Amphion played his lyre, and saw the stones move by rhythmic masonry to the places where they were wanted. And the beauty is the more effective by contrast with the sternness and vigor of the lines of Adams and Madison."

Randolph Hill is a locality on the Jefferson-Hill road, from which one of the noblest views of the White Mts. is obtained. It is about 5 M. from Gorham by a well-constructed road which passes over the crest of the hill, 600 ft. above the village. Therefrom is afforded a grand * view of Mt. Madison (on the l.) and Mt. Adams (r.), with the tremendous gorge of King's Ravine opening into the heart of the latter.

"After the first mile the summits are in view all the way. As the sides of the mountains are more and more clearly seen, attention is arrested by the correspondent lines that run N. W. from Adams, and S. E. from Madison. They are alike in almost all their details. These earthquake rhymes are more interesting for the intellect than the granite physiognomy in Franconia. And the lower outworks and braces of Mt. Madison repeat, in reduced form and reverse order, the shapes of the two great hills. There is no drive more valuable than this for the close study of the multitudinous details that make up the foreground of a vast mountain, — the abutments, the water-lines, the ravine walls and edges, the twistings of rock beneath the soil, that give character to a view ten miles off, which almost every eye feels, but which only a critical one can explain. . . . . And then the general aspect of these mountains during this drive. How proud and secure! What weight and what spirit! They are not dead matter, — they live. So solid, yet soaring! They seem to lift themselves to that glorious height. . . . . Here we see the N. E. wall of the White-Mt. chain declining sharply to the valley. From Randolph Hill we look down to the lowest course of its masonry, and up to the two noblest spires of rock which the ridge contains. How lonely and desolate it looks, aloft there! And yet those pinnacles, that are scarcely fanned by a breath of summer, and that feel such storms as the valleys never know and could not bear, — is it not wholesome to look at them and think what they undergo for the good of New England?" (STARR KING.)

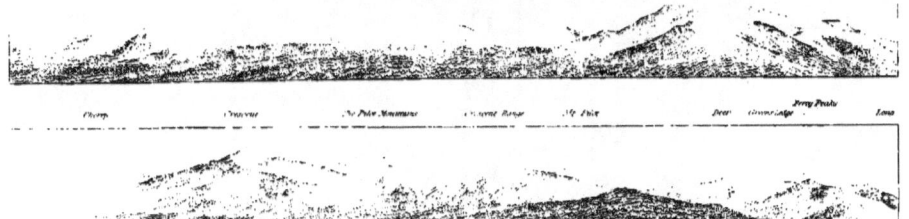

Cheops.　　　　Uncompah.　　　No Palm Mountains　　Saguache Range　　Mt Ouis　　　　Deer　　Greenwledge　　Jerry Peaks　　Lone

the
fron

T
abo
noo
Mor

M
hair
fron
poir
the
the
heig
Fab

Fr
ft. al
Hou
lines
They
flutt
more
by a
more
such
woul
the
And
lines

R
of tl
Gorl
hill,
Mt.
of K

" A
mou
lines
all tl
than
of Mi
great
multi
ment.
the s
but w
these
what
seem
the V
down
whicl
those
storm
at th
KING.

Pine Mountain, or *Camel's Hump,* is the most northerly of the White Mts., and lies S. W. of Gorham, occupying the area which is bounded by the Moose, Androscoggin, and Peabody Rivers, and the Old Pinkham-Notch road (which separates it from Mt. Madison). It is a low and broad-based mountain, on whose ridge are three distinct crests partially covered with woods. Hamilton Willis had the trees cut away from one of the lower peaks so that views might be obtained of Mts. Madison and Adams. This point was called *Willis's Cut,* and was about ¾ M. from Gorham, whence it was approached by a good path.

A recent scientific visitor thus records his march to the middle peak of Pine Mt.: Glen Road (1 M. from Gorham) to end of logging-road, 1 hr., 2¼ M.; thence to first peak, 2 hrs., 1½ M.; thence to the second peak, 1 hr., ³/₅ M.

29. Mount Hayes.

The easiest, and in some respects the most profitable mountain-excursion to be made from Gorham is the ascent of this eminence, from which is obtained one of the grandest views of the White Mts. The mountain here approaches the river, and is faced with rocky ledges and cliffs. The summit is a little over 2 M. from the village, and the path is entered from the N. end of the suspension-bridge. For many years Mt. Hayes has been celebrated for " bears, blueberries, and views."

The distance from the suspension-bridge to the house on the ridge is nearly 1½ M., and the summit is about ⅓ M. beyond. The path is easy and well defined, leading through a forest of second-growth birches; and has plenty of water in its vicinity. On account of its easy grade this mountain is much visited by tourists. The most favorable time is in the morning, when the Presidential Range is lit up from the E., and the ravines show distinctly. The summit is a broad plateau, covered with outcropping ledges of granitic gneiss, and its highest part is overgrown with trees. The best view of the Androscoggin Valley is gained from the ledges near the house ; but the peaks to the W. are best seen from a high rock farther up on the plateau. The Guide-Book party was 80 min. in making the ascent.

"Mt. Hayes is the chair set by the Creator at the proper distance and angle to appreciate and enjoy Mt. Washington's kingly prominence. All the lower summits are hidden, and you have the great advantage of not looking along a chain, but of seeing the monarch himself soar alone, back of Madison and Adams and seemingly disconnected with them, standing just enough to the S. to allow an unobstructed view of the ridges that climb from the Pinkham road up over Tuckerman's Ravine, to a crest moulded and poised with indescribable stateliness and grace. It completely dimmed the glory of Mt. Adams. The eye clung ever fascinated and still hungry upon those noble proportions and that haughty peace. We were just far enough removed to get the poetic impression of height which vagueness and airy tenderness of color give. . . . . It was satisfactory, artistic, mountain-eminence and majesty that we were gazing upon."

* *The View.* — Toward the S. is the great and massive ridge of Mt. Moriah, falling away on the r. towards the lower peak of Imp, which rises sharply over the Peabody Glen. Farther to the r. is the symmetrical and lofty cone of Mt. Carter, partly concealing Carter Dome and its formidable southern outworks; and adjoined by the lower and more distant Mt. Wild-Cat, which falls off on the l. into the Carter Notch, and on the r. toward the Pinkham Notch. Over the foot of the Imp's long slope is the Glen-House clearing; and at the end of the valley the Rocky-Branch ridge trends up on the r. to the rugged crags on the end of Boott's Spur. The S. wall of Tuckerman's Ravine is on the side of the Spur, and then the high crest of Washington is seen, with its white hotel. In the foreground, on this line of vision, is the low and heavily wooded Pine Mt. On the r. of Washington are the long and broken rocky ridges of Madison, ascending on the r. to the shattered peak, on whose r. are the fine crests of Adams, concealing Jefferson from view. The Moose-River Valley opens away on the N., by Randolph Hill, and at its end is the long blue Cherry Mt., with the Owl's-Head peak on the N. end. Thence to the r. extend the long and monotonous hill-ranges of Randolph, Crescent, and Pliny, culminating in Mt. Starr King. Toward the N. W. are the higher and more picturesque Pilot Mountains, showing fine outlines and covering a great area, with the twin white mamelons of the Percy Peaks far away on the r., seen over a nearer wooded hill. On the l. of the line to the Percies are the crags of Green's Ledge, falling off sharply to the S.; Deer Mt., with a knob in the centre; and the more distant heights of Sugar Loaf and Stratford Mt. More to the r., and nearly over Mt. Forist, is Long Mt., in Odell.

The view to the N. and N. E. is closed by the woods on the summit of Hayes. When standing on the ledge near the old house, the visitor sees the Moose-River Valley on his r., the Peabody Glen in front, and the Androscoggin Valley on the l. This scene is one of rare beauty. "The rich upland of Randolph, over which the ridges of Madison and Adams heave towards the S., first holds the eye. Next the singular curve in the blue Androscoggin around the Lary Farm, arching like a bow drawn taut. Directly beneath us lay two islands in the river, — one of a diamond shape, the other cut precisely like a huge kite, and fringed most charmingly with green. Down the valley, Shelburne, Gilead, W. Bethel, and Bethel were laid into the landscape with rich mosaics of grove, and grass, and ripening grain, needing a brush dipped in molten opal to paint their wavering, tremulous beauty. Directly opposite, seemingly only an arrow-shot's distance, were the russet ravines of Moriah and the shadow-cooled stairways of Carter." (STARR KING.)

30. Mount Surprise.

This peak is on a lower ridge of Mt. Moriah, and is about 2¼ M. from Gorham, whence it may be reached in about 2 hrs. The path was formerly a good one, adapted for the ascent of saddle-horses; but it has been neglected, and is now obscured in some places by bushes and wood-choppings. The entrance of the path may be found through the pastures near J. R. Hitchcock's, although it formerly passed in from the E. bank of the Peabody, near the highway-bridge. The services of a guide will be useful at least in getting the visitor fairly started in the woods. The route to

Surprise is the lower half of the bridle-path to Mt. Moriah, and was formerly a broad, plain, and well-travelled way. It would be well if the people of Gorham would have it once more opened and put in order, — a work which would involve but little outlay in comparison to the benefits accruing.

Mount Surprise is a ledgy knoll rising boldly from the long flank of Moriah, and commanding a very noble * view. On the S. S. E. is the main peak of Moriah, over a green wooded spur, with the Imp on the r. Then comes Mt. Washington, flanked by Boott's Spur on the l., and by the high humps of Mt. Clay on the r. The noble peaks of Jefferson, Adams, and Madison are next seen, with Pine Mt. below, on the r. Farther away are the Crescent and Randolph Ranges, Mt. Starr King, and the broken lines of the Pilot Mts., over which are the white domes of the Percy Peaks. The Stratford, Bowback, and Deer Mts. are in the distant N. W. Close at hand on the N. is the Androscoggin Valley, with the rocky ridges of Mts. Hayes and Baldcap over it, and Gorham's white houses below. Farther away towards the N. N. E. are the mountains about the Grafton Notch, and the pointed peak of Goose-Eye; and more to the r. is the long and beautiful valley of Shelburne.

Mt. Surprise was a favorite resort of Thomas Starr King, and no description of its view can equal that which he wrote: " Looking up the valley of the Peabody, we see the five highest peaks of the Washington range, but a full view is given of two only, — Madison, the Apollo of the Highlands, and the Herculean structure of Washington, with his high, hard shoulders and stalwart spurs. There are very few hills of moderate height accessible by bridle-paths, from which a good view of any portion of the great range can be gained, — positions near enough to reveal the extent and freshness of the forests, and yet far enough to allow the effect of light and symmetry. We know of none so favorable in both these respects as Mt. Surprise. It ought to be to Gorham what Mt. Willard is to the Notch. Certainly after several visits to Mt. Willard, when the senses have become used to the impression, at first so startling, made by looking over the cliff into the awful gulf of the Notch, the view gained there of the summits of the Washington chain, especially of Mt. Jefferson, is more fascinating to an artistic sense. And Mt. Surprise gives a still more striking spectacle. Plain prose, however eloquent, is no fit medium to describe that proud smooth swell of Madison from the Peabody Valley to a peak that pricks the sky. It needs rhythm; it needs the buoyant surge of a blank-verse like that of Coleridge, to ensoul the fascination of that soaring beauty, which spires at last into granite grandeur. There is no point among the New-Hampshire hills where the ' hymn in the valley of Chamouni ' breaks from the lips so readily as here. And if one wants to see forest-costume in the utmost richness of folds and retinue, let him look at the broad miles of wilderness that flow down the opposing sides of Carter and Madison. . . . . One is tempted to believe that these two points — the tops of Carter and Madison — were lifted up gradually from the level land at first, and held off from each other just far enough to let the forests droop in the most graceful folds from them, and meet with trails soft as velvet upon the valley.

" Ah, and what intensity of expression in the ragged crest of Adams, which starts out, it may be, from a melting fog, and overtops the gentler slopes of Madison; and what energy in these far-running southward braces of Washington, engraved perhaps upon a white cloud-background, — each worn to the rocky bone by the torrents of summer, and the slower but more penetrative wrath of winter cold! It is indeed rich music for the eye that is afforded by the quintette of summits seen from Mt. Surprise; and one who can detect some dim analogy between tones and forms will find increased delight here in seeing how, in the mountain choir, the sharp soprano of Madison is brought into contrast and balance with the heavy bass of Washington, and how the body of the harmony is filled up by the tenor of Adams, the baritone of Jefferson, and the alto of Clay, whose bulk and lines are merely suggested by their crests that jut into view.

"But a sweeter melody still is offered to the eye that turns from the great hills to the Androscoggin intervale. It is the strength that 'setteth fast the mountains' which appeals to us on the W. ; on the E we have the smile of the landscape, the fluent curves of the river moving 'like charity among its children dear,' the sweet phrases which man has added to the wild natural music, the colors vivid and tender that glow upon winding miles of shorn grass and ripening grain. No mountain as high as Washington can offer, in its comprehensive pageant, any one passage so lovely as this nearer view from Mt. Surprise of the farms that border the Androscoggin. Here the infinite goodness responds by appropriate symbols to the infinite majesty which is represented by the barren hills."

31. Mount Moriah.

This fine mountain is on the N. part of the great eastern range, between Mt. Carter and the unnamed peaks over Shelburne. It attains a height of 4,653 ft. above the sea, or 3,841 ft. above Gorham. On account of the impracticable nature of the country in this vicinity, the greater part of the range remains unexplored, and the name Moriah is applied to nearly all the peaks and ridges between the Androscoggin, Wild, and Peabody Rivers, over an area of nearly 20 square miles. The crests N. of Moriah are of commanding altitudes and fine shapes, and deserve individual names. It is said that the present title was given to the range by an early settler on account of its supposed resemblance to the hill of that name in the city of Jerusalem. The good pioneer probably thought that the height and extent of the Hebrew Moriah must have been in some degree proportionate to the greatness of the events which transpired thereon.

At one time there was a house on the summit, and a smooth and well-worn bridle-path led up to it. But since the occupation of Gorham by the railway, the burning of the hotel, and the construction of easy roads up Mt. Washington, this route has been nearly abandoned, and is now in bad repair. The worst part of it is on the lower half, and the route from Surprise to Moriah is comparatively plain and easy, — after traversing the tangled hollow S. E. of the former It is about 2½ M. from Surprise to Moriah, the summit of the first ridge being 1 M. from Surprise. Beyond this point the trail leads among and over the bold nubbles on the lofty crest-line, traversing a fine forest and passing some picturesque ledges. Water is found at various points, and during their season berries are abundant. Pleasant views are occasionally given through or over the trees, and the beautiful woodland scenery tempts the visitor to frequent rests. During the last 1½ M. the weary pedestrian is disposed to think that each of the rocky knolls which he is forced to ascend is the final peak. When the latter is attained, after passing the ruins of a small house on the r., the visitor stands on the summit of a bare ledge which overlooks a vast area of country in a wide and unbroken prospect. "The forest-path itself — unequalled as far as we know in the whole mountain tour — is lovely enough to tempt the visitor, independently of the prospect from the crown."

** *The View.* — Toward the S. W. is Mt. Washington, looming up in the centre of a line of vast mountains, and flanked on the l. by the long and lofty terrace of Boott's Spur, which ends in a pile of crags. Below the Spur are the striped walls of Tuckerman's Ravine, with the gorge of Raymond's Cataract and the upper part of Huntington's Ravine on the r. On the r. of the white Summit House, and below, is the Half-way House, over which are the rugged humps of Mt. Clay, at the head of the Great Gulf. More to the r. is the well-marked and smoothly lined crest of Jefferson, succeeded by the stately pyramid of Adams. Just across the Peabody Glen, on the r. of Adams, is the graceful summit of Madison, whose N. flank flows off into the Moose-River Valley. Pine Mt. is on the r. of Madison, and far below it. Beyond, over the flank of Madison, is the Randolph Range, with the Crescent Mts. more to the E. Mt. Starr King is over the Randolph Mts.; and the clustered peaks of the Pilot Mts. are over the Crescents. Over the r. slope of Pine is the long Deer Mt., with the white Percy Peaks beyond, and the Bowback and Stratford Mts. in the background, far to the N. W. Green's Ledge is nearer and more to the r., with Long Mt., in Odell, behind it.

Gorham is seen below the mountain, its white houses strongly contrasted with the rich green of the Androscoggin meadows and the dark hues of the rearward heights. A little W. of N. are seen the pleasant levels of the valley, as it passes through Berlin and Milan, flanked on the l. by Mt. Forist, the Berlin Heights, and the Milan Hills. The Dixville and Crystal Mts. are far beyond, low down on the horizon.

Across the Androscoggin Valley is the bold rocky ridge of Mt. Hayes, with a house on the S. of its plateau-summit. The ridge stretches away to the r. to Mt. Baldcap, a round-topped and ledgy peak over the bright glens of Shelburne. Beyond this ridge, a little E. of N., and far away on the horizon, are the Magalloway Mts., with the conspicuous peak of Mt. Carmel looming out of the blue distance and Mt. Dustin somewhat nearer. Farther S. is Mt. Aziscoös, which is near the N. shore of Lake Umbagog.

The N. peaks of Moriah are now seen, covered with white rocks, and rising nobly beyond a deep ravine. In this direction are seen the high peaks of Mt. Ingalls and Goose-Eye; and between their crests are the mountains toward the Grafton Notch, Speckled Mt., the Bear-River White Cap, and the Sunday-River White Cap. Then comes a third N. peak of Moriah, reaching a great height and surrounded with steep gray cliffs. A cluster of the chief peaks of Maine is now seen towards the N. E., the distant Mt. Bigelow being over the Grafton Notch, with Saddleback on the r., and Mt. Abraham nearly in a line over the Sunday-River White Cap. Nearer Moriah are the fair glens of the Androscoggin, stretching through Gilead and Bethel, with the shaggy ledges of Tumble Down Dick in their midst, and on the r. Mt. Calabo, a double ridge with a bare top. Over this valley, and nearly in line with Puzzle Mt., is the graceful cone of Mt. Blue, in Avon.

On the E. is the great Wild-River Forest, stretching down the long valley, across which, S. of E., is the heavy double-headed ridge of Mt. Royce, about 5 M. distant, and on the Maine border. Over Royce is the rounded top of Speckled Mt., and through the gap on the r. is a part of Upper Kezar Pond, beyond which are Bear and Hawk Mts., in Waterford. More to the S. are Upper Moose Pond, Highland Lake, and part of Long

Pond; and then Mt. Pleasant comes into view, showing a long ridge rising from apparent levels and crowned by a white house. The ocean is seen in this direction for a long distance, but can hardly be separated from the sky, except at early morning, when the sun is reflected from it. With a powerful glass, on a clear day, the city of Portland is visible to the l. of Mt. Pleasant. Over a bare spot on the first ridge S. of and close to Moriah are the two sharp peaks of Baldface, covered with white rocks, 5 – 6 M. distant across an untrodden wilderness. Beyond and on the r. of Baldface are Mts. Eastman and Sable, rising from the heart of the Wild-River Forest; and farther to the r., nearly over Sable, is the symmetrical cone of Kiarsarge, crowned by a house, and flanked on the r. by Mt. Bartlett. Due S. over the highlands about the Perkins Notch is Double-Head, the S. peak nearly eclipsed by that on the N., over which the rich Saco Valley is seen, stretching away to the remote distance. Walker's Pond and other localities in Conway are discernible. On the r. of Double-Head is Thorn Mt., with the clear-cut ridges of Moat Mt. over its r. flank.

The view next falls on the Carter Range, near at hand, and bristling with forests. The nearest summit is Mt. Imp, which is comparatively low, and lies to the r., towards the Peabody Glen. Farther away, and much higher, are the double peaks of the true Mt. Carter, S. of which is the lofty crest of the Carter Dome, looking down into the Carter Notch.

32. Shelburne.

Boarding-houses. — The Winthrop House and T. J. Hubbard's (40 guests each; $6 – 8 a week), near the station; Gates Cottage (30 guests), at the foot of Baldcap, and commanding a noble view; G Philbrook's (30 guests), N. of the Androscoggin and in the E. part of the town; Charles Philbrook's (25 guests) and D. Evans's (15 guests), towards Gorham.

Shelburne has 259 inhabitants, and covers 18,140 acres of land, of which 15,000 acres are rugged and mountainous, and the remainder is along the river, and forms the richest of intervales. The present population is but about half that of the town in 1850. Farming is the chief occupation of the people, the best estate being Judge Burbank's White-Mt. Stock-Farm. The town is divided into two parts by the Androscoggin River, which is crossed here by three bridges, and receives the waters of ten brooks, the chief of which is Rattle River. The chief mountains of the town are Ingalls, Baldcap, and the N peaks of Moriah, which are sometimes ascended by the ravine of Rattle River. The hamlet of Shelburne is near the railway station, and has a small church. It is 723 ft. above the sea.

Mr. King says of the road through Shelburne, down to the Gilead Bridge on the S. bank and back to the Lead-Mine Bridge on the N. bank: "No drive of equal length among the mountains offers mo e varied interest in the beauty of the scenery, the historic and traditional associations involved with the prominent points of the landscape, and the scientific attractions connected with some portions of the road."

Shelburne was granted to Mark H. Wentworth and others in 1771, and was occupied by settlers in 1770. In the autumn of 1776 a party of American soldiers emerged from the forest into the town, having deserted from the army which lay before Quebec and crossed the wilderness amid terrible sufferings. One of their number, named Hall, was left near the frontier, exhausted and famine-stricken. He was found by a searching party, dead, with his face in the water of the brook, which has since been known as Hall's Stream. Early in the history of the town an Indian passed through the settlement, and was soon afterwards attacked by a pack of wolves, and devoured, after an heroic struggle. Not long afterward his bones and clothing were found, with the carcasses of seven huge wolves alongside. In 1781 the Canadian Indians ravaged the town.

Moses Rock is a ledge near the road E. of the village, which is 60 × 90 ft. in area, and inclined at an angle of 50 degrees. During the early surveys, the best lot in the town was offered to whoever would climb it, whereupon Moses Ingalls, removing his shoes, ran up its face and won the reward. It is said that a hound once pursued a moose over this cliff, and both fell to the bottom dead.

Granny Starbird's Ledge is near this point, and once supported an enormous boulder (since blasted for railway work), under which Granny Starbird, the ancient doctress of this region, remained all night, during a tremendous mountain-storm, standing and holding her horse.

The Shelburne Basins are a series of not very remarkable potholes, occurring in the upper glens of Pea Brook, over 1 M. from the road, on an old logging-track.

Mt. Winthrop is in the town of Shelburne, and occupies a favorable position for overlooking the Androscoggin Valley. The summit is a little over 1 M. from the Winthrop House, and is reached by a wood-road which leads to the r. from the main road E. of the hotel, crosses a pasture and ascends the slope. The summit is composed of uneven ledges, and is partly occupied by trees.

The View from the top includes a long sweep of the valley, whose verdant meadows are banded by the gracefully winding Androscoggin River. Nearly N., across the valley, is Baldcap, beyond which are the high ridges towards Mt. Ingalls. More distant, and farther to the r., are the many mountains of Newry and other ranges of Western Maine, sweeping down by the glens of Gilead to the broadening plains of Bethel.

The best view of the White Mts. is gained from a point below the woods on the farther slope of the summit. The great ranges of Mts. Moriah and Carter are near at hand, across a dark wooded valley; and the N. E. crest of Moriah towers on the r. Beyond these are seen the northern presidential peaks, massively outlined and boldly advanced.

Hark Hill is in the E. part of the town, overlooking the glens of Gilead, and was so named because the inhabitants of the town took refuge on its summit during the Indian foray of 1781, and heard thence the whoopings of the enemy throughout the night. It was a war-party of Canadian Indians who had made an attack on Bethel, and, on their return-march, sacked several houses in Shelburne, and killed one man and captured another. The next morning Hark Hill was evacuated, and the people fled through the wilderness to Fryeburg, 59 M. distant.

6

Near the point where the Lead-Mine Bridge road diverges from the Shelburne road is a small hill whence is obtained the noble view described by Starr King: "Mt. Madison sits on a plateau over the Androscoggin meadows. No intervening ridges hide his pyramid, or break the keen lines of his sides. He towers clear, symmetrical, and proud against the vivid blue of the western sky. And as if the bright foreground of the meadows, golden in the afternoon light, and the velvety softness of the vague blue shadows that dim the desolation of the mountain, and the hues that flame on the peaks of its lower ridges, and the vigor of its sweep upwards to a sharp crest, are not enough to perfect the artistic finish of the picture, a *frame* is gracefully carved out of two nearer hills, to seclude it from any neighboring roughness around the Peabody valley, and to narrow into the most shapely proportions the plateau from which it soars."

The Lead-Mine Bridge is about 4 M. from Gorham, and 2 M. from Shelburne village. From its centre a noble * view is obtained of Mt. Madison, with Adams over it and Washington to the l., the river forming a beautiful foreground.

"The best time to make the visit is between five and seven of the afternoon. Then the lights are softest, and the shadows richest on the foliage of the islands of the river, and on the lower mountain sides. And then the gigantic gray pyramid of Madison with its pointed apex, back of which peers the ragged crest of Adams, shows to the best advantage. It fills up the whole distance of the scene. The view is one of uncommon simplicity and symmetry. The rolling slopes upon the base of Mt. Moriah on one side, and the jutting spurs of Mts. Hayes and Baldcap on the other, compose an effective avenue through which the eye roams upward to the higher mountain that sits back as on a throne. . . . . [It is] a view which at once takes the eye captive, and not only claims front rank among the richest landscapes that are combined in New Hampshire out of the White Mts. and the streams they feed, but impresses travellers that are fresh from Europe as one of the loveliest pictures which have been shown to them on the earth. . . . For eye-landscape, to be enjoyed without reference to the demands of the canvas, it would be difficult to conceive a scene where greater beauty of river and islands is crowned with a mountain so bold and yet so tenderly tinted, so symmetrical and yet so masculine, so satisfactory in height without losing on the surface clearness and vigor of detail." (KING.)

The lead-mine was about 2 M. from the bridge, on the N. side of the river, in a deep mica-slate ravine, where are found veins of zinc, copper, and silver-bearing lead-ore, sometimes in heavy masses of fine quality. Three pounds of silver were derived from a ton of lead-ore. The mine has long been abandoned.

The *Gates Cottage* is about 2 M. E. of the Lead-Mine Bridge, on the N. side of the river, and near the base of Mt. Baldcap. It is on high ground, and a rich * view is enjoyed either from the house or from the arbor on an adjacent knoll.

"How grand and complete is the landscape that stretches before us as we look up the river 7 – 8 M. to the base of Madison and to the bulk of Washington, whose majestic dome rises over two curving walls of rock, that are set beneath it like wings! Seen in the afternoon light, the Androscoggin and its meadows look more lovely than on any portion of the road between Bethel and Gorham, and more fascinating than any piece of river-scenery it has ever been our fortune to look upon in the mountain-region. . . . . The Shelburne view is superior in simplicity, largeness of feature, and bold picturesqueness. In graceful picturesqueness it must yield to Conway, but the mountain-forms in Madison and the crest of Jefferson are more spirited and decisive." (STARR KING.)

33. Mount Baldcap.

Mt. Baldcap is N. of the central part of Shelburne, and is 2,952 ft. above the sea. It is a vast pile of ledges, heaped above each other line over line, and abounds with blueberries in their season. The top is wooded, but a fine view is commanded from adjacent ledges, marked by piles of stones. The best route of attack is from the Peabody farm, about 1 M. E. of the Lead-Mine Bridge, whence the ascent may be made in 1½ - 2 hrs. A plain path leads in from the house to a secluded pasture-lot, and runs out again towards the mountain from its r. upper corner. When it reaches the crest of the first ridge, the path should be left, and the visitor ascends to the l. oblique, following up the spur through a pleasant second-growth forest, and then over far-viewing ledges. The dubious route from the Gates Cottage is not to be recommended.

* *The View* is in some respects superior to that from Mt. Hayes. Mt. Ingalls is a little N. of E., and is flanked by the more distant hills of Gilead and Bethel, prominent among which is Tumble Down Dick. The beautiful Androscoggin is seen in this direction for many leagues, winding gracefully through the rich meadows and under the shadow of lofty heights. To the S. of Bethel is Mt. Calabo, a long ridge with a little peak on the N.; and on its r. is Mt. Royce, round-topped and massive. Over the bend of the river below Shelburne is the N. extension of Baldface, about which are the dark highlands of the Wild-River Forest. Mt. Winthrop is the low eminence E. of S. across the valley, with pastures on its sides. Near its base is the hamlet of Shelburne, with the bare ledge of Moses' Rock on the r., over which and on the r. are the many peaks and lofty ridges of Mt. Moriah, indented by the ravines of Rattle River and Pea Brook, and bearing Mt. Surprise on its r. flank. The crest of Mt. Carter is further to the r.

The view now rests on Mt. Washington, on whose l. is Boott's Spur, with the wall of Tuckerman's Ravine and the crags of the Lion's Head. The Half-Way House and portions of the carriage-road are seen, leading up to the white hotel on the last summit. The next view-line is up the entire length of the Great Gulf to Mt. Clay, whose uneven rolling crests are seen at its head. Somewhat nearer, and to the r., are the long stony ridges of Madison, surmounted by a pyramidal peak, behind which is the higher crest of Adams, of graceful shape and imposing height. In this direction, but nearer, is the low and leonine shape of Pine Mt.; and the rich intervales of the Androscoggin fill the foreground, traversed by the bright sinuous band of the river.

Over the N. flank of Madison is the distant blue ridge of Cherry Mt., on whose r. and nearer are the Randolph and Crescent Mts., the Pliny Range, and the Pilot Mts., all mingled in a wide wilderness of wavy crests and wooded slopes. A picturesque and irregular ridge runs W. from Baldcap, dotted with low sharp cliffs and bristling knobs.

Dream Lake is nearly N. of the mountain, and is seen from ledges on that side, far below in the forest.

34. Berlin Falls.

Hotel. — The Mt.-Forist House (kept by M C. Forist) is a small country-inn overlooking the river. The Grand Trunk Railway passes the village. Stages run N. to Milan daily.

The town of Berlin has 722 inhabitants, and covers 31,000 acres of land, of which only 1,340 are improved, the remainder being occupied by mountains and forests. It is traversed by the Upper Ammonoosuc, Dead, and Androscoggin Rivers; and contains parts of the Pilot, Pliny, and Crescent Mts. Berlin Falls is the name of the only village, which is at the confluence of the Dead and Androscoggin Rivers, and has a railway station and a small church. It is the site of the great mills of the Berlin Lumber Company, which saw from 20,000,000 to 25,000,000 ft. of lumber every year, valued at about $350,000. The logs are rafted down the Androscoggin from the northern forests and the Umbagog country.

This town was granted in 1771 to Sir William Mayne, of Barbadoes, and others of his family, and bore the name of *Maynesborough* until 1829. The population has more than trebled within 20 years, on account of the rapid development of the lumber business.

The * **Berlin Falls** are 5 – 6 M. from Gorham, and are within a few rods of the Berlin-Falls railway station. They are visited by a path which turns in near a squalid shanty on the E. of the road, and soon reaches a foot-bridge over the gorge. Just before descending to this point, the path crosses a line of ledges whence is gained a view of the Presidential Range in the S. S. E. Starr King's much-lauded view from the bridge itself has been spoiled by the growth of a curtain of trees on the adjacent shore. The vicinity of the falls, though so near an uncomely village, is fortunately surrounded by trees, and remains in a comparatively primitive condition. The Androscoggin River here descends nearly 200 ft. in 1 M. of its course, in a series of powerful falls and rapids, and is confined between high and curving walls of dark schist. The main fall is just above the bridge, and is noticeable for massiveness and power rather than for altitude. From this point the stream dashes down a line of vexed rapids until it comes out on the widenings below. Beyond the bridge is a high wooded island, traversed and encircled by paths, and separated from the E. bank by a deep gorge, filled with fantastically piled boulders and water-worn rocks. The river formerly poured vast bodies of water through this chasm during the season of floods.

" We do not think that in New England there is any passage of river passion that will compare with the Berlin Falls . . . . Here we have a strong river that shrinks but very little in long droughts, and that is fed by the Umbagog chain of lakes, pouring a clean and powerful tide through a narrow granite pass, and descending nearly 200 ft. in the course of a mile. . . . . How madly it hurls the deep transparent amber down the pass and over the boulders, — flying and roaring like a drove of young lions, crowding each other in furious rush after prey in sight! On the bridge, we look down and see the current shooting swifter than 'the arrowy Rhone,' and over-lapped on either side by the hissing foam thrown back by each of the rock walls. Above the bridge, we can walk on the ledge of the right hand bank, and sit down where we can touch the water and see the most powerful plunge of all, where half the river leaps in a smooth cataract, and, around a large rock which, though sunken, seems to divide the motion of the flood, a narrow and tremendous current of foam shoots into the pass, and mingles its fury at once with the burden of the heavier fall." (STARR KING.)

The **Alpine Cascades** are 3½ M. from Gorham and 1½ M. from Berlin Falls, and are interesting during seasons of full water, though at other times they are rather weak. Carriages are left near the road, on the W. bank, and visitors cross the Androscoggin by two light suspension-bridges, suitable only for pedestrians. Just above the main bridge is a cataract in the river, which is preferred by many to the cascades. On the knoll above the bridge is a small house where refreshments are sold and where a toll of 25 c. is taken from each visitor. A good path of about ¼ M. long leads thence to the foot of the Alpine Cascades, whose course may be followed for a few rods by means of stairways and ropes on the ledges. The precipitous ledges and sharp crags in the bed of the brook afford every facility for a picturesque fall, when enough water is supplied by heavy rains.

Mount Forist is over the village of Berlin Falls, and may be ascended in 45 minutes by following a path which runs from the houses opposite the railway station to the foot of the ledges, and then clambering up over the rocks. The mountain is 915 ft. high above the railway, and is covered with trees on top, the sides being surrounded with high bare cliffs of imposing dimensions. According to Prof. Hitchcock, Forist is the E. end of a long chain which includes the Pliny, Randolph, and Crescent Mts.

35. The Milan Road.

The roads between Berlin Falls and Milan Corner afford a favorite drive for summer-visitors in this section. The distance between the two villages is about 8 M., and the roads are good, following the course of the Androscoggin and giving panoramas of pleasant valley-scenery and farming-lands. Views are also given of the Percy Peaks and the other mountains of Stratford, on the N. W., and of Goose-Eye, Chickwolnepy, and the border ridges on the E. The road W. of the river is usually taken for the north-ward drive, and the E. road for the return, since this arrangement enables one to see the Presidential Range at its best.

Berlin Heights are reached by a road which diverges from the W. road near an old school-house, about 3 M. from Berlin Falls. The ascent by carriage is somewhat difficult, on account of the steepness of the way; but there is a succession of pleasant mountain-views from the upper courses of the hills. The road is about 1½ M. long, and terminates at a farm-house.

The road from Milan Corner S. along the E. bank of the Androscoggin crosses Chandler's Brook about 2⅓ M. below the village. Soon afterwards a point is reached whence one of the best views of the Presidential Range may be gained, and thenceforward until Berlin Falls are approached the noble prospect is continued. This * view may also be obtained by driving 4 M. from Berlin Falls up the E. bank of the river, and then returning slowly. The highest peak on the l. is Washington, on whose r. are the cones of Madison and Adams, with the notched mound of Pine Mt. in the valley below. This view has been painted by one of the most eminent artists of America, and is now in England. Starr King laments that so few tourists have yet taken this ride and enjoyed its superb prospect. The excursion may easily be made in a two-hours' drive from Gorham or a shorter route from Berlin Falls. The tranquil beauty of the river on this reach is also an attractive feature, and " on a still afternoon it sleeps here as though it had not been troubled above, and had no more hard fortune to encounter below."

Of the main view from the E. road, Starr King says : " Mt. Washington has lifted his head into sight beyond Madison, and has pushed out the long outline of the ridge that climbs from the Pinkham forest, and by all the stairways of his plateaus, to his cold and rugged crown. What a majestic trio ! What breadth and mass, and yet what nervous contours ! The mountains are arranged in half-circle, so that we see each summit perfectly defined, and have the outline of each on its characteristic side lying sharp against the sky, — Adams as it is braced from the N., Madison from the S. E., Washington from the S. They hide the other summits of the range entirely.

" The Milan view is superior in symmetry to Shelburne, but is not equal to Conway in variety and proportion. The arrangement gives three distinct distances almost ungraduated. First, the river and its meadow borders, suddenly cut off by the dark joining of the long flattened spur of the Pilot Hills, and the abrupt and higher base of Mt. Hayes ; beyond these nothing but a wide space of gray air ; while far away in this arise the great mountains, grouped in a triple-peaked pyramid, admirable in proportions, and strangely beautiful in the afternoon light."

36. Milan

(pronounced *Mile-un*) has a small inn, and is reached by daily stages from the Grand Trunk trains at Berlin Falls. The town has 710 inhabitants, and covers 31,154 acres, of which 5,512 acres are improved. It was granted to Sir William Mayne in 1771, and retained the name of *Paulsbury* until 1824. The chief occupation of the people is farming, though there are several lumber and starch mills. *Milan Corner* is a small hamlet on high ground W. of the Chickwolnepy Mts., whence a road runs W. N. W. over the Milan Hills to the hamlet of W. Milan, on the Grand Trunk Railway. Grand views of the mountains are obtained from W. Milan, from the road E. of the Androscoggin S. E. of Milan, and from the heights N. of Milan. The latter view shows the broad and peaceful valley of the Androscoggin in the foreground, running towards the White Mts. In the centre of the group is Mt. Adams, with the immense gorge of King's Ravine cut into its side, over which is Mt. Washington. Jefferson is on the r., and Madison rises boldly on the l.

Green's Ledge is 2½ M. from W. Milan, and is reached by following the road to Higgins's Mill, 1½ M. from the village, whence a logging-road ascends almost to the summit. It is nearly 2,000 ft. above the sea, and is lined on the S. by remarkable cliffs, at whose foot are piles of boulders, one of which is 70 ft. long and 50 ft. high. The view includes the Presidential Range, which is nearly S., to the r. of which are the Crescent Mts. and Mt. Forist. S. W. and W. are the long lines of the Pilot Range, with their many wooded crests; and about N. W. are the white domes of the Percy Peaks. Mill Mt., near Stark, and certain of the Stratford mountains are next seen on the r. The Ragged Mts. in Odell are about N. N. W., beyond which are the ranges in Millsfield, with many of the chief peaks of Western Maine, including Goose-Eye.

Farther around toward the White-Mt. range are the wooded lines of Mt. Hayes, near Gorham, beyond which are the broken ridges of Mt. Moriah and the Carter Range.

Success is the pathetic name of the township E. of Milan, which, though granted in 1773, had but 5 inhabitants in 1870. It covers 30,000 acres, and is traversed by the Chickwolnepy Mts.

Cambridge is another wilderness-town, on the N. E., covering 23,160 acres, granted in 1793, and containing but 28 inhabitants in 1870. The people live near the S. end of Lake Umbagog, beyond the Hampshire Hills, and formerly had a forest-trail to Milan.

Dummer adjoins Milan on the N., in the Androscoggin Valley, and covers 23,041 acres. It has 317 inhabitants, engaged in farming; and much of its area is covered with cold highlands.

37. Upper Bartlett.

Hotels. — The Bartlett House (by Frank George) is a large country-inn near the station. The Carrigain-Hotel Company was chartered by the State recently, designing to expend $75,000 in the erection of a hotel at this point, but nothing has yet been done.

Distances (on the highway). — Upper Bartlett to Lower Bartlett, 7¼ M.; N. Conway, 12¼; Bemis Station, 5½; Willey House, 11; Crawford House, 13¾. These distances were obtained by odometer surveys.

Upper Bartlett is a station on the P. & O. Railroad, near which a small hamlet has arisen. It is in the centre of a picturesque amphitheatre of mountains, having Carrigain, the Nancy Range, Tremont, and Haystack on the W.; Hart's and Willoughby Ledges, Mts. Parker, Crawford, Resolution, Langdon, and Pickering on the N.; Kiarsarge and Moat on the E.; and Table and Bear Mts. on the S. Numerous excursions may be made from this point over the adjacent peaks; and there is rich trouting in Albany Brook and other tributaries of the Saco. If the Carrigain Hotel is built, and the railroad does not locate too many laborers here, Upper Bartlett may become one of the best centres of forest-tours in the mountain-district. Mt. Carrigain is seen to the best advantage hence, and Champney made his celebrated painting of it from near the old mill. The formidable and frowning peaks which surround the hamlet are finely contrasted with the rich and narrow intervales of the Saco. A bridge here crosses the river to a group of farms on the N., and through the low pass on the S. a road is projected to Albany and the Swift-River Intervale.

The Bartlett Land and Lumber Company own 40,000 acres of woodlands, and have a large steam-mill at Upper Bartlett. During the winter of 1874 – 5 they hauled out 7,000,000 ft. of lumber. Near the hamlet is the Chapel of the Hills, a little church which was opened in 1854.

The *Willoughby Ledge* is about a half-hour's walk from Upper Bartlett, across the Saco, and beyond a succession of sheep-pastures. In some places its dark cliffs overhang so as to form ample shelters, where the sheep take refuge during stormy weather. From the summit a beautiful view of the Saco Valley is afforded, with the dark mountains on the S.

Sawyer's Rock is a dark ledge on the l. of the road, about 2 M. W. of Upper Bartlett, which derives its name from the old hunter who helped the discoverer of the Notch to get the first horse through. During the laborious transit the refreshing bottle was often called into requisition, so that when he reached this point it was quite empty, and was dashed against the ledge.

Bear Mountain is an immense line of ridges, extending through considerable portions of Bartlett and Albany, between the Saco and Swift Rivers. Its crests are wooded, and dense forests cover all its long flanks, which are cut into by deep ravines. The height is estimated at 3,000 ft. But few persons have visited these tangled wilds, and their topography is yet unsettled. Good woodsmen may reach the summit most easily by going S. from Upper Bartlett along a logging-road, 1½ M. long, which will carry them within 1½ M. of the crest. From this point the ascent is steep, through dense forests and tangled thickets. Mr. Ripley says that the upper part of this mountain affords "a most perfect view of Mt. Washington and the whole range."

Table Mountain is S. E. of Bear Mt., and is a flat-topped and forest-covered ridge, deriving its name naturally from its level upper plateau. It is separated from Moat Mt. by the approaching ravines of two brooks. It is nearer the Swift-River road than the Saco, and may be visited more easily on that side.

38. Mount Langdon

is nearly N. of Upper Bartlett, and 2 M. distant, the ascent requiring about 2 hrs. It is in the centre of the range of rocky summits N. of the Saco, and has ledgy sides, draped with woods and abounding in berries. The summit is of a sandy character, and is 2,460 ft. above the sea. The ascent is easy, though without a path, and is entered by crossing the Saco bridge and traversing the pastures beyond, then striking through the woods toward the straight flanking ridge. A bridle-road is promised for the year 1876, and there is talk of building a hotel on the summit. If the small thickets near the top could be cleared away, the view would be greatly improved. This peak was at one time ascended by Mrs. Lucy Stone Blackwell, who named it in honor of her husband; but the Appalachian Mountain Club has given it the better name of *Mt. Langdon*, in honor of Samuel Langdon, D. D., an early president of Harvard University, and joint-author of the first map of New Hampshire (published in 1761). The Langdons were one of the ancient patrician families of this State.

Mr. H. W. Ripley says: "It is one of the most perfect gems in White-Mt. views, and its easy ascent will invite many who love to look upon so charming a landscape as is seen from its summit."

* *The View.* — Nearly due W. is the lofty and imposing Mt. Carrigain, with Vose's Spur and the long ridges running to the S., and the sharp points of Mts. Lowell and Anderson on the r., nearly eclipsed by the nearer Mt. Nancy. Over and to the r. of this range are Mts. Bond and Guyot and the Twin Mts., forming a long and massive line of heights in the distance. Toward the N. W., within 2 M., is the well-marked peak of Mt. Parker, beyond which Crawford is seen on the l. and Resolution on the r. From the N. of Resolution the dark Montalban Ridge runs off toward Mt. Washington, and over it is seen the round swell of Mt. Pleasant, with Franklin and Monroe above on the r. Almost due N. is the wild valley of the Rocky Branch, filled with dense and tangled forests, and beyond its solitary leagues rises the noble cone of Mt. Washington, with the alpine hamlet on its summit and the lofty terrace of Boott's Spur on the r. Farther away through the Pinkham Notch is Pine Mt., near Gorham; and on the r. is Mt. Wild-Cat, overtopped by the higher peak of the Carter Dome, with a portion of the Moriah Range beyond.

Iron Mt. lifts its white and ledgy sides across the glen of the Rocky Branch, 3 - 4 M. away, beyond which is the dark Mt. Sable, standing nearly in front of the high crest of Baldface. The singular peaks of Double-Head are about W. N. W., in front of which are the densely wooded crests of Mts. Eastman and Slope. Farther to the r., and nearer, is Thorn Mt., back of which are the resembling spires of Mt. Gemini. Down the Rocky-Branch glen, and almost over Lower Bartlett, is the hotel-capped summit of Kiarsarge, flanked on the r. by the Green Hills of Conway. The elongated sierra of Mt. Pleasant, in Maine, is next seen, over the Green Hills, and marked by a hotel on its central height.

The view now rests on Mt. Attitash and the ponderous braces and foot-hills of Moat Mt., whose white and beautiful N. peak is about S. E. To the r. of the chaotic ridges of Moat is the flat-topped Table Mt., over which is the gable-like sharp peak of Chocorua, flanked by thin white ridges. On the r. of Table is Bear Mt., across the Saco Valley, forming a heavy rounded mass, which is covered with dark forests and flanked by a desolate defile. Passaconaway and Whiteface are far away over this mountain, and on their r. is the three-pointed mass of Tripyramid. To the r., across the Saco Valley, is Haystack, showing a sharp dark peak, with the tawny crest of Tremont beyond. The bold cliffs of Hart's Ledge are still nearer, at the great angle of the Saco River; and over them appear the distant but stately crests of Tecumseh and Osceola, of which the latter is the higher and nearer. Farther to the r. are the nameless mountains of the inner Pemigewasset Forest; and Mt. Hancock is visible up the valley of Sawyer's River, closing the circuit and falling behind the noble Carrigain.

39. Mount Tremont

is not far from Upper Bartlett, the journey to the top requiring but about 3 hrs. Darius Cobb is competent as a guide on this mountain. The slopes are covered with dense forests, among which are a few windfalls; but parts of the upper ridge have been burnt over, giving opportunity for a broad view. There is no path, and visitors usually strike in from the road beyond Sawyer's Rock.

Tremont is 3,393 ft. high, and its upper peaks are of a light-colored stone, presenting the appearance of colossal white knobs when seen from a distance. In many respects the view is one of a high degree of grandeur, and especially is this true in the direction of the lower presidential peaks and towards Carrigain. It is to be hoped that a good path may be cleared up to this peak by the people of Upper Bartlett, in order that this view may be more generally enjoyed. It forms the subject of one of the finest of Morse's panoramas, printed by the N. H. Geological Survey.

* *The View.* — Nearly N. W., and about 5 M. distant, is the immense mass of Mt. Carrigain, with the long S. ridge sweeping up from Sawyer's River to its rounded peak, below which, on the r., is the dome-like Vose's Spur, enclosing the great ravine of the W. Fork. On the r. is the profound gorge of the Carrigain Notch, through which are seen portions of Mts. Bond and Guyot, with the S. Twin to the N. Over the ledgy sides of the adjacent Camel's Hump is the symmetrical cone of Mt. Lowell, with Anderson over it, on the r., and the N. Twin Mt. far beyond. Mt. Nancy is next seen, close at hand, continuing the Lowell-Anderson range to the N. E., and running off in long terraces to the Saco Valley. Over its r. shoulder is the sharp peak of Willey, with Mt. Field behind it on the l.; and a portion of the round disk of Willard appears over a lower terrace, with the Deception range beyond. Then comes the steep side of Webster, striped with brilliant hues, and with a portion of the Frankenstein Cliff below its r. flank, girded by the P. & O. Railroad. The view now runs up the narrow and winding Saco Valley, whence the great gorge of the Mt.-Washington River is seen diverging to the r., near the Frankenstein Cliff.

On the r. of Webster, and apparently at the head of the Saco Valley, is Mt. Jackson, showing a low point against the sky; and farther to the r. is the higher and flatter top of Clinton. The prominent rounded crest of Pleasant is next seen, followed by the level plateau of Franklin and the high crags of Monroe, over which peers the summit of Jefferson. Then Mt. Washington comes into view, pre-eminent above all the others, bearing houses on its summit, and overlooking the deep chasm of Oakes's Gulf. On the r. Boott's Spur projects to the S. S. E., whence the shaggy Mont-alban Ridge descends directly towards Tremont, well foreshortened, to the nearer Crawford group, whose highly colored ledges are plainly visible. Mt. Crawford is on the l.; over it are the two upper rock-faced terraces of the Giant's Stairs; and to the r. is the high and unmarked swell of Mt. Resolution. Mt. Hope is nearer, rising over Hart's Ledge towards Mt. Crawford; and Mt. Parker is on the r., just S. of Resolution. On the r. of and far beyond Resolution, the sky-line sinks down into the Pinkham Notch, on whose r. are the remote blue ridges of Wild-Cat and the Carter Dome, with a small upper section of Moriah. Portions of the Eagle and Black Mts. of Jackson are seen, about N. E., nearly over Mt. Langdon; and then the ledgy flanks of Iron Mt. appear, 7 M. distant, over which rise the distant white peaks of Baldface, cutting the horizon. Farther away is a part of Speckled Mt.

The next conspicuous points are the similar peaks of Double-Head, finely relieved against the sky, with Sable behind them on the l., and Slope on the r. The view then passes over the low ledges of Mts. Pickering and Stanton, at the confluence of the Saco and the Rocky Branch, to the long-flanked crest of Thorn Mt., the second connected swell on the l. of which is Tin Mt. A line of remote blue mountains in Maine meets the horizon above Thorn. The Saco Valley now opens to the E., and is visible for 8 – 10 M., with the winding river on the l. and the straight band of the railroad on the r. At the end of this beautiful vista is the tall and graceful pyramid of Kiarsarge, with the sister peaks of Mt. Gemini on the l., and Humphrey's Ledge below. On the r. of Kiarsarge are the Green Hills of Conway, much lower, and laid in wavy lines against the blue background of the Waterford mountains and the highlands of Norway and Sweden. Farther to the r., and over the high dark ridge of Bear Mt., are the white peaks of Moat Mt., over whose l. flank is Mt. Pleasant, near Sebago Lake. Farther to the r., and but 3 M. distant, is the flat top of Table Mt. The Freedom hills and the Green Mt. in Effingham are seen beyond and very distant. Nearly S. S. W. is the stately white crest of Chocorua, supported by lofty piles of rocky ledges, and flanked on the r. by the lowly curved top of Paugus, over which a part of the Ossipee Range is seen. Nearly S. is the white mound of the Potash, beyond which looms the black dome of Passaconaway, with the peaks of Whiteface still farther away. The sierra of Tripyramid is nearer and more to the r., and a portion of the Mad-River country opens beyond, with the high peaks of Tecumseh and Osceola on the r. The abrupt heights of Green's Cliff are about 3 M. distant, on the S. W.; and farther to the r., a little S. of W., is the high blue ridge of Moosilauke. The view now passes over the lower part of the Pemigewasset Forest, and rests on the confused ridges and tall peaks of Mt. Hancock, back of Carrigain. Farther away, between Hancock and Carrigain, is a portion of the thin serrated crest-line of Lafayette.

40. The Bartlett Haystack

is the fine conical peak S. W. of Upper Bartlett, near Mt. Tremont. The summit is 2½ M. from Upper Bartlett, and is reached with very little fatigue by a march through the woods. There is no path, at present.

The View. — To the N. of Haystack is the Saco Valley, bending around Hart's Ledge in its great curve from W. to N., and visible in either direction for several miles. On its l. side, above Hart's Ledge, is the Franken-stein Cliff, pushing out from the vicinity of the Arethusa Falls and approaching the sinuous river. The long defile of the White-Mt. Notch extends far up to the N., and near its farther end are Mts. Willey and Webster, respectively on the l. and r. sides. Nearly N. of Haystack across the valley are the high and ledge-crowned Mts. Crawford and Resolution, the sharp peak of the former contrasting with the even plateau of the latter. Below Crawford is Mt. Hope and below Resolution is Mt. Parker, with Giant's Stairs rising behind and above. Over the flanks of Crawford are Mts. Jackson, Clinton, and Pleasant, with the rugged crown of Monroe on the r.; and the cone of Washington is over Resolution. Portions of Jefferson and Adams are seen on either side of Washington, nearly eclipsed by the central peak; and the upper heights of Mt. Wild-Cat and the Carter Dome are to the r. Across the valley and rising from the narrow intervales of Upper Bartlett are Hart's Ledge (on the l.), Willoughby Ledge, Mt. Langdon, Mt. Pickering, and Mt. Stanton. The yellow ledges of Iron Mt. are about N. E., beyond which are Black Mt. in Jackson, and Baldface, with its white crest. More to the r. and nearly as far are the twin mamelons of Double-Head and the wooded swell of Thorn Mt. Below the similar cones of Mt. Gemini is the sharp and symmetrical crest of Kiarsarge, buttressed by Mt. Bartlett on the W.

Nearly E. is the great Bear Mt., exhibiting several heads and over-looked by the brightly colored N. peak of Moat Mt. Chocorua shows its white spire, a little E. of S., and a part of the blue Ossipee Range is seen far beyond. Farther to the r. is the dark dome of Passaconaway, nearly concealing the high slopes of Whiteface. Still to the r. is the thin sierra of Tripyramid, with Sandwich Dome in the background; and either Osceola or Tecumseh appears to the W. Nearly W. is the straight-sided plateau of Green's Cliff, and close at hand in the N. W. is Tremont, whose bare white ridge shuts out the view in that direction over a considerable arc of the circle of vision.

41. Mount Lowell and the Nancy Range.

Mount Lowell was formerly known as *Brickhouse Mt.*, and received its present name in 1868, in honor of a Portland gentleman who has been an enthusiastic mountain-explorer. It is 3,850 ft. high, and falls into the Carrigain Notch with remarkable cliffs, apparently in a half-ruined condition, and showing ledges of bright colors. "The slopes of these two mountains in Carrigain Notch are more imposing, both on account of their exceeding steepness and of their great height, than any others yet described in the White Mts." Lowell is the farthest W. of the four peaks of the Nancy Range.

Mt. Lowell is ascended from the Carrigain Notch by following up its S. spur, the route being steep and badly obstructed. The summit is exceedingly sharp, and is only reached after a breathless and exhaustive scramble. This is the point that looks so much like a black spire when viewed from the distant mountains.

Mt. Nancy is 3,800 ft. high. It is at the E. end of the range, over the Saco Valley, and occupies a conspicuous position in the views from the N. and E. On its inner slopes the forests are large and easily passable. This peak received the name of *Mt. Amorisgelu*, some 20 years since, in commemoration of the fate of the unfortunate maiden who died at its foot (see page 138). The word is compounded of two Latin words, meaning "The Frost of Love," and is accented on its third syllable, in order to resemble an Indian name.

Mt. Anderson is between Mts. Nancy and Lowell, attaining a height of nearly 4,000 ft. It was named in honor of the chief engineer of the Portland & Ogdensburg Railroad.

The Editor did not visit the peaks of the Nancy Range, nor has any one else, so far as he knows. The following list of mountains visible therefrom has been formed by comparing the views from other mountains, and finding those which looked upon the Nancy Range. It is necessarily incomplete, but is correct as far as it goes; and will answer, with slight changes, for either of the peaks, though Mt. Lowell is more particularly its subject.

Nearly N. N. W. the line of the Presidential Range is seen, foreshortened because viewed along its axis. Webster, Jackson, Clinton, Pleasant, Franklin, and Monroe are visible, beyond which are the tall peaks of Jefferson and the humps of Clay. At the head of the Mt.-Washington-River Ravine is the lofty crest of Mt. Washington, with its conspicuous houses, flanked on the r. by Boott's Spur. The Montalban Ridge runs thence to the S, nearly hiding the great Carter Range, terminating in the red-capped cluster of Giant's Stairs, Resolution, and Crawford. Toward the E. is Mt. Langdon, with Iron Mt. beyond, and Double-Head still farther away; and Thorn Mt. is almost over the near Mt. Parker. Over the ledgy sides of Mts. Pickering and Stanton is the high-pointed Kiarsarge, with its summit-house; on whose r., far away over Middle Mt. and the Green Hills of Conway, is Mt. Pleasant, which is near Sebago Lake, and is marked by a hotel near the centre of its elongated ridge. The view next rests on the high white peaks of Moat Mt., nearly in a line towards which are the rocky knobs of Tremont, the lofty ridge of Bear Mt., the black cone of Haystack, and the plateau of Table Mt. Nearly S. E. is the splendid spire of Chocorua, with a wide area of Maine beyond; and then the Sandwich Range appears, — the lowly Paugus, the rounded Passaconaway, and the double-pointed black Whiteface. To the l. of Passaconaway is Copple Crown Mt., near Wolfeborough. Farther to the r. is a part of the massive Sandwich Dome.

The view is now closed by the immense Mt. Carrigain, which looms up close at hand across the gulf of the Carrigain Notch. From the peak of Lowell the singular structure of this mountain and its outlying spurs may be studied to the best advantage ; and the wonderful steepness of the sides of the two ridges is here perceived. To the W. N. W., over the r shoulder of Carrigain, the eye wanders over a wide wilderness, the unbroken Pemigewasset Forest, to the sharp and lofty crags of Mts. Lincoln and Lafayette. Nearer, and looming massively from the central wilds, are the peaks of Mts. Bond and Guyot, with the Twin Mts on the N. Farther to the r., a little E. of N., are the crests of Mts. Field and Willey, over the White-Mt.Notch.

42. Mount Carrigain

is in the Pemigewasset Forest, between Sawyer's River and the East Branch, the loftiest in a long range of summits, and overlooking many leagues of unbroken wilderness and stately mountains. It is 4,678 ft. high, and its E. spur is 4,419 ft. high. The bold and remarkable architecture of this peak makes it an object of great interest, but its remoteness from the roads has rendered it nearly inaccessible to ordinary tourists.

Prof. Vose says : " Mt. Carrigain stands almost exactly in the centre of the vast group of the White and Franconia Mts., and, rising as it does to a height of nearly 5,000 ft., is a marked feature in the landscape from almost every point of view. Conversely, the view from Carrigain must embrace the whole mountain mass, and must sweep around over all the principal summits. . . . . Ranges and notches, huge mountains and broad valleys, never seen from the points commonly visited in this region, are spread all around. From its central position a better idea of the arrangement of the White and Franconia Mts. is had than from any other point, perhaps, in the whole group."

It is about 8 M. in a straight line from Upper Bartlett to the summit of Carrigain, but so many and formidable are the difficulties in the way that it should be made a two-days' trip, the camp being well up in the ravine of the W. Fork of the Carrigain Brook. A third day would give time to enter and explore the noble Carrigain Notch.

The summit of Carrigain consists of a narrow ridge several rods long, covered with a forest of low spruce-trees, which spoil the view except for such as are willing to climb among the branches. To the E., 300 ft. lower, is a knob-like peak whose summit is more clear, and affords a broad view. More to the S., and running at right angles with the main ridge, is a long stony spur 240 ft. below the peak, bending to the W. of the great ravine, and marked by a signal-pole set in a pile of stones. The vicinity of this beacon is the best point for obtaining a general view to the N., E., and S., the W. being covered by the main ridge. The depressions between the forest-covered peak and the knob to the E. and the Signal Ridge on the S. are inconsiderable, and may be easily traversed. Surveyor Crawford intends to have the summit-ridge cleared, and that would open one of the noblest view-points in the mountains. Vose's Spur makes out to the E., partially enclosing a singular trilateral ravine, which is conspicuously seen from Mt. Washington. On the W. is a long ridge, running towards Mt. Hancock, with lofty cliffs on its sides, looking down into the deep gorge on the S. W.

The favorite route of ascent has hitherto been up the course of Sawyer's River, leaving the Saco road at Lawrence's farm, and going up Sawyer's for about 2 M. to Duck-Pond Stream, the first brook coming in on the r. Ascending the latter for about 1½ M., the course is next laid due W. for ¾ M., by a line of blazed trees, to the Carrigain Brook, whose bed is followed upward for about 2 hrs. An old camp is here reached, where the heavier baggage may be left. It is about two hours' march from this point to the summit, and the route is exceedingly steep, rising over 3,600 ft. The early parties ascended the W. fork of the Carrigain Brook, over successive steps of granite, where every mu cle was called into play; but now a less dangerous route is followed, through the woods. The ascent is long and arduous, on whichever side it may be attempted. A railroad has been built 2 – 3 M. up the valley of Sawyer's River, to bring out lumber from the forest; and much of the labor of the approach may be obviated by walking up the track to the Carrigain Brook. Darius Cobb has been a competent guide for previous parties. He lives in Upper Bartlett, not far from Sawyer's Rock.

The Editor has made the ascent by another route, which, however, cannot be recommended. Leaving his camp, in the ravine of the brook which flows between Carrigain and Hancock, on the S W. of the former peak, the stream was followed until the main ridge was encountered, and its W. slope was ascended to the summit. The journey occupied about 4 hours, and 2 hours were taken for the descent from the Signal Ridge. The ascent is for the most part through a forest of tall spruce and fir trees, affording good walking-ground and plenty of water well up the mountain. The long ridge which runs S. from the main peak to Sawyer's River would afford an easier mode of access but for the lack of water along the route.

Mr. Warren Upham gives the following directions for the ascent from the Pemigewasset Forest: " Leave Cedar Brook at its fork 1 M. from the East Branch, and steer straight for the summit. The ascent by this way is an almost straight, regular slope; no undergrowth exists for any part of the way; and under foot for most of the distance is the abundant mountain-moss and wood-sorrel. This is the easiest ascent, for the height reached, that we found in all our mountain-climbing."

The mountain was named in honor of Philip Carrigain, who was born at Concord in 1772 and died there in 1842. He was a graduate of Dartmouth College, and a lawyer; and was Secretary of State from 1805 to 1810. His most notable work was the map of New Hampshire, which he published in 1816. The White-Mt. Club, of Portland, has made this peak one of its chief objects of attack.

* *The View.* — The valley of the Saco is a little S. of E. from Mt. Carrigain, and the yellow sands of the river-bed form a wide band through the green plains. The clearings of Upper Bartlett open down the valley, and Hart's Ledge crops up on the l. Well down the opening is the high gable of Kiarsarge, and the saddle-like summit of Tremont is nearer at hand. Over it is the ragged sky-line of Moat Mt., with Mt. Table and Bear Mt. to the r. over Haystack, and the Green Hills over the N. slope. The high, long crest of Chocorua looms up on the S. E. over the farms in the Albany Intervale. S. S. E. about 5 M. is the plateau of Green's Cliff, over which is Mt. Paugus. To the r. of Paugus successively follow Passaconaway (with the Ossipee Range beyond), Whiteface, and the saw-edged Tripyramid, next to which is the tall blue ridge of Sandwich Dome. To the r. of the deep Mad-River Notch, in which the Greeley Ponds are hidden, are the peaks of Osceola and Tecumseh, running out to the S. W. into Green Mt. and Fisher's Mt. Close at hand in the W. are the heavy and characterless mounds of Hancock, near which in the field of vision,

but really far beyond, is Moosilauke. Then come the Coolidge and Potash
Mts., near the confluence of the Pemigewasset and the E. Branch, with
the formidable line of peaks called Mts. Flume, Liberty, and Lincoln.
Between Flume and Lincoln is the more distant crest of Mt. Kinsman;
and the serrated crags of Lafayette are seen over Owl's Head, which is
between the Franconia and Lincoln Branches. In the foreground, and extending for many leagues in every direction, is the broad wilderness of the
Pemigewasset Forest, including the valleys of the E. Branch and its tributaries. This trackless and virgin forest is the cradle of the Merrimac
River, and contains neither road nor house nor clearing.

Beyond, and to the r. of Lafayette, is the well-proportioned Haystack,
and then come Mts. Bond and Guyot and the Twin-Mt. group, with
Mt. Hale on their r. down the New-Zealand Notch. Farther down the
Notch is the blue swell of Cherry Mt., and then come Mts. Tom, Field,
and Willey, the latter bearing steep ledges on its S. slope. A section of
the striped flank of Mt. Webster is seen W of Willey, over which is Mt.
Jefferson, with Pleasant at the l. slope. W. of Jefferson, up the White-
Mt. Notch and Ammonoosuc Valley are Mts. Mitten and Dartmouth,
lowly in comparison. Mt. Washington looms up pre-eminently in the N. E,
and is succeeded by the blue crown of Carter. Farther S. and nearer are
the reddish-yellow ledges on Mt. Resolution and Giant's Stairs, over which
are Mts. Eastman and Baldface. Below is the slide-scarred slope of Mt.
Lowell, walling on the E. the profound chasm of the Carrigain Notch. A
little N. of E., above the Saco Valley, are the white ledges on Iron Mt.,
with the sister-knobs of Double-Head beyond. Still nearer is Mt. Langdon, over Upper Bartlett, with white ledges near the top, and then the
eye follows again down the shaggy valley of Sawyer's River to the fairer
plains of the Saco.

The **Carrigain Notch** is a deep pass between Mts. Carrigain and Lowell, 2,465 ft. above the sea, offering the best route for a road from the Saco
Valley to the Pemigewasset Forest. In some respects this is the finest
pass in the White Mts., the peaks on either side being lofty and well
marked, while the falling lines of Carrigain are full of grace and beauty.
Prof. Vose says: "The slopes of these two mountains in Carrigain Notch
are more imposing, both on account of their exceeding steepness and of
their great height, than any others yet described in the White Mts." Unfortunately, the gorge is filled with tall and sturdy trees, which obscure
the view of the adjacent ranges; so that it is better comprehended from
the bare ledges above. A fine distant view is enjoyed from Mt. Chocorua;
and Mt. Tremont also commands a good prospect into this wild pass.

The Carrigain Notch is visited by following up the main stream of the
Carrigain Brook, which rises in the upper part of the great hollow and
flows S. E. into Sawyer's River. It is not more than 3 M. from the rail-

road up Sawyer's River to the centre of the Notch, but the route is difficult to traverse.

43. The Crawford Glen.

Bemis Station.

This rich centre of mountain scenery lies nearly in the centre of *Hart's
Location,* a political division which includes the Saco Valley from Sawyer's Rock to the Crawford House, a distance of 12½ M., with an average
width of about 1 M. (population, 26). The glen is between the mountains
of the Crawford group on the E., and the Nancy Range on the W., and
was formerly occupied by valuable intervales which have been nearly
ruined by slides and avalanches from Mt. Crawford. Nancy's Brook here
enters the Saco from the W., and Sleeper Brook from the N. E. The
railroad crosses the highway on grade, at Bemis Station, just S. of which
is the gabled cottage of Dr. Bemis, the patriarch of the mountains, who
left Boston many years ago and settled in this remote glen, around which
he owns thousands of acres of woodlands.

Near the station is the old Mt.-Crawford House, formerly one of the
chief hotels in this region, but long since closed to the public. While N.
T. P. Davis kept this house, he opened the long path to Mt. Washington
over the Montalban Ridge (see Route 89).

The site of the Mt.-Crawford House was occupied before the year 1800 by Abel
Crawford, who had married Capt. Rosebrook's daughter. He died at the age of 85.
" He had been so long accustomed to greet travellers in the summer, that he longed
to have his life spared till the visitors made their appearance in Bartlett, on their
way to the Notch. He used to sit in the warm spring days, supported by his daughter, his snow-white hair falling to his shoulders, waiting for the first ripple of that
large tide which he had seen increasing in volume for 20 years. Not long after the
stages began to carry their summer freight by his door, he passed away."

" Here the mountains assumed the form of an immense amphitheatre ; elliptical
in its figure ; from 12 to 15 M. in length ; from 2 to 4 in breadth ; and crowned with
summits of vast height and amazing grandeur. Compared with this scene, all human
works of this nature, that of Titus particularly, so splendidly described by Gibbon,
are diminished into toys and gewgaws. Here more millions could sit than hundreds
there ; every one of whom might look down with a full view of the valley beneath."
(DWIGHT's *Travels in New England.*)

There are picturesque bits of water-scenery (difficult of access) on the
Davis Brook, 1 M. N. of Bemis Station; and 1 M. farther N. is the Bemis
Brook, on which are the superb Arethusa Falls (see Route 49).

Bemis Pond is 6 – 8 M. S. S. W. of the station, in the wilderness beyond
Mt. Tremont. It was formerly frequented by trout-fishers, who found
good sport there. A spotted line leads in to the pond from the Saco road.

Nancy's Bridge is on the highway ½ M. S. of Bemis Station, and crosses
Nancy's Brook, which here traverses a rocky cañon, 200 ft. long, 16 – 20
ft. wide, and 25 – 35 ft. deep. This fine gorge was formed by the decomposition of a trap-dike and the subsequent disintegration of the enclosing

granite walls by water freezing in its seams. Nancy's Brook comes from
a small pool 2 – 3 M. distant on the side of Mt. Nancy, and forms several
pretty cascades on the way.

The bridge, brook, and mountain derive their names from a sad incident in the
early history of the country. In the autumn of 1788, a young woman by the name
of Nancy was employed in Col. Whipple's family at Jefferson, and became engaged
to one of the men on the farm. She gave him her two-years' wages when they were
about to depart for Portsmouth to be married, but he started away during her absence, leaving no explanation. On her return at night she set out after him, hoping
to catch the recreant lover in camp at the Notch, before the dawn. The ground
was covered with snow, and the route for 30 M. lay through the forest, marked only
by a line of spotted trees. She reached the camp, but it was abandoned, and after
vainly striving to rekindle the smouldering fire, she pressed on down the Notch,
fording the icy Saco in several places, and at last sank down in utter exhaustion on
the S. bank of Nancy's Brook, where she was speedily chilled to death. The bitter
northwest wind had driven blinding masses of snow upon her; her clothing had
become saturated in fording the streams; and she was found stiff and cold, with her
head resting on her staff. The men at Col. Whipple's had doubted that she would
face the storm, but, becoming alarmed at her long absence, they followed the trail
and found her, not long after her death. On learning of her dauntless faith and
terrible fate, her lover became insane, and died, a few years later, in fearful par-
oxysms; and there is a tradition that long afterwards these valleys resounded
on still nights with the weird and agonizing shrieks of his restless ghost.

44. Mount Crawford

is 3,134 ft. high, and is the lowest but most alpine of the peaks of the
Crawford group. It rises boldly from the Saco Valley, near the inflowing
of the Mt.-Washington River and Sleeper Brook, and exhibits a broad gulf
on the side towards Bemis Station. A ridge runs S. from Crawford along
the E. side of the Saco Valley to Hart's Ledge, around which the river
bends to the E. The peak of *Mt. Hope* is on this line, and is over 2,000
ft. high. The upper parts of Crawford are covered with broad red ledges,
which render it easily recognizable from afar. The rock is rapidly disin-
tegrating and forming beds of gravel. The peak is high and steep, especially on the E., where it is nearly precipitous, though by no means so
phenomenally beetling as the picture in *The White Hills* (page 15) would
indicate.

The peaks of the Crawford group are reached by means of the wreck
of the old Davis bridle-path. This route has been so long abandoned that
Nature has reclaimed considerable portions of it, and the visitor will
sometimes be perplexed by patches of jungle and rank bushes. It is,
however, more difficult to follow the path on its higher grades, because
there it ran over a long series of ledges, leaving but slight traces, and the
rocks themselves have since become disintegrated. The outlet of the
trail on the Saco meadows is very obscure, and can only be found by local
assistance or by a keen-eyed forester. It is reached by fording the river
on the stones, the Saco being here wide and shallow, but after heavy rains
it is impassable for the pedestrian. If the path can be found and retained
through the woods it will be of great assistance, not only as showing the

most direct way, but also as affording a clear track through the thickets, and an even grade over the ledges. At some points it was constructed at considerable expense, being terraced out along the sides of steep banks. The path passes to the r. of and about ⅛ M. from the peak of Crawford; traverses the depression towards Resolution; winds around the W. and N. W. flanks of the latter; passes through the ravine to the Giant's Stairs; and ascends their terraces on the S. W. side.

In ordinary weather this is a dry mountain, having no water on its upper slopes, and visitors will therefore need to carry potables with them. Just after rains, enough water may be found in the hollows of the rocks near the summit. The mountain was burnt over about the year 1815, and even the soil was destroyed.

Mts. Crawford, Resolution, and Giant's Stairs received their names from Dr. S. A. Bemis, who has named more of the mountains than has any other man.

* *The View.* — Nearly N. W., and about 5 M. distant, is the high and alpine peak of Mt. Willey, fronted with crags, and falling steeply to the S. A small portion of Willard is seen to the r., and then the eye crosses the broad ravine of the Mt.-Washington River to the S. W. peaks of the Presidential Range, of which Webster is on the l., and the low crests of Jackson and Clinton follow. The bulging dome of Pleasant is due N., and is followed to the r. by Franklin and Monroe, beyond which the range culminates in the lofty crest of Washington, with the flat Boott's Spur on the r. Under the r. flank of Washington, and about 1½ M. from Crawford, are the poorly outlined terraces of the Giant's Stairs, on whose r., and adjoining Crawford on the N. E., is the higher and flattened top of Resolution, which shuts out the view in that direction. On its S. is the dark Mt. Parker.

Mt. Kiarsarge is finely seen on the E. S. E., below the highlands of Jackson, and down the Rocky-Branch glen. On its r. are the Green Hills of Conway, over which, far away, is Mt. Pleasant, in Maine, with its summit-hotel. The view next includes Mts. Langdon and Pickering, beyond which, on the r., is the high red peak of Moat Mt. S. S. E, over the plateau of Table Mt., is the stately white spire of Chocorua, reaching the sky-line. Nearly S., and close at hand, is the low cone of Mt. Hope, and the glens of Upper Bartlett are seen to the l., with the little hamlet near their centre. Farther away is the long dark ridge of Bear Mt., far beyond which, due S., is a part of the blue Ossipee Range. Passaconaway is a high black hemisphere, about S. S. W.; and over the near white knobs of Tremont is the distant Whiteface, nearly obscured by Passaconaway. To the r. is the serrated ridge of Tripyramid, uplifting three well-marked peaks, on whose r., far distant, are the crests of Tecumseh and Osceola.

In the foreground is the pleasant Crawford glen, with the buildings about Bemis Station, and the winding river still nearer. The dark-green Mt. Nancy rises just across the glen, and behind it are Anderson and Lowell, with the triple head of the lofty Carrigain still more to the

W. S. W. To the W. is the high plateau which borders the Pemigewasset
Forest on the E., over which looms the distant sierra of Lafayette, beyond
the monotonous Twin Range. Across the Saco Valley is the abrupt front
of the Frankenstein Cliff, beyond which rises Mt. Willey.

"On the top of Mt. Crawford the spectator, without moving from his station,
commands the whole circumference of the horizon, and a series of views the most
varied and interesting. On the E., a little S., is the conical summit of Kiarsarge,
in the S. the rough Chocorua, with its remarkable four-toothed summit, the peak to
the r. sharply pyramidal, and much higher than the others. To the W., the great
ranges of the unbroken wilderness. To the N. W., the fine view of the Willey Mt.
and the Notch. To the N., the whole S. W. range of the White Mts., their summits,
ridges, and sides clear and distinct, Mt. Washington being about 10 M. distant. To
the N. E., at a short distance, the curious and most striking Stair Mt., with its two
immense and regular steps. At the E., close at hand, are the bare and most desolate sides of Mt. Resolution, the brown crumbling granite wearing away so fast that
no vegetation can obtain a hold upon its surface, which is strewn here and there
with a dreary chaos of fallen timber, the effect of the fires which have laid bare the
mountain. On the W., beneath your feet, in the valley below, is the Mt.-Crawford
House, and the clearing, with its orchards and meadows, with the line of the road,
and the shining river." (Oakes's *White-Mt. Scenery.*)

45. Mount Resolution

is 3,400 ft. high, and consists of a lofty plateau, flanked by decomposing
red ledges, and separated from Crawford and the Giant's Stairs by narrow
and shallow ravines. The summit is covered with dense thickets of dwarf
trees, through which are occasional lane-like openings, not broad enough,
however, to give connected views. In some places are beds of red gravel,
which have been formed by the rapid disintegration of the ledges. Among
the thickets many bear-signs are seen, and water is found on the N. slope.
The mountain was named by Dr. Bemis, because when Davis had completed his path to this point he became discouraged, but afterwards ascended hither and made a *resolution* to carry it through to Mt. Washington.

The march from Crawford to Resolution takes about 1¼ hrs., and may be entered
upon by descending the precipitous side of Crawford to the E., striking the old
bridle-path, if possible, in the deep hollow beyond, toward the slope of Resolution.
The easiest way to get to the top is to keep along the path to a point where a long
red slide comes down on the r., and then ascend the line of rocks and gravel.
The walking is not so good as on the wooded slopes, but there are no annoying
bushes.

The only satisfactory view-point on Resolution is the rocky ridge to the
N. N. E., over the Rocky Branch, and fronting towards the Giant's Stairs.
A pole has been erected here, on a pile of stones. This point may be
reached by keeping around on the half-obliterated path beyond the traversing slide, to the ridge where it descends towards the Giant's Stairs.

The View. — Across the ravine to the N. are the fine terraces of the
Giant's Stairs, all of which are seen, with high precipices fronting them,
over which the pre-eminent cone of Washington appears. To the r.
is the depression of the Pinkham Notch, W. of which are the blue

highlands of Mts. Wild-Cat, Carter, Carter Dome, and the distant Moriah. Farther to the r. is the crest of Royce, before which are the white crags of Baldface. In this direction are the wooded heights of Mts. Sable and Eastman; and the twin mamelons of Double-Head are about E., to the r. of which is Thorn Mt. The view now passes down the Rocky-Branch glen to Iron Mt., whose light-hued ledges are under the double pyramids of Mt. Gemini, on whose r. is the house on queenly Kiarsarge. Still more remote is Mt. Pleasant, in Maine, with a hotel on its ridge; and the Moat range is next seen.

The view to the S. is closed by the woods on the main ridge of Resolution, which rise up close at hand. Beyond this intrusive curtain, portions of Carrigain and the Nancy Range are seen in the W. S. W., and farther to the r., over the edge of the Pemigewasset Forest, are the long and massive Mts. Bond and Guyot and the Twin Mts., with Lafayette beyond. Farther to the r. is the sharp point of Willey, beyond which the S. W. presidential peaks are seen, Webster on the l., then the low point of Jackson, Clinton's level top, Pleasant's gray hemisphere, the narrow terrace of Franklin, and the double crest of Monroe, resting against Washington.

46. Mt. Giant's Stairs

is 3,500 ft. high, and derives its name from two remarkable step-like terraces near its summit, which present the semblance of colossal stairs when seen from distant points. They appear to be cut with great regularity and sharpness of outline, the uppermost being 450 ft. high, and the second 350 ft. They fall off on the S. E. to the Rocky-Branch glen.

Giant's Stairs may be reached from Bemis Station by the old Davis path (if it can be found and followed), which ascends to the summit from the S. W., and is thence prolonged over the Montalban Ridge to Mt. Washington (see Route 89).

The mountain has been visited from Jackson by way of Littlefield's, 4–5 M. distant, over Iron Mt. The ascent of the lower stair is along a steep gully, over loose and friable rock; and the clamber up the top stair is at an angle of 60°, requiring one to pull himself up by the shrubs. The following are the times made on this route by a recent party: leave Littlefield's at 9 A. M.; at confluence of Rocky Branch and Stair-Mt. Brook, 9.50; at foot of stairs, 12.30); on summit, 1.20; left summit, 2.15; reach Littlefield's, 5.45. Paul Hayes was the guide.

The View to the N. is shut out by dense spruce woods, and on the S. is the high swell of Mt. Resolution, hiding most of the Bartlett glen. But fine prospects open out on the E. and W., extending from Mt. Pleasant, near Sebago Lake, to Moosilauke, near the Connecticut River. The Guide-Book party were within 1 M. of the summit of the Stairs, when it was forced to retreat, on account of a deficiency in its commissariat. So that the view cannot be described here, and it can only be said that the Giant's Stairs are visible from Mts. Wild-Cat, Carter Dome, Moriah,

Baldface, Double-Head, Thorn, Kiarsarge, Iron, Pleasant (Maine), Moat, Chocorua, Tremont, Passaconaway, Whiteface, Tripyramid, Osceola, Moosilauke, Carrigain, the Twins, Lafayette, Willey, and the S. W. peaks of the Presidential Range.

47. The Willey House

is about 3 M. from the Crawford House, through the Notch, and is much visited on account of the tragedy of 1826. The house inhabited by the Willeys is the low building attached to the N. end of the main white house, which is kept as a cheap tavern. Visitors are escorted through the old house, on payment of a small fee, but they will see nothing of interest. In the rear is the remnant of the great rock which sheltered the house by splitting the avalanche in its course toward it; and the track of the slide may easily be ascended for a considerable distance beyond, through a scattered forest of birch-trees. Below the house is a pile of stones which shows where the bodies of several of the Willeys were found. There is now but little danger of slides from Mt. Willey, since its side is stripped nearly to the bed-rock; but on the opposite side of the valley is Mt. Webster, whence immense rocks dash downward during storms, with a terrific roaring and crashing. True lovers of nature will find more of interest in the majestic mountains which environ this glen than in the sensational element attaching to the house. The view of the splendid cliff on the sides of Mt. Willard (to the N.) is of extreme beauty, the Notch being apparently blockaded by its heavy mass.

The view from near the Willey House is thus described in Oakes's *White-Mt. Scenery:* "The Willey Mt. is the highest, but its summit is not seen from below; and although gloomy and grand, with its high ledges and deep slides, it is less striking than Mt. Webster, which is among the most unique and magnificent objects of the White Mts. This vast and regular mass rises abruptly from the plain below, to the height of about 2,000 ft.; its shape is that of a high fort with steep scarped sides, its immense front apparently wholly inaccessible. Its top, nearly horizontal, and rough with precipitous crags, juts over with heavy and frowning brows. So mighty a mountain wall, so high, so wide, so vast, and so near the spectator that all its gigantic proportions are seen with the utmost distinctness; it fills at once the eye and the mind with awe, admiration, and delight. In a bright day, when its outline at top is seen sharp and distinct against the blue sky, its gray granite cliffs and ledges colored with iron brown, or stained with darker shades, its sides seamed with long gullied slides of brown gravel, its wide beds of great loose rocks, black with lichens, contrasted with the summer green or varied autumnal colors of the trees, make it as beautiful and interesting in its various hues and parts, as it is great and sublime in its total impression."

The Willey House was built in 1793 (some say in 1820), as a public house on the Coös road; and in 1825 it was occupied by Samuel Willey, Jr., and his family. In June, 1826, two slides fell off the flank of Mt. Willey, near the house, premonitory of the coming disaster. A long drought ensued through the months of July and August, followed by a S. wind which heaped immense masses of clouds upon the mountains. On the night of the 28th of August a deluge of rain fell, washing out the sides of the ridges, flooding the valleys, and inflicting great damage in all the adjacent towns. Bartlett, Conway, and Gilead alike suffered; all the Saco bridges were swept away; and the Ammonoosuc was swollen to ten times its usual width.

The first traveller who afterwards forced his way through the chaotic ruin in the Notch found the Willey House deserted, with the doors unclosed and the Bible lying open on the table. He gave the alarm in Conway, and the people who came up found the bodies of Mr. and Mrs. Willey, two of their children, and two hired men, buried in the slide and sadly mutilated. The bodies of the other three children were never recovered. It is supposed that the family left the house in apprehension of the rising floods of the Saco, and retreated to a point farther up on the mountain, where they were overtaken by the avalanche and swept away to a fearful and united death. Had they remained in the house, they would have been safe, for it was not moved by the water, and the slide parted at a great rock behind it and reunited below, leaving the house intact. A theory has been advanced to the effect that the fatal slide was caused, not by a heavy rain-storm, but by the breaking of massive clouds on the ridges of Mt. Willey. Three chapters are devoted to this catastrophe in Rev. B. G. Willey's *History of the White Mts.*; and T. W. Parsons has commemorated it in a ballad of 40 stanzas. See also *N. H. Hist. Colls.*, Vol. III.

48. The White-Mountain Notch

is a deep pass through the mountains, dividing the great New-Hampshire group near its centre. The Notch proper is somewhat less than 3 M. long, extending from the Gate to a little below the Willey House; and lies between Mts. Willard and Willey on the W. and Mts. Webster and Jackson on the E. The valley from Lower Bartlett to the Willey House is narrow, and mountain-ranges rise boldly on either side, thus forming an appropriate approach to the narrower gorge beyond. The highest point in the Notch is 1,914 ft. above the sea, and the massive walls are visible for 2,000 ft. above, the peaks being still higher, but generally out of sight. The Willey House being 1,323 ft. high, the road rises nearly 600 ft. in about 2 M. The bottom of the defile is occupied by the impetuous Saco River, which maintains a long and steady descent, through and about masses of boulders and the rent rocks of old avalanches. The turnpike is on the E. side of the river for 1½ M., where it crosses the Black-Camp Bridge. The railroad crosses the turnpike at grade near Bemis Station, and then commences the ascent, being elevated on artificial terraces along the W. walls. At the Willey House it is several hundred feet above the turnpike. The magnificent scenery of the Notch can be seen to much better advantage from the railroad than from the old highway, which is generally overhung with trees and masked in thick woods.

One of the finest view-points in the Notch is a little over 1 M. from the Crawford House, where the bases of Mts. Willard, Webster, and Willey approach each other. The descent through the Notch is said to give a more marked impression of its grandeur than the ascent. Dwight says: " The first ¼ M. is a mere chasm between ruptured cliffs. The remainder is a vast ravine." The course of the road is nearly E. of S.

" When we entered the Notch we were struck with the wild and solemn appearance of everything before us. The scale, on which all the objects in view were formed, was the scale of grandeur only. The rocks, rude and ragged in a manner rarely paralleled, were fashioned and piled on each other by a hand operating only in the boldest and most irregular manner. As we advanced, these appearances increased rapidly. Huge masses of granite, of every abrupt form, and hoary with a moss which seemed the product of ages, recalling to the mind the *Saxum vetustum* of Virgil, speedily rose to a mountainous height. Before us, the view widened fast to the S. E. Behind us, it closed almost instantaneously; and presented nothing to the eye but an impossible barrier of mountains." (Dwight's *Travels in N. Eng.*)

"Descending the river, the mountains in some places seem to close before you, and meet together. In other places their bare sides, scarred with avalanches, rise perpendicularly at first, then, receding, swell into rugged pinnacles, with projecting crags on either side, which nod over the bleak ridges underneath, threatening to burst from the gigantic mounds and crush the lower walls that surround them. The Saco has now swelled to a maddening torrent, and thunders down the chasm with a fierce roar and a wild echo. . . . . After struggling through the mountains, the river issues, with a calm flow, upon the plain below; and scarcely can the country furnish a more pleasant vale than that which borders the slow-winding current of the Saco in the towns of Conway and Fryeburg." (Barstow's *Hist. of N. H.*)

"The craggy sides of these giant-hills are seamed and furrowed by innumerable avalanches, which, during the last few years, have hurled headlong down their destructive masses of earth, stones, rocks, and trees, into the terrific-looking glen below. With these the river was literally choked up, exhibiting altogether such a picture of universal devastation as I never beheld even among the very wildest mountains of Switzerland. Had I wished to behold the most striking emblem of the general deluge that once swept over the earth, I could not have witnessed anywhere such fearful traces of ravage and appalling chaos as the scene displayed." (Tudor.)

Autumn in the Notch.— The splendor of the autumnal scene portrayed here by President Dwight in 1797 is still annually presented to tourists through this pass.

"The darkness of the evergreens was finely illumined by the brilliant yellow of the birch, the beech, and the cherry, and the more brilliant orange and crimson of the maple. The effect of this universal diffusion of gay and splendid light was to render the preponderating deep-green more solemn. The mind encircled by this scenery irresistibly remembered that the light was the light of decay, autumnal and melancholy. The dark was the bloom of evening, approximating to night. Over the whole, the azure of the sky cast a deep, misty blue; blending, toward the summits, every other hue; and predominating over all. As the eye ascended these steeps, the light decayed, and gradually ceased. On the inferior summits rose crowns of conical firs, and spruces. On the superior eminences, the trees, growing less and less, yielded to the chilling atmosphere, and marked the limit of forest vegetation Above, the surface was covered with a mass of shrubs, terminating, at a still higher elevation, in a shroud of dark-colored moss"

Hear also Starr King : "The only way to appreciate the magnificence of the autumnal forest scenery in New England is to observe it on the hills. I never before had a conception of its gorgeousness. The appearance of the mountain-sides as we wound between them and swept by, was as if some omnipotent magic had been busy with the landscape. It was hard to assure one's self that the cars had not been switched off into fairy-land, or that our eyes had not been dyed with the hues of the rainbow. No dream could have had more brilliant or fantastic drapery. Now we would see acres of the most gaudy yellow heaped upon a hillside; soon a robe of scarlet and yellow would grace the proportions of a stalwart sentinel of the valleys; here and there a rocky and naked giant had thrown a brilliant scarf of saffron and gold about his loins and across his shoulders; and frequently a more sober mountain, with aristocratic and unimpeachable taste, would stand out, arrayed from chin to feet in the richest garb of brown, purple, vermilion, and straw-color, tempered by large spots of heavy and dark evergreen. It did not seem possible that all these square miles of gorgeous carpeting and brilliant upholstery had been the work of one week, and had all been evoked, by the wand of frost, out of the monotonous green which June had flung over nature."

Geology of the Notch. — The Notch "has been excavated almost entirely out of granite. It lies near the E. border of the vast sheet of Labrador granites heretofore described, perhaps on the line of eruption. This deep valley exists for the reason that the denuding agents have excavated it out of the softest materials occurring in this vicinity. The summits of Mts. Webster and Willey consist of flinty slates, which resist decomposition much more steadfastly than the intervening granite. A climb up both these mountains shows that the granite extends nearly to their summits. In descending, one finds an abundance of loose, friable rocks, inclined at the greatest angle possible for such materials. These fragments accumulate gradually through the action of frost, and, under favorable circumstances, when rendered pasty by abundant rains, make a kind of plastic material, which slides to the bottom of the valley, where the river disintegrates it still further, and carries it towards Conway. The plains below Bartlett are largely composed of the fragments brought down from this narrow valley. The Saco Valley below Mt. Webster is lower, because the walls are composed entirely of this softer rock, and have yielded readily to the forces of disintegration." (PROF. C. H. HITCHCOCK.)

History of the Notch. — The White-Mt. Notch was known to the Indians of the adjacent valleys, but was probably rarely used by them on account of their superstitious dread of the mountains. It is claimed, however, that certain war-parties of Canadian Indians, returning from successful forays on the New-England coast, carried their captives and plunder through this pass. In the spring of 1746 an Indian war-party fell upon Gorham, Me., and took several prisoners, one of whom described their march to Canada by the way of the Saco River and through the White-Mt. Notch (*Hist. of Gorham*, p. 51). It was first made known to the English in 1771, by Timothy Nash, a border hunter, who, being in pursuit of a moose which eluded him, climbed a tree on Cherry Mt., in hopes to see his game, and was surprised to discover a deep pass cloven through the mountains. He speedily reconnoitred the Notch, and passed down the Saco River through the gorge, going on to Portsmouth, where he informed Gov. Wentworth of his happy discovery. Wishing to test the value of the pass as a route of commerce, Wentworth requested him to bring a horse through it from Lancaster, offering as a reward in case of success the tract now called Nash and Sawyer's Location, extending from the Gate of the Notch to a point beyond the Fabyan House, and including 2,184 acres. Nash associated with himself a fellow-pioneer by the name of Sawyer, by whose aid he lowered the unfortunate horse over the cliffs and drove him through the rocky river until they emerged at Conway. A road was soon built here " with the neat proceeds of a confiscated estate," and a direct route was formed between the coast and the upper Coös country, which had previously been accessible only by a long detour around the S. side of the mountains. The first article of merchandise carried through from Lancaster was a barrel of tobacco; and the first freight up from the coast was a barrel of whiskey, most of which was consumed on the way, " through the politeness of those who helped to manage the affair." In 1803 the Tenth N. H. Turnpike was built through the Notch, at an expense (for 20 M.) of $40,000. Nash and Sawyer, with the usual improvidence of hunters, speedily squandered the proceeds of their grant, and were forced to seek the forest again.

The ancient county-road crossed the Saco 32 times during the ascent of the valley, and was a singular specimen of highway engineering. The Tenth Turnpike was the avenue of an immense amount of travel, until the construction of the railroads into the mountain-region, and received profitable tolls. It is possible that portions of it will now be discontinued, since the adjacent railroad serves all purposes of freight and passengers, and the rattling six-horse stages, with their loads of merry tourists, have vanished from the Notch road. The Tenth Turnpike was more skilfully constructed than its predecessor, having but four bridges between the Crawford House and Conway, — the Black-Camp Bridge, 1½ M. S. of the Crawford House; the Deep-Hole and Pleasant-River Bridges, 2-3 M. S. E. of the Willey House; and the bridge near the inflowing of the Rocky Branch.

49. The Cascades in the Notch.

The **Flume Cascade** is about ¾ M. from the Crawford House, where a small brook descends from the mountain on the E. and passes downward to the Saco. Its name is derived from the singular trench through which the stream flows near the bridge, where it is deeply sunken in the channelled ledges. The main cascade is on the l. side, descending the road, and is about 250 ft. high. This point, as well as the Silver Cascade, should be visited just after heavy rains, when the enlarged mountain-torrent leaps over the cliffs in vast white sheets of foam.

See the quaint and minute description of the Flume Cascade, given by President Dwight in 1797 : " At the distance of ¾ M. from the entrance, we passed a brook, known in this region by the name of *the Flume;* from the strong resemblance to that object, exhibited by the channel, which it has worn for a considerable length in a bed of rocks : the sides being perpendicular to the bottom. This elegant piece of water we determined to examine further ; and, alighting from our horses, walked up the acclivity, perhaps a furlong. The stream fell from a height of 240 – 250 ft. over three precipices ; the second receding a small distance from the front of the first, and the third from that of the second. Down the first and second, it fell in a single current : and down the third in three, which united their streams at the bottom in a fine basin, formed by the hand of Nature in the rocks, immediately beneath us. It is impossible for a brook of this size to be modelled into more diversified or more delightful forms : or for a cascade to descend over precipices, more happily fitted to furnish its beauty. The cliffs, together with a level at their foot, furnished a considerable opening, surrounded by the forest. The sunbeams, penetrating through the trees, painted here a great variety of fine images of light, and edged an equally numerous and diversified collection of shadows ; both dancing on the waters, and alternately silvering and obscuring their course. Purer water was never seen. Exclusively of its murmurs, the world around us was solemn and silent. Everything assumed the character of enchantment ; and had I been educated in the Grecian mythology, I should scarcely have been surprised to find an assemblage of Dryads, Naiads, and Oreads, sporting on the little plain below our feet. The purity of this water was discernible, not only by its limpid appearance, and its taste, but from several other circumstances. Its course is wholly over hard granite ; and the rocks and the stones in its bed, and at its side, instead of being covered with adventitious substances, were washed perfectly clean ; and by their neat appearance added not a little to the beauty of the scenery."

The *** Silver Cascade** is about 1 M. S. of the Crawford House, and is one of the most graceful falls in the mountains. It is the brightest jewel on the route of the railroad, and is seen with fine effect from the cars. The descent of the brook within 1 M. of advance is over 1,000 ft., the most remarkable falls occurring towards the road. Much of its downward course is in long slides over the smooth surfaces of highly inclined ledges, and these reaches are succeeded by short and nearly perpendicular leaps over steeper rocks. The course of the stream may be ascended as far as the visitor's time and strength allow, but the route is arduous, and in some places perilous. Mt. Webster has been ascended along this stream; and from the midway cliffs good views are given of Mt. Willard and the Hitchcock Flume. The portion seen from the road is about 300 ft. high. Close at hand it falls 20 ft. sharply, and then rushes through a deep flume under the bridge. It is a pleasant place to spend a moonlight hour, when the significance of its name is more manifest. The brook dwindles to puny proportions in dry weather, and dribbles over the rocks in weak threads of water. The ponderous massiveness of the mountains on either side helps to increase the pleasing effect of the scene, when the stream is thundering down whitely, with full banks.

The *** Ripley Falls**. — The stream which crosses the road about 1¼ M. below the Willey House is *Cow Brook*, which has retained its homely name, though Starr King and Mr. Ripley attempted to christen it *Avalanche Brook*. A cart-road turns in to the r. near the point where it crosses the highway, leading to a platform on the railroad, across which and a little way to the r. is the beginning of the path to the Ripley Falls. The path is well made and is practicable for ladies. It winds around the upper side of a densely wooded ravine, in which frequent glimpses of the brook are gained. In less than 1 M. the falls are reached, and several good view-points may be chosen in their vicinity, according to the time and wood-craft of the visitor. The cliffs at the head of the ravine are of imposing height and grand proportions. Before taking the final leap the brook falls over several rocky steps, and then passes down the cliff at a high angle, its breadth varying widely according to the rains which have fallen or been withheld. Starr King estimated its height at 150 ft., but a careful meas-urement with an aneroid barometer makes it 108 ft. It is not so high nor so nearly perpendicular as the Arethusa Falls, but has about the same quantity of water.

"Mr. Champney, who visited these falls about a fortnight after their discovery, is inclined to ascribe to them a nobler beauty than any others thus far known among the mountains. He describes the picturesque rock-forms as wonderful, and their richness in color and marking, in mosses and lichens, as more admirable than any others he has had the privilege of studying in the mountain region."

These falls were discovered by a fisherman, and were visited, in 1858, by two gen-tlemen named Ripley and Porter, who named the main fall the *Sylvan-Glade Cat-arart*, and the smaller one above, the *Sparkling Cascade*. With questionable taste, Starr King had the former name changed to *Ripley's Falls*.

The stream may be ascended for more than 1 M. beyond the Ripley Falls, through primitive woodlands, and by several pretty cascades. Far up its course, near the inflowing of a tributary brook, is another beauti-ful fall, which has been named the *Sparkling Cascade*. From this point woodsmen can follow the ridge to the S., behind the Frankenstein Cliff, and strike the Bemis Brook, near the Arethusa Falls.

The *** Arethusa Falls** are on Bemis Brook, which rises N. of Mt. Nancy and passes under the roads about 2 M. N. of Bemis Station, or 6½ M. S. of the Crawford House. They are only 1 M. from the railroad, though the visitor will think that the distance is nearer 5 M., so rugged and arduous is the route. The only way of approach is by following the stream up-ward, sometimes among the dense thickets on the banks, and sometimes over the rocks near the water. Several pretty cascades are passed, and the forest scenery is beautiful. For long distances the brook glides over

nearly level sheets of granite, where the visitor can walk as if on a pave-
ment, and then he is forced to cross a chaotic mass of rocks, flanked by
bristling thickets.   When within about ⅛ M. of the end, there is a charming fall of about 40 ft., over wide granite ledges.

The Arethusa Falls were discovered by Prof. Tuckerman, many years ago, but
have not been visited by a dozen people since, and are wellnigh forgotten.   They
were visited and measured by the Editor and Prof. Huntington, in September, 1875,
and then (being nameless) received the provisional name of the Arethusa Falls, in
allusion to Shelley's lines : —

" Arethusa arose
From her couch of snows
In the Acroceraunian mountains, —
From cloud and from crag,
With many a jag,
Shepherding her bright fountains.
She leapt down the rocks,
With her rainbow locks
Streaming among the streams :

" Her steps paved with green
The downward ravine
Which slopes to the westward gleams ;
And gliding and springing,
She went, ever singing,
In murmurs as soft as sleep.
The Earth seemed to love her,
And Heaven smiled above her,
And she lingered towards the deep."

The falls are 176 ft. high, and are among the finest in the mountains.
They are surrounded by rich and luxuriant forest-scenery, among large
old trees and rugged groups of water-worn boulders.   The best point of view
is about 100 ft. below the falls, where their long white line is seen through
the foliage.   A nearer approach brings the visitor too much under the
cliff to realize its height.   It will be seen that the brook leaps almost perpendicularly over a lofty precipice, and plunges heavily into a deep pool
below.   An impressive view down the ravine and over the dizzy brink
may be obtained by bearing to the l. through the forest and ascending to
the top of the falls.

Remembering how the power and sensational effect of the mountain-
cascades are enhanced by a large volume of water, tourists will do well
to visit the Arethusa Falls soon after heavy rains, — although the labor of
reaching them then is greater, because the line of the stream is not then
available as a route.

The Guide-Book party passed from the Arethusa Falls to the Ripley Falls by taking a northerly course through the forest, leaving the Frankenstein Cliff about ¾ M.
to the r.   The march occupied about 2 hrs. and crossed several low ridges, through
open and easily travelled woods.   Care must be taken not to deflect too far to the r.
and strike Cow Brook *below* the falls.   It is better to strike it above and descend its
madcap downward course.

50. The Crawford House

accommodates 300 – 400 guests, charging $ 4.50 a day, with considerable reductions
for regular boarders, especially in June and September.   It is managed by Mr. C.
H. Merrill, A. T. & O. F. Barron being the proprietors.   The post-office is in the
large central hall, which serves as a rendezvous and exchange for tourists.   The
hotel has a barber-shop, bowling-alley, telegraph-office, and a large livery-stable ;
and is lighted with gas.   Railroad and stage connections are made between this and
all the other mountain-hotels, so that transient visits may be planned from this as a
centre.   The house opens about June 1st.

Distances (from the hotel-list). — Crawford House to Gibbs's Falls, ½ M. ;
Elephant's Head, ¼ ; Beecher's Cascades, ½ ; Gate of the Notch, ¼ ; Profiles, ½ ;
Pulpit Rock, ½ ; Flume Cascade, ¾ ; Silver Cascade, 1 ; Devil's Den, 2 ; Mt. Wil-

bird, 2; Hitchcock Flume, 2; Willey House, 3; Ripley Falls, 6; Ammonoosuc Falls, 5½; Mt. Washington, by bridle-path, 8½; Mt. Washington, by railroad, 13. Mountain-wagons are frequently despatched from the hotel to Mt. Willard, the Willey House, and other points of interest, and passengers are taken for a small sum each.

Routes. — From Boston by the Eastern R. R., leaving at morning, and arriving late in the afternoon. Or by the Eastern R. R. to Portland, and thence by the Portland & Ogdensburg R. R. By the Boston, Concord & Montreal R. R. from Boston to the Fabyan House, and thence by the P. & O. R. R., the same afternoon. From New York in about 24 hours, by various combinations with the Sound steamers and the Connecticut-Valley lines.

The Crawford House is on a little plateau 1,900 ft. above the sea, occupying a tract which the geologists maintain was the bed of an ancient lake. On the W. are Mts. Field and Tom. Mt. Clinton is on the E., and in front are Mts. Webster and Willard, with the majestic portals of the Notch. A few rods in front of the house is the pretty little building of the railroad station. There are more interesting short excursions in this vicinity than near any other of the mountain-hotels, except the Profile House. The environs of the house are neatly kept, and embellished with fountains and costly lawns. Two springs which rise near by seek different courses, one flowing into the Ammonoosuc, the other into the Saco. A short distance from the house, on the l. front. is **Saco Lake**, from which the young Saco River flows, entering its impetuous course through the Notch. This sheet of water has been enlarged and deepened by artificial means, and is provided with boats and little piers. On its E. shore, at the foot of the mountain, and approached by a short path, is the pleasant bit of disciplined forest called **Idlewild**, which overlooks the lake and glen, and is provided with many rustic seats. Pretty views are enjoyed from the high bluff in Idlewild, where seats and tables are placed.

The **Elephant's Head** is about ½ M. from the hotel, and derives its name from the resemblance of its shape, as seen from the piazzas. It is a rocky bluff on the E. side of the Notch, and commands a good view of the upper part of the defile and of the Crawford-House plateau. The path diverges to the l. from the road below the lake, and winds around through the forest to the top of the rock.

The double * **Gate of the Notch** is about 80 rods in front of the hotels, from which it is well seen, and consists of the narrow pass (26 ft. wide) through which the road leads and the Saco flows, between two huge piles of rock. The Gate has not been enlarged to admit the passage of the railroad, since a new passage has been made for the track between the W. portal and the rocky side of Mt. Willard. It has been proposed to span these two gates by a double triumphal arch, to commemorate the ancient victory of nature, aided by water, and the recent victory of man, aided by fire, over the flinty barriers in their way.

A short distance below the Gate of the Notch, and about ½ m. from the hotel, guide-boards are raised by the roadside, telling where to look on the adjacent cliffs to see the profiles of the Old Maid of the Mountain, the

Infant, the Young Man, the Sentinel, the African Face, and the Grandmother. The labor of hunting out these profiles, whether successful or otherwise, is doubtless a prolonged and valuable discipline of the imagination. The Old Maid and the Infant are seen from nearly the same point, the former being on a spur of Mt. Webster, and the latter on the side of Pulpit Rock.

* Beecher's Cascades are about ½ m. from the hotel, and are reached by a good path turning to the r. from the front of the house, and crossing the railroad-cut on a foot-bridge. It soon enters the forest and crosses and ascends on the l. bank of the brook, being provided with stairways and rustic seats. There are several beautiful falls on the brook, amid picturesque rock-scenery, and surrounded by a primitive forest. Passing all these cascades the path crosses the brook at the head of the uppermost, ascends a shelving ledge, and terminates at a rustic seat above, from which the visitor gains a fine view of the peaks toward Mt. Washington, down a long vista of foliage. The pools and falls along this glen have been pleasantly described by Henry Ward Beecher, in honor of whom they were named.

Gibbs's Falls are about ¼ M. from the hotel, by a path entering the woods on the l. They were named in honor of a former landlord of the Crawford House. The brook falls 30 – 40 ft., around a projecting central ledge. This path is the same as the old bridle-path to Mt. Washington for a part of its course. There are pleasant bits of water-scenery all along the course of the brook.

There is a legend to the effect that the hardy old Scotchman who founded the Crawford family of New Hampshire erected a rude log-house near the Notch, and began to attack the forest. The royal Governor Wentworth occasionally took solitary *incognito* journeys through the Province, and on one of these jaunts he found Crawford's hut, and began to make himself agreeable to Mrs. Crawford. But she stoutly rebuffed him, and complained to her husband on his return, all vainly, however, for the jovial old Scot saw in Wentworth a kindred soul, and bade him stay all night. The evening was passed in drinking and merry-making, until Crawford protested that his guest was " the best fellow he had met wi' sin the day o' the baillie o' Glasgow, who was aye fou' six days out o' the seven, and ended his life at last ae drifty night amang the snaw." Before leaving, in the morning, the Governor got the frontiersman to promise to visit him at Wolfeborough, where he was well known as " Old Wentworth." Here he was royally received, and the dismayed old woodsman, repentant for his wassail with the king's representative, was finally dismissed with a deed of 1,000 acres of land about his farm.

The Crawfords opened a hotel called the *Notch House*, close to the Gate of the Notch, many years ago. It was in its glory in 1840, when Thomas J Crawford was the proprietor.

51. Mount Willard

stands in the mouth of the Notch, between Mts. Field and Jackson, its E. base forming one side of the Gate of the Notch. It is about 2,570 ft. above the sea, or 670 ft. above the Crawford House. The N. side is covered with woods, but on the S. is a vast precipice of brilliantly colored rocks, falling off sharply to the Saco glen below. Mts. Field and Willey tower over

Willard, on the W., 1,600–1,900 ft. higher; and Mts. Jackson and Webster, on the E., are 1,600–1,700 ft. higher. The summit is reached from the hotel by a good carriage-road, 2 M. long, which crosses the track below the station. Mountain-wagons are often sent up from the Crawford House; and the journey on foot is easy and pleasant, leading through picturesque woods. The road leads out to the edge of the cliff, from which a noble view is gained. It is not a mountain panorama, such as may be enjoyed from Mt. Willey or Mt. Clinton, and its horizon is narrowed by the adjacent ranges; but it has a singular beauty and quaint individuality which no other view possesses. It is preferred by many frequenters of the hill-country to any other prospect in this region. According to some accounts this mountain was named for Prof. Sidney Willard, of Harvard University; but others say from Joseph Willard, of Boston, an enthusiastic admirer of the view, in the days of the Crawfords. In 1844 Prof. Tuckerman ascended it and named it *Mt. Tom* (after Tom Crawford).

** *The View.* — Nearly N. E. is the high peak of Mt. Jefferson, rising rapidly to the r. from the insignificant and chaotic hills of the Mitten and Dartmouth ranges. On its r. are the uneven humps of Mt. Clay, from which the rugged slope of Mt. Washington ascends to the hamlet on the summit. Mt. Clinton's shaggy sides fill up much of the horizon in this direction, but are overlooked, on the l., by the curving crest of Pleasant. Farther to the r. are the formidable masses of Mts. Jackson and Webster, just across the Notch, cut into by ravines and jewelled with cascades, — the brilliant stripes on Webster's W. front showing but in part. Beyond the r. flank of the latter are the wooded ridges below the Saco, Table Mt., and the W. spurs of Moat Mt.; and about S. S. W. the weird white crown of Chocorua peers over the dark ridge of Bear Mt. More to the r. is the long and gracefully outlined Mt. Nancy, rising from the Notch to a narrow plateau (over which Mt. Tremont is seen), and thence to a rounded crest. The mighty sweep of Mt. Willey then cuts off the view, rising to a great height, and trending to the r. in the line which includes also Mt. Field (nearly W.) and Mt. Tom (N. N. W.). The great gorge in which the Saco flows is, however, the chief feature of interest, filled, as it is, with dense woods, and traversed by the narrow and often-obscured bands of the turnpike and railroad. The white walls of the Willey House are far below, a mere dot on the ruin-swept lowlands.

" Much of this scenery, I say, is superior to the famed and classic lands of Europe. I know nothing, for instance, on the Rhine equal to the view from Mt. Willard, down the mountain pass called the Notch." (ANTHONY TROLLOPE'S *North America.*)

" Under our feet yawned the tremendous gulf of the Notch, roofed with belts of cloud, which floated across from summit to summit nearly at our level; so that we stood, as in the organ-loft of some grand cathedral, looking down into its dim nave. At the further end, over the fading lines of some nameless mountains, stood Chocorua, purple with distance, terminating the majestic vista. It was a picture which the eye could take in at one glance; no landscape could be simpler or more sublime. The noise of a cataract on our r., high up on Mt. Willey, filled the air with a far, sweet, fluctuating murmur, but all around us the woods were still, the harebells bloomed, and the sunshine lay warm on the granite. I had never heard this view particularly celebrated, and was therefore the more impressed by its wonderful beauty. As a simple picture of a mountain pass, seen from above, it cannot be surpassed in Switzerland. . . . . A portion of the effect, of course, depends on the illumination, but no traveller who sees it on a day of mingled cloud and sunshine will be disappointed." (BAYARD TAYLOR.)

"But to know the Notch truly, one must take the drive from the Crawford House to the top of Mt. Willard, and look down into it. A man stands there as an ant might stand on the edge of a huge tureen. . . . . The road below is a mere bird-track. The long battlements that, from the front of the Willey House, tower on each side so savagely, from this point seem to flow down in charming curves to meet at the stream. . . . . The view of the summits of the Mt. Washington Range, too, ... is a reward for the short excursion, almost as valuable as the view of the Gulf of the Notch. And let us again advise visitors to ascend Mt. Willard, if possible, late in the afternoon. They will then see one long wall of the Notch in shadow, and can watch it move slowly up the curves of the opposite side, displacing the yellow splendor, while the dim green dome of Washington is gilded by the sinking sun 'with heavenly alchemy.'" (STARR KING.)

The **Hitchcock Flume** is on the E. flank of Mt. Willard, and was discovered by Prof. C. H. Hitchcock, in 1875. It is reached by a good path which diverges to the l. (in ascending) from the carriage-road, about ¼ M. below the summit. The flume consists of a narrow cañon between high perpendicular and parallel walls of rock, its bed having a rapid slope, and being so damp as to be unsafe to venture through. This singular gorge is 350 ft. long, and its walls are 30 – 60 ft. high, approaching in some places to within 6 – 8 ft. of each other. In the lower portion it bends to the r. around a tall crag; far below which is the railroad, with many a cliff between. The *Butterwort Flume* is another interesting locality on the mountain.

The **Devil's Den** is a black-mouthed cavern on the S. side of the cliffs of Mt. Willard, and is plainly seen from the road. It is reported that Crawford visited it, many years ago, and found the bottom strewn with bones and other ghastly relics; and marvellous stories of the supernatural were afterwards told about it. In 1856, however, the Den was explored by Dr. Ball, who was lowered down the cliff by ropes. He found it to be 20 ft. wide and deep, and 15 ft. high, very cold and damp, but containing nothing remarkable or interesting During the State Survey of 1870, explorers were lowered to the Den by a rope 125 ft. long, but their discoveries did not repay the peril of the descent.

52.   The Mount-Willey Range.

This conspicuous and interesting ridge commences near the White-Mountain House in Carroll, and runs nearly S. for 8 M., to the ravine S. of Mt. Willey. It shows a steady rise from the low granitic peaks on the N. to the bold crest of Willey; and in its southerly section it forms the W. wall of the White-Mt. Notch. It is covered with unbroken forests, which, however, afford no worse obstacle than light underbrush, easily parted or trodden down, save about the thicket-fringed ledges and on the densely wooded crests. Strong clothing should be worn during the exploration of this range, since any other would soon be riddled by the spruce boughs. No water is found between the base of Mt. Field and the farther slope of Mt. Willey, and the visitor would do well to carry a bottle of cold tea or some other refreshing beverage.

Mt. Tom is just to the westward of the Crawford House, and is about 3,200 ft. high. It was named in honor of Thomas J. Crawford, the founder of the Crawford House. There is but little to interest tourists on this summit, as it is covered with trees which prevent a view from being obtained. The ascent is made by following the stream on which Beecher's Cascades are situated, up to its source, and then gaining the ridge to the W., which is followed to the r. to the jungle on the crest.

Mt. Avalon is the name given by the Editor to the peak above Beecher's Cascades, because it had no name of its own and on account of its resemblance to certain bold hills in Avalon, the great S. E. peninsula of Newfoundland. It is ascended by striking off into the forest in a left oblique direction from the head of Beecher's Cascades, and the upward journey may be achieved in 1½ hrs. The first peak is low and wooded, and is separated by a shallow trough, abounding in thickets, from the main peak, which consists of a high pile of rugged rocks.

This is a fine watch-tower from which to reconnoitre the main Mt.-Willey Range, whereof Mt. Tom stands on the r., Mt Field in front, and Mt. Willey on the l. On the other side, and far below, is the round and wooded summit of Mt Willard, with the Crawford-House clearing on the N. and the long gorge of the White-Mt. Notch to the S. To the S. E. is the striped front of Mt. Webster, beyond which the peak of Mt. Crawford cuts the sky ; and at the foot of the Notch are the Swift-River peaks, terminated by the white gable of Chocorua. To the N E. are the colossal stairs leading to Mt. Washington, — woody Clinton, rounded Pleasant, and spire-like Monroe ; and beyond the great peak is the massive bulwark of Jefferson. The northern view includes a section of the Ammonoosuc Valley, the Mt Deception, Cherry-Mt , and Starr-King ranges, and a portion of the valley of Israel's River. Farther to the N. W. are the Essex-County hills and the cloven mountains at Willoughby Lake.

Mt. Field was formerly called *Mt. Lincoln,* in honor of the martyr President, but the Mt. Lincoln in Franconia claimed priority of nomenclature, so the State Geologist, in order to avoid confusion of localities, named this peak in honor of Darby Field, who first ascended Mt. Washington (in 1642). It is 4,070 ft. high, but possesses no interest for tourists, the top being covered with dense thickets which shut out the view. The ascent is made only as a part of the attack on Mt. Willey, and is usually conducted by way of Beecher's Cascades and Mt. Avalon. It takes a little longer time to visit Avalon, but the detour is advisable as affording opportunity to plan the route of ascent. The best line is found by striking directly up the N. flank of Mt. Field to the dark-green foliage, which indicates heavy timber and less undergrowth.

Mt. Willey attains a height of 4,330 ft., and is the chief peak and S. end of the Mt.-Willey Range, which forms the W. wall of the White-Mt. Notch. It was first ascended in 1845, by Prof. Tuckerman. The views from its summit are broad and beautiful, and include one of the best general prospects over the Pemigewasset Forest. The ascent is difficult, being through pathless forests and over several tracts of formidable ledges; and the excursion requires the better part of a day.

A skilful woodsman can reach the summit from Mt. Willard by crossing the deep ravine on the S. W. ; but the simplest routes of ascent are by the Kedron ravine and by Mt. Field. The former (and by all odds the best) is traversed by going down the Notch road to the Brook Kedron, below the Willey House, and ascending its steep and rugged ravine to the S. flank of the mountain, which is reached within less

than 1 M. of the crest. The second route is entered by way of Beecher's Cascades,
Mt. Avalon, and over Mt. Field (see page 153), whence the ridge is followed to the S.
for about two hours. There are two slight eminences on this line of heights, and the
woods are comparatively open, being free from fallen timber and encumbered only
with light underbrush. The time necessary for climbing from the Crawford House
to the summit of Mt. Willey is about five hours. The crest is covered with a dense
thicket of low trees, but is surrounded by a ring of open rocky land, whence the
outlook is gained.

* *The View.* — Mt. Webster is nearly E., across the White-Mt. Notch,
scarred by numerous long slides and made brilliant by contrasting bands
of red, yellow, and green Its long thin crest droops toward the S., and is
faced by bright-red rocks. From this point the Presidential Range trends
to the N. E., including Jackson's gray summit, the meadow-like cap of
Clinton, the symmetrical dome of Pleasant, the slide-striped slopes of
Franklin, and the craggy top of Monroe. Over all is the alpine hamlet
on Washington, with the level lines of Bigelow's Lawn and Boott's Spur
running off to the r. On the l. is the railway-station, in the valley of the
Ammonoosuc; and portions of the track are also visible. On the l. of
Washington are the zigzag crest-lines of Clay, the rocky pyramid of Ad-
ams, and the high peak of Jefferson, farthest to the l.

Nearly N. E., over the range beyond the Notch, is the Carter Dome,
with a portion of the Moriah Range; on whose r. are the sky-meeting peaks
of Baldface, capped with white ledges and flanked by a section of Royce.
Eastman and Sable are seen next on the r., and then come the distant
crests of Double-Head and Mt. Gemini, over the Montalban Ridge and
below the horizon. Thorn Mt. also rises beyond this ridge, and far beyond
are the highlands of Waterford and Sweden, scarcely rippling the level
sky-line. Over the r. of Thorn is the lofty cone of Kiarsarge, indicated
by a house on the summit, with the upper cliff of Giant's Stairs below,
and Resolution close alongside the latter. On the S. of Kiarsarge, the
Green Hills of Conway are seen, over which rises Mt. Pleasant, near Sebago Lake, with a hotel near the middle of its long rolling ridge. Nearly
S. E., below the Green Hills, and across the Notch, within a few miles,
are the massive ridges of Mts. Crawford, Resolution, and Giant's Stairs,
whose upper portions are flecked with brilliantly colored ledges of light-
red rock. Crawford is distinguishable by its sharply cut peak; Resolution is crowned by a broad plateau, 266 ft. higher; and the upper terraces
of the Giant's Stairs close the group on the N. Farther to the r. is a portion of the Notch, traversed by the white band of the P. & O. Railroad;
and N. Conway is seen on the widening plain beyond, with the bold peaks
of Moat Mt. on the r. Over the section between Kiarsarge and Moat extends a vast area of Western Maine, diversified by low mountains, highland lakes, and white villages. Between Pleasant and Moat, about S. E.,
the city of Portland and the ocean beyond can be seen with a good glass.
To the r. of Moat, and in the same range, S. of the Saco, are the flat tops

of Table Mt., the undulating ridge of Bear Mt., the dark cone of the Haystack, and the whitish knobs of Tremont. Beyond these, to the S., is the white and storm-beaten ridge of Chocorua, with a singularly regular pyramid as its crowning point; over which, on the r., and far to the S., is a portion of the blue Ossipee Range, with Copple Crown and Moose Mt., below Lake Winnepesaukee.

The view now passes over the long plateau on the S. to Mt. Nancy, whence a continuous range runs to the S. W., including a nameless second peak and then the dark and pointed tops of Mts. Anderson and Lowell. Beyond these are the peaks of the Sandwich Range, Passaconaway being due S. and nearly over Mt. Anderson, with the white mound of Potash at its base. The double-crested Whiteface is just to the r., over the peak of Lowell; and Tripyramid is a triple-headed blue ridge still farther to the r.

The view now rests on the vast and imposing form of Mt. Carrigain, about 8 M. distant, W. of S., rising directly from the wilderness, with step-like spu s on the N. and a sharp descent on the S. In front of the great ravine in its N. E. side is Vose's Spur, below which is the profound Carrigain Notch; and the Lowell-Anderson-Nancy Range extends thence to the l. to the White-Mt. Notch. A portion of Sandwich Dome is visible through the Carrigain Notch; and Osceola is farther to the r., through the gap between Carrigain and the wooded mass of Hancock. On the S. W. is the quiet and secluded sheet of Ethan's Pond, beyond which the valley of the true Merrimac (East Branch of the Pemigewasset) extends for many leagues, covered with unbroken forests and without signs of habitation or civilization. This desolate region is diversified by low ridges and is bounded irregularly by bold mountains. 8–10 M. distant across this wilderness is the Twin Range, whereof the S. Twin is due W. of Mt. Willey, with the N. Twin on the r. and the high and massive ridges of Mts. Guyot and Bond on the l.

The stately blue ridge of Moosilauke appears over the l. flank of Mt. Bond, and on its r., over the Twins, are the sharp crests of Mts. Flume, Liberty, and Lincoln, terminating with the thin and lofty serrated ridge of Lafayette. The N. Twin is marked by a long white slide, and slopes down toward the New-Zealand Notch, through which a vista of the Ammonoosuc Valley is gained, with Mt. Hale and the Sugar Loaves closing in on the W. Nearly N. W. are the sharply notched mountains at Willoughby Lake, beyond the Victory Hills, in Vermont; while more to the r., and much nearer, is Cherry Mt , with the Owl's-Head peak on its N. end. Between Cherry and Jefferson are the low and monotonous ranges of Mt. Deception, with long dark slopes; Mt. Dartmouth, with three well-marked peaks; and Mt. Mitten, showing a knob-like summit. Above these are the Starr-King and Pilot Ranges, with the village of Jefferson Hill; and the valley of the upper Connecticut opens above Lancaster, in which Cape Horn stands out conspicuously. In the remote distance,

E. of N., are the Dixville and Magalloway Mts.; and beyond all (over Mt. Mitten) is the blue crest of Mt. Carmel, between the Connecticut and Parmachene Lakes, and within 7 M. of the Canadian frontier.

53. Eastern Pemigewasset.

The term *Pemigewasset* has been applied to the great wilderness which surrounds the East Branch and its tributaries, including (according to Prof. Hitchcock) the Mt.-Willey and Nancy Ranges, Mts. Carrigain, Hancock, Tripyramid, Osceola, Tecumseh, Welch, Black, Loon-Pond, the Twins, the Franconia Range, the Haystack, Mt. Hale, and many nameless peaks. The boundaries therefore seem to be the Saco River on the E., the Ammonoosuc and Gale Rivers on the N., the Pemigewasset on the W., and (partly) the Mad and Sawyer's Rivers on the S. This broad region is still in a condition of primeval wildness, and has not been invaded by clearings, roads, or trails. "Clear to the Franconia Notch extends this untracked and unvisited realm of Nature, who yet holds one fastness in the heart of busy New England, with its glorious falls not yet harnessed as 'water-powers,' and its stately trees yet undeveloped into sashes and blinds." \

This forest may be entered to better advantage on the W. side, from the town of Lincoln, but certain localities toward the E. are reached more easily from the Saco Valley. The inner solitudes should be entered only under the guidance of experienced foresters; and travelling will be found very slow and arduous. The scenery is simply that of a vast primeval forest, most of the environing mountains being hidden by the foliage or by intervening ridges. Trout increase and multiply almost undisturbed in the brooks and ponds; and during the winter many deer are shot here.

Ethan's Pond is S. W. of the precipitous sides of Mt. Willey, which is seen to great advantage from its shores. It covers a small area, and is surrounded by dense forests on all sides. The pond is probably over 2,500 ft. above the sea; and is visited only by occasional parties of trout-fishers, who camp on the shores. The route is through the *Willey Notch*, whose summit is 2,800 ft. above the sea, or about 1,500 ft. above the Willey House. The Saco Valley is left near the Willey House, and the great slide on the S. of Mt. Willey is ascended to the plateau S. of the mountain, whence a line of spotted trees leads in to the pond, 3 M. from the Saco turnpike. This secluded mountain-tarn is probably the fountain-head of the Merrimac River, whereof Whittier and Thoreau write: —

> " Go, child of that white-crested mountain whose springs
> Gush forth in the shade of the cliff-eagle's wings,
> Down whose slopes to the lowlands thy wild waters shine,
> Leaping gray walls of rock, flashing through the dwarf pine.
>
> " From that cloud-curtained cradle so cold and so lone,
> From the arms of that wintry-locked mother of stone,
> By hills hung with forests, through vales wild and free,
> Thy mountain-born brightness glanced down to the sea."

"At first it comes on murmuring to itself by the base of stately and retired mountains, through moist primitive woods whose juices it receives, where the bear still drinks it, and the cabins of settlers are far between, and there are few to cross its stream; enjoying in solitude its cascades still unknown to fame; by long ranges of mountains of Sandwich and of Squam, slumbering like tumuli of Titans, with the peaks of Moosilauke, and the Haystack, and Kearsarge reflected in its waters; where the maple and the raspberry, those lovers of the hills, flourish amid temperate dews; flowing long and full of meaning, but untranslatable as its name, Pemigewasset, by many a pastured Pelion and Ossa, where unnamed muses haunt, tended by Oreades, Dryads, and Nereids, and receiving the tribute of many an untasted Hippocrene.

> "Such water do the gods distil,
> And pour down every hill
> For their New England men.
> A draught of this wild nectar bring
> And I'll not taste the spring
> Of Helicon again."

Where it meets the sea is "Plum Island, its sand ridges scalloping along the horizon like the sea-serpent, and the distant outline broken by many a tall ship, leaning, *still*, against the sky. . . . . Standing at its mouth, looking up its sparkling stream to its source, — a silver cascade which falls all the way from the White Mts. to the sea, — and behold a city on each successive plateau, a busy colony of human beavers around every fall. Not to mention Newburyport and Haverhill, see Lawrence, and Lowell, and Nashua, and Manchester, and Concord, gleaming one above the other."

Many years ago, when Ethan Allen Crawford was ranging the woods, he encamped on these shores for a night. While catching trout for supper, he saw two large brown moose among the lily-pads, and by quick action he succeeded in killing them both. After a glorious feast of trout and moose-meat, he retired to sleep between the skins of the fallen animals, regardless of the wolves that were howling on every side. Since that night the pond has been known as *Ethan's Pond.*

The **Thoreau Falls** may be reached by descending the Merrimac from Ethan's Pond for about four hours' march (4–5 M.). At this point the river has as much water as the Saco shows at Bemis Station, and descends over 200 ft. in ½ M., the latter part of which contains several nearly vertical leaps. These falls were named by the Editor in memory of Henry D. Thoreau, the poet-naturalist, who has so often written lovingly of the Merrimac River and its fountains in the wilderness; and also in allusion to the polyglot meaning of his name, "Thunder-Water."

The march from the Notch to the Thoreau Falls will take 6–7 hrs., and by encamping at their foot the visitors can descend to the Pemigewasset Valley the next day. Another, but more difficult route of exit is by crossing the forest to the N. W. to the outlet of the New-Zealand Pond, ascending to and beyond the pond, and up the stream which enters it from the N. A short journey from the head of this brook leads through the New-Zealand Notch, and to the upper part of the New-Zealand River, which may be followed down to the Ammonoosuc, between the White-Mt. and Twin-Mt. Houses. Prof. Huntington estimates the distance from the Thoreau Falls to the Ammonoosuc highway at 5–6 M., — a good day's journey.

Western Pemigewasset, see Route 111.

Mt. Hancock is close to Mt. Carrigain, on the W., and covers a great

area of the Pemigewasset Forest. Its extent and shape are vaguely represented on the maps because much of the adjacent country is still unexplored. It is the *Pemigewasset Peak* of Prof. Guyot's map, according to
which it reaches a height of 4,420 ft.

54. The Fabyan House

accommodates over 500 guests, charging $ 4.50 a day, and $ 21 - 25 a week. Its external architecture is unattractive, but the halls and rooms are high-studded and
airy. Within the house are telegraph and post offices, billiard and bowling rooms,
a news-stand, and other conveniences. The building is lighted by gas. There is a
large livery-stable : and a farm is connected with the hotel. During the season an
orchestra is kept here to play on the piazzas and in the halls. Visitors experience
immunity from the attacks of rose-cold, or hay-fever. The house is kept open from
June 20th to Oct. 1st. It is within a stone's-throw of the junction of the Portland
& Ogdensburg Railroad with the Mt.-Washington Branch of the Boston, Concord &
Montreal Railroad. The main front of the hotel is 320 ft. long ; the parlor is 100
× 35 ft. in area ; and the dining-hall is 130 × 45 ft. The great central hall is 60 ft.
square, and contains the offices. It is the rendezvous of the guests at all times.

The *Mt -Pleasant House* is a new and well-located hotel, between the railroads,
and about ½ M. from the Fabyan House. It will be open in 1876, at about $2
a day.

The *White-Mountain House* is about 1 M. W. of the Fabyan House, and is an ancient inn whose rates are $ 2.50 a day ($ 7 - 15 a week). It is at the intersection of
the Bethlehem and Cherry-Mt. roads, and is a station on the railroad. The building was constructed about the year 1845, by one of the Rosebrooks.

Distances. — (The excursions from the Fabyan House are made on the railroads, or on the highways parallel thereto. The following distances are from the
measurements of J. F. Anderson. Engineer of the Portland & Ogdensburg Railroad.)
The Fabyan House to the White-Mountain House, 1 M. ; Lower Ammonoosuc Falls,
1½ M. ; Twin-Mountain House, 5 ; Crawford House, 4 ; Upper Bartlett, 18½ ; N. Conway, 29. (From the B., C. & M. R. R. Survey.) To the base of Mt. Washington, 6 ;
the summit of Mt. Washington, 9. (From Walling's odometer survey.) To Jefferson Hill, 18 ; Bethlehem, 13 ; the Profile House, 22½.

Routes. — Through trains (with parlor-cars) leave Boston at 8 A. M., and run
over the Lowell road and the B., C. & M. R. R., reaching the Fabyan at 4 P. M.
(see Routes 2 and 3). Passengers on the afternoon trains from Boston pass the
night at Plymouth, and reach the Fabyan the next forenoon. Tourists can also go
through in one day from Boston by the Eastern and Portland & Ogdensburg Railroads. From New York by the evening boats on the Sound, reaching the Fabyan
at 4 P. M. the next day ; or by the 10 A. M. train, via Springfield, passing the night
at Plymouth. From Saratoga and the Lake-Champlain region by the Montpelier &
Wells-River Railroad or the Portland & Ogdensburg Railroad. From Montreal in
the morning by the Southeastern Railway, dining at Lake Memphremagog, and taking supper at the Fabyan.

The air-line distance from the Fabyan House to the summit of Mt.
Washington is about 7½ M., and the Summit House is 4,722 ft. higher
than the Fabyan, — the latter being 1,571 ft. above the sea. Most of
the summits of the Presidential Range are visible from this point, rising
over the wide forests which shade the upper streams of the Ammonoosuc,
and exposed nearly to their bases, scarred with slides and gullied by deep
ravines. The view is greater in extent than that from the Glen House,
but far less majestic, though Mt. Washington shows more nobly. The
Fabyan cupola is lifted above the dull foreground and commands an unobstructed view. On the N. is a low and wooded spur of Mt. Deception,
on whose r. are the flanks of Mt. Jefferson, with Clay extending toward

Washington. Barren gray rocks cover the latter nearly half-way down, over which, on the l. ridge, is the railroad-track, and on the summit is the white hotel. A bold spur descends to the r. to the craggy line of Monroe, below which is a long and slide-scarred ravine. Next to the S. is the flat crest of Franklin, and then, over the iron bridge, is the white dome of Pleasant, in whose ravine the yellow stripes are thought to resemble in outline an Indian warrior. Next to the r. is the long green crest of Clinton, whose triple head is over Crawford's monument. Then comes the little peak of Jackson, and at the end is Mt. Webster, which falls off into the White-Mt. Notch. Mt. Rosebrook is nearer, on the S.

The small spur of Mt. Deception which lies opposite the hotel is sometimes ascended for the sake of its view of the White-Mt. Notch and Mt. Washington. There is no good path. It is called *Mt. Prospect* or Mt. Peabody. The **Giant's Grave** was a mound of river-gravel, or a sand-bar formed by the reaction of the ocean-waves against the adjacent hills. It was 300 ft. long, 75 ft. wide, and 50 ft. high. At one time a cannon was kept upon its summit, with which the echoes of the mountains were often aroused. The grave and monument of Ethan A. Crawford are seen from the hotel, near the road to Mt. Washington.

N. P. Rogers spoke thus of the view from near the White-Mountain House, in 1841 : " We descried through the thick atmosphere a gloomy range of mountain, — its summit, or summits, hid in thick clouds, and its awful breast gashed and lacerated with the mighty slides. We at once recognized it as the high object of our journey. Nothing could exceed its awful majesty and vastness. Everything around us had for some time betokened that we were in the suburbs of one of the capitals of nature. The majestic woods, the tremendous elevations of the mountain-ranges, and the vastness of the forest, — the stillness in the air, and its altered temperature ; and the majestic roar of the Ammonoosuc along its bed of precipices, spoke of its mountain descent, and that its fountains could not be far distant. It was a glorious hour."

The *Lower Ammonoosuc Falls* are alongside of the road, a little way below the White-Mountain House. So far as the falls are concerned, they are not worth visiting, because their beauty and power have all been taken away by the erection of a saw-mill above. The ledges and rock-walls on the sides are curious, and are of considerable interest to geologists. In their regularity they resemble fine masonry, and show the results of the long polishing of the water.

The *Upper Ammonoosuc Falls* are 3½ M. from the Fabyan House, on the road to Mt. Washington. They are worthy of a long visit. The stream descends over rapids for some distance above, and then makes a fall of nearly 50 ft., through a narrow gorge whose walls are polished ledges of granite. Below the plunge it whirls in white and billowy masses through a sinuous chasm between massive cliffs. This portion of the stream may be visited by a circuit through the forest, and reveals a fine view of the falls above. The ledges near the stream, even at points high above its level, are indented by large potholes, 6 – 8 ft. in diameter, through some of which the transparent water still whirls.

The *Twin-River Farm* is so called because a tributary stream here flows parallel with the Ammonoosuc. It is nearly 500 ft. above the Fabyan House. Farther on, and close to Mt. Washington, is the station of *Marsh-field*, the headquarters of lumbermen during the winter and spring. There is a small hotel at this point, where visitors to the mountain often sojourn. Fine near views of Mt. Washington are obtained from various points along the line of this highway, as Starr King says: " A very noble view of Mt. Washington itself is gained by approaching near its base, on this area, and seeing it separated from the rest of the ridge." Good trout-fishing has been found on Jefferson Brook and other adjacent streams.

Mt. Deception is N. W. of the Fabyan House, which is overlooked by one of its low spurs, cut off from the main peak by a deep ravine. It is 2,449 ft. high, and is covered with dense woods. Harriet Martineau says that the name was derived " from its real being so much greater than its apparent height." An old legend claims that it was named by some people who ascended it under the belief that it was Mt. Washington. A line of rugged ridges extends E. from this point to the base of Mt. Jefferson, through an unbroken and unvisited wilderness, the summits being wooded and uninteresting. The first range E. of Deception is *Mt. Dart-mouth*, named probably from the town adjacent, which formerly bore that title. The range between Dartmouth and Jefferson is *Mt. Mitten* (called Mitten Hill on the map of 1791), so named because one of its early visitors lost a mitten there. So puerile is the origin of many of our mountain names !

Eleazar Rosebrook was one of the earliest of the pioneers, having removed in 1772 from Grafton, Mass., to Lancaster, N H., and thence to Colebrook. In 1792 he settled on Nash & Sawyer's Location, and built an extensive pile of mills, stables, etc., at the base of the Giant's Grave. Here he died, 25 years later, and was succeeded by his grandson, E. A. Crawford. Ethan Allen Crawford, " the White-Mountain Giant," is almost the only resident of the hill-country in whom any interest centres. He was born at Guildhall, Vermont, in 1792, and was carried to the mountains when a child, afterwards inheriting and occupying the house at the Giant's Grave. He was of large stature and powerful frame, and became famous for skill in hunting and woodcraft. His singular adventures with bears, deer, and wild-cats are even now remembered and chronicled. He was one of the first and best of the mountain-guides, and made the Crawford-House bridle-path and the first summit-house.

In 1803 the first public-house was erected here, and it was burnt in 1818, when occupied by Ethan Allen Crawford. Two other hotels on this site have since been destroyed by fire, helping to confirm (or perhaps giving rise to) the old tradition that an Indian once stood on the mound at night, waving a torch and crying, " No pale-face shall take deep root here ; this the Great Spirit whispered in my ear." Some time afterward a new hotel containing 100 rooms was erected on this site, and was kept by Mr. Fabyan, until its destruction by fire, about 20 years since. The present Fabyan House was erected in 1872–3, and its constructors committed a needless act of vandalism in levelling the mound of the Giant's Grave.

" We have heard the chants in Westminster Abbey, and the breath of the mighty organ towering up from its chancel like a little church, as it reverberated away among the arches, and along its interminable aisles. But we never heard mortal sounds to be named with *the echoes of Fabyan's tin horn !* " This famous horn was 6 ft. long, and was often sounded from the Giant's Grave.

55. The Twin-Mountain House

is on a little terrace N. of the Ammonoosuc River, and in the vicinity of a group of saw-mills. The house accommodates 300 guests, charging $ 4.50 a day. It is provided with post and telegraph offices, billiard and bowling rooms, and croquet-grounds; and keeps a band during the summer. It is equal to the Glen House or Bethlehem as a refuge for people affected by hay-fever, though it is much lower than either of those points. The house was built in 1869-70. The routes are the same as to the Fabyan House, so that passengers can leave Springfield, Montreal, or Burlington in the morning, and reach this house at afternoon

Distances (from the hotel-list). — From the Twin-Mountain House to the Fabyan House, 5 M.; the Profile House, 17; the Crawford House, 9; Mt. Willard, 11; Glen House, 30; Whitefield, 8; Lancaster, 16; Jefferson Hill, 11; Gorham, 28; Willey House, 12.

The view from the Twin-Mountain House includes on the l. the symmetrical knolls known as the Sugar Loaves (and sometimes locally called the Baby Twins), on whose r. is the massive Mt. Hale. Farther to the r., and more distant, is the North Twin Mountain, the South Twin being invisible. Then come the sharp cone of Haystack and the serrated ridge of Lafayette, towards the S. W.; and more to the W. are the low but conspicuous hills at Bethlehem, — Mt. Agassiz and Round Hill. The view to the E., N., and W. is completely obstructed by low wooded spurs of Cherry Mountain and Beech Hill. From the flag-staff mound in front of the house a pleasant prospect of the upper peaks of the Presidential Range may be gained, and a better one is found on the pasture-hill above the railway station. The only really striking panorama of mountains to be found in this vicinity is seen by ascending 2-3 M. on the Jefferson road, whence the Franconia Range appears in stately proportions.

The situation of the Twin-Mountain House was not well chosen, with reference to mountain-views, but its *cuisine* is justly celebrated; and to the usual amusements of a summer-hotel it adds the pastime of boating, on an adjacent mill-pond of considerable length.

The people about the house stoutly maintain that the two mountains nearly opposite are the Twin Mountains, but they are not, since the Twins lie in a line nearly N. and S., and the N. Twin only is visible, Mt. Hale being the other "Twin" (as regarded by the hotel-people). The Twin Mountains are seen from Mts. Washington and Lafayette, and other points E. or W. of their line, but not from the Twin-Mountain House. (See *Geology of New Hampshire,* Vol. I. p. 197.)

56. The Twin Mountains.

This ponderous and extensive range lies between the Ammonoosuc and the true Merrimac (East Branch), running at r. angles to both streams, and drained by Little River on the N. E. and the Franconia Branch on the W. It is also between the Field-Willey and Franconia Ranges, being 4-5 M. from the latter, and 6-7 M. from the former. The range is 8-10 M. long, running nearly N. and S. Prof. Hitchcock says that "scarcely any mountains are more difficult to reach than these, owing to the stunted growth near their tops." The group consists of the N. Twin, S. Twin, Mts. Guyot and Bond, and the adjacent ridges and foot-hills. The geologists maintain that the Twins were once continuous with the Field-Willey Range, and that the present valleys of Little and New-Zealand Rivers were excavated by atmospheric agencies.

The topography of this range is not understood nor described. The only map which ventures to be definite is Prof. Guyot's *Karte der Weissen Berge,* which places six peaks on the Twin ridge, of which the second from the N. is given a height of 4,700 ft. ; the third, 4,900; the fifth, 4,900 ; and the sixth, 4,800.

The Editor has not visited the Twin Mountains. Some of the following items were furnished him by Prof. Huntington, who has ascended them. Parties who intend to visit either of the Twins should advance to the head of Little River, and encamp, a short but wearisome day's journey. The next day should be devoted to the ascent and the return to camp ; and the hotel would be reached on the third day. Axes will be found of material service, and the clothing should be of the most sturdy character.

The **N. Twin Mountain** is 5,000 ft. high, and is the one which is visible from the Twin-Mountain House. On its N. E. side is a spire of coarse granite which rises about 150 ft. from the slope, and is called *The Nubble,* being conspicuous from several points in the Ammonoosuc Valley. The mountain-top is nearly covered with woods, and is therefore of but little interest to tourists. If the proprietors of the Twin-Mountain House would have a path cleared to this summit and an observatory of logs made upon it, they would confer a favor on their guests and on all lovers of alpine adventure. The mountain is recognizable from the peaks to the E. by a great white slide on its W. flank.

The **S. Twin Mountain** is of about the same height as its brother-peak, from which it is separated only by about 1 M. and a depression of 200 ft., a like slight descent intervening between it and Mt. Guyot. On the highest point of the ridge is an uplifted ledge, from which a view is obtained on all sides. This peak is reached from the Twin-Mountain House by ascending Little River and the branch which comes in from the W. near its head. The journey from the hotel to the foot of the mountain is 6 M., and it is about 2 M. thence to the summit. On the ascent and along the ridge are wide areas of dwarf spruces, 4 – 6 ft. high, through which it is almost impossible to force a passage, the most energetic progress being but slow and painful. Frequently the most rapid mode of advance is found by lying flat on the ground and crawling under the bristling boughs. But few persons have ever attempted the attack on the Twins, and most of them have been repulsed by these unflankable bayonet-lines of shrubbery.

The view from the Twins cannot be given, but we submit a list of peaks from which they are visible, as a basis on which visitors can work on backward lines of view. Mt. Willey is about E., across the wide New-Zealand valley, and the minor peaks of Field and Tom extend thence to the N. Over this ridge is the Presidential Range, high and round-topped Pleasant, flattened Clinton, rugged Monroe, and supreme Washington, with Clay, Jefferson, and Madison on the N. ; then Double-Head appears, more distant ; and S. of Willey is the red-crested group of Crawford,

Resolution, and Giant's Stairs.   To the r. of Crawford is the house-crowned Kiarsarge, with Langdon nearer and Moat on the r.   Far out in this direction is Mt. Pleasant, near Sebago Lake.   The Nancy Range is to the r , and nearer ; and Chocorua is visible from N. Twin, through the Carrigain Notch.   Mt. Carrigain lifts its immense bulk nearly S. E., with Hancock adjoining it on the W.   These two must shut out most of the Sandwich and Waterville peaks from the view ; but Osceola and Tecumseh appear on the r.   About S. W. is Mt. Flume, beyond which is the high plateau of Moosilauke ; and the Franconia Range extends thence to the N.

There is a profile in the *Geology of New Hampshire* which shows a fine view from N. Twin toward the W.   Beginning on the S. with the flank of S. Twin, the peak of Osceola is seen far away beyond ; and a considerable distance to the r., over the valley of the Franconia Branch, is the Big Coolidge Mt.   In the centre of that valley is the wilderness-ridge called Owl's Head, over which are the sharp peaks of Mts. Flume and Liberty, with Lincoln's pyramid on the r.   Then the high, thin, and serrated ridge of Lafayette is seen, about W S. W., on whose r., and much lower, is the bold conical crest of the Haystack.

On the N. the N. Twin is visible from Mts. Agassiz, Starr King, the Pilot Range, the Percy Peaks, and Dalton Mt.

The most southerly peak of the range has recently been named **Mt. Bond**, in honor of the late Prof. G. P. Bond, of Harvard University, who made the best and most accurate map of the White Mts.   The first peak to the N. of Mt. Bond has been named **Mt. Guyot**, in honor of Prof. A. Guyot, of Princeton, the eminent geographer, author of another map of the White Mts.   Mr. Warren Upham has visited these remote peaks, and has furnished the Editor with the following clear description: "I visited Mts. Bond and Guyot in 1871, for geological exploration.   The easiest access to them is from Mr. Pollard's, in Lincoln; and the journey is enough for a day. Follow the East Branch to Franconia Branch, then up the latter (on which there are interesting falls) to Red-Rock Brook (so named by Prof. Hitchcock because of abundant pebbles and boulders of red compact feldspar); follow up the Red-Rock Brook a mile or so, and you come to a little opening of swampy land *without trees*.   From this point we made our ascent, going up the western spur.   The ascent near the top we found quite difficult, because of the thick growth of black spruce; on the S. side this would be in part avoided.

"The name *Mt. Bond* I understand to be given to the southern peak of the whole range.   This is probably 300–400 ft. (appearing less, rather than more) lower than the highest of the Twin Mts.   At ½ M. N. from Mt. Bond is a depression of 75–100 ft.; and at ⅔–¾ M. N. (somewhat E. of N.) from Mt. Bond is the summit which Prof. Hitchcock understands to be *Mt. Guyot*.   This point is perhaps 100–125 ft. higher than Mt. Bond. Northward from this summit (Guyot) there is only a slight depression (not 50 ft. probably), and the ridge continues all the way to N. Twin of very nearly uniform height.   This feature of the Twin Range (a long, comparatively level-topped, high ridge) is well seen from Lafayette.   The summits of Mts. Bond and Guyot are destitute of forest, and have the characteristic alpine vegetation of our highest mountains.   The dwarf spruce in the hollow between these summits rises only 2–3 ft. (sometimes 4–5 ft.), and the route across may be selected so as to avoid this.   On the N. W. side, the granite of which Mt. Bond is composed presents a very noticeable precipice, almost vertical for 100 ft. or more. (This caused the name *Craggy Mt.* to be employed by us to distinguish this in our explorations.)

"The view is wholly of forest-covered mountains on every side, and I think no evidence is seen of the works of men, except the small settlement on the summit of Mt. Washington. From Mt. Bond especially the view of the whole valley of the East Branch is fine, bordered by irregular mountain ridges, sloping in varying curves to the stream. From this point we get the best acquaintance with the form of the range S. of the East Branch, and between this and the Hancock Branch; Mts. Carrigain and Hancock, the most conspicuous peaks, appearing as massive rounded summits, similar in form and height."

Mt. Hale is a fine wooded peak E. of Little River and S. of the Ammonoosuc. It is plainly seen from the hotel, and is usually called there "one of the Twins." The present name was given it by Prof. Hitchcock, in honor of the Rev. E. E. Hale, formerly one of the most active of the explorers among these mountains. N. of Mt. Hale are three bold granite hills called the *Sugar Loaves.* This group and the connected nameless peaks as far as the New-Zealand River are considered as a part of the Twin Range, separated from the main summits by the erosive action of Little River.

57. Bethlehem.

Hotels and Boarding-houses. The Sinclair House is at the intersection of the roads to the White and Franconia Mts. It has been greatly enlarged, and now accommodates over 200 guests, at $3.50 a day, and $15-20 a week. The situation of this house is good, and it commands broad views to the N. and N. E. The Maplewood House is about 1½ M. from the village, towards the White Mts., and was remodelled and enlarged during 1875-76. It accommodates 80 guests, at $3 a day, and $14-18 a week. This house is isolated from all others, and commands the Mt.-Washington range in a series of grand prospects. It is 1½ M. from the Bethlehem station. The Strawberry-Hill House is just W. of the Sinclair, and accommodates 75 guests, at $10-14 a week. The Bellevue is a new and high-studded house on the W. edge of the village, accommodating 60 guests, and having a broad view toward Littleton and Lancaster. The Prospect House is on the lowlands ½ M. N. of the village, and views the White Mts. It cares for over 80 guests, charging $7-10 a week. The Mt.-Agassiz House (65 guests) is near the Sinclair, on the Franconia road; the Avenue House (60 guests) is also in the village; and J. N. Turner's (50 guests) is towards the Maplewood. Among the other boarding-houses are those kept by Elisha Swett (40 guests), W. G. Simpson (35), W. G. Bunker (35), J. Plummer (28), S. H. Thayer (25), T. J. Spooner (20), W. H. Bean (18), H. W. Wilder, S. F. Winch, Myron Bailey, and many others. The rates at these houses vary from $6 to $10 a week, according to the location of rooms and the number in them. On Aug. 6th, 1875, there were 621 city boarders in this little village.

Distances. — Bethlehem to the summit of Mt. Washington, 22 M.; Crawford House, 17; N. Conway, 42; Bethlehem station, 3; Littleton, 5; Whitefield, 8; Lancaster, 15; Jefferson Hill, 15; Gorham, 32; Profile House, 10; Plymouth (by stage), 39; Crufts' Ledge, 2; Wallace Hill, 3½; Kimball Hill, 5; Sugar Hill, 7½.

Routes. — The usual way for passengers from the S. is by the Boston, Concord & Montreal R. R. and its Mt.-Washington Branch, to Bethlehem station, where stages from the hotels are in waiting. The morning trains from Boston, Springfield, etc., reach this point after mid-afternoon. Passengers from N. Conway to Bethlehem station go through on the Portland & Ogdensburg R. R., by the Notch and the Crawford and Fabyan House. Daily stages run between Bethlehem and Littleton; and to and from the Profile House.

The town of Bethlehem contains 998 inhabitants, who are settled in three neighborhoods, the Street (or Heights), the Bridge, and the Hollow. Along the course of the Ammonoosuc there are large lumber-mills, and the farms produce good crops of grain, potatoes, and hay. But the chief business is the care of summer-boarders, from whom there is an annual receipt of $160,000. The number of these who remained here for periods longer than a week was over 4,000, in the year 1872, and this number is steadily increasing. The village of Bethlehem Street is on a high plateau, 1,450 ft. above the sea, and 263 ft. above the adjacent Ammonoosuc Valley. It is said to be the highest village E. of the Rocky Mts.; it is probably the highest in New England. The Street is composed of a church, several shops, and a large cluster of boarding-houses and hotels, built on the N. side of an elevated ridge which rises higher on the S. and hides the Franconia Mts. It looks out to the N. and W. for many leagues, viewing the mountains of Lancaster, the Percy Peaks, the Pilot Mts., and the great Presidential Range, whose majestic summits are finely seen. The view of the White Mts. is broader and in some respects more imposing than that from N. Conway; though the beautiful environs of the latter village give it an overbalancing advantage, the surroundings of Bethlehem being comparatively uninteresting. The superior altitude of this ridge and its exposure to the N. renders it a very cool place during the summer, even when the other mountain-resorts are visited by intense heats. For the same reason people who are afflicted with hay-fever or rose-cold find immunity from their attacks in this high pure air. A village-improvement association is now at work to provide proper drainage and lighting for the streets, and otherwise to better the appearance of the village. There is but one church here, in which, however, clergymen of various sects officiate during the season.

"Bethlehem is about as far from Mt. Washington as N. Conway is, and lies on the opposite side. The drives in the neighborhood, commanding as they do, within short distances, both the Franconia and White-Mt. Notches, and the meadows of the Connecticut, are very varied and delightful. The town lies, also, at the favorable landscape-distance from the hills. . . . . No village commands so grand a panoramic view. The whole horizon is fretted with mountains." (STARR KING.)

Around the Heater is a favorite drive from which a succession of fine views is obtained. It passes out by Mt. Agassiz and around the adjacent heights to the Littleton road, on which the return is made. *Wallace Hill* is about half-way to Littleton, and is visited for the sake of the views therefrom. The *Montgomery Pond* is 5–6 M. N. of the village. The *Cherry-Valley Drive* is 4–5 M. long, leading N. by the Prospect House, then around by the roads to the l., and returning by the Littleton road. *Crufts' Ledge* is 2 M. from the village, and is reached by a path which enters the fields near the great barn beyond the Maplewood House. It commands a noble view of the White, Twin, and Franconia Mts., the latter being seen across the primitive forests in the valley of Gale River. An impressive view of the Franconia Range may also be obtained by ascending the Franconia road for about 1 M., to the crest of the ridge.

Bethlehem was formerly known as *Lord's Hill*, and was settled in 1790. A rude trail was cut to it from the Giant's Grave, and a log-bridge was thrown across the Ammonoosuc, which was soon succeeded by a timber-bridge. Famine frowned on the settlement frequently, and in 1799 the people were obliged to make a load of potash and send it to Concord, Mass. (170 M.), for sale, living on roots and plants until their envoy returned with provisions, four weeks later.

During his passage through Bethlehem, in 1803, President Dwight found only log-huts, "recent, few, poor, and planted on a soil singularly rough and rocky. There is nothing in Bethlehem which merits notice, except the patience, enterprise, and hardihood of the settlers, which have induced them to venture and stay upon so forbidding a spot; a magnificent prospect of the White Mts.; and a splendid collection of other mountains in their neighborhood, particularly on the S. W. . . . . A nobler group cannot be imagined than that, which is seen from Bethlehem. . . . . Their form, their extent, their height, their position, and all the circumstances of their appearance, are so varied through the several gradations of beauty, boldness, and splendor; and so happily related to each other; that the mind finds, here, everything, which can gratify its wishes in rude, wild, and magnificent scenery."

58. Mount Agassiz.

Mt. Agassiz (formerly known as *Peaked Hill*) is near and S. E. of Bethlehem, and affords a very favorable point of view over the White, Franconia, and Green Mts. The visitor ascends the long hill on the Franconia road (running S. from Bethlehem), and on the crest of the slope, about 1 M. from the village, enters a path which crosses the field to the l. This route leads through low second-growth thickets and over white ledges to the summit, which is marked by a beacon of the U. S. Coast Survey. The path is about ¾ M. long, and is well cleared and of easy grade. The peak is 2,042 ft. high; and a carriage-road is now being built to the summit. The present name of the mountain was given in honor of Prof. Louis Agassiz, whose researches on the glacial remains in this vicinity were of great value.

Prof. Agassiz was much interested in the glacial remains about Bethlehem, to which he devoted careful study. "The Littleton road from Bethlehem, and the roads to Franconia Notch from both these towns, frequently intersect terminal moraines. Those familiar with the topography of the Franconia Range, and its relation to Picket Hill [now Mt. Agassiz] and the slope of Bethlehem, will at once perceive that the glaciers which deposited the front moraine to the N. of Bethlehem Village must have filled the Valley of Franconia to and above the level of the saddle of Picket Hill, making it 1,500 ft. thick, if not more, — thicker, in short, than any of the present glaciers of Switzerland. It will be observed, also, that as soon as the N. portion of that glacier had retreated to the wall which encircles the Franconia Valley on the N., the glacier, occupying henceforth a more protected valley within the range, must have made a halt, and accumulated at this point, that is, S. and W. of the saddle of Picket Hill, a very large terminal moraine. The finest lateral moraines in these regions may be seen along the hillsides flanking the bed of the S. Branch of the Ammonoosuc, N. of the village of Franconia. The best median moraines are to the E. of Picket Hill and Round Hill. The latter moraines were formed by the confluence of the glacier which occupied the depression between the Haystack and Mt. Lafayette, and that which descended from the N. face of Lafayette itself. . . . The lane starting from Bethlehem Street, following the cemetery for a short distance, and hence trending N , cuts sixteen terminal moraines in a tract of about 2 M. Some of these moraines are as distinct as any I know in Switzerland." (*Amer. Assoc Advan. Sci., Proc. XIX*)

The View. — About 7 M. distant, across the densely wooded valley of Gale River, a little E. of S., is the thin and craggy crest of Mt. Lafayette, whose immense spurs and foot-hills run far out over the wilderness. On its r. flank is the rugged mass of Eagle Cliff, with the deep gorge of the Franconia Notch adjoining, and apparently blocked up by the low Bald Mt. Next comes the massive and round-topped ridge of Mt. Cannon, flanked by the sharp and rolling summit of Kinsman. Nearly S. S. W. up the long Landaff valley is the high top of Moosilauke; and on the r., much nearer, are Ore Hill and Sugar Hill. More to the W. is the long and monotonous range of Mt. Gardner, which occupies parts of four townships. Far beyond, along the horizon, is a line of blue peaks in Vermont, extending for scores of leagues down the Green-Mt. range. Among these Camel's Hump is seen, nearly W. N. W., and the high plateau of Mt. Mansfield is more to the r., with Mt. Elmore apparently adjoining it. In this direction, but close at hand below, are the hotels of Bethlehem, beyond and above which is the larger village of Littleton, with the high school and the Oak-Hill House on the heights. The Scythe-Factory Village runs to the E. from Littleton along the Ammonoosuc, and Mann's Hills are seen beyond, to the r., over which is a distant mountain which may be Jay Peak. About due N., across the Ammonoosuc Valley, is Dalton Mt., at whose foot is a bright lake shining among the forests. Portions of the great ranges towards Willoughby Lake are seen far beyond, with the mountains near the Connecticut River.

The view now passes over the plains of Whitefield, bordered on the N. by the Martin-Meadow Hills, and the round summits of Mts. Pleasant and Prospect. Over the saddle between the latter two runs the Lancaster highway; and Cape Horn is seen beyond. Nearer at hand is Kimball Hill, in Whitefield; and over the r. flank of Prospect are the two white domes of the Percy Peaks, backed by the long Stratford and Bowback Mts. The blue lines of the Pilot Mts. extend to the r.; and towards the N. E. is the white village of Jefferson Hill, at the foot of Mt. Starr King. The black mass of Cherry Mt. is more to the r. and much nearer, and fills a long section of the horizon. Then, nearly E. N. E., the view rests on the sharp and lofty pyramids of Mts. Jefferson and Adams, flanked by the rough ridge of Clay; and then the imposing crest of Washington appears, with portions of the railroad, the white station-buildings, and the Summit House. To the E., down the long Ammonoosuc Valley, is the Twin-Mountain House, while beyond rise the peaks which run from Mt. Washington to the Notch, Monroe, Franklin, Pleasant, and Clinton. Then come the huge and far-extending Twin Mts., 6-8 M. distant, and not far from S. E. Close to Agassiz is the wooded top of Round Hill, over which appears the boldly cut peak of the Haystack, resting to the r. on Lafayette.

59. Whitefield.

Hotels. — The old Whitefield House has been closed. The Mossy-Brook House (Ira M. Aldrich) is a large summer boarding-house pleasantly situated on the edge of the village, accommodating 35 guests, and charging $ 7 – 10 a week.

Distances (from Walling's odometer-surveys). — Whitefield to Dalton, 6½ M. ; Jefferson Hill, 11 ; Lancaster, 8 ; Northumberland, 14 ; Gorham, 26. Distances by railroad, — to Lancaster, 10 M. ; Groveton, 20 ; Littleton, 12 ; Fabyan House, 19 ; Crawford House, 23 ; summit of Mt. Washington, 28. Distances by roads, — to Bethlehem, 8 ; the Profile House, 18 ; the Twin-Mountain House, 8.

The town of Whitefield lies to the N. W. of the White Mts., and is traversed by John's River. Its surface is hilly, and the land is valuable for farming purposes. The growth of the town has been rapid, since in 1810 it had but 51 inhabitants, while in 1870 there were 1,196, and the number is now much greater. The village of Whitefield is near the railroad station, and consists of two parts, the section about the lumber-mills and the houses of the workmen being close to the station and river, while the village proper is on the plateau beyond and is not seen from the railroad. In the centre is a public square, from which diverge the highways. The street leading to the S. E. has several large villas upon it, among which is the Mossy-Brook House; and pleasant mountain-views are enjoyed from other points on the plateau. The village has about 700 inhabitants, three churches, and a large graded school. Of the various manufactures carried on at this point, the lumber-business is the most important, and the Brown Brothers have the chief mills.

The immense lumber-mills of Messrs. A. L. & W. G. Brown are located at this point, and have a railroad 12–15 M. long running through their forests, up the John's River and towards Cherry Pond. This firm owns 30,000 acres of spruce and pine woodland, most of which is in Carroll, Jefferson, and Lowe & Burbank's Grant, and employ nearly 300 men, many of whom are French Canadians. 12,000,000 ft. of logs have been cut and sent out from their camps in a single year, being drawn on long cars and unloaded into the pond above the mills. The mills are on the John's River, at the Whitefield station, and are of great extent. They are run by steam, and lit by gas, and are supplied with water by a long and costly aqueduct. The danger of fire is averted by stand-pipes and by a large steam fire-engine, with a disciplined fire-department. The railroad has two locomotives, and in some places rises 215 ft. to the mile. It is intended to prolong it to Jefferson Hill. Estimates have been made to show that it will take 40 years to clear the timber from the company's lands, by which time the tracts which are now being stripped will have been covered with new forests.

The *Mountain-View House* (kept by Wm. F. Dodge) and the *Cherry-Mountain House* (J. S. Fitch) are 3½ – 4 M. from the Whitefield station, 6 M. from Lancaster, 4 M. from Mt. Prospect, 10 M. from Jefferson Hill, 10 M. from Bethlehem, 2½ M. from Bray Hill, and 12 M. from the base of Cherry Mt. They accommodate respectively 50 and 30 guests, and charge $ 7 – 10 a week. These houses occupy positions on a plateau which slopes downward towards the mountains, and command grand views. They are far isolated from other houses, and are consequently very quiet and retired. At Dodge's the Rev. Dr. E. S. Gannett spent the last summer of his saintly life. The Rev. Dr. Waterston, of Boston, owns an estate near this house, which he has highly improved by skilful landscape-gardening.

The **Howland Observatory** is situated on Kimball Hill, which is distant from Lancaster, 10 M.; from Bethlehem, 5 M.; from Littleton, 7 M.; from the Twin-Mountain House, 10 M.; and from the Sumner House, 8 M. It is 70 ft. high, and the upper platform rises above the tops of the tall birch-trees which surround it. The basement has a refreshment-stand, and the top is provided with a telescope. The entrance-fee is 25c. The observatory is about 2 M. S. of Whitefield, and may be reached either by the road which curves around it and ascends on the S., or by the nearer road out from the village. In either case the visitor can drive over the summit of the hill, and near the tower.

* The View. — The ponderous mass of Cherry Mt. is comparatively near the observer, and extends over most of the horizon between E. and S. E. On its l. is the peak of Owl's Head, and over the long r. flank is the bare crest of Jefferson. The S. ridge of Cherry is prolonged for miles, and over it are seen the clear peak of Washington, the broken outlines of Monroe, the unrecognizable flat of Franklin, and the hemisphere of Pleasant, forming a continuous ridge. Beyond the end of the ridge of Cherry, on the r., a vista opens up the Ammonoosuc Valley, over the plains of Carroll, and terminated by the low sharp peak of Jackson, and the flattened top of Webster, which falls off sharply to the r. into the White-Mt. Notch. About S. S. E., and a few miles distant, is the wooded mass of Beech Hill, over which appears Mt. Hale, beyond the broad clearings near at hand. Over the r. flank of Beech is the well-marked peak of the N. Twin Mt., rising high over the horizon, with a part of the S. Twin beyond, over the farms and fields in the foreground. The bold mass of Haystack next appears, and extends on the r. to the high sharp peak of Lafayette, whose minor N. crags appear in front. The mountain sweeps away to the r. in graceful falling lines, broken by Eagle Cliff, on whose r. is the cleft of the Franconia Notch, with Mt. Cannon on the r. over the clearings towards the S. Still more to the W. the sharp crest of Kinsman is seen, dropping away into the glens of Landaff. Farther to the W. are the hills of Bethlehem and Landaff, with Mt. Agassiz conspicuously shown, and the Benton Notch far beyond.

About E. is the clear sheet of Burns Lake (Long Pond), over which are the high hills of Littleton, with portions of the Green Mts. of Vermont beyond; and then Dalton Mt. rises on the N. W., across the narrow valley, with its long wooded flanks, beyond which, on the r., are Mts. Tug, Cow, and Burnside, in Lunenburg and Guildhall. Over the adjacent road are the rounded and connected hills of Mts. Orne, Pleasant, and Prospect, near Lancaster. A road runs toward Prospect, on whose r. are seen the crest of Cape Horn and the white Percy Peaks, flanked on the r. by Long Mt., in Odell. The great assemblage of the Pilot Mts. then comes into

view, extending along the horizon for miles, with a massive line of dark
peaks. Nearly N. E. is the hamlet of Jefferson Hill, over which rises Mt.
Starr King. The space between this point and Cherry Mt. is filled with
the dark ridges of the Pliny, Randolph, and Crescent Ranges, forming a
great undulating forest amid which rise dull heights and characterless
crests innumerable.

60. Dalton.

Hotel. — The Sumner House (75 guests) is a summer-resort near the
hamlet, standing on a bluff by the river, and provided with extensive
grounds, boats, billiards, bowling, croquet, and a livery-stable. The
rooms are remarkably high and airy, and a pleasant view is enjoyed to
the N. E., including the Pilot Mts. and the Percy Peaks. The hotel opens
on June 20th, and its rates are $ 12 – 18 a week. Mr. Prime speaks of
this house as " beautifully situated on the bank of the Connecticut. It
looked like a pleasant and quiet place to do summer loitering." The
favorite drives are over the river to the Lunenburg Heights (see Route 62),
to Whitefield and Lancaster, to Mt. Prospect, and to Dalton Mt. The
Fifteen-Mile Falls flow along most of the N. W. boundary of the town,
and are a chain of wild rapids in a narrow valley which has been excavated through the Mt. Gardner Range. Between Dalton and the
foot of McIndoes Falls the distance is 24 M., and the river falls nearly
400 ft.

The hamlet of Dalton consists of but a few houses, and is situated on
the Connecticut River, at the head of the Fifteen-Mile Falls, 8 M. from
Lancaster and 6 M. from Whitefield. It is 1½ M. from Dalton station, on
the B., C. & M. R. R. (by a pleasant forest-road), and less than ¼ M. from
the station of the P. & O. R. R., where that line crosses the Connecticut
River.

Dalton is a hilly town, with fertile uplands and a long front on the Connecticut
River. It has 773 inhabitants, and produces annually 3,500,000 ft. of lumber, 125
tons of starch, 9 tons of maple sugar, 2,292 tons of hay, 12 tons of butter, 50,000
bushels of potatoes, and 6,500 bushels of corn, oats, and wheat. It contains 16,455
acres, of which 6,913 are cultivated. The town was settled during the Revolutionary
era, and was incorporated in 1784 and named after Hon. Tristram Dalton, one of the
grantees, the earlier name having been *Apthorp.*

61. Dalton Mountain.

Dalton Mt. is a long and lofty wooded ridge, running across the town of
Dalton from S. W. to N. E. Its main peak has been cleared of trees, for
the convenience of visitors, and commands a broad view. It is 3½ M. from
Dalton village, and is reached by the Round-Pond road, passing the
Ruggles farm. Where this road reaches the crest of the high pass, a disused road is seen diverging to the l. through the woods. The ascent along
this road and the connecting path is about ¼ M. long, and must be made

on foot, the cleared peak being generally visible in advance. The upper slopes are seamed by deep sheep-paths, but the road is unmistakable, and soon leads to the bare and rocky summit. The road from the Sumner House to the peak is an almost continuous ascent, but the excursion up and back may easily be made in half a day. It is 4 M. from the peak to Whitefield village, and the road affords a series of fine views after the forest has been passed.

The View. — A high wooded spur of Dalton Mt. shuts out the view to the N. E. On its r. is seen the great group of the Pilot Mts., with their clustering blue peaks and desolate ravines, terminating, on the r., at Mt. Starr King, at whose base is the village of Jefferson Hill, nearly over Bray Hill, and beyond the valley of Israel's River. The Pliny and Randolph Ranges are next seen, over Cherry Pond; and the Randolph Notch is nearly due E. The long dark line of Cherry Mt. rises beyond the wooded plains of Whitefield, with the Owl's-Head peak nearly over the village; and beyond, with their lower slopes concealed by Cherry, are the stately White Mts., with the rounded top of Madison on the N., flanked by the immense pyramids of Adams and Jefferson, and the stately cone of Washington, distinguished both by its altitude and the white house upon the crest. To the r. are the craggy peaks of Monroe, the level top of Franklin, the high curves of Pleasant, and the rounding swell of Clinton. Beyond the succeeding summits the range drops off into the White-Mt. Notch, which is nearly S. E., across the valley of Carroll. On the r. is a portion of the Field-Willey Range, with Mt. Hale on the r.; and then the massive Twin Mts. and Mts. Guyot and Bond are seen, beyond Beech Hill, filling a great area with their dark and monotonous crests. Next comes the sharp peak of the Haystack, which is flanked on the W. by the lofty sierra of Lafayette, gray and rugged in the distance. The village of Bethlehem is nearly S., with Mt. Agassiz just over it, far back of which is the Franconia Notch. The lower plains and forests of Whitefield are in the foreground, lighted by the bright waters of Burns Lake and Round Pond. On the r. of the Franconia Notch is Mt. Cannon, flanked by the long ridge of Mt. Kinsman, which drops away into the glens of Landaff. Far down in this direction, about S. S. W., is the high crest of Moosilauke, with the Benton Black Mt. on the r. The highlands of Littleton and Lisbon are nearer at hand, in the S. W.; and the long Mt.-Gardner range hides the Connecticut River, Bald Ledge being about W. S. W. The view between S. W. and N. includes a vast area of Eastern Vermont, with bold lines of mountains rising all along the horizon, marked by stately blue peaks. The highlands just across the river, to the W., are in Concord and Waterford; and it is claimed that Jay Peak and Owl's Head are visible in the N. W. Perhaps it will be as interesting to leave these long ranks of the Green Mts. undesignated in this catalogue, allowing a wide field for imaginative work. The Editor ascended Dalton Mt. on a hazy day, with an ignorant guide, and was at that time comparatively unfamiliar with the Vermont mountains. He was unable to reach this point again, and now prefers silence to doubtful hypotheses about the view.

62. Lancaster.

Hotels. — The Lancaster House is a large and well-built hotel at the church end of the village, pleasantly located among the trees. It accommodates 150 guests, charging $ 2.50 - 3 a day, or $ 8 - 16 a week. The Williams Mansion, low down in the village, and the Hillside Cottage, on the slopes beyond, are summer boarding-houses, accommodating about 40 guests each, at $ 7 - 10 a week. The American House is not recommended. Small boarding-houses are kept by several of the villagers.

Railroads. — Lancaster is on the Boston, Concord & Montreal R. R. (Route 2), and may be reached from Boston in 8½ hrs. ride. It is but 10 M. from Groveton Junction, where connection is made with trains on the Grand Trunk Railway (Route 7) for Montreal and Quebec, Gorham and Portland. About the same distance to the S. is the junction of the Portland & Ogdensburg R. R., whereon the tourist can pass W. to Lake Champlain, or E. to the Notch and N. Conway.

Stages run from Lancaster to Jefferson Hill on the arrival of the mail-trains from the S., and hotel-coaches often at other times. Glen-House or Gorham stages frequently return empty from Lancaster over the Jefferson-Hill road, and will take return-fares, if possible. The Lancaster House has a livery-stable.

Distances by road (from Walling's odometer surveys). — Lancaster to Dalton, 8¾ M.; Whitefield, 8; Jefferson Mills, 3⅔; Jefferson Hill, 6⅔; Gorham, 24; Groveton, 10; Lost Nation, 5⅔. Northumberland, 6; N. Stratford, 22. Distance to Lunenburg Heights, 8 M.; to Bray Hill, 8 M.; to the top of Mt Prospect, 2 M.

Lancaster is the largest, and in many respects the most beautiful, of the villages near the White Mts. The town contains 2,248 inhabitants, most of whom are in the village, although the rural roads which interlace the plain in every direction are lined with farms. It is the best town in Coös County, and the twelfth in the State, in the value of its agricultural products; and the intervales at this point are the most extensive on the Connecticut River. They extend back for nearly 1 M., and on their rich alluvial soil grow large crops of corn, oats, and hay. The uplands are prolific in wheat. The climate is mild and delightful, the E. winds being turned by the mountain-wall on the E. Lancaster is one of the shire-towns of Coös County.

The village is on Israel's River, not far from its confluence with the Connecticut, and occupies a level plain bordered by low hills. It contains six churches (Cong., Meth., Bapt., Epis., Unit., Cath.), 20 - 30 stores, an academy, a public library, a savings-bank, and graded schools; and has two weekly papers (Republican and Democratic). The Masons and Odd Fellows have large societies here. Among the articles manufactured are various kinds of lumber, paper, potato starch, carriages, sashes and blinds, harnesses, etc., to the value of $ 403,000 annually. Main Street is the chief thoroughfare, on which are the shops and several lawyers' offices. At the N. end of the village is the ecclesiastical quarter, where five churches are clustered together, near the county-buildings. The Episcopal church is opposite the Lancaster House, and was built in 1875. The population of the village is about 1,500. The influx of the farmers from the adjacent rural districts imparts an air of bustle and activity to the village, which supplies an extensive district of Northern New Hampshire. The chief streets are lined with fine old trees, which give an embowered appearance to the houses. The neatness and cleanly appearance of the village also render it pleasant as a summer-home. It is 870 ft. above the sea, or 350 ft. higher than N. Conway.

Israel's River is crossed by two bridges in the village, and affords a good water-power. This stream rises near the base of Mt. Adams, and flows N. W. for 15 M. Israel and John Glines were two noted beaver-trappers of the last century, and Israel's and John's Rivers were named in their honor; though Capt. Powers, the first explorer of this region (in 1754), named the first *Powers's River* and the second *Stark's River*, — after Ensign Stark, who was captured by the Indians hard by. It would have been better to have retained the old and melodious Indian name of Israel's River, *Singrawack*, meaning " the Foaming Stream of the White Rock." Sir Charles Dilke says that " the world can show few scenes more winning than Israel's-River valley, in the White Mts. of N. H., or N. Conway, in the S. slopes of the same chain."

The Connecticut River runs for 10 M. along the W. and N. W. borders of Lancaster, with a width of 20 – 30 rods. 6 – 8 M. distant, over pleasant meadow-roads, is the commencement of the Fifteen-Mile Falls.

That Lancaster has not become more famous as a summer-resort is due to the fact that its citizens have not prepared sufficient boarding-house accommodations for tourists. Several new houses should be opened (not in the village but) on the line of heights to the S. E., whence the mountains are visible. The rich and fruitful meadows, the bright rivers, the peaceful and prosperous village, and the distant environing mountains form elements whose ever-varying combinations produce the most beautiful scenic effects. The good roads which run out on every side give a succession of pleasant drives; and the railroad connections favor longer excursions. The society of the village is cultured and refined, being largely formed by the members of the three learned professions. There are five private libraries of over 1,000 volumes each, and one of 4,000 volumes.

The view from the large cupola of the Lancaster House is one of the best to be gained from the village. It includes the imposing forms of the White Mts. on the S. E., 16 – 20 M. distant, yet plainly seen up the deep valley of Israel's River. A few miles to the S. are Mts. Prospect and Pleasant and the Martin-Meadow Hills, with their graceful round tops, and then the view passes to the r. over the Lunenburg Heights and beyond the broad green meadows. All along the W. are the mountains of Vermont; and to the N. are Cape Horn, the weird white Percy Peaks, and Stratford Mt. Toward the E. is the long wall of the Pilot Mts., rising from the Lancaster valley for leagues, and famous for its exquisite dolphin-like changing tints at sunset. One of the chief charms of a sojourn at Lancaster is the superb effect of the evening light on these mountains.

Broad views are afforded from *Bunker Hill* and others of the low eminences which surround Lancaster. A much-admired prospect is obtained near the bridge over the Connecticut, where a rich expanse of meadows extends toward the N., terminated by Cape Horn and the Percy Peaks and flanked on the r. by the Pilot Mts. This is a pleasant objective point for a short evening stroll. A favorite drive is along the high ridge of *Stebbins Hill*, on the flank of Mt. Pleasant, whence a succession of broad views is obtained to the N. and E. The drive around Mt. Prospect is also much admired. It leads S. through the pass between Pleasant and Prospect, and then passes around the latter to the E. and N., opening wide and noble vistas towards the Presidential Range, and finally coming out on the Jefferson road, whence the route is to the N. W. into Lancaster.

Lunenburg Heights are beyond the Connecticut River, in one of the picturesque hill-towns of Vermont. They are easily reached by the road which runs out to the S. W. from Lancaster and crosses the river on a covered bridge. Pleasant views are given of the serpentine course of the stream, whose annual inundations enrich the broad and fruitful intervales on either side. The road ascends the heights by easy stages, traversing a broad belt of forest, until it reaches a road which runs along the upland, 1,200 ft. above the river. Beautiful prospects of the mountains are now obtained, the best of which is from the Smith farm, 8 M. from Lancaster. To the N. E. are the white Percy Peaks, towering over the Connecticut Valley, beyond which are the blue ranges of the upper townships. Somewhat nearer, and in the same direction, is Cape Horn, near Groveton, from which the imposing and unbroken line of the Pilot Mts. sweeps around to the E., overlapping the Starr-King group. In the S. E. are the bold blue masses of the Presidential Range, strongly defined and lifted well N. of the long, dark ridge of Cherry Mt. Over the S. slope of the latter is the low crest of Mt. Willard, at the White-Mt. Notch, to the r. of which are Mt. Hale and the Twin-Mt. group. Beyond the massive cone of Haystack, and nearly due S., is the pyramid of Mt. Lafayette. W. of S. are several very distant peaks, the chief of which is said to be Moosilauke, beyond Mt. Agassiz, in Bethlehem. In the foreground of this vast semicircle of mountains are the valleys of the Connecticut, the Ammonoosuc, Israel's and John's Rivers, with the villages of Lancaster and Whitefield. The rich intervales present a pleasing view, extending for leagues N. E. and S. W., and dotted with clumps of graceful trees.

The return drive to Lancaster is often made by way of the quiet little hamlet of **Lunenburg** (*Chandler House*), high above the valley. This place is about 1 M. from the S.-Lancaster station, and has several houses in which summer-boarders are taken (J. M. Lucas, Levi Barnard, Chester Dodge, etc.).

"Grand combinations, too, of the river and its meadows with the Franconia range and the vast White-Mt. wall, are to be had in short drives beyond the river, upon the Lunenburg Hills . . . . where the bright blue of the river and the embowered homes of the village are set in the relief and under the protection of the long White-Mt. wall, tinged with the violet of departing day." (STARR KING.)

Mount Prospect is a round-topped summit about 2 M. S. of Lancaster and the easternmost of a line of similar eminences which runs W. to the Connecticut River. It is separated from Mt. Pleasant by the high pass which is traversed by the Whitefield road. There is a comfortable path leading up the mountain from the old Legro farm, which is reached by the road diverging to the l. from the Whitefield road, nearly 2 M. S. of Lancaster. The summit of Prospect is partly cleared, and commands a broad and noble mountain panorama.

* *The View.* — Lancaster is plainly seen, on the N., on whose l. and beyond extends the broad Connecticut Valley, with the winding river, the fair intervales, and hundreds of farms. The massive bluff of Cape Horn appears, farther to the N., with the dark ridges of Bowback Mt. over it. The white crests of the Percy Peaks are about N. N. E., on whose l. and r. are the distant peaks of the upper townships. Across the valley of Israel's River, from N. E. to E., extends the long line of the Pilot Mts., within a few miles and nobly conspicuous. Mt. Starr King closes their ranks on the r. (E.), and the village of Jefferson Hill is seen at its foot, with the Pliny and Randolph Mts. beyond. About S. E., up the Israel's River valley, is the lofty peak of Mt. Washington, with its hotel and railroad. On its l. are Mts. Clay, Jefferson, Adams, and Madison; and on the r. are Mts. Monroe, Franklin, Pleasant, and Clinton. This is one of the best possible panoramas of the Presidential Range, because the view-line from Mt. Prospect meets its line near the centre, at right angles. The great ravines which deeply indent the chief peaks on the W. are plainly seen, with the formidable cliffs at their heads. King's Ravine is deeply sunken into the side of Mt. Adams, whose three peaks rise airily over its shadowy recesses. On its l. is the Madison Ravine, and the Ravine of the Castles is on the r. The massive ruins of the Castellated Ridge run out from the flank of Jefferson, hiding the deep gorges beyond. Mt. Prospect is one of the most advantageous points from which to reconnoitre the Presidential Range, notwithstanding its distance, and it is to be hoped that the long-promised carriage-road to its summit may soon be built. The latter part of the afternoon is the best time to enjoy this superb view. Between Prospect and the Presidential Range are the low wooded peaks of Mitten, Dartmouth, and Deception. The bold ridge of Cherry Mt. is S. S. E., hiding the Notch, and overlooking the plains of Whitefield; and the Twin Mts. are nearly S. Farther to the r. are the stately peaks of Haystack, Lafayette, Cannon, and others of the Franconia Mts.; with Mt. Agassiz and Bethlehem village nearer and more to the r. Still more distant, nearly over Bethlehem, are the blue heights of Moosilauke, looming up over the glens of Benton. The narrow basin of the Ammonoosuc Valley opens away to the S. W., with the hills of Littleton and Mt. Gardner enclosing its lower section, and the Vermont highlands beyond. Nearly S. E , across the wooded valley of John's River, is the long ridge of Dalton

Mt., darkened by great forests. On the r. of this range is a section of the Connecticut Valley, with the heights of Concord and Waterford on its farther side. Several of the high crests of the Green Mts. are seen beyond.

To the W. is the wooded crest of Mt. Pleasant, which is close to Prospect, and is separated from it by a high pass, through which the Whitefield road is carried. Beyond and adjoining Pleasant are the dull slopes of the Martin-Meadow Hills, falling off on the W. to the Connecticut Valley. Across this broad depression are the farm-strewn terraces of the Lunenburg Heights, with the Victory Hills looking over them, and parts of still more distant peaks. The Westmore Haystack and several mountains toward Willoughby Lake are farther away, in the N. W., towards the dark and picturesque Lowell Mts. More to the r., and comparatively near, is the idyllic village of Lancaster, resting on the borders of rich and fruitful meadows. The highlands of Guildhall are over and beyond Lancaster.

Lost Nation is a prosperous farming neighborhood, 5¾ M. N. E. of Lancaster, at the base of the Pilot Range. It is in the town of Northumberland, but so far remote from the chief village thereof that but little account was made of it, and the people were forced to go around through Lancaster when they wished to visit the riverward parts of their town. The name of *Lost Nation* was attached to the neighborhood at an early date. A short distance beyond is *New France*, a farming district inhabited by French Canadians, surrounding the estate of Father Isidore Noiseaux, the priest of Lancaster.

"The drives about Lancaster for interest and beauty cannot be surpassed. . . . . There is no single meadow view in Lancaster equal to the intervale of N. Conway. But the river is incomparably superior to the Saco ; and in the combined charm, for walks or rides, of meadow and river, — the charm not of wildness, such as the darker and more rapid Androscoggin gives, but a cheerful brightness and beneficence — Lancaster is unrivalled." (STARR KING.)

The Lancaster meadows were the camp-ground of Capt. Powers's rangers, who advanced hitherward in 1754, searching for hostile Indians, whom having found, they speedily retreated to the Merrimac Valley. This region was known as the Upper Coös, Haverhill and Newbury being the Lower Coös, and Colebrook "the Coös above the Upper Coös." The period between its first exploration and the settlement was short, since it was granted in 1763 and in 1764 it was occupied by a group of families from Petersham, Mass. In 1775 most of the people abandoned the town, in fear of Indian incursions, but the Stockwells fortified and stood their ground. In 1790 there were but 161 inhabitants here. The town was named after Lancaster, Mass., whence several of its early settlers came.

Early in the present century, Lancaster found its principal market at Portland, where the produce was carried in wagon-trains, the rates of freight being $1.00 a hundred-weight in winter, and $2.00 in summer. At this time large amounts of whiskey were here distilled from grain and potatoes, and sent to Portland.

63. The Pilot Mountains.

The Pilot Range is about 16 M. long and 13 M. wide, occupying portions of the townships of Stark, Kilkenny, Jefferson, and Berlin. It consists of a great congregation of low mountains, covered, for the most part, with forests, and diversified by several high, bold peaks. The range presents an especially interesting appearance from the Connecticut Valley on account of the abruptness with which it rises from the lowlands, and also by reason of its massive and wall-like aspect. The prevailing stone is porphyrite, of a dark brown or green hue when freshly fractured and weathering to a whitish color. Although occupying such a commanding position with reference to the populous towns of the upper Connecticut Valley, the Pilot Range has never been thoroughly explored, and parts of it have never been visited. The forests are dense and continuous, but the practised woodsman can gain much assistance from the rude logging-roads which follow the ravines. The upper sections of the brooks in this region contain many small trout, but can only be effectually fished by parties in camp.

In earlier times the Pilot Hills were also called *Little Moosehillock*, since Dwight locates a range so named as just S. of the Upper Ammonoosuc ; ending on the W. in

Northumberland; bounding the plains of Lancaster and Jefferson on the N ; 20
M long; and overlooking the town of Durand (Randolph). The town of Kilkenny
is covered with these peaks, and is almost uninhabited. A half-century ago John
Farmer said of the people of Kilkenny: "They are poor, and for aught that appears
to the contrary, must always remain so, as they may be deemed actual trespassers
on that part of creation, destined by its author for the residence of bears, wolves,
moose, and other animals of the forest."

" A range of mountains lying to the N. E., in the townships of Northumberland,
Percy, and Kilkenny, are distinctly seen from this place (Lancaster), and form the
background of a very beautiful picture. They formerly served to guide the hunters
to Connecticut River, and are called ' land pilot hills.' " (*Mass. Hist Colls.*, 1815.)
The popular tradition avers that these mountains were anciently the resort of a
hunter and trapper named Willard, who on one occasion got lost among their defiles,
and wandered vainly for days, until he was nearly starved. At a certain time every
day his dog left him and remained away for a short space; and finally, when Willard
had lost all hope, he concluded to expend the last remnants of his strength in following the dog. The intelligent animal soon led him out to his well-provisioned
camp beyond the hills. Thenceforward the dog was called *Pilot*, and the great range
of hills thus came to be named in his honor.

The chief summit of the Pilot Range is **Mt. Pilot**, which reaches the height of
3,640 ft. It is reached from Lancaster by driving out to Lost Nation (see page 175)
and turning to the r. from the main road where it makes a sharp turn near the district school-house, to the old Hutchins farm, now occupied by Mr. Cummings.
There was formerly a path from this point to the summit, but it is now obliterated
by the growth of shrubbery. The visitor should take a guide, unless he is an experienced woodsman, for the distance is about 4 M., and the route is not direct. The
remains of old logging-roads aid one in reaching a point far up on the main ridge.
The summit is nearly bare, and commands a wide extent of country to the S.

The View. — About S. S. E. are the commanding forms of the Presidential Range, Madison on the l., the spire of Adams in the centre, and
the massive bulwark of Jefferson on the r., over whose flank is the higher
range and culminating peak of Washington. Farther around to the S. W.
are portions of the line of peaks which sweep from the chief summit
toward the Crawford House. A portion of the Mt. Willey Range is also
seen down the White-Mt. Notch, beyond which are certain of the remoter
heights S. W. of Carrigain. In the foreground, E. and S., are the numerous wooded hills of the Pilot Range and the long highland forests of Randolph, with Mt. Starr King culminating on the S. Farther to the r. is Mt.
Hale, beyond the Ammonoosuc Valley, with the shapeless masses of the
Twin Mts. and Mts. Guyot and Bond on the W., filling a great area with
their ponderous spurs. Nearer, in the foreground, are the populous plains
of Lancaster and Jefferson, apparently terminating at the blue wall of
Cherry Mt. The valley is covered with fertile farms and hamlets, and is
gemmed by bright lakelets. About S. S. W. are the Franconia Mts., with
the sharp Haystack peak next to the Twins, and then the boldly defined
and high-placed crags of Lafayette, back of which are portions of the
more southerly peaks. To the r. of the deep-cut Franconia Notch is the
ungainly form of Mt. Cannon, overtopped by Mt. Kinsman, beyond which
a portion of Moosilauke is visible. Farther to the S. W. is a long vista of
the rural towns of the Lower-Ammonoosuc Valley; and the beautiful
village of Lancaster is nearly S. W. Beyond Lancaster, on the S., are the
round-topped hills which are known as Mt. Prospect, Mt. Pleasant, and

the Martin Meadow Hills, over whose r. flank is the dark wooded ridge of Dalton Mt. The hills of Littleton and Bethlehem are partially revealed over the nearer Lancaster range.

A great extent of the Connecticut Valley is seen both below and above Lancaster, with occasional reaches of the broad blue river, strongly relieved in the midst of the level and verdant meadows, and numerous farms and houses. Beyond the valley are the highlands of Vermont, gradually rising from the foot-hills near the river to the tall crest of Mt. Mansfield, many leagues distant. Over Lancaster, on the r., are the far-viewing Lunenburg Heights; and Cow Mt. and Mt. Burnside, in Guildhall, are across the Connecticut Valley on the W, while between them are the Victory Hills and Umpire Mt., in Burke. Nearly N. W. are the East-Haven hills, Haystack Mt. in Westmore, and the sharp-cut peaks about Willoughby Lake. To the r. of the hills of Maidstone and Ferdinand are the cold highlands about Smuggler's Notch, in Brunswick, and the long and monotonous ridge of Bowback Mt., in Stratford. Nearly N. are the noble yellow domes of the Percy Peaks, with the shaggy sides of Stratford Mt. beyond, passing towards the Long Mt. of Odell. About N. E., over the lower portion of the Mill-Brook Valley, is Mill Mt., near the village of Stark, beyond which are the hills of Dummer and Millsfield. The view to the E. is limited by the nearer summits of the Pilot Range, and gives but slight glimpses of the valleys of Milan and Berlin, and portions of the ridges in the vicinity of Gorham and Shelburne.

64. Northumberland.

This town contains 955 inhabitants, and has 6,555 acres of improved land. It is bordered on the W. by the Connecticut River, and is crossed by the Upper Ammonoosuc. Its extensive intervales are prolific in corn, and the uplands are used for pasturage. During the spring floods the meadows are overflowed, and the Connecticut resembles a long lake.

Northumberland village (small inn) is on the Connecticut River, 5 M. from Lancaster, and is connected by a bridge with the Vermont hamlet of *Guildhall Falls* (Essex Summer Boarding-house). There are a few small mills at this point; and at Guildhall Falls are the public buildings of Essex County, while W. of the village are Mt. Burnside and Cow Mt. Cape Horn (Mt. Lyon) and Bellamy Mt. are close to this hamlet, on the E. and N. E. 6 – 8 M. N. W. is *Maidstone Lake*, famous for its lunge-fishing and highland scenery.

Groveton (*Melcher House*, $2 a day) is on the Upper Ammonoosuc, not far from its confluence with the Connecticut, between Bellamy and Morse Mts. It is the chief village of Northumberland, and contains a church, a graded school, and several stores and mills. It derives its importance from being the point where the Boston, Concord & Montreal

8 *   L

Railroad intersects the Grand Trunk Railway. Prof. Huntington says: "These is some remarkable scenery in the vicinity of Groveton. Coming from the S. towards the village, the Percy Peaks will attract the attention of the most indifferent observer, on account both of their symmetrical form and color. . . . . The village itself is surrounded by mountains."

Cape Horn is about 3 M. from Groveton, and may be reached by taking the Lancaster road for 2 M. to the old Richey tavern, and ascending ½ M. through the pasture and then ¼ M. through the woods. From the summit the ridge extends E. of S. and has a precipitous face to the N. E. for ½ M. The rock of the mountain is porphyritic. There are thick woods along the summit, with occasional vistas through the trees, — but the view is limited on the E. and S. E. by the Pilot Range. On the S. is a broad expanse of the Connecticut Valley, with Lancaster in the middle and Moosilauke apparently at the S. end. On the W. are Mt. Burnside and Cow Mt., in Vermont, and the Smuggler's Notch, in Brunswick. On the N. and N. E. are the mountains of Stratford and the Percy Peaks.

This mountain is 2,735 ft. high, or 1,834 ft. above Groveton. Its ancient name was *Cape Horn*, and this apparently irrelevant (but quaint) name is the one which is always used by the inhabitants of the adjacent towns. The new name of *Mt. Lyon* was given recently, in honor of a prominent railroad official; but has been rejected by the Appalachian Mountain Club.

In 1758 Rogers's Rangers marched up from the Merrimac Valley, and built a spacious palisade-fort in Northumberland, just below the confluence of the Connecticut and Ammonoosuc Rivers. They then had a great feast, and named the work *Fort Wentworth*. The Rangers were soon ordered to Lake George, and "left the ungarrisoned fort to slowly rot away under the shadow of the white summits of Percy Peaks." The town was occupied by the pioneer settlers in 1767; and in the war of the Revolution it was defended by a fort N. of Cape Horn and near the river.

65. The Percy Peaks.

These remarkable domes of rock are in the S. E. corner of Stratford, near the Upper Ammonoosuc Valley. The S. peak is 3,149 ft. high, and the N. peak 3,836, the two being separated by a deep and tangled ravine. They are visible from the valley as far S. as Dalton, and thenceforward are landmarks for the traveller bound N. The name is derived from *Piercy*, the ancient title of the town of Stark; and some of the people who live near call them the *Stratford Peaks*. A recent painting gives them the pretty name of the *Twin Sisters*. The peculiar whitish color of the peaks and their curving symmetry cause them to be prominent objects in the landscape, even when seen from a great distance. The round parts of the domes are of light granite, partly disintegrated, and sometimes degenerating into beds of new-made gravel. The peaks were burned **over** many years ago, and so thorough was the destruction wrought by the fire that even the soil was consumed. Now, every rain that falls sweeps over the broad curves of rock and gravel, keeping them clear of vegetation and life. In ascending the slopes long low cliffs, too steep to scale, are succeeded by smooth ledges inclined at a high angle, over which the climber must pass on hands and knees, if he pass at all. This peculiar character of massive and clean rocky architecture gives a great charm to the attack on the Percy Peaks, as it does also at Welch Mt. and Chocorua.

The Percy Peaks are visited from Groveton, and Richey is the guide. The traveller first drives out about 4 M. N. E. to the foot of the mountain, crossing the Ammonoosuc 3 M. out, and then Nash's Stream, beyond which a side-road is taken to the l., which leads in to a farm at the foot. A large area of tangled woods lies between this point and the ledges, though the S. peak is not more than 1 M. distant. Unless one has a skilful guide he should take a careful compass-line, to prevent wandering in the forest. There is a path from the farm-house to the ledges, which saves the visitor from a hard battle with the bushes, provided he can keep the faint trail of the berry-pickers. If the traveller is in a hurry and desires to visit the higher peak only, it is best to bend around the E. slope of the S. peak, about half-way up, and pass into the saddle between the two, whence the ascent is readily made. But if time is not limited, both crests should be ascended, since from the lower a magnificent view of the N. peak is obtained, worth all the labor of the climbing. The S. peak is ascended directly from the edge of the forest, the slope being visible ahead, frequently impassably steep, and sometimes requiring the greatest caution. The summit is perfectly clear, and circumscribed in area, but commanding a pleasant series of views, somewhat similar, of course, to those from the N. peak, except that its chief element is the impressive dome of that crest, belted with cliffs and gracefully curved lines.

The march from the S. peak to the N. peak takes more than an hour, and leads down through a sharply cut ravine, wherein bushes abound. By following this ravine a long way down to the r., the thirsty voyager can find water, hidden under the rocks. The march up the N. peak is fatiguing on account of the steepness and smoothness of the ledges, but it has elements of inspiration in the rapidly opening view and the swinging upward curves of the mighty rock. It seems as if one were climbing up over the dome of the United-States Capitol. Rarely is a comparatively low mountain found where there are so few bushes and jungles; where the way is so clear, free, and open; and where the dashing fervor of invigorative exercise is so little mingled with the vexations incident to areas of fallen timber and patches of dwarf trees. The climber had better take some cooling beverage with him, for sparing use, as both of the peaks are dry in ordinary weather, and the descent of the ravine to the spring is long and arduous. The top of the N. peak is nearly flat, and covers about an acre, the out-cropping ledges having been decomposed into gravel-beds.

Dwight said of the N. Peak (in 1803): "The second is the most exact and beautiful cone which I ever beheld. It is not improbable that most, if not all of them, have been derived from volcanic explosions, which have long since ceased. This mountain is a noble object to the eye, and is seen on the road with the utmost advantage."

The View. — About S. E., beyond Potter's Pond and up the pleasant valley of the Ammonoosuc, are the low but abrupt eminences of Devil's Slide and Mill Mt., near Stark village. Back of and to the r. of these, and conspicuously seen from the Percies, is the great assembly of the Pilot Mts., culminating on the S. in Mt. Pilot and Mt. Starr King. Still farther away and nearly S. is the majestic Presidential Range, — Washington in the centre, with Clay, Jefferson, Adams, and Madison on the l., and parts of the lower ranges on the r. Mt. Deception is seen on the r. of Pleasant, and to the r. of Mt. Starr King is Cherry Mt. Far beyond these are the dim blue crests of the Twin Mts., Mts. Guyot and Bond, and other remote and unvisited heights of the Pemigewasset region. To the r. of Cherry Mt., and much farther away, is the Haystack Mt. of Franconia, with the noble serrated crest of Lafayette just to the r. The Franconia Notch is seen to the W., cutting down deeply between Lafayette and Mt. Cannon, with the low knoll of Bald Mt. at its entrance. Somewhat nearer is the white summer-village of Bethlehem Street, with Mt. Agassiz rising close behind it. Still farther S. are the massive heights of Mt. Kinsman. A few miles distant, down the broad valley of the Connecticut, is the beautiful village of Lancaster, on fertile plains which extend far down the river and along the Vermont shores. Below Lancaster are Mts. Prospect and Pleasant and the Martin-Meadow Hills, rising boldly from the meadows to symmetrical rounded summits. The fortress-like eminence of Cape Horn towers over the Northumberland meadows, 5 – 6 M. from Percy, and the hamlet of Groveton lies still nearer, in the S. W. Farther down the valley are Dalton Mt., Sugar Hill, and the high ridges of Landaff; while the Mt.-Gardner Range is at a long distance, nearly over Groveton. Just to the r. of Mt. Kinsman and nearly over Mt. Bellamy and Cape Horn is the blue crest of Moosilauke, low down on the horizon. About S. W., and on the Vermont side of the Connecticut Valley, are the Lunenburg Heights (opposite Lancaster), and Niles, Burnside, and Cow Mts. Farther to the r. are the Victory Hills, N. W. of which are the bold mountains of Burke. The hill-ranges of Ferdinand and Brunswick are more to the r., and a considerable section of the Connecticut Valley is seen to the W. Toward the N. W. is the Bowback Mt. of Stratford, with Sugar Loaf toward the E. end; and several fine ridges are seen in the N. and N. E., including the chief mountains in Odell and Millsfield. The narrow valley of Nash's Stream is cloven through this rugged and desolate region, running down for leagues from the remote Trio Ponds in Odell. The character of the view in this direction is wild and primeval, the mountain-forests being as yet unbroken by tilled clearings or roads, and no vestige of civilization is visible.

D\
ful c
have
mou
adva

7

vall
Slic
and
Pil
far
ing
and
of
the
oth
of
wi
Nc
Ca
is
cl
A
be
ri
ar
o
H
tl
v
t
t
t
s
l
l
l

To the E. is the near ridge of Long Mt. in Odell, which shuts out much of the view in that direction. Farther to the S. and along the horizon are the mountains about the Grafton Notch. Farther out and more to the r., beyond the plains of Milan, is the sharp spire of Goose-Eye, across the Maine border. Toward the S. E. are portions of the pleasant Androscoggin Valley, on the N. edge of which are Mts. Hayes, Baldcap, and Ingalls, in Gorham and Shelburne. The high crests of Moriah are more to the S , below the Androscoggin, and apparently resting to the r. on Mt. Madison.

Sugar Loaf is a tall and picturesque detached mountain in the E. part of the town of Stratford. It is not ascended by tourists as yet, and perhaps the excursion might not be found profitable. The road which runs from Stratford 6 – 7 M. N. E. up the course of Bog Brook leads to the Hall farm, whence it is a journey of about 2 hrs. through the woods to the top. The summit is composed of a great sheet of granite, 170 ft. long by 60 ft. wide, with nearly precipitous sides. Through the thin trees which fringe the edge pleasant prospects may be obtained down the Connecticut Valley and of the Percy Peaks and the Pilot Range.

66. Jefferson Hill.

Hotels. — The Waumbek House and the adjacent cottage accommodate over 150 guests, at $ 3 a day, and $ 10 – 17.50 a week. It is open from June 15 to Oct. 15. The house is on the flank of Mt. Starr King, and commands a superb view. It has a croquet-ground, reading-room, telegraph-office, and livery-stable. The Plaisted House is below the Waumbek, and is more modern in its appointments. It accommodates 100 guests, charging $ 3 a day, or $ 10 – 18 a week. It is surrounded with piazzas, whence broad views are gained. The Jefferson-Hill House and the Starr-King House are close to the Plaisted, and accommodate 60 – 70 guests each, at $ 9 – 15 a week. The Sunnyside House cares for about 25 guests; Mrs. Bowles's Maple House, opposite the Plaisted, has 18; Levi Stalbird's Pleasant-Valley House has about 18; and there are other smaller houses. On the road to the E., fronting the noblest mountain scenery, are the following houses: W. H. Crawford's, 2 M. out, 20 guests; J. M. Pottle's Highland House, 3 M. out, 30 guests; E. A. Crawford's, 4 M. out, 30 guests; and Crawshaw's Mt.-Adams House, 5 M. out, 75 guests.

Routes. — Lancaster is the railroad station for Jefferson Hill, and is about 8 M. distant. Stages are in waiting on the arrival of trains from the S. Passengers from Boston in morning trains reach Jefferson Hill at supper-time. The nearest way from Montreal and Quebec is by the Grand Trunk Railway (Route 7) to Groveton Junction, and thence by the B., C. & M. R. R. (Route 2) to Lancaster. The *John's-River Railroad* (from Whitefield) will soon be prolonged to the Meadow, 2 M. from the village, and will then be the best route, if it runs passenger-cars.

Distances. — (The figures in heavy type are from Prof. Walling's odometer surveys, and are reliable. They are in miles and tenths, on the nearest highways. The other figures are from the hotel-lists. The appended sums are from the carriage-tariff of the Plaisted House, showing the expense of the rides, on the basis of a single person with a buggy.) Jefferson Hill to Bray Hill, 6 M., $ 1.50; to Stag Hollow, 6 M., $ 1.50; to Blair's Mill and the Valley Road, 8 M., $ 2.25; the Gore Road, 8 M., $ 2.25; Jefferson Mills, 3 M., $ 1.50; Lancaster, 8 M., $ 2.50; Whitefield, **10.8** M., $ 3; Dalton, **14.5** M.; Lunenburg Heights, 19 M., $ 5 50; Fabyan House, **12** M., $ 4.50; Crawford House, 16 M., $ 5; Willey House, 19 M., $ 5.50; summit of Mt. Washington, by railroad from the Fabyan House, 21 M.; summit of Mt. Washington, by carriage, via the Glen House, 27½ M., $ 8; Cherry-Mt. path, 6 M.; Twin-Mountain House, 13 M., $ 4.50; Bethlehem, 18 M., $ 5; Profile House, 28 M., $ 7; Mt.-Adams House, 5 M., $ 1.50; Gorham, **17** M., $ 4.50; Shelburne, **20.5** M.; Berlin Falls, 23 M., $ 6.50; Glen House, 19 M., $ 5; summit of Mt. Starr King, 2½ M.; around Mt. Prospect, 16 M.; North-Road drive, 8 M.

Jefferson Hill is a hamlet which consists almost entirely of summer boarding-houses, and is situated on a high spur of Mt. Starr King, over the valley of Israel's River. Its height above the sea is over 1,430 ft , but no precise measurements have been made. The Jeffersonians claim that their village is considerably higher than Bethlehem (which is visible from Jefferson Hill). It would be desirable to have this question settled by careful levelling, for the Bethlehemites are equally convinced of the superior altitude of their village. Sufferers from hay-fever and catarrhal complaints receive relief here. The village churches are of the Baptist and Methodist persuasions. The absence of trees gives a bare appearance to the street, but there are fine forests back of the Waumbek House; and in the remoter mountain-brooks many trout are found. The roads to the Twin-Mountain and Fabyan Houses pass down from the flank of Mt. Starr King, and out over the Jefferson Meadows, which are not far below the village. A peaceful and pastoral scene may be enjoyed by a stroll in this direction, in the afternoon. There is a road to Lancaster, somewhat longer and harder than the usual route, which passes up over the W. spurs of Mt. Starr King, leaving Jefferson Mills far to the l., and coming in on the Lost-Nation road. This route is called the *North Road*, and is superior in scenery to the stage-road, since it crosses higher ridges. The *Gore Road* leaves the North Road 2¼ M. from Jefferson Hill, running to the N. into the region of Lancaster Gore, and then bending to the W. and rejoining the North Road, 2 M. from Lancaster. This route is hilly, but reveals much romantic scenery in the retired glens. A shorter drive is out to Jefferson Mills, and then across to and back on the North Road.

The View from Jefferson Hill. — On the r. of the Randolph Notch begins the long slope of Mt. Madison, succeeded by the higher peak of Mt. Adams, with King's Ravine opening in its l. side. Mt. Jefferson comes next, with a noble elevation, and a deep ravine indenting its flank; and then the stately peak of Washington is seen, with its hotel and railroad. Under its r. slope is the wilderness-peak of Mt. Mitten, and to the r. is the double-cragged crest of Monroe. Mt. Dartmouth is in the same range with Mitten, and reaches the sky, hiding Franklin; but the curving swell of Pleasant comes out on its r. Next to Dartmouth rises the long and wooded ridge of Deception, reaching the horizon, and flanked by the Field-Willey Range, which is nearly S. Then comes the great mass of Cherry Mt., near at hand and covering several miles, the peak of Owl's Head being N. of the highest point. Over the r. slope of Cherry appear three sharp peaks, Haystack, Lincoln, and Lafayette, with a long apparent plateau to the r., flanked by the low crest of Kinsman. The curved tops of Round Hill and Mt. Agassiz are seen, back of Bethlehem, beyond which the view passes over the Whitefield and Littleton hills to the highlands of Vermont. Dalton Mt. is the long wooded ridge about W. S. W., on whose r. and far beyond are the mountains of Essex and Caledonia Counties, in Vermont, sweeping around by Mt. Tug and Niles Mt. Mts. Prospect and Pleasant, below Lancaster, are about W. N. W., and their bases are girdled with clearings. The view is now cut off by Mt. Starr King, on whose slope the village stands.

"Jefferson Hill may without exaggeration be called the *ultima thule* of grandeur in an artist's pilgrimage among the New-Hampshire mountains, for at no other point can he see the White Hills themselves in such array and force. This view has other qualities to justify such a claim. The distance is happily fitted, not only to display the confederated strength of the chain, but also to reveal in the essential marks of form and texture the noblest character of the separate mountains. As we have said also, the smaller Franconia group rises farther away in front, separated from them by the dark bulk of Cherry Mt. in mid-ground; and on the r. hand the savanna that stretches along the Connecticut presents a landscape contrast of a magnitude and distinctness rarely met with.

"The White-Mt. range is so much grander when seen from Jefferson than from any other point where the whole of it is displayed, and yet is set at such a distance as to show the richest hues with which, as one feature of the landscape, it can be clothed, that we must award to this village the supremacy in the one element of mountain majesty. . . . . The light and rich purple-olive at Jefferson Hill, with the greater delicacy and variety displayed there, adds beauty to majesty. . . . . Here Lafayette, with other Franconia mountains, comes into view. From no point is a better landscape-picture of him to be gained than he offers here with that long serrated summit, and the precipices of his sides reduced to dimples. And here, too, the color of the White Hills is richest in the late afternoon light." (STARR KING.)

Bray Hill is a low eminence, covered with pastures, about 6 M. S. W. of Jefferson Hill, on the road to Whitefield. It is easily ascended in a few minutes, and from its position with reference to the valley of Israel's River it commands one of the noblest possible views of the White and Franconia Mts. Says Starr King: "We would especially speak of the spectacle from Bray Hill, on the edge of Whitefield, around which Nature spreads, about five in the afternoon, as gorgeous a feast of color on the meadows and cultivated uplands that lie within the dark circle of larger mountain guards, as New Hampshire can supply."

Blair's Mills are in Randolph, 7 M. S. E. on the Gorham road. It is a favorite drive for visitors at Jefferson, on account of its magnificent mountain-views. The return is usually effected by turning to the r. into the *Valley Road*, which follows the course of Israel's River down to the Jefferson Meadows. The Cherry-Mt. road is here met, and is followed to the N. up to Jefferson Hill. This circular drive is 18 M. long.

Stag Hollow is 3 M. from Jefferson Hill, on the Gorham road, where the *Potato Road* diverges to the r. and enters the Valley Road. This circuit is about 9 M. long, and is one of the favorite minor excursions.

Mount Pliny is the ancient name for the long wooded range in the N. E. part of Jefferson. The name was used by Dr. Bigelow, in 1816, and was perhaps given by Dr. Belknap, in 1784. It was formerly famous for the number of moose that were found in its forests. The culm'nating point of the range, toward the N. W., has received the name of *Mt. Starr King*. 3 – 4 M. back among these dark ridges is the *Pond of Safety*, 2,000 ft. above the sea, from which the Upper Ammonoosuc River flows.

Cherry Pond is in the W. part of the town, about 2 M. from the Meadow, and is on the John's-River Railroad, by which great quantities of lumber are taken out from the surrounding forests. It is about 1 M. around, and from its lower shores the adjacent mountains are seen. Dr. Belknap (in 1784) spoke of it as a place "where yᵉ Moose at this feafon go to bathe to get clear of yᵉ flies, & are sometimes shot in yᵉ Water."

The town of Jefferson has 825 inhabitants, and covers an area of 26,675 acres, of which nearly 20,000 are unimproved, being occupied by rugged mountains and primitive forests. It is traversed from S. E. to N. W. by Israel's River, which separates the Deception and Cherry ranges from the Pliny and Starr-King groups. There are a few small lumber and starch mills, but the chief occupation of the people is farming, the town producing about 80,000 bushels of potatoes yearly.

This town was granted by the name of *Dartmouth* to Col. John Goffe, in 1765, and was settled in 1772 by Col. Joseph Whipple, who for many years exercised a semi-feudal and patriarchal sway over the adjacent country, owning, as he did, a vast area of the northern valleys. He visited Portsmouth once a year and brought up supplies for his tenantry, carrying down the surplus products of the valley. During the Revolutionary War, he was captured here by a war-party from Canada, but escaped while they were plundering the place, rallied his people, and drove them away. The summer-village of Jefferson Hill may almost be considered as the creation of the Rev. Thomas Starr King, who first brought its magnificent scenery into wide and worthy notice, and directed the march of tourists in that direction.

Among the defiles to the S. E. at the head of Israel's River tradition locates the destruction of a detachment of Rogers's Rangers, under circumstances weirdly horrible. In October, 1765, Major Rogers led several hundred of his veteran rangers from New England to the St. Lawrence River by an arduous secret march, and made a dashing night-attack on the Indian village of St Francis. The savages were surprised while engaged in a martial feast in honor of their victorious forays on New England, and the village was plundered and destroyed after its people had been slain or scattered. The church-plate, candlesticks, and a massive silver image were carried away by the conquerors. But large Indian forces rapidly gathered and hung on the line of their retreat by Lake Memphremagog, while heavy snows impeded the march. The rangers soon broke up into small parties, each of which made its independent way southward. Nine of them attempted to carry the silver image through the Notch, but were led astray by a perfidious Indian guide, who piloted them up Israel's River into the trackless gorges near its head, and then eluded them, after having poisoned one with a rattlesnake's fang. Bewildered among these dark ravines they sank, one by one, under terrible deprivations, and but one reached the settlements, bearing a knapsack partly filled with human flesh. The golden candlesticks of the church of St. Francis were found near Lake Memphremagog in 1816, but the most earnest quest has not availed to discover the silver image.

Numerous legends have been developed on this romantic background, among which are those of the hunter and the fawn-skin parchment, the skeleton Indian in the speaking storm, the magic stone, the fortune-teller and the midnight quest, and the screeching of lost spirits. The most beautiful of these traditions is that of a lonely hunter, camping at night far up towards Mt. Adams, before whose astonished vision the mountain mist rolled back and showed "a great stone church, and within this was an altar, where from a sparkling censer rose a curling wreath of incense-smoke, and around it lights dispersed a mellow glow, by which in groups before the altar appeared a tribe of savages kneeling in profound silence. A change came in the wind; a song loud and long rose as a voice-offering to the Great Spirit; then glittering church-spire, church, and altar vanished, and down the steep rock trailed a long line of strange-looking men, in solemn silence. Before all, as borne by some airy sprite, sported a glittering image of silver, which in the deep shadows changed to fairy shape, and, with sparkling wings, disappeared in the rent rocks."

67. Mount Starr King.

Mt. Starr King is properly a portion of the Pilot Range, of which it forms the most southern point, being partially separated from its brethren by a deep ravine on the N., and connected with the Pliny Range. The flanks of the mountain meet the plain of Israel's River at the village of Jefferson Hill, whence the ascent is usually undertaken. The mountain is composed of feldspathic rock, resembling the material of the other Pilot Mts.; and is covered with woods up to the summit. The name of the peak was bestowed in 1861, in honor of the author of *The White Hills*, who extols the scenery of this vicinity with so much eloquence and feeling. There has been a good path from Jefferson Hill to the summit, but it has sadly deteriorated of late years, owing to the neglect of the hotel-keepers to cut out the fallen trees and intrusive bushes. The trail cannot be lost, being well trodden by hundreds of visitors every year; but the ascent is awkward, at least for ladies, and is not comfortable after rains have wet the bushes. The distance from the Waumbek House to the summit is about $2\frac{1}{2}$ M. (by chain measurement), and the ascent may easily be accomplished in $1\frac{1}{2}$ - 2 hrs. The lower section of the path is not made as plain as it should be, but it may be found by following the l. bank (r. bank in ascending) of the brook which comes out near (and W. of) the Waumbek House. Traversing a grove of sugar-maples and then an open pasture (crossing a fence to the r. oblique), the path is seen entering the woods near the gully of the brook. After fairly getting into the forest, the route is plain. A more direct, but less agreeable way, is to follow the cow-path which turns off opposite the Waumbek Cottage through a field, a maple-grove, and a pasture, to the point where the brook and path emerge from the woods. The summit is clear and is covered with large weathered rocks, on one of which is a beacon of the U. S. Coast Survey. A path leads down from this point to the ridge on the N., where, at a few rods from the summit, is a rock-arched spring of clear cold water. The best time for visiting the mountain is at afternoon, when the great ravines of the Presidential Range are filled with light. The height of the beacon above the sea is 3,800 ft.; and above Jefferson Hill it is about 2,400 ft.

   * *The View.* — On the E. S. E., beyond the higher adjacent Pilot Mts., the long and distant ridges of Moriah, Imp, and Carter are seen over the Peabody Glen. Farther S. are the imposing peaks of the Presidential Range, Madison, Adams, Jefferson, the craggy wall of Clay, and the crest and long W. flanks of Washington, traversed by the railway. The profound chasm of King's Ravine opens in the l. side of Adams, with its enwalling cliffs heading up near the fortress-like ridge towards Madison. Between Adams and Jefferson are the shadowy depths of the Ravine of the Castles, in which Israel's River rises. To the r. of Washington is Monroe, which assumes an unusually conspicuous appearance; Franklin

is scarcely perceptible; Pleasant resembles a flattened dome; and Jackson lifts a low but marked peak against the sky-line. The ravines in the sides of this range are well defined and clearly cut. In the foreground are the low and monotonous lines of Mts. Mitten, Dartmouth, and Deception, a dark and wooded ridge which runs from the foot of Mt. Jefferson nearly to Cherry Mt. Beyond, and over the middle of Mt. Deception, is Mt. Willey, with Mts. Field and Tom in a continuous line on the r.; and a group of lower summits runs N. W. towards the White-Mountain House. Over the r. of Mt. Deception is the high round top of Mt. Carrigain, just to the r. of which is Mt. Hancock. The New-Zealand Notch is over the middle of the gap between Mts. Deception and Cherry, and through it is seen the distant cone of Osceola, with Tripyramid on the l. In the foreground is the vast mass of Cherry Mt., often highly colored by the evening lights, and nowhere reaching the sky-line. Beyond Cherry are Mt. Hale, W. of the New-Zealand Notch; the formless flanks of the North Twin Mt.; and the conspicuous cone of Haystack.

About S. W. is the foreshortened crest of Lafayette, with the tent-shaped peaks of Lincoln and Liberty on the l. The bold outlines of the Franconia Notch are somewhat diminished by the intrusion of the insignificant Bald Mt. W. of the Notch is Mt. Cannon, beyond which is Mt. Kinsman; and part of Moosilauke is seen far down the Landaff Valley.

The pond-strewn plains of Whitefield are towards the S. W., above which Mt. Agassiz and Round Hill (at Bethlehem) are seen, nearly over the forest-environed Cherry Pond. More to the W. are Dalton Mt. and the bold hills S. of Lancaster, — Mts. Prospect and Pleasant, and the Martin-Meadow Hills. Lancaster is seen on its verdant meadows, nearly N. W.; and a great arc of the western landscape is occupied by the fertile valley of the Connecticut, beyond which rise the Green Mts., in number sufficient to defy arrangement. Camel's Hump and Mansfield are seen, far away; and nearer, below and beyond Lancaster, are the Lunenburg Heights, the Victory Hills, and the well-defined mountains of Burke, N. of which are the peaks towards Willoughby Lake. Beyond Lancaster, and on the edge of the Connecticut Valley, are Cow Mt. and Mt. Burnside, in Guildhall, and farther to the N. is Cape Horn, looming up boldly from the Northumberland plains.

The remaining section of the prospect is chiefly occupied by the shaggy knolls of the Pilot Range, though several interesting vistas may be gained between them. Nearly due N. is the summit of Mt. Pilot, which is visited from Lost Nation, beyond which are the light-hued Percy Peaks, standing before the Stratford mountains. But little can be seen of the plains of Berlin and Milan, though portions of the ranges in Shelburne and Gorham are visible, including Baldcap and Mt. Ingalls. Several conspicuous peaks cut the horizon beyond the Maine border, among which are Speckled Mt., Puzzle Mt., and the picturesque cone of Goose-Eye (N. of E.).

68. Cherry Mountain

is a long and massive ridge which runs nearly N. for several miles in the town of Carroll. Most of its area is covered with dense woods, giving it a peculiar dark appearance when seen from distant points. The ridge was formerly known as *Pondicherry Mt.* The highest part is 3,670 ft. above the sea; and on the N. end of the ridge is the sharp peak of Owl's Head, which is conspicuous in the view from the E. or W.

The town of Carroll, in which this mountain stands, was granted in 1772, under the name of *Breton Woods* (probably in allusion to the then recent conquest of the French fortresses on Cape-Breton Island). It now has 378 inhabitants, and covers an area of 24,640 acres, of which only 2,915 are improved. There is no village. It is entered by the John's-River R. R. on the N. W., and by the B., C. & M. R. R. on the S.; and contains the Twin-Mountain and White-Mountain Houses.

Cherry Mountain is usually ascended from John M. King's farm, which is 6 M. from Jefferson Hill, 10 M. from Whitefield, 8 M. from the Twin-Mountain House, and 10 M. from the Fabyan House. Mr. King has made a good path from his fields to the summit, and keeps it well cleared, charging each person who uses it 30 c. The distance from the road to the top of the Owl's-Head peak is about 1¼ M, and the ascent may be made in 1 – 1½ hrs. The path is steep, and passes directly up the long slope and the sharp cone, without crossing intermediate levels. Somewhat more than half-way up is a pleasant resting-place, where a vista has been cut through the trees, opening out on the plains of Whitefield and Jefferson. The summit is covered with ledges, among which are low thickets. The Owl's-Head peak is N. of the highest point of the mountain, and is the only satisfactory view-point. A path 1 M. long has been bushed out over the S. edge of Owl's Head, through the intervening ravine, and up on the highest peak of Cherry Mountain; but it does not pay to follow this route, since the whole ridge is thickly wooded, and only occasional outlooks can be obtained. There are no springs or water-courses on the mountain. Owl's Head is 1,800 ft. above the King farm, and over 3,000 ft. above the sea.

The View. — The Presidential Range begins a little S. of E. with the far-reaching flanks of Madison, to which succeeds the dark spire of Adams, with the castellated head of King's Ravine below. Beyond the shadowy depths of the Adams Ravine is Mt. Jefferson, then the low walls of Mt. Clay, and then Mt. Washington, with the railway curving up its side. The range continues to the r. with the ragged crest of Monroe, the level plateau of Franklin, the bulging hemisphere of Pleasant, and the long apparent ridge in which are blent Clinton, Jackson, and Webster, the latter falling off sharply into the White-Mt. Notch. Between Cherry Mt. and the Presidential Range is a long, low, secondary ridge, of which Mt. Mitten is under Jefferson, Mt. Dartmouth is under Washington, and the highest part of Mt. Deception is to the r. of Pleasant. Under the pitch-off of Webster, and nearer, is the low rounded top of Mt. Willard, on whose l. is the Crawford-House clearing, with the Elephant's Head and the Gate of the Notch beyond. Far down the Notch, over Willard, is the sharp peak of Chocorua, with the lower and nearer summits of the Bartlett Haystack and Tremont, the one on the l., the other on the r.

To the r. of Willard is Mt. Willey, cutting the sky-line with its bold head, on whose r. is Mt. Field, in an apparently continuous ridge, with the rocky knoll of Avalon near its flank. Farther to the r. is the round top of Mt. Tom. The long ridge of Cherry Mt., which has occupied the foreground, now heaves up towards the main peak, which lies nearly due S. and shuts out the view of the Twin and Franconia Ranges. Over the woods on the W. flank of Cherry Mt. is seen the village of Bethlehem, with Mt. Agassiz in the rear. Still farther S. W. are Landaff Hill and Sugar Hill, in Lisbon; and the village of Littleton appears among the surrounding highlands, with Mt. Gardner far beyond. N. of W. is the great forest of the John's-River region, and Whitefield village appears in line with the more distant Dalton Mt. To the l. is Kimball Hill, with its observatory rising through a dense grove, above wide farm-lands. Over the clearing at the foot of Owl's Head, and many miles away, are the Lunenburg Heights and Mt. Niles, beyond which are the Victory Hills of Vermont and the Burke mountains. Among the extensive forests which run N. W. from Cherry Mt. Cherry Pond is seen, to the l. of and beyond which are the triple hills S. of Lancaster, — Mt. Prospect, Mt. Pleasant, and the Martin-Meadow ridge. Still farther to the N. W. is Cow Mt., between Pleasant and Prospect, and Mt. Burnside (in Guildhall), to the r. of Mt. Prospect. Lancaster is nearly hidden by an elevation on the E., but more to the N. are Cape Horn and the Bowback Mt. of Stratford, with the ridges about Smuggler's Notch (in Brunswick) to the l. Towards the N. is the general assembly of the Pilot Mts., Mt. Pilot being marked by a white ledge on its side. Across the valley of Israel's River is the village of Jefferson Hill, back of which rises Mt. Starr King, its peak being marked by a beacon (l. of two small slides). From N. W. to W. N. W. are the Pliny, Randolph, and Crescent Ranges, with Goose-Eye over the latter. The Moose-River Valley runs out about W. N. W., with Pine Mt. and Mt. Moriah near its end; and down the continuous opening of the Androscoggin River are several bold peaks of Western Maine.

69. The Mt.-Adams House

is on the Cherry-Mountain Road, near Israel's River, in the S. E. corner of Jefferson, and 5 M. from Jefferson Hill. Its railway-stations are Gorham (12 M.) or Lancaster (13 M.). The house accommodates about 60 guests, at $7 – 8 a week.

Boy Mountain is a spur of the Pliny Range, N. W. of the hotel, and about ½ M. distant, to whose summit an easy path has been made. It commands an admirable * view of the valley and the vast mountain-wall beyond; and is often ascended by sojourners in this vicinity. The *Cold-Stream Falls* are on the Moose River, and are frequently visited from the hotel. A path leads down to Israel's River, ½ M. from the hotel. The Crawford House and Notch, Gorham and Berlin Falls, and the Glen-Ellis Falls are within a half-day's ride; and the Glen House is 14 M. distant. The old Lancaster path up Mt. Washington is hence accessible; and the new route up Mt. Adams commences 4 M. distant. Chas. E. Lowe is a first-class guide for this region.

To the majority of travellers the greatest interest will centre on the stately peaks of Adams and Jefferson, that culminate within 4 – 5 M. of the house, showing their long firm lines and ponderous ridges to the best advantage. The pyramidal apex of Adams cuts sharply into the air over the shadowy recesses of profound ravines, and forms the noblest object in the landscape.

" Emerging upon the road at Martin's, where now stands the Mt.-Adams House, you see the whole great chain of the chief peaks, their forests speckled with light, and apparently so near that one almost feels like putting his hand upon their flickering sides across the densely wooded ravine which winds up and up till it is lost in the gray distance of the heights of Mt. Washington." (*Picturesque America.*)

Starr King thus describes the view from this place : " There is no point in New Hampshire where its monotony is so poetic and sublime, where the wilderness, miles and miles in extent, unenlivened by a clearing or the smoke of a cabin, unravaged by the axe and unspotted by fire, flows off in such noble lines and folds from the shoulders of the bleak hills. The forms of the mountains are nobler on this side than on the side towards the old Fabyan place near the Notch. The largest members of the chain are the most prominent here. The ridge is not so lank, and its braces run out with more vigor; the ravines are more powerfully furrowed; and Mt. Washington is far better related to the chain."

70. The Cherry-Mountain Road.

Distances. — Lancaster to Jefferson Hill, 8 M. ; Fabyan House to Jefferson Hill, 12 ; Jefferson Hill to the Mt.-Adams House, 5 ; to Wood's, 11 ; to Gorham, 17 ; to the Glen House, 19. *In reverse.* Glen House to Wood's, 8 ; to the Mt.-Adams House, 14 ; to Jefferson Hill, 19 ; to Lancaster, 27 ; to the Fabyan House, 29.

" There is as much beauty to be enjoyed on other routes ; but for grandeur, and for opportunities of studying the wildness and majesty of the sovereign range, the Cherry-Mountain road is without a rival in New Hampshire." (STARR KING.)

The Cherry-Mountain Road skirts the N. and N. W. sides of the main range, running from Jefferson through Randolph and Gorham and over spurs of Cherry Mt. to the Fabyan House and the Twin-Mountain House. The road is easy and smooth, except for 2 – 3 M. of the passage of Cherry Mt. toward the Fabyan House, where it is steep and rugged. Tourists who do not wish to travel over the latter route can drive from Lancaster to Jefferson Hill over an admirable road, 8 M. long. There are no public conveyances on this route, but carriages and mountain-wagons can be hired at the Glen House or Gorham, Lancaster or the Fabyan House. 5 – 7 hours is ample time for the journey, in either direction. By passing from Lancaster to Gorham the advantage is given of a drive of 12 M. to the S. W., in which the mountains are steadily approached, until they grow from the remote gray peaks seen at Lancaster to the vast elevations which tower close at hand over the Mt.-Adams House. The best teams, however, are to be obtained at the Glen House ; and when the journey is taken in this direction, from E. to W., the advantage is gained of the drive from Jefferson Hill to Cherry Mt., facing and approaching the main range for several miles.

Having thus stated that this route is rich in scenery and interest, in whichever direction it is travelled, the following description applies to it as seen by riding from Lancaster to the E., and may be read also in reverse order by tourists going from E. to W.

The Roads over Cherry Mountain.

The Cherry-Mt. Road from the Fabyan House leaves the Ammonoosuc Valley near the old White-Mountain House and ascends the banks of a tributary stream into the high pass between Cherry Mt. (W.) and Mt. Deception (E.). The upper section of this road leads through thick forests, and is usually very muddy after heavy rains. Beyond the crest of the pass, it descends to the Israel's-River Valley, having the Owl's-Head peak of Cherry Mt. conspicuous on the l. The tourist then keeps straight to the N. to Jefferson Hill, avoiding the side-road to the r. which passes under the minor ranges of Deception and Dartmouth. Israel's River is crossed, and then for 5 M. the road runs over the Jefferson Meadows, in plain view of the main range, whose great peaks loom up not far away, on the S. E., with Mt. Washington proudly pre-eminent.

The route from the Twin-Mountain House ascends the Alder-Brook glen, between Beech Hill and Cherry Mt., and opens a noble panorama of the Twin and Franconia Ranges. It then swings around the long W. and N. bases of Cherry Mt., through the farming neighborhoods of the town of Carroll, with pleasant views to the W. and N. Passing the house whence the Cherry-Mt. path diverges, it soon descends to the Fabyan-House road, on which it passes N. to Jefferson Hill, over the meadows of Israel's River.

The Road from Lancaster.

The road from Lancaster to Jefferson Hill (7 M.) is a pleasant one, affording a succession of broad and attractive views. The Pilot Range and Mt. Starr King are constantly in sight on the l., and frequent prospects are afforded of the Presidential Range. At 2½ M. from Lancaster the road crosses a bold hill, 300 ft. above the village, from which is given a panoramic view of the White, Twin, and Franconia Mts., and several minor ranges. This is one of the best locations for a hotel in all the mountain district, but it has not yet been so utilized. Leaving Mt. Prospect on the r., the road descends to Israel's River, which is crossed at the hamlet of Jefferson Mills. The direction is S. E., and Bray Hill is seen on the r. The mountain-view seems almost without limit, and includes portions of three counties, with the majestic forms of the chief peaks of the White Hills.

The road from Jefferson Hill to the Mt.-Adams House gives a succession of rich and splendid views. From the summit of the long hill that forms the height-of-land, a brilliant westerly view is gained. "At first sight there is something grander than the range before us in the long lines crowned with forest that sweep with even slope towards the Connecticut. And what breadth of prospect! At the r., the Cherry Mt. heaving out of a vast plain attracts us; then at the l., the Pliny ridge, on which, far up towards the summit, the wilderness has been displaced by smiling farms; the cultivated hills of Bethlehem glow like huge opals on the S. W.; and more northerly than these, and far beyond them, summits of the Vermont mountains peer dim and blue. The view is as vast as from many a mountain-top."

From the little red school-house on the r. of the road George L. Brown painted one of his most notable pictures, "The Crown of New England," now in Windsor Castle.

In driving to the E. over this road we see the peaks to the S. W. of Mt. Washington, Monroe, Franklin, and Pleasant, peering over the long forests on the r., and above the Deception and Dartmouth ranges. One by one they are eclipsed by nearer mountains, until Washington alone appears on the S. Several summer boarding-houses are passed on this line, and at length the *Mt.-Adams House* is reached, with its magnificent views to the S. E. (See page 188.)

Beyond the Mt.-Adams House the road crosses the water-shed and descends to the valley of Moose River. Noble views are obtained on the r., including the Castellated Ridge of Mt. Jefferson, "whose rocks rising over a deep ravine seem to be the turrets of decaying fortifications." Then King's Ravine opens off to the r., far into the mass of Mt. Adams, high up from the valley, and heading into lofty precipices. A succession of impressive panoramas is now witnessed, as the route passes between the dark mountains of the Randolph Range, on the l., and the far-reaching spurs of Adams. The town of Randolph was granted to some London people in 1772, and long bore the name of *Durand.* It has only 138 inhabitants, on 26,680 acres of land.

As the valley of Moose River is entered the massive flanks of Mt. Madison appear to close in on the r. and the low Crescent Mts. lie to the N. At Wood's tavern the Glen-House road diverges to the S. E. between Mt. Madison and Pine Mt. The Gorham road now leaves the river and passes over *Randolph Hill,* whence is gained a rich and widely celebrated view. (See page 114.) 5 M. beyond this point, after a long descent, the visitor enters the brisk village of Gorham. (See page 112.)

The following extract describes a portion of this route as taken in a reverse direction: "Leaving Gorham, and following the stage-road to the W., you soon emerge on a hillside, leaving the Androscoggin Valley behind, and when about a mile up this little valley, at a turn in the road, you suddenly find yourself gazing up at the steep side of Mt. Madison, which rises in a clear sweep from its base, washed by the rocky Moose River, and its flanks clothed with huge forest-trees to its gray and rocky summit. Now we see one slope of the mountain, and now another, as the road winds along, till at length the trim peak of Mt. Adams, very like in form to

Madison, peeps over one of the immense shoulders of Adams, and soon its sides rise into view. Mt. Jefferson, in its turn, comes into sight, and the deep gullies in its sides and its rocky flanks present the same unbroken and satisfactory slopes which had made Madison at first seem quite the ideal mountain of one's imagination. From the moment the journey is commenced at the hill-top in Gorham, it is interesting, but, to be fully enjoyed, it should be taken with the afternoon light purpling the mountain-sides, and when the large, picturesque trees, twisted and bent, stand, like sentinels, profiled against the broad, soft light of the hills. Driving along, one flank after another comes into view, shutting off the previous one, filling one with an ever-new surprise at the number and variety of these mountains, which yet are always immense in their sweep and grand in curve. The mountains on this side are much more abrupt than when seen on their W. declivity, and the rocky structure of their formation is more conspicuous." (*Picturesque America.*)

A MAP
of the
PRESIDENTIAL RANGE
WHITE MOUNTAINS
N.H.
"Osgood's White Mountain Guide"

THE PRESIDENTIAL RANGE.

71. The Ancient Paths.

THE Jackson people used to ascend Mt. Washington by climbing over the New-River cliff, near the road, following the brook into the hollow beyond, and ascending a slide at its upper corner, coming out in the sub-alpine region. Prof. Tuckerman says of this route: "My own impression of it was that it was shorter and easier than any other foot-path." About the year 1840 there was another route, consisting of a foot-path through the woods and dwarf shrubbery from the Elkins Farm, 3 M. N. of New River. Both these routes have long been forgotten.

In 1784 Cutler's and Belknap's party ascended the Shelburne road, embarrassed by windfalls, from Adams (now Jackson) to the height-of-land in the Pinkham Notch. Here they encamped, near a beaver-dam, with the peak of Mt. Washington hidden by its great E. spur. They ascended to the first summit (out of the ravine) in 3 hrs. and 52 min., and to the peak in 1½ hr. more. Their route was N. W., with the stream then named Cutler's River on their l.; and Belknap's map (1791) makes Cutler's River issue from the present Tuckerman's Ravine. According to Belknap, also the New River was the first stream to the S. running nearly parallel with Cutler's River, and less than ¾ M. distant. Bigelow's account (1816) is to the same effect, — Cutler's River leading directly towards the principal summit, and being met ¼ M. therefrom, — passing through low black spruces, then into a place surrounded with vast cliffs, then a gentle slope of ½ M. and a laborious scramble of ¼ M. up the rocks, reaching a second flowery plain, where a slight declivity descended on the l. to a brook.

The Editor therefore concluded not to accept the recent maps, in so far as they relate to New River and Cutler's River. He was confirmed in this decision by a letter from Prof. Edward Tuckerman (Dec. 11, 1875), in which the case is stated clearly as follows: "He [Belknap] says that the fountain-head of the Ellis River is so near that of the Peabody River that you might stand in one and touch the other. And his map shows this as he understood it. The head of the Ellis River is a little stream coming in from the N. from a point where the Peabody River is made to head. And the main stream from Tuckerman's Ravine (which is the ' New River ' of Bond's map, and regarded the head-water of the Ellis in Scudder's[1]) beyond all question is, as you say, the original Cutler's River. . . . . About ½ M. S. of the stream called Cutler's River, another runs down, says Belknap, from the same mountain into the Ellis, which originated in the great flood of 1775, and was called New River And where this stream falls into the Ellis River, he further says, are noble falls of 100 ft., which, again, must be surely the Glen-Ellis Falls. . . . . New River was not far S. from the stream from Tuckerman's Ravine, we may be sure, and also that the way by it led up Boott's Spur, and gave a gradual way of ascent to Mt. Washington. It was the way by which Cutler, Boott, Bigelow, and other early visitors ascended, — I was informed in 1840, — and was also the way used by the Jackson people. . . . . The ascent of Cutler's River to Mt. Washington takes one through the whole of Tuckerman's Ravine, and is surpassed by no walk in the mountains for grandeur and interest."

[1] The names of New River and Cutler's River were not located by Mr. Scudder.

When the Editor visited this locality, he was informed by reliable men of Jackson that the rivulet which falls over the lofty cliff l. of and near the road, N. of the path to the Glen-Ellis Fall, was the remnant of the old New River. The present New River crosses the road just above, and was reported to have formed this bed during the cataclysm of 1826. If, as this would certainly indicate, the New River ran in the course which lies over the cliff, before 1826, we have an explanation of Belknap's "noble falls" near the mouth of New River. Probably Belknap did not see nor hear of the Glen-Ellis Falls, since they are so far from the road; but the New-River cascade is plainly visible from the road, even now in its diminished estate.

72. Tuckerman's Ravine.

This wonderful gorge is on the E. side of Mt. Washington, N. of Boott's Spur, and is divided into two sections, the broad vestibule in which are the cascades and Hermit Lake, and the inner and higher chasm of the ravine itself. It is the most remarkable piece of scenery of this character in all New England, for though it is neither so deep nor so long as King's Ravine, it surpasses it in the steepness and sweep of its cliffs, and in its close relation to the supreme summit. It is, moreover, much easier to traverse, being free from dangerous crevasses, and requiring only a powerful exertion of the thews and sinews. Ladies have frequently traversed the ravine, and encamped in its depths. In view of the possibility of accidents it is not prudent for people to make this excursion alone. Tuckerman's Ravine is plainly seen from the mountains of Jackson and Conway, and is distinguishable even from Portland. Its S. wall is very conspicuous from the Glen House, as well as the crags of the *Lion's Head*, which form its N. portal.

This ravine was probably the route of Gorges's party, in 1642. It was traversed by Capt. Evans in 1774; and ten years later by Dr. Cutler's party. Its name was given in honor of Prof. Edward Tuckerman, of Amherst College, an eminent botanist and author of several works on American lichens, who has, moreover, been for many years a tireless explorer of this mountain region. When Thoreau was botanizing here, he fell and sprained his ankle, but when rising he discovered, for the first time, the leaves of the *arnica mollis*. He was alone, and was made helpless by his injury, so that he might have died here but that some one chanced into the ravine, and was attracted by his shouts. In 1852 the gorge was carefully explored by Messrs. Beckett and Hall, who gave their present names to the Hermit Lakes, the Mountain Colosseum, and the Fall of a Thousand Streams.

The cliffs on the sides of Tuckerman's Ravine are composed of andalusite slates, dipping at a moderate angle into the mountain northwesterly. The genesis of the gorge is probably referable to a line of depression left at the time of the upheaval of the great ridges, and afterwards deepened and widened by the action of frost and the water. Through thousands of years successive surfaces of the rocky walls have been split off by frost, and afterwards have been pulverized and washed away by the streams, finally producing the present stupendous chasm.

There are three ways of visiting the ravine. The first (and probably the best for gaining a comprehensive idea of the scenery) is to go in from the Jackson road to the Crystal Cascade, cross the stream at its foot, and ascend on the S. side. The estimated distances on this route are as follows: from the road to the Crystal Cascade, ¼ M.; to Hermit Lake, 2¼ M.; to the Snow Arch, 3 M.; to the Summit House, 4¼ M. (4 – 6 hours). There is a rude path along this line, but it is usually difficult to find and keep it. The Glen-House people would add a great attraction to their vicinity by clearing out both this and the trail from the mountain-road. Nearly ½ M. above the Crystal Cascade, but not visible from the path, is a pretty columnar fall, which has cut a bowl 6 ft. in diameter in the level floor of granite. A short distance below is a beautiful basin, filled with transparent water. 20 rods above is the cascade called the *Birch Pitch.* Frequent views of the overhanging heights on the S. are gained as the visitor passes up through the woods; and the high bluff on the S. shore of **Hermit Lake** is soon reached. This is a favorite place for camping, because of the magnificent views obtained on every side, although on account of its bleak and exposed situation, it will be found less comfortable than a nest in the adjacent thickets. The lake itself is a small black tarn among the primitive forests, and under the immense cliffs of the Lion's-Head spur. On the S. are seen still more formidable precipices, and the tall environing walls of the ravine are in front. This vicinity is prolific in *odonata.* Not far beyond this point the visitor is obliged to take to the bed of the stream, in order to avoid the thickets on either side, and if the water is low he will find but little trouble in advancing over the rocks. After a hard scramble up an ascending steep, Tuckerman's Ravine is entered, and its innermost part is seen to be semicircular, the outer cliffs rising over 1,000 ft. The majestic curve rests on the S. against projecting crags of Boott's Spur, and on the N. are the beetling rocks of the Lion's-Head. From this point up to the Snow Arch, the massive walls are seen on either side, towering to a great height.

The following items may be of interest: The Guide-Book party encamped at Hermit Lake, and walked thence to the Snow Arch in 45 min., whence it took 1½ hrs. to reach the Summit House. The following altitudes were taken with an aneroid barometer: Hermit Lake to the Snow Arch, 615 ft elevation; the base of the arch to the top of the ravine-wall, 650 ft; the edge of the ravine to the Summit House, 1,220 ft. The usual time for walking from the Glen to the Summit House by this route is 4½ - 6 hrs.

The Snow Arch. — The snow remains in the head of Tuckerman's Ravine until the latter part of the summer, but disappears before the autumnal storms set in. At the time of the Editor's last visit (August 20, 1875), the length of the bank was about 120 ft., and the arch remained intact, a solid mass of dripping ice, with a side corridor leading to the r. During the subsequent week tourists found snow here, and traversed the arch. But in ordinary seasons it disappears by the third week in August. Throughout the winter immense masses of snow, hundreds of feet in depth, are accumulated in the ravine, from the heavy mountain-storms, and by being blown down from the adjacent ridges. The lower portion of this incipient glacier becomes solidly congealed from the great pressure and partial meltings above, and is thus fitted to resist the summer sun when it pours into the gorge. It has been visited in winter by men on snow-shoes, and the most brilliant effects have then been noticed as produced by the sun shining on the immense stalactitic icicles which hung from the cliffs, and the great sheets of translucent ice over the Fall of a Thousand Streams.

The tiny rivulets which rise on Bigelow's Lawn and trickle over the cliffs at that Fall form one of the upper tributaries of the Ellis River, and their coalescent brook wears an archway under the snow, through which it flows down.   July 16, 1854, when the officers of the new carriage-road dined here, the arch was measured, and was found to be 266 ft. long, 84 ft. wide, and 40 ft. high (to the roof).   "The roof is always uneven, the irregularities resembling the conchoidal fracture of cannel coal, flint, and other minerals."

The Fall of a Thousand Streams is over the Snow Arch, and in ordinary weather is but a mere dribble of weak threads of water over the face of a lofty cliff.   The route of ascent from the arch to the Summit House is on the r. of the head of the ravine, no other way being practicable, — and thence to and up the cone, — a sharp and breathless scramble over the rugged rocks.

The second route to the ravine is not feasible unless it has been recently cleared.   It consisted of a bridle-path which was cut by Mr. Thompson, many years ago, from the Mt.-Washington road, about 2 M. from the Glen House.   It ran for about 2½ M. through the woods and entered the previous route near Hermit Lake.

The third way is by descending from the Summit House to the Snow Arch, and returning over the same course.   This is the favorite route with visitors to the ravine, and is, of course, much easier than the others, though it does not give so good an idea of the locality.   Guides may sometimes be found at the Summit House, but it is to be desired that some one will so mark out the route that it may be easily followed by strangers.   Some prefer the descent by this route and down Cutler's River to the Crystal Cascade and the road.

"The rock-ribbed organization of the hills is grandly revealed by it; while the spirit of mountain strength, the enormous vitality that is compressed into the resisting power of a great ridge, is suggested there more intensely, perhaps, than in any other mountain line or feature of this region. . . . . The stupendous amphitheatre of stone would of itself repay and overpay the labor of the climb.   It is fitly called the *Mountain Colosseum.*   No other word expresses it, and that comes spontaneously to the lips.   The eye needs some hours of gazing and comparative measurement to fit itself for an appreciation of its scale and sublimity. . . . . It seems as though Titanic geometry and trowels must have come in to perfect a primitive volcanic sketch.   One might easily fancy it the Stonehenge of a Pre-adamite race, — the unroofed ruins of a temple reared by ancient Anaks long before the birth of man, for which the dome of Mt. Washington was piled up as the western tower.   There have been landslides and rock-avalanches as terrible in that ravine as at Dixville Notch, — the teeth of the frosts have been as pitiless, the desolation of the cliffs is as complete, but the spirit of the place is not as gloomy as at Dixville, — is sublime rather than awful or dispiriting. . . . In Tuckerman's Ravine there is a grand battle of granite against storm and frost, a Roman resistance, as though it could hold out for ages yet before the siege of winter, and all the batteries of the air." (STARR KING.)

The following quaint account is extracted from a MS. now in the archives of the Mass. Historical Society, — Dr. Jeremy Belknap's *Tour to the White Mountains* (1784) : —

" Having rifen many very steep and extreemly difficult precipices, I found my breath fail & the Compy having been obliged to make many paufes on my accᵒ and yᵉ Pilot supofing we were not more yⁿ ½ way to yᵉ Plain — a Confultation was held & it was tho't beft yᵗ I shᵈ return before we proceeded further . . . . I arrived I fuppofe abt 10 hᵒ much fatigued — took some refrefhmt & went to sleep " . . . . After I left yᵐ yᵉ afcent became much more steep & difficult yᵉ growth shorter and shorter till it came to shrubs then short bufhes then a sort of grafs called Winter-grafs mixed w mofs. The bufhes are either fir or fpruce, a sort of red berry — & blue berries grew on small Vines. *The Plain* is composed of stones, covered w mofs, mixed w this winter grafs — yᵉ mofs of a light grey colour (yᵗ below is *green*) — & so spread over yᵉ stones & their interftices as to look like yᵉ furface of a dry Pasture or Common — (in some parts yᵉ interftices of yᵉ Rocks were filled w mofs — in others open & dry) — in some openings water appeared. The area of this plain is an irregular figure suppofed near a mile from yᵉ Edge to yᵉ bottom of yᵉ Pinnacle or Sugar loaf [, which] is a pile of loose dark grey rocks suppofed abᵗ 100 feet perpendicular height — & not so difficult of afcent as yᵉ precipice below yᵉ plain wʰ in some places is inaccefsible efpʸ on yᵉ S. E. sides."

73. Huntington's Ravine

heads in near the S. angle of the mountain carriage-road, near the fifth milestone, and runs toward the S., where it meets the outer section of Tuckerman's Ravine. It was inspected in 1871 by the scientific expedition then wintering on the mountains; and was named in honor of Prof. J. H. Huntington, one of the members of that arctic conclave. It is less a ravine than a gulf, which heads into the great N. E. spur of Mt. Washington, and is terminated by the finest line of cliffs in the mountains, surpassing, in altitude, the rock-walls of Tuckerman's Ravine. These immense precipices are so steep that in some places they seem to overhang the bottom and present an impassable barrier to the ascent. The best point from which to view the walls of this chasm is its upper edge, along the Alpine Garden, or by diverging to the l. from near the fifth mile-stone on the mountain-road. It is nearly impracticable to ascend by this gorge to the summit of Mt. Washington, the distance and the obstacles are so great; but Mr. Raymond succeeded in getting through the forests, and ascended the cliffs beyond, under circumstances of considerable peril. Another party of visitors has ascended to the base of the cliffs, by following the bed of the brook which flows out of the ravine, but no thorough exploration of its course has yet been made. The cliffs are visible from many points on the S. E. and S., and even from Portland, whence they present an impressive appearance in the spring-time, the snow assuming the semblance of a vast marble cathedral, whose spire extends far up toward the zenith.

Raymond's Cataract is about ½ M. from the Glen-House path to Tuckerman's Ravine, and is reached by turning to the r. along the third brook (the first comes from the E. spur, the second from Huntington's Ravine), and ascending through a line of windfalls. It is also about ½ M. from Hermit Lake to the place where the path should be left. The cataract is named in honor of one of its early explorers; and the enormous cliff of seamless granite on the l. is called Grinnell Rock, in honor of the naval officer who took the British Arctic ship *Rescue* back to England, and who, at one time, ascended the side of this rock. The cataract is caused by the fall of a brook which rises on the Alpine Garden, and here passes over the edge of the spur in a long white line, sliding down at a high angle. Its height has been variously estimated at from 300 ft. to 1,000 ft. The white and steady line of this fall is visible from a point on the Jackson road, from Mt. Wild-Cat, and from various peaks in Jackson. This locality has long been known, since it was alluded to by John Farmer, in 1823, as "a most elegant cascade," falling from near *St. Anthony's Nose.*

74. The Great Gulf

(called also the *Gulf of Mexico*) is the profound gorge between Mt. Washington and Mts. Jefferson, Adams, and Madison, in which rises the West Branch of the Peabody River. Its depth is about 2,000 ft., and its narrowest point is opposite the Ledge, above which it has a trough-shaped character, and rises rapidly. The deep and narrow lateral ravines on each side of Mt. Adams open into the Gulf, and are traversed by small brooks.

According to Prof. Huntington, the Great Gulf is the most monotonous and uninteresting place about the mountains. It has been entered several times by naturalists and trout-fishers, but no one has yet visited it to study the scenery. Prof. Tuckerman and Dr. Robbins have encamped in its depths, carrying on botanical explorations; and "Josh Billings" passed many hours of the summer of 1875 in fishing for trout in its lower sections. The forests which fill the bottom are so dense as to prevent the adjacent peaks from being seen; and every obstacle known among the White Mts. is met with here. During stormy weather vast masses of mist roll into and move about the Gulf, until it seems like an immense caldron. The best view-point for the Gulf is the gap between Mts. Washington and Clay.

The solitary and secluded pond known as Spaulding's Lake is near the head of the Great Gulf, and is partly filled by the slides from the adjacent slopes. It was first visited (in 1853) by J. H. Spaulding, author of *Historical Relics of the White Mts.* The ascent of Mt. Clay has been made from this point, amid many perils and discouragements; and Mt. Madison has been climbed from the lateral ravine on the r.

Prof. Hitchcock maintains that the Great Gulf is a gorge cut out of an ancient plateau, which extended from Boott's Spur to Mt. Madison. The excavation has been formed by the powerful erosive agency of water, acting through a vast space of time.

"It balances on the N. the deeply cut Tuckerman's Ravine on the S. Seen from the bottom, the last is the grander. Looking down from the top, the Great Gulf is the more terrible. And while the first has only even walls to rim its desolation, the last is crowned by Mt. Jefferson, and commands a grand view also of Adams and

Madison. And the afternoon is the time to see it, — when the sun pours a sheet of light . . . . up the whole southerly wall of chippy rock, tinged with pale-green lichens, and the bottom of the pit lies dark under the grim guard of the three peaks that bend around it on the N. E., and have lost the sun. Those that have not seen this view of the Gulf . . . . are unacquainted with one of the grandest spectacles which the summit of Mt. Washington affords." (STARR KING.)

75. King's Ravine.

The ascent of Mt. Washington by way of King's Ravine may be effected, by stout climbers and skilful woodsmen, in 1½ days. The best way is to advance into the ravine early in the afternoon, and to ascend it as far as it is possible to find a good supply of wood and water. By starting very early the next morning the head of the gorge may be scaled, and the party can cross Mts. Adams, Jefferson, and Clay, and perhaps reach the Mt.-Washington Summit House by dark. The march in the reverse direction would be easier, but far less rich in progressive inspiration and forward views. The Guide-Book party descended from a camp near the top of Mt. Adams to the mouth of the ravine on the Randolph road in 7 hrs.; Starr King's party, the first that entered this gorge, ascended from the road to the Gate in 9 hrs. There is no path, and the ravine is not visited once in a year. The best way to see it is to look down into its depths from the ridges of Mt. Adams, for soon after the visitor reaches the bottom he is enveloped in dense and almost impenetrable woods. The lower part of the ravine is uninteresting, being a long narrow valley, flanked by heavily wooded ridges, between which flows the Moose River. A rude logging-road follows this stream for about 1 M., entering on its true l. bank. Beyond the end of this trail the winding stream must be ascended, now along the ridges on either side, and again on its rocky floors, alternating forest-clambering with rock-leaping. After winding about in this manner for 2 – 3 M., the stream becomes so small as to render it of little service, and its course is often covered with stupendous rocks. Absolute silence rests in all the vast amphitheatre, whence even the winds are shut out.

" For miles the floor of the gorge is overlaid with immense boulders, often two or three deep, between which are cavernous holes which are in some places 30 ft. deep. A slip from the damp moss in leaping over these crevasses would land the unfortunate mountaineer with broken bones in a pitfall from which egress would be difficult, if not impossible. We took to the forest on the side, and forced our way for miles through thickets whose bristling boughs were almost as sharp and as thickly set as the bayonets of the British squares at Waterloo. Low-lurking stubs tripped the most careful feet ; shaggy and stinging branches swept our faces, to the utter confusion of eyesight ; we were lanced by dead pines, impaled on fallen timber, snared by moose-bushes, and pummelled by rolling rocks. Wide detours around piles of ragged rocks ; cautious fordings of deep-cut torrents ; breathless clambers over intrusive ridges ; and long crackling conflicts with obstinate thickets consumed our time and strength, and gave many a tear to garments and shoes and hands." When this chaos of boulders is reached, the party must take to the forests on the sides, for it would require a brigade of engineers, with fascines and scaling-ladders, to fight their way over the rocks.

The visitors at last emerge from the higher black-growth woods, and see the vast walls towering on either side. The best points for observation are gained by working cautiously out into the middle of the ravine, and clambering upon a high pile of boulders. In front are the immense shelving cliffs at the head of the chasm; on the r. is a less abrupt but loftier slope, whose green walls bristle with dwarf firs and are striped with waterfalls. On the l. is the sharper N. wall, which runs off on the r. to a series of wonderful spires and domes of rock. All around are thousands of boulders, tumbled upon each other in utter confusion, and forming a picturesque but impassable labyrinth. At the head of the ravine, visitors should work in close to the l. (N. E.) side, and clamber up over the crags, taking care lest those in the rear receive injury from stones dislodged and set rolling by the foremost men. On the r. are long and perfectly smooth ledges, with an area of acres, tilted at a high angle; and on the l. the wall rises hundreds of feet sheer, exhibiting great variety of rock-sculpture, in fantastic forms of towers, spires, and domes. These marvellous formations, which surround the crest-line of the head of the ravine, and appear also on the S. side, are seen to best advantage from near the middle of the narrow ascent, since on account of the great depth of the bottom they are scarcely discernible from below.

Let us now borrow the vivid word-painting of Mr. King: "The last part of our path out lay up the eastern wall, just where it joins the left-hand cliffs; and here we had the excitement of grand rock-scenery overhanging and threatening us as we climbed; while the opposite rampart, covered with green, and channelled by streams into very graceful lines, responded to the blasted cliff like Gerizim to Ebal, the hill of blessing to the mount of cursing. One could not turn the eye from side to side, without repeating mentally the passage, 'strength and beauty are in his sanctuary.' The last few rods of the passage out of the ravine led us up a narrow and smooth gateway, quite steep, and carpeted with grass. We sat some time in it, looking at the rocky desolation and horror just about us, balanced by the lovely lines into which the verdure of the western ramparts was broken, — not knowing what a splendid view was in reserve for us when we should step out on the ridge. The huge cone of Mt. Madison rose before us, steep, symmetrical, and sharp, with more commanding beauty of form than any other summit of the White Hills has ever shown to my eye. . . . . The cone of Madison, seen from the gateway of the ravine, is not only steep, regular, and pointed, but, all other mountains being shut out, it looks immensely massive. . . . . And its color was even more fascinating than its form. It puzzled us to understand how the rounding lines of the summit, as seen from the road to Randolph, could have been conjured into the lance-like sharpness here revealed to us."

76. The Ravine of the Mt.-Washington River.

The Mt.-Washington River heads in Oakes's Gulf and runs S. S. W. for 7 M. to the Saco River, which it enters at a slight angle, and after a longer course than the stream from Saco Lake, — wherefore it is claimed that this is the true Saco. Its ancient name was *Dry River*, on account of its tendency to dwindle away in times of drought, and also because there is a broad expanse of gravel near the mouth of its valley. The ravine lies between the S. W. peaks of the Presidential Range, on the W., and the Montalban Ridge on the E.; and numerous tributaries fall into the brook during its downward course. There is but little of interest in this long glen, except for the trout-fisher, who finds rare sport here. There are two small cascades, with a fine intervening basin, well up the river; and in the higher parts geologists find ledges of ossipyte rock. The scenery is seldom interesting, because the bottom and sides are so clothed upon with large trees as to hide the adjacent mountains.

The best point to enter is about 3½ M. above Bemis Station, whence the distance is about 12 M. to Bigelow's Lawn. There is no fallen timber, but the density of the woods and the rugged rocks near the brook prove sufficient obstacles. The journey from Bemis Station to the top of Mt. Washington by this ravine cannot be accomplished in one day.

Oakes's Gulf is the profound chasm on the E. of Mt. Monroe, towards the head of the Montalban Ridge, and S. of Bigelow's Lawn. It is visited by descending from Monroe, but is interesting only to enthusiastic naturalists. It is supposed that at least one of the missing men who have been lost from the Crawford-House Path has been killed by falling into this gorge. The Gulf was named in honor of William Oakes, of Ipswich, Mass., an eminent botanist and explorer of this region, and author of an illustrated folio volume on *White-Mountain Scenery* (published in 1848).

On a branch of this stream which comes in from under Mt. Pleasant, an old hunter claimed to have found two immense ledges covered with diamonds, whence the sun flashed so brilliantly as nearly to blind him. Specimens which he carried out were sold at a good price; and several parties afterwards made diligent but unavailing search for the diamond ledges. Fine quartz crystals have been found along the Mt.-Washington River; and some of its tributaries are said to be so strongly impregnated with iron that fish cannot live in them. An old tradition locates in this ravine the great carbuncles whose marvellous brilliancy was the wonder of early explorers. "Hearing that a glorious carbuncle had been found under a large shelving rock, difficult to obtain, placed there by the Indians, who killed one of their number that an evil spirit might haunt the place, we went up Dry River, with guides, and had with us a good man to lay the evil spirit; but returned sorely bruised, treasureless, and not even saw that wonderful sight."

77. The Alpine Garden and the Lion's Head.

The Alpine Garden is the nearly level terrace on the E. of and below the peak of Mt. Washington. It is perhaps 1½ M. long and ½ M. wide, and is covered with rocks and patches of verdure, among which are the springs which flow down to Raymond's Cataract. A long half-day may be spent here with profit; or the pedestrian, on the way down the carriage-road to the Glen House, may make a detour in this direction. The best point to approach it from above is from the mountain-road below the Chandler Ridge, near the seventh mile-stone, whence one bears to the S. around the base of the cone, and along the high plateau. It is about ½ M. from the road to the edge of Huntington's Ravine, whence is afforded a view of the stupendous cliffs which surround the dark gorge below. (See page 197.)

9 *

The **Nelson Crag** is finely seen from the Alpine Garden, towering sharply over the N. end of Huntington's Ravine, and reaching an acute point. It is easily ascended from the Garden, or from the sharp bend of the mountain-road near the fifth mile-stone; and consists of a bold peak of rugged rocks reared on the N. E. spur of the mountain.

The View from the Nelson Crag includes a vast area to the N., E. and S. Toward the S. W. are parts of the upper cone and the frowning Chandler Ridge, flanked on the r. by the clustered humps of Mt. Clay, beyond which are parts of the valleys of Carroll, Jefferson, and Whitefield, with Cherry Pond, and Mts. Prospect and Pleasant, near Lancaster. Then comes the vast peak of Jefferson, looming out of the Great Gulf, and flanked on the r. by a profound ravine and the long and rugged slope of Adams, over which appear Mt. Starr King, the blue Pilot Range, and the white domes of the Percy Peaks. Next appears the noble pyramid of Adams, with a sloping subordinate spur on the r., running out towards the formidable crest of Madison, which is nearly N. Then the view passes up the Androscoggin Valley, including the white hamlet of Berlin Falls, with the low Mt. Forist on its l., and the Berlin Heights beyond; the Chickwolnepy Range being on the r. Over the lower r. flank of Madison is the dark and wooded Pine Mt., hiding Gorham, with the rocky sides of Mt. Hayes beyond, on the r. About N. E., down the Peabody Glen, are the rich meadows and farms of Shelburne, on the Androscoggin, with Mt. Baldcap beyond. Still farther away are the mountains about the Grafton Notch and in Newry, with the sharp blue peaks of Goose-Eye and the Sunday-River White Cap. Just across the Peabody Glen is the immense mass of Moriah, with Mt. Surprise scarcely discernible on its l. flank, and Imp apparently before it on the r. Then come the slide-striped peaks of Carter, and its long ridge, over which are Mt. Royce and Speckled Mt., with broad areas of Western Maine. In the deep green hollow below are the white walls of the Glen House. Then the entrance of the Carter Notch is seen, and Mt. Wild-Cat comes before the Carter Dome, which is flanked on the r. by two white peaks of Baldface, with Mts. Sable, Eastman, and Slope below. Beyond are the clearings of N. Fryeburg and Lovell, with the long rampart of Mt. Pleasant, crowned by a hotel near the centre, and surrounded by ponds. Just to the r., and about S. E., a powerful field glass shows Portland and the sea, over the nearer waters of Sebago Lake.

Black Mt. is next seen, about 8 M. distant, a long, low ridge diversified with light-green clearings, over which are seen the double peaks of Mt. Gemini and the house-crowned cone of Kiarsarge, with the twin mamelons of Double-Head in line with and below Gemini. Over the near range of the Eagle Mts., which runs S. from Wild-Cat, is the high mound of Thorn Mt., backed by the Green Hills of Conway. The view now runs nearly S. S. W. down the garden-like Saco Valley, including N. Conway, Conway, and the bright sheet of Walker's Pond; and to the r., over Humphrey's Ledge, is the distant Green Mt. in Effingham. The S. wall of Tuckerman's Ravine is now seen to great advantage, showing its formidable cliffs for a long distance. Nearly over its l. slope are the yellow ledges of Iron Mt., beyond which rise the bold white peaks of Moat Mt. Then the crests of Pickering and Langdon are seen, with Table and Bear Mts. beyond, and the white needle of Chocorua still more distant between them. Paugus is on the r. of and continuous with Chocorua; and the blue Ossipee Range is seen far beyond.

The view now includes the rugged plateau of Boott's Spur, over which appear the cloven peaks of Whiteface, with parts of Tripyramid and Sandwich Dome. Over the gentle slope of Bigelow's Lawn, and nearer the cone of Washington, is the peak of Carrigain. The distant views from the Nelson Crag are, however, less important than the display of the grand architecture of the E. side of the mountain, Boott's Spur, and the near peaks to the N. A short descent to the S. W. (one needs to be very careful), leads to the edge of the amazing cliffs of Huntington's Ravine.

The **Alpine Garden** affords a comparatively easy walk of 1½–2 M. from the road to the edge of Tuckerman's Ravine, over a rock-strewn sedgy plain. Care should be taken to keep well up to the r. in order to avoid the deep hollow which falls away towards Raymond's Cataract. It is perilous and unprofitable to attempt the descent of this gully, as the cataract must be seen from below (see page 197), and the bottom cannot be reached by this route, on account of high and continuous cliffs. There are no thickets on the plain; and plenty of water can be found. The views on every side are full of interest and wildness.

At the S. E. end of the Alpine Garden, at the point between Tuckerman's Ravine and Huntington's Ravine, is a fortress-like pile of rocks, with broad outworks, standing on the extreme point of the spur. This jutting crag has long been known as the **Lion's Head** (though its earlier name was *St. Anthony's Nose*), from the resemblance which it assumes when viewed from the Glen House. It commands one of the best views of Tuckerman's Ravine and its approaches, with the vast S. wall and the rocky mounds on Boott's Spur. It also looks upon the cone of Washington and the Summit House, the Chandler Ridge, and the sharp Nelson Crag. The great N. peaks are hidden by the N. E. spur, which runs out to the Crag.

The View. — Over the declining ridge to the N. is seen a section of the Androscoggin Valley, with the mountains near Gorham, the highlands of Shelburne and Success, and the woody Baldcap. Moriah and Imp are nearly over the Glen, on whose r. are the high and slide-marked Carter and the Carter Dome, filling this part of the horizon with their huge masses. Only about 3 M. distant, to the E., is Mt. Wild-Cat, over which are the white caps of Baldface. Then comes Mt. Eastman, on whose r., and far beyond, is the long ridge of Mt. Pleasant, with its central hotel, and a bit of Sebago Lake is seen beyond to the r. (possibly also the city of Portland). Nearer are the twin round peaks of Double-Head, under the twin sharp peaks of Gemini, and over the dark and light green plaid of Black Mt. Then comes the high pyramid of Kiarsarge, with its summit-house, flanked by the Green Hills of Conway; and still more to the r. is the beautiful valley of the Saco, with the white hamlets of Conway and the distant gleam of Walker's Pond, the vista being closed by the Green Mt. in Effingham, and flanked by Moat Mt. The view then rests on the S. wall of Tuckerman's Ravine, and explores the profound depths below, resting on the dark Hermit Lakes.

78. Boott's Spur

is the great ridge running S. E. from Mt. Washington between Tuckerman's Ravine and Oakes's Gulf, and is nearly 3 M. long. It is about 5,000 ft. high, and has several masses of rugged rocks drawn across its line. The nearly level expanse between the base of the cone and the crags on the plateau was formerly called Carrigain's Lawn, and is now known as *Bigelow's Lawn* (in honor of Dr. Jacob Bigelow, an early explorer in this region). It consists of a narrow sedgy plain, rich in rare alpine plants, and strewn with worn rocks.

The excursion out on Boott's Spur may be made from the **Summit House**, to which the return should be effected. Strong pedestrians can do it as a digression from the Crawford-House path. The distance from the hotel to Bigelow's Lawn is about 1 M., and from that point to the outer crags it is over 1½ M. The Lawn is reached by descending the cone on a line about ½ M. W. of the route to Tuckerman's Ravine. This excursion is one of the most interesting that can be made from the summit of the mountain, and is easy of accomplishment, the only severe climbing being the ascent of the cone of Washington on the return. Care should be taken to choose a clear day, since fogs often settle down on the Spur, which would be likely to bewilder and imperil the visitor. It would be advisable, in threatening weather, to find and follow the old Davis Path (see Route 89), in order to secure a line of retreat.

Where the path turns suddenly to the r. the visitor should leave it and advance upon the crags in front, the second line of rocks which crosses the plain. The noble view described below is enjoyed from the central ledge in this line. Below this line of crags the Spur drops rapidly away toward the Ellis Glen, and a broad stony terrace is seen about 15 minutes' walk from the crest. From the edge of this terrace one can look into the ravines on the E. and S. E. of the Spur. On the N. end of the terrace is a lofty crag, which is easily accessible from the N. end of the rocks, and commands the best view which can anywhere be obtained of the cone of Mt. Washington, rising in impressive majesty over the darkness of Tuckerman's Ravine. Below and in front are the Hermit Lakes, the N. wall of the ravine, the outer portions and entrance, the fortress-like rocks of the Lion's Head, the expanse of the Alpine Garden, and the sharp-angled Nelson Crag, rising over the stupendous gray cliffs of Huntington's Ravine. The view from this point is one of the most grand and awe-inspiring in the mountains. The rock should be approached with caution, as a single slip would precipitate the careless visitor upon the sharp stones, a thousand feet below.

* *The View.* — The most conspicuous object is the imposing peak of Washington, which looms boldly over the high walls of Tuckerman's Ravine. The N. wall of the ravine is seen for a long distance, falling precipitously downward, and heading on the r. at the rocky promontory of the Lion's Head. Farther to the r. are the lofty gray cliffs of Huntington's Ravine, running up to the Nelson Crag. To the r. a broad view opens down the Peabody Glen, including the white Glen House, Pine Mt., and, beyond the Androscoggin, the rugged crests of Mt Hayes, Baldcap, and Ingalls, with the sharp peak of Goose-Eye far away. Nearer, and along the E. wall of the Peabody Glen, are Mts. Moriah, Imp, and Carter, the latter showing two fine peaks, that to the S. being the Carter Dome. Just across the Pinkham Notch to the E., about 2 M. distant, is Mt. Wild-Cat. Mt. Royce and Speckled Mt. are over Carter ; and Eastman is on the r., over Wild-Cat, with Sable next in line. Then the clearings on Black Mt are seen, under the similar and rounded crests of Double-Head and the twin spires of Mt. Gemini. Farther to the r., over the clearings of Jackson, are the mounds of Thorn Mt. and Thorn Hill. In the distance is Mt. Pleasant, near Sebago Lake ; and the view then rests on the high pyramid of Kiarsarge, and the Green Hills beyond. The Saco Valley then opens toward the S. in a long and beautiful vista, in which are seen the white village of N. Conway, the Ledges, and Walker's Pond, with the Green Mt. in Effingham beyond. Much nearer, down over a few miles of wilderness, are the pastured sides of Iron Mt., over which are the two peaks of Moat Mt. On the r. are the wooded ridges of Table and Bear Mts., beyond which rises the weird white peak of Chocorua. Under Bear Mt. is Mt. Parker.

About S. S W. are the red-ledged summits of Resolution (a long flat) and Crawford, over the former of which is the Bartlett Haystack, with Paugus above and to the l., to the r. of and continuous with Chocorua. In the same distant range are seen the dark dome of Passaconaway and the double crest of Whiteface, with Tremont nearer, over whose long flank is Green's Cliff. Above the latter is the serrated ridge of Tripyramid, with a part of Sandwich Dome on the r. The view-line now passes down the long valley of the Mt.-Washington River to the Nancy Range, whose peaks are overshadowed by the mighty mass of Carrigain (S. W.), behind which is a portion of Hancock. Over the upper flank of Carrigain the peak of Osceola is seen; and on the r. extends a broad expanse of the Pemigewasset Forest. About W. S. W., 7–8 M. distant, is the sharp peak of Willey, with Mts. Field and Tom on its r. Over Willey are the Potash and Coolidge Mts.; and over Field are Mts. Bond and Guyot, looming out of the central wilderness. Nearer, and forming the W. wall of the Mt.-Washington-River Valley, are Mts. Jackson, Clinton, and Pleasant, the latter lifting its hemispherical top within 3 M. Between Jackson and Clinton, far out beyond the Mt.-Willey Range, are the Twin Mts., beyond which, on the horizon, are the sharp and well-marked peaks of the Franconia Range, — Mts. Flume, Liberty, Lincoln, and Lafayette, the serrated summit of the latter being just to the l. of Pleasant. The Ammonoosuc Valley is partly hidden by the adjacent crags of Monroe, but the Sugar Loaves are visible, and to the r. are the wooded crests of Deception and Dartmouth.

79. Mounts Jackson and Webster.

These forest-bound peaks are but rarely visited, being difficult of access and possessing less landscape interest than their brethren to the N. They form the lower end of the Presidential Range, and, according to some, of the White Mts.

Mount Jackson is 4,100 ft. high, and is surmounted by a singular small peak, covered with bushes, which is visible from many distant points. The greater part of the summit is masked with trees, a circumstance which renders it of little interest to the lover of scenery. The longer way up is the easier, consisting of the ascent of Mt. Clinton by the bridle-path from the Crawford House, and then traversing the low ravine S. of Clinton. The distance is 1–1½ M., and the transit is very laborious, the way being frequently obstructed by thickets of dwarf spruce. Mt. Jackson was named by Mr. Oakes, the botanist, who sent his guide to its summit and had a bonfire kindled there to celebrate the christening.

Mount Webster is 4,000 ft. high, and is the point where the White-Mt. Range falls off sharply into the Notch. It was formerly known as the *Notch Mt.* The summit is wooded, and so possesses but little interest for scenery-seekers. The great feature of this mountain is its W. side, where it slopes to the Saco Valley in a steep wall, free from foliage, and striped with brilliant colorings by the slides which have laid bare the bed rock. It is rarely ascended, the best route being up the course of the Silver Cascade and its first tributary on the r., which should be followed to the edge of the cliff, whence a fine view of Mts. Willey and Carrigain is obtained. Prof. Hitchcock descended Webster along the steep track of one of the slides, near the Willey House. Mt. Webster was ascended by Prof. Tuckerman, in 1844, and he advanced hence along the entire ridge, botanizing.

80. Mount Clinton.

The summit of Mt. Clinton is 4,320 ft. above the sea, and 2,400 ft. above the Crawford House. It is reached from the Crawford Path (Route 88) by turning to the r. above the Mt.-Clinton guide-board, and ascending for a few rods. The summit is marked by a rude cairn, and a considerable plateau extends on either side. From the S. E. part of the crest is obtained a view of the shallow ravine to the S., with the bristling approaches to Mt. Jackson.

The View. — About N. N. E. is the fine peak of Jefferson, rising to a sharp point, and flanked on the r. by the saw-edged ridge of Clay, which runs into Washington, whose Summit House is plainly seen. The l. flank of Washington falls upon the graceful round dome of Pleasant; and its r. flank encounters the crags of Monroe, below which is the flat-topped Franklin. Oakes's Gulf is on the r., and its sides are scarred by long slides. The wider valley of the Mt.-Washington River approaches from the Gulf, across which is the long and darkly wooded Montalban Ridge. Over this the first peaks to the l. are Wild-Cat and the Carter Dome, with the twin flat crests of Double-Head to the r., followed by the noble pyramid of Kiarsarge, with its summit-house, and the distant Mt. Pleasant, near Sebago Lake. Mt. Clinton is seen from Portland over the S. shoulder of Kiarsarge. Close on the r. of Kiarsarge are the Green Hills of Conway, with a part of the Saco Valley and the distant ponds. The sharp peak of Moat Mountain is next seen, on whose r., and within 5 – 6 M., S. S. E., are the clustered crests of the Giant's Stairs, the flat-topped Resolution, and Mt. Crawford, the latter having a remarkably sharp summit. On the r. of Resolution is Table Mt., and over Crawford is the heavy ridge of Bear Mt., with the splendid white spire of Chocorua beyond. Over the r. flank of Crawford is the low Mt. Hope, with the Bartlett Haystack over it and the blue Ossipee Range beyond. On the r. of Chocorua and the Haystack is Paugus, low and rocky, below which is Tremont. Over the r. of the latter is the black hemisphere of Passaconaway, on whose r., and equidistant, is the cloven crest of Whiteface, with the ridge of Tripyramid, farther to the r., over the Nancy Range.

In this direction, about S. S. W., and within 1½ M., is the dark peak of Jackson, flanked by the rounded top of Webster. Directly over the former, on the distant horizon, is Sandwich Dome, with the Sachem Peak on its r. slope; and the majestic crest of Carrigain is on the r. of this line, about 10 M. distant. On the r., still more remote, is the crest of Osceola, with Tecumseh breaking off on the S. Then comes the massive Hancock, rising over the wilds of Pemigewasset. Nearly S. W., and 3 – 4 M. distant, is the alpine peak of Willey, falling off sharply on the S., with the rounded swells of Field and Tom in the same ridge, toward the r. Over the gap between Willey and Field are the distant Franconian peaks of Potash and Big Coolidge. Mts. Bond and Guyot are over Field; the S. Twin is over Tom; and farther to the r. is the N. Twin, marked by a long slide, and Mt. Hale is to the r. and below. Still farther out, over the ravine between Field and Tom, is the sharp spire of Lincoln; and Lafayette lifts its high, serrated ridge over the S. Twin. To the r. of and nearly over Hale are

Mt. Agassiz and Round Hill, back of Bethlehem, beyond which, to the r., are Mann's Hills and other summits in Littleton, with the Gardner range far beyond.   The white walls of the Fabyan House are now seen, down the Ammonoosuc Valley, beyond which is Beech Hill, with the long dark line of Cherry Mt. nearly N. W.   Dalton Mt. is farther away, on the l., over the plains of Whitefield.   Nearer, and running E. to Mt. Jefferson, are the wooded heights of Deception, Dartmouth, and Mitten.   Nearly over Dartmouth is Mt. Starr King, marked by a slide; and farther to the r. is the Pilot Range, filling the N. with its blue and tumultuous summits, among which the South Peak stands out prominently.

81. Mount Pleasant

is 4,764 ft. high, the peak rising 714 ft. above the gap on the S., and 364 feet above the N. gap.   As seen from distant points it always presents the form of a symmetrical dome, of a darkish hue, and is the most conspicuous peak on the S. W. range.   The summit consists of a slightly curving plateau of about six acres' area, covered with tufts of short grass, amid which are found rare flowers.   It slopes but slightly from the centre, and is "smooth enough for a parade-ground."   Its W. flanks are scarred by great slides, most of which occurred in 1826; and one of them is ½ M. wide near its base.   At the centre is a pile of stones, whence the view is best enjoyed.   Near the N. E. base of the cone is *Red Pond*, a dull puddle of bad water, whence it is said that the water flows both to the Ammonoosuc and the Saco, in seasons of heavy rains.   It was named on account of the red moss which is found in the vicinity.

The ways to Mt. Pleasant are described in Route 88.

* *The View* from Mt. Pleasant is one of the finest along the range, and should be studied carefully if that from Mt. Washington has been or threatens to be obscured by clouds.   Nearly N. N. E. is the massive peak of Jefferson, flanked on the r. by the humps of Clay.   On the long slopes of these two are the white stripes of several ancient slides.   Then the stately cone of Washington is seen with grand effect, with its houses on top, and the high trestles along the railroad, over which, at train-time, the puffing locomotives are seen toiling upward.   Under the r. flank of Washington, and close at hand, are the two rugged rock-piles which form the peaks of Monroe, towering over the still nearer level of Franklin, from whose E. side long slides fall away into Oakes's Gulf.   The long and massive plateau of Boott's Spur, which runs out to the r. from Washington, is higher than Pleasant, and hides all the mountains to the N. E.   The Mont-alban Ridge sweeps thence to the r., and is seen across the valley of the Mt.-Washington River, extending well to the S., covered with woods except at one burnt place, and maintaining nearly an equal height.   The first notable elevations seen over this ridge are parts of the Eagle and

Black Mts., in Jackson, over which appear the flat-topped twin crests of Double-Head and the pointed peaks of Gemini. The long ridge of Mt. Pleasant, in Maine, is seen on the S. E., crowned by a white hotel, and on its r. is the bright gleam of Sebago Lake. Tin and Thorn Mts. are nearer at hand, to the r. of Double-Head; and beyond is the graceful pyramid of Kiarsarge, with its summit-house. Over its l. flank the city of Portland can be seen with a strong field-glass, and the ocean beyond; and below it, on the r., are the Green Hills of Conway. A part of the Saco Valley is nearer at hand, with the yellow ledges on Iron Mt. still nearer, across the Rocky-Branch glen, and the Cathedral and White-Horse Ledges to the r. The rocky peaks of Moat Mt. are next seen, over the nearer ridges of Mts. Stanton and Pickering, and the sandy top of Mt. Langdon.

Within a few miles, across the valley of the Mt.-Washington River, is the Crawford group, with Giant's Stairs and the flat top of Resolution in the centre, the crest of Parker on the r., and the sharp red peak of Crawford on the l., Hope being below and beyond. Table Mt. is over and beyond Parker. Bear Mt. is over the r. of Resolution, and the Bartlett Haystack is over Crawford and Hope; while Tremont stands at the foot of the Mt.-Washington glen, over a part of the Saco Valley and a section of the P. & O. Railroad. Still farther away rises the Sandwich Range, Chocorua lifting its line of splendid white peaks over the l. of Bear Mt.; Paugus adjoining it on the W., over the r. flank of the Haystack, with Belknap far beyond; and the round dome of Passaconaway is over the r. of Tremont. Farther out in this direction is the cloven peak of Whiteface, flanked by the fine sierra of Tripyramid, far beyond which is Sandwich Dome, showing a rounded top, with a sharp point on the W. end (about S. S. W. from Pleasant). Under these and much nearer, across the Saco Valley, are the peaks of the Nancy Range.

The view-line now passes over the bold little peak of Jackson, within 3 – 4 M., and rests on the vast pile of Carrigain, which rises by successive terraces from the S. to a high rounded peak. To the r. and more distant are the crests of Osceola and Tecumseh; and Hancock adjoins Carrigain on the W. On the r. of Jackson is Webster, whose r. flank runs into the brown top of Clinton. To the r., beyond this ridge, is the Field-Willey Range, Willey showing a fair peak, falling slowly to the S.; Field marked by a rounded crest on the r , with Willard below; and Tom farther to the r. Moosilauke is seen far away toward the S. W. Over the l. flank of Field are the distant Franconian peaks of Potash and Coolidge; and over its r. flank is the high crest of Mt. Bond. Over the r. of Tom is the S. Twin, on whose r., and continuous, is the N. Twin, with Hale on the r. and below. The remote spire of Liberty is over the r. of Bond and Field; Lincoln is over Tom; and the noble sierra of Lafayette is over the r. of Tom and beyond the S. Twin. Above the ridge which runs N. from Field are the Sugar Loaves, near the Twin-Mountain House; and

Agassiz is beyond. The view now extends down the Ammonoosuc Valley, beyond the white walls of the Fabyan and White-Mountain Houses, to the hills and houses of Littleton. On the r. of the valley are Beech Hill and Cherry Mt., the latter being marked by the bold peak of Owl's Head, on its N. end. Dalton Mt. is beyond, across the plains of Whitefield.

About N. W., and to the r. of the near and forest-covered crest of Deception, are Prospect and Pleasant, near Lancaster; and Dartmouth and Mitten lift their woody heights on the r. of Deception and continuous with it. Jefferson-Hill village is over the gap between Deception and Dartmouth, with Mt. Starr King to the r. and over it, followed by the general assembly of the Pilot Range, Deer Mt., and Green's Ledge.

82. Mount Franklin

is 4,904 ft. high, or 504 ft. above the gap on the S., while to the N. there is but little depression. On the latter account, and also by reason of its flat and unmarked summit, Franklin has but little individuality, and can hardly be recognized from distant points, appearing as a spur of Monroe. It is connected with the latter mountain by a long and narrow ridge, over which the bridle-path passes. The level summit of Franklin is traversed near its middle by the winding and well-marked path. (See Route 88.)

* *The View.* — The stately peak of Jefferson is about N. N. E., 4 M. distant, and is flanked by the lower and broken ridge of Clay, over which towers the ponderous mass of Washington. Close at hand are the two craggy crests of Monroe; and the high plateau of Boott's Spur sweeps away on the r., with the dark Montalban Ridge passing from it to the S. Between these formidable heights and Franklin is the deep and wooded ravine of the Mt.-Washington River. Over the downward trend of the Montalban Ridge the twin crests of Double-Head and of Gemini are visible; and Mt. Pleasant is far away toward Sebago Lake, over which Portland may be seen. Kiarsarge lifts its fine pyramid S. of S. E., nearly over Thorn Mt., and is flanked on the r. by the Green Hills of Conway, with N. Conway at their base. Lower Bartlett is at the head of a long reach of the Saco Valley, in which the clear mirror of Walker's Pond is seen, with the Ledges; and the highlands of Eaton and Freedom are far beyond. The yellow ledges of Iron Mt. next appear, much nearer, and then Moat Mt. lifts its fine rocky peaks, beyond Mts. Stanton and Pickering, on whose r. stands Mt. Langdon. The cone of Parker, the flat top of Resolution, the wooded crest of Giant's Stairs, and the red peak of Crawford are then seen, nearly S. and 5-6 M. distant, forming a compact group towards the end of the Montalban Ridge. Over these, and running to the r. from Moat, are the flat-topped Table Mt., the long ridge of Bear Mt., the black cone of the Bartlett Haystack, and the crests of Tremont; and again, over this line appear the high white peaks of Chocorua, the curving ledges of Paugus, and the dark hemisphere of Passaconaway. Portions of a fourth line appear in the remoter distance, consisting of the blue Ossipee Range and the twin Belknap summits. From Passaconaway to the r. extend the cloven crest of Whiteface, the triple heads of Tripyramid, and the great swell of Sandwich Dome.

The view now rests on the nearer Nancy Range, across the Saco Valley, beyond which loom the majestic peaks of Carrigain and Hancock, with Osceola and Tecumseh between. Hancock rises over the nearer crests of Jackson and Webster; the bold mass of Willey is over Clinton; and Moosilauke is far away in the S. W. Over the adjacent dome of Pleasant is the rounded top of Field, with the Coolidge and Potash Mts. far beyond. Over the r. flank of Pleasant are Mts. Bond and Guyot, and over Tom is the S. Twin, adjoined by the slide-striped N. Twin. The Franconian peaks of Liberty, Lincoln, and Lafayette appear over the Twin range. Over the r. of the Field-Willey chain are the Sugar Loaves and Mt. Hale, beyond which appear Mt. Agassiz and a part of Bethlehem, with the white Fabyan House in the foreground. Beyond Beech Hill Littleton's highlands are visible; and farther to the r. is Cherry Mt., with its Owl's-Head peak. Over Deception, Dartmouth, and Mitten, in the foreground, are the more distant summits of Prospect and Pleasant, with the rich plain and village of Lancaster. Many of the Vermont hills are seen in this direction. Farther to the r. one sees the white hamlet of Jefferson Hill, Mt. Starr King, and the blue Pilot Range, terminating on the E. in Deer Mt. and Green's Ledge.

83. Mount Monroe

is 5,384 ft. high, or 284 ft. above the gap on the N., and 480 ft. above Mt. Franklin. The main peak lies to the N. of the ridge, and on the S. is a minor crest, 5,204 ft. high, which was called *Little Monroe* by Prof. Guyot. The bridle-path formerly passed between these two peaks, but is now on the E. of both of them (see Route 88). On account of its sharp and massive crags Monroe presents a fine alpine appearance to the distant observer, — which is, however, somewhat lessened by its vicinity to Mt. Washington, whose greater altitude dwarfs it. From points near at hand, where the latter is not visible, Monroe has a formidable aspect, and the noble symmetry of its craggy walls excites the most lively interest.

A little way N. of the pool on its E. flank, and up the steep slope to the W., is seen a natural portal in the rampart-like rocks of Monroe's crest, 12 ft. wide and with walls 12 – 15 ft. high. The detour from the path to the summit and back need not take more than ½ hr., and will richly repay the tourist.

Near the N. E. base of Monroe is one of the " tiny pastures " of the high ridges, so highly esteemed by botanists, and abounding in rare and delicate alpine grasses.

On the stony plateau near by are found alpine cinquefoils, and other unusual plants. The cloudberry, the mountain rattlesnake-root, and the alpine bearberry are also seen in this vicinity, and along the S. W. peaks generally. The descent into *Oakes's Gulf* is sometimes made from this mountain, and many rare plants and flowers are found therein.

* *The View.* — The peaks of Jefferson and Clay are seen close at hand towards the N., very much foreshortened; and then the visitor gets one of the best near views of Washington, including its long W. flank, with the railroad-station in the valley, the trestles on the slopes, and some of the houses above. Just over the l. flank appears a part of Adams. The rocky ridge of Monroe runs out towards the Lakes of the Clouds, and the winding bridle-path is seen ascending the high cone of Washington. Farther to the r. the great craggy plain of Boott's Spur runs to the r., beyond the dark Oakes's Gulf, which lies below, within pistol-shot. The twin crests of Double-Head are toward the S. E., over the descending forests of the Montalban Ridge, with Mt. Pleasant (in Maine) beyond; while far away in this direction is Portland, at the gates of the sea. The sharp point of Kiarsarge comes next, nearly over Thorn, and with N. Conway on the r., below the Green Hills, then Conway, the Ledges, and Walker's Pond, well down the fair Saco Valley. Iron Mt. is much nearer in this direction; and over Mts. Stanton and Pickering is Moat Mt., near N. Conway. Then comes the Crawford group, near at hand, at the end of the Montalban Ridge, with the sharp slope of Giant's Stairs falling towards Resolution, Parker being on the l. and the acute peak of Crawford on the r. Beyond this cluster is the range which runs W. from Moat, — the plateau of Table Mt., the broken crest-line of Bear Mt., the dark peak of the Bartlett Haystack, the burnt ridge of Tremont, and the square-sided Green's Cliff. Chocorua is over the gap between Table and Bear, showing stately white peaks; Paugus lifts its slow curve over the Haystack; black-domed Passaconaway is over Tremont; Whiteface is on the r. of Passaconaway; and the sierra of Tripyramid is at the apparent end of the Saco Valley, over Green's Cliff. The blue lines of the Ossipee Range are farther away. The Nancy Range is at the foot of the Mt.-Washington-River glen, and over it is Sandwich Dome, indenting the horizon with its slow curve, and bearing the Sachem Peak on its r. end. Then come the noble peaks of Carrigain and Hancock, with Osceola and Tecumseh through the gap between them. Clinton and Jackson lie to the S. W., in the Presidential Range, with the sharp head of Willey nearly over them; and to the r. the adjacent plateau of Franklin lies toward the round top of Pleasant, over which is Field, continuous with Willey and overlooked by Mts. Bond and Guyot, whence the long ridge of the S. and N. Twins runs to the r. Nearly over the l. of Mt. Bond is Moosilauke, with the Coolidge, Potash, and Flume Mts. to the r.; and farther N. are Mts. Liberty, Lincoln, and Lafayette, lifting their sharp and distant peaks over Mt. Tom, — Lafayette being over the Twins. The rugged peak of Little Monroe lies close below, towards the

Ammonoosuc Valley. in which are seen the Fabyan and White-Mountain Houses, the knolls called the Sugar Loaves, and Mt. Hale, N. of the Twins. Mt. Agassiz and part of Bethlehem and Littleton villages are beyond; and the Dalton Mt. is farther away, towards the long line of the Vermont mountains. On the r. are the dense woods of Beech Hill and Cherry Mt., on the latter of which a dark peak rises towards the N.; and the shaggy sides of Deception are nearer. Farther towards the N. are Mts. Prospect and Pleasant, with the village of Lancaster; and then come the white houses of Jefferson Hill, under Mt. Starr King, and with the long lines of the Pilot Range on the r.

The Lakes of the Clouds are two tiny tarns in the depression between Mts. Washington and Monroe, 5,000 ft. above the sea. Their basins were excavated by the great drift current, and the glacial scratches are found in their vicinity and for 200 ft. above. The Ammonoosuc takes its rise here, and falls over 2,000 ft. in the first 3 M. of its course. The larger pond covers about ¾ of an acre, and is of an oval shape, with deep, cold, transparent, and sweet waters, and a bottom of rock and gravel. There is a small spring at the S. E. angle. The other pond is smaller, and is somewhat boggy.

The Lakes of the Clouds were observed by Thomas Gorges, in 1642. and he thought that they formed the source of the Connecticut River. They were visited by the Lancaster explorers in 1820, and then the larger received the name of *Blue Pond*. When the Rev. S. J. May made the ascent, he found the name of *Washington's Punch Bowl* attached to this lake. In the vicinity of the lakes may be found the willows and alpine birch, the alpine violet and bistort and cranberry, *Linnæa borealis*, the common harebell, and numerous other species of plants.

There are pretty cascades on the young Ammonoosuc River, about ½ M. below the lakes, but the route thither is difficult. They have also been visited by leaving the railroad train on Jacob's Ladder, crossing the slides, and ascending the first brook. The lower part of the Ammonoosuc Ravine still awaits an explorer.

84. Mount Madison

is 5,365 ft. above the sea, and 3,733 ft. above the Glen House. It occupies a prominent position in the views from the Glen House and Gorham, and other points on the N. and W. Starr King bestowed upon it the epithets of "the Narcissus of the range" and "the Apollo of the highlands," calling it also "beautiful, clear, symmetrical, proud, charming, gigantic," and "of feminine symmetry."

The summit consists of a narrow ridge of rocks 3 – 4 rods long, from which long and ponderous flanks descend toward the Peabody and Moose Rivers and the Great Gulf. Five ravines head into its sides, cutting in deeply on either flank of the ridges. The upper parts of the mountain are above the forests, though below the district known to naturalists as the alpine region. They are covered with ragged fragments of weather-beaten rock, among which nestle a few rare flowers.

** *The View.* — Nearly S. across the wide chasm of the Great Gulf is the Ledge, on Mt. Washington, above and below which are the long lines of the spur which trends downward from the Summit House towards the Glen House. Parts of the winding white stripe of the carriage-road are seen, and to the r. of and below the summit is the head of the Great Gulf. Next comes the rough ridge of Clay, with the deep ravine on the N. The noble pyramid of Adams is about 1½ M. distant, hiding Jefferson, with serrated cliffs descending towards the Great Gulf, and a long battlemented ridge approaching Madison. Above the r. flank of Adams is the long black ridge of Cherry Mt., over whose highest point are the highlands of Fayston, Vermont, with the Adirondack peak of Mt. Marcy beyond, and a trifle to the r. To the r. of Cherry is the village of Whitefield, beyond Island Pond, with the long low ridge of Dalton Mt. over it. Over Whitefield, and but a trifle N. of W., is the stately and distant peak of Mt. Mansfield, the chief of the Green Mts. Cherry Pond is to the r. of and near Whitefield, and the Martin-Meadow Pond and Hills are on the r. of Dalton Mt., with Mt. Niles and Mt. Tug beyond, and the rugged line of Sterling Mt. low down on the horizon. The eye now follows the course of the Israel's-River Valley to the low rounded tops of Mts. Prospect and Pleasant, below Lancaster, over which are the Umpire and Burke Mts., and the Victory Hills; and Jay Peak lifts its dim point far away on the N. W.

The white houses of Jefferson Hill are seen much nearer at hand, at the foot of Mt. Starr King, over whose long and wooded S. slope is a part of Lancaster. Nearly over Lancaster, out in Vermont, are the Cow and Burnside Mts. of Guildhall, Mt. Seneca, and East-Haven Mt.; and then the sharply cut-off Mt. Hor, at Willoughby Lake, with the Westmore Haystack on its r. Between the two last is the distant peak of Owl's Head, at Lake Memphremagog. To the N. W., across the Moose-River Valley, are the dark Randolph and Pliny ranges, covered with forests, and culminating at Mt. Starr King, over which is Owl's Head. The Pilot Range is to the r., and runs far to the N. with a long line of dark and wooded peaks. The wilderness-bound lake of the Pond of Safety is but a few miles away, above the Randolph Mts.; and nearly over them, far away, are the mountains about the Smuggler's Notch in Brunswick, to the r. of which is the distant Pinnacle Mt., in the Eastern Townships of Canada. Over the three chief peaks of the Pilot Range is the long Bowback Mt., in Stratford. The round white domes of the Percy Peaks are about N. N. W., with the Stratford Mt. behind, and Long Mt. extending to the r., back of which is the peak of Mt. Whitcomb. Mill Mt., near Stark is on the r. of the Percy Peaks, and farther to the r. are Green's Ledge and Deer Mt. A little E. of N., over the nearer Crescent range, is Cedar Pond, near W. Milan, over which is Patience Peak, in Millsfield, with the Dixville Mts. farther out, on the l. Head Pond, in Berlin, is next seen, far over which are Mt. Pisgah and Mt. Carmel, the latter being on the Maine border, near the Canada line.

Farther to the E. is a long extent of the Androscoggin Valley, in which appear the white villages of Berlin Falls and Milan, with Mt. Forist and the Milan Hills near them, and the Chickwolnepy range beyond. In the far distance, in the narrow arc of the horizon between Mt. Carmel and the view-line prolonged over Milan, are the Diamond Peaks, the Half-Moon Mt., the sharp spire of Mt. Dustan (N. W. of Umbagog), and the bare and rounded summit of Aziscoös, with the Magalloway Mts. beyond and on the W. Over Milan and Berlin Falls is the distant glimmer of Lake Umbagog, with Lakes Welokenebacook and Moosetocmaguntic beyond, over which appears a long line of lofty nameless peaks in the wilderness of Maine. Nearly N. E., at a great distance, is Mt. Bigelow, showing a sharp peak near the centre of a ridge. Farther to the r. are the Pierce-Pond Mts., and then the view rests on the cloven top of Saddleback and the uneven peaks of Mt. Abraham. Nearer than these, but in the same direction, are the Bear-River White Cap and Speckled Mt., at the Grafton Notch. Following over the rocky ridges, beyond a part of Gorham which is visible close at hand, the view-line reaches Goose-Eye, its highest part cutting the sky, and the mountains of Byron appearing over its ridges.

The eye now rests gladly on the beautiful green meadows of the Androscoggin, through which the river winds in a series of exquisitely graceful curves. Back of white Gorham, beyond Pine Mt., is the rugged crest of Mt. Hayes, over which are the Sunday-River White Cap and the fine distant peak of Mt. Blue, in Avon. The farms of Shelburne and Gilead are farther down the Androscoggin Valley, with Baldcap and Mt. Ingalls just to the N., and Puzzle Mt. and the Rumford White Cap beyond. A little N. of E., down the valley, are the white houses of Bethel, around which rise many hills.

The view now rests on the neighboring summits across the Peabody Glen, that to the N. being Moriah, flanked on the r. by Imp. Then come the high and formidable peaks of Mt. Carter and the Carter Dome, marked by a slide on the S.; and across the deep ravine of the Pinkham Notch is Wild-Cat, with a long and curving ridge. The white walls of the Glen House rise at the apparent junction of the Carter and Pinkham Notches. Over these ranges stretches a vast area of Western Maine, extending to the mountains of Waterford, Norway, and Paris Hill. The rocky top of Royce is over Imp, and the white crests of Baldface are over the Carter Dome. Much farther away in this direction are long reaches of the ocean. It is thought that a part of Sebago Lake is seen in this direction, towards which are several mountains of Brownfield and the Saddleback Mt. in Baldwin. Over the r. of Wild-Cat is the N. peak of Double-Head, on whose r. is the graceful pyramid of Kiarsarge, with its crowning hotel.

About S. S. E. is N. Conway, standing whitely out in the pleasant Saco Valley, and flanked by the Ledges on the W., with Walker's Pond beyond. Farther to the r. are the low ledges of Iron Mt., with the fine peak of, Moat Mt. beyond, and a part of the Green Mt. in Effingham far away on the horizon. Nearly S., over the E. flank of Washington, it is thought that the white crest of Chocorua may be seen.

85. Mount Adams

is 5,794 ft. high, or about 500 ft. lower than Mt. Washington. Although it is the second mountain in New England in respect to altitude, it is exceeded by no other in picturesque grandeur and bold alpine character, on account of its sharp and slender peak and of the profound ravines which traverse its flanks. As seen from most points it presents the appearance of a symmetrical pyramid, rising from a high and narrow ridge and flanked by bold crags. There are no ledges on its summit, but a heap of frost-broken fragments of rock. It is doubtful whether its summit was ever attacked by the grinding and erosive power of the glaciers, the evidence being in favor of its immunity. The main peak is flanked on the N. by a minor summit, and on the S. W. by a long rocky ridge.

This peak may be reached either by the route over Madison (see Route 90), or by its reverse, as taken to the N. from Washington. The march from Washington to Adams and back will require a long and clear day, because, though the distance is short, a line of high peaks must be crossed on the way. It may also be reached by ascending King's Ravine, a very difficult mode of access (see Route 75). The new path opened by Mr. Lowe, in 1876, is beyond all comparison the best way to reach the crest of Mt. Adams (see Route 90).

There are three distinct ridges running N. W. from Mt. Adams, of which the first forms, with Madison, the Madison Ravine; the second (called *Nowell's Ridge*) and the first enclose King's Ravine; and the third, with the Castellated Ridge of Jefferson, encloses the Ravine of the Castles, across whose outlet is the bold foot-hill of Mt. Bowman.

Starr King gave the name of *John Quincy Adams* to the rampart-like crest which lies between the main peak of Mt. Adams and Mt. Madison, at the head of King's Ravine ; and Prof. Hitchcock applies the name of *Mt. Quincy Adams* to a lesser peak on the S. of the highest crest. It is doubtful whether these minor crags merit separate names, but if they do the similarity of those aforementioned, as applied to different points, is liable to create confusion.

Considerable confusion has been caused by the fact that the names of Mts. Adams and Jefferson were transposed on Bond's map and on Walling's Coös-County map, and many still believe that Adams is next to Washington. The report of the Lancaster party that named the peaks (published in 1823) distinctly states that "Mt. Adams is known by its sharp terminating peak ; and Mt. Jefferson is situated between" Washington and Adams.

"From the top of this pyramid of Adams, whose rocks are so huge and lawless that it would be scarcely possible to make a horse-path to it from the plateau, we gained glorious views of the northern country, — the beautiful Kilkenny Range, the lovely farms and uplands of Randolph and Jefferson, the long unrolled purple of the Androscoggin, making a right angle at the Lary Farm, the Pond of Safety, on the N. side of the Pilot Hills, and Umbagog, Richardson Lake, and Moosetocmaguntic, whose dreamy waters, framed by the unbroken wilderness, are stocked with portly trout, and haunted by droves of moose." (STARR KING.)

** *The View.* — W. of S., about 4 M. distant, is the Summit House on Mt. Washington, with the Chandler Ridge below on the l., and the long N. E. spur trending away to the Glen House, with parts of the carriage-road. Below are the dark depths of the Great Gulf, from which long and shallow ravines run up to Washington. On the r. slope parts of the railroad and the tanks are seen; and the low humps of Mt. Clay close in at the head of the Great Gulf. About S. W., and 1½ M. distant, is the ragged and massive crest of Jefferson, with narrow slides towards Adams, and vast brown and gray ledges along its upper courses. Portions of the S. W. peaks may be visible, though they are not alluded to in the Editor's field-notes from Adams. Over the r. flank of Washington, and far away, is the ridge of Tripyramid, on whose r. is the round swell of Sandwich Dome, with the Sachem Peak on its r. flank. Then comes the imposing peak of Carrigain, with gracefully rounded crests and their sharp slopes to the enclosed ravine. Hancock appears from behind Carrigain; and the rolling peaks of Osceola and Tecumseh are far beyond. Through the gap between Clay and Jefferson is the brown-sided Willey, sloping to the S., with Mt. Field on the r. and continuous. Far away over Willey is Mt. Cardigan; and several distant and unrecognizable peaks loom up beyond Field.

On the r. of Jefferson is the distant sharp spire of Liberty, with the high and house-crowned ridge of Moosilauke still beyond. Mts. Bond and Guyot and the Twins are nearer, the N. Twin being marked by a white slide; and over them appear the pinnacle of Lincoln, the yellowish ledges of Lafayette (falling to the r.), the rounded heads of Mts. Cannon and Kinsman, and the sharp nearer crest of Haystack. Mt. Hale is still nearer, and the Landaff and Lisbon hills are clustered beyond. The Deception and Dartmouth ranges of wooded highlands are below the ridge towards Jefferson. The hotels of Bethlehem are plainly seen over Beech Hill, with Mt. Agassiz behind. In the W. are the double lines of the Green Mts. of Vermont and the Adirondacks of New York, seen nearly in the same relations as from Mt. Washington (see Route 95). Due W., and but a few miles distant, is the long black ridge of Cherry Mt., with small clearings on its sides, rising over the lower knolls of Adams. Over its l., and r. of Bethlehem, is Littleton, flanked by rolling hills; and Camel's Hump is over its higher part (with Marcy on the l., and farther away). Beyond the glimmer of the Cherry and Island Ponds is the white village of Whitefield, back of which rises the long, low, and dark ridge of Dalton Mt., with Mt. Mansfield looking over it far away. A part of the Connecticut Valley is seen just to the r., with Mt. Niles beyond, and Mt. Tug over the Lunenburg Heights. Down the populous valley of Israel's River are the rounded Mts. Prospect and Pleasant, near Lancaster, over which are the Victory

Hills and Burke Mt. To the l. of the Martin-Meadow Ponds is the high hamlet of Lunenburg, with the lofty Lowell and Belvidere Mts. far beyond. To the r. of Prospect, in the distance, is the sharp Willoughby Notch, between Mts. Hor and Annanance, in which lies Willoughby Lake. To the l. is the remoter point of Jay Peak, and the Westmore Haystack is on the r. About N. W., not many miles away, is Mt. Starr King, marked by a small diagonal slide near the central peak; and on its l., over the near white building of the Mt.-Adams House, is the hamlet of Jefferson Hill, above and beyond which is the large village of Lancaster. Far away over Lancaster is the rounded top of Owl's Head, on the Canadian shore of Lake Memphremagog, with Mt. Orford to the r.; and nearer, on this line, are Mt. Seneca and East Mt. Over the r. of Starr King is the Smuggler's Notch, in Brunswick, with high Canadian peaks beyond.

The view now passes over the near and wooded Randolph range and on to the blue assemblage of the Pilot Mts., over whose highest peaks appear the Bowback and Sugar-Loaf Mts. The white domes of the Percy Peaks are about N. N. W., over the Pilots, with Stratford Mt. behind and the Long Mt. of Odell to the r., overlooked by Mt. Whitcomb. Parts of the Stark and W. Milan valleys are seen in this direction, on the r. of the Pilots, and connected with them are the heights of Green's Ledge and Deer Mt. Due N. is Patience Peak, in Millsfield, beyond which are the Dixville and Magalloway Ranges, with the remote crest of Mt. Megantic, near Megantic Lake. The Crescent range is now seen, near at hand below, under Head Pond; nearly over which, far away over Odell and the Dartmouth-College Grant, are Mts. Pisgah and Carmel, with Saddle Hill and Mt. Nicolet still farther off, on the Maine-Canada border. The weary eyes now rest on the adjacent Androscoggin Valley, running N. from Gorham (with Mt. Hayes behind) to the white villages of Berlin Falls (with Mt. Forist on the l.) and Milan (with the Milan Hills on the l.). Through these pleasant meadows the river flows peacefully, in long bright curves. Over the l. of Milan is Mt. Dustan; and over its r. is Lake Umbagog, with the Rangeley Lakes beyond, and the great Mt. Aziscoös over it. The view now passes over Mt. Madison, and is nearly identical with the view from that peak (see page 212) as far around as Mt. Wild-Cat. Nearly S. E. through the Carter Notch (near the Glen House) is Mt. Pleasant, a long ridge crowned by a hotel, directly over which the city of Portland is visible, beyond the bright sheen of Sebago Lake. The Upper Moose Pond and Long Pond are over the l. of Pleasant, and Lovewell's Pond is on the r. near Fryeburg and in line with Saddleback Mt. in Baldwin. A great expanse of Western Maine, bordered by the dim sea, is visible in this direction. Nearly over a little notch in Wild-Cat is Double-Head, with the fine pyramid of Kiarsarge on its r., crowned by a house, and looming over Thorn Mt. The Green Hills of Conway appear beyond; and the Eagle Mts. run down from the r. end of

10

Wild-Cat. The rich Saco Valley next appears, with the white villages of Lower Bartlett, N. Conway, and Conway, the Ledges, and Walker's Pond, the vista being closed by the Green Mt. in Effingham. On the r. is Moat Mt., over which is a part of Ossipee Lake; and Copple Crown Mt. is far away toward the S.

86. Mount Jefferson

is 5,714 ft. high, or 80 ft. lower than Adams, 735 ft. above the gap on the S., and 775 ft. above the gap on the N. The dull gray of the rocks on the N. is relieved by many blocks of white quartz; and on the S. is the long grassy slope of the Monticello Lawn. There are two peaks of nearly equal height, lying W. N. W. and E. S. E., with a minor nubble towards Adams. The E. side falls sharply off into the Great Gulf, but on the W. and N. W. longer spurs make out, the chief of which is the *Castellated Ridge*, whose vast crags resemble the walls of a battered fortress. In the depressions S. E. of the main peak snow remains until July. The profile of the Sentinel is to the W. of the main peak.

The old Lancaster trail formerly led over the flanks of Jefferson, leaving the Jefferson road a little way E. of the Mt.-Adams House. It has been long discontinued, and is now overgrown. Messrs. W. F. Channing and E. E. Hale ascended this peak in 1842, by the ravine of Israel's River. The present ways of access are described in Route 90. The top of Jefferson may be visited in little more than a half-day from the Summit House (including the return).

* *The View.* — Toward the S. E. is the vast mass of Washington, with its roads and hotel, and the sides and bottom of the Great Gulf. Clay is below, over which is the distant Ossipee Range; and above the slope between Clay and Monroe are the round-headed Paugus and the black-pointed Bartlett Haystack, with Tremont and dark-domed Passaconaway nearly over Monroe. The white knoll of Potash is at the foot of Passaconaway, and on the r. is the split summit of Whiteface. The lower terrace of Franklin is on the r. of Monroe, with Green's Cliff over and beyond it, and the serrated ridge of Tripyramid to the r. Farther away, to the r., is the great curve of Sandwich Dome. Over the near dome of Pleasant is the Nancy Range, and the peaks of Jackson and Webster are on the l. of Pleasant, with Clinton on the r., over which looms the lofty Carrigain, beyond the Nancy Range. Hancock is on the r. of Carrigain, and over it peer the distant crests of Osceola and Tecumseh. Above the dark forests towards the Notch towers the sharply cut-off Mt. Willey, with Mts. Field and Tom descending to the N. To the r. of Willey is the remote Mt. Cardigan; and the rounded Mt. Willard is over against Willey. The Loon-Pond Mts. appear over Willey; Mt. Carr is far away over Field; and the round ridges of Mts. Bond and Guyot are over Tom. On the r. of Bond is Mt. Flume, to the r. of which are the high points of Liberty, Lincoln, and Lafayette, with Moosilauke's long ridge over Liberty, the S. Twin under

Lafayette, and the N. Twin on the r., with a slide-striped side. Mt. Hale and the Sugar Loaves are just to the S. of the Ammonoosuc Valley; and the view then passes out over the white Fabyan House and the village of Bethlehem to Mt. Agassiz, Sugar Hill, the heights of Landaff and Lisbon, Mt. Gardner, and the village and hills of Littleton. Camel's Hump is far beyond, due W. The wilderness-ranges of Mts. Deception and Dartmouth then fill the foreground, with the long dark Cherry Mt. over them, marked by the peak of Owl's Head on its r. end, while beyond are the ponds and village of Whitefield, backed by Dalton Mt. Over the l. of Dalton are Elmore Mt. and Mt. Mansfield, far away on the horizon; and over the r. are Mt. Niles, the Umpire and Burke Mts., and the Lowell Mts. Nearly over Cherry Pond and close to the r. of Dalton Mt. are Mts. Prospect and Pleasant, with Lancaster village to the r., and the Cow and Burnside Mts. beyond. Still farther away in this direction are the peaks about Willoughby Lake, the Westmore Haystack, the remote Jay Peak (to the l.), and Owl's Head and Orford Mts., near Lake Memphremagog. The view now rests on the nearer Mt. Starr King, with the village of Jefferson Hill at its foot, below Lancaster; and to the r. extends the great group of the Pilot Mts., over which are seen the white Percy Peaks, the Bowback and Stratford Mts., and the mountains about the Smuggler's Notch, in Brunswick. In the foreground, towards and r. of Starr King, are the densely wooded Pliny and Randolph ranges; and on the r. of the Pilots are Deer Mt. and Green's Ledge.

(The Editor was prevented from getting the view between N. and E. by a massive cloud which came down on Jefferson during his visit, and remained throughout the rest of the day. The view from Jefferson in these directions is not essentially different from that from Adams, except in so far as the noble peak of Adams itself, rising near at hand, eclipses Madison and many of the mountains in that direction) To the S. E. the view includes Long Pond and Sebago Lake, with the long ridge of Mt. Pleasant, surmounted by a hotel. Over the r. end of this ridge the city of Portland is visible. A vast area of Western Maine, bordered by the ocean, is seen in this direction, and thence N. to the Canadian frontier. The view towards Kiarsarge, Double-Head, and the Saco Valley is closed by the massive N. E. flanks of Mt. Washington, which cannot be overlooked.

The * **Castellated Ridge** runs about N. N. W. from Mt. Jefferson, and may be visited in about 4 hrs. from the main peak and return. But little underbrush is met on the way, the course lying over rocky slopes. The ridge is but 5 – 20 yards wide along its crest, falling away gradually on the W., and faced with formidable precipices on the E. The Castles are about 1½ hrs. march from the peak of Jefferson, and are three in number, the central one being the largest, and the northern one the most sharply cut. They consist of lofty piles of rock, with nearly perpendicular sides; and support minor crags which resemble turrets and battlements (shown in a picture on page 381, in *The White Hills*). They can be scaled, though with difficulty. *The Ravine of the Castles* is wild, steep, rocky, and well watered.

87. Mount Clay

is close to Mt. Washington on the N. W. and forms the precipitous head of
the Great Gulf. It is 5,553 ft. high, or only 136 ft. higher than the gap
towards the S.; and consists of a long ridge on which rise three well-
marked hummocks, whereon is exhibited a fine regularity of stratification.
"The humps of Clay" are formidable to travellers on the long march from
Randolph, since they have to be ascended in order over steep acclivities.
They may easily be visited in a half-day from the Summit House. The
view of the Great Gulf from this ridge is very impressive. The peaks are
low and steep-sided, and the hollows between them are carpeted with
tough sedges. The N. peak is the highest, and from it the following
view is taken. The route to Clay diverges from the second tank on the
railway.

* *The View.* — Mt. Washington is seen close at hand, in all its vastness,
with the Summit House, the great N. W. and N. E. flanks, and parts of
the carriage-road and railway. Between the S. peak of Clay and Monroe are the dark cone of the Bartlett Haystack, the round head of Paugus,
and the blue Ossipee Range. The dark hemisphere of Passaconaway appears over Monroe; and the cloven crest of Whiteface is between Monroe
and Little Monroe. On the r. of the latter is the sierra of Tripyramid,
with the long swell of Sandwich Dome on its r. Below Monroe is Franklin's level top, over which is the Nancy Range, and far away to the l.
(nearly above Paugus) is Mt. Belknap, over the visible gleam of Lake
Winnepesaukee. A few miles away, on the Presidential Range, is the
high dome of Pleasant, flanked by Jackson and Webster on the l. and
Clinton on the r. Over its l. is the imposing crest of Carrigain, with Hancock over the r. flank of Clinton, and the low peaks of Osceola and Tecumseh far away beyond. Low down on the horizon is Cardigan, over the r.
of Hancock. The sharp S. flank of Willey is beyond the Notch, with
Field and Tom continuous to the r., Willard below the l. of Field, the ravine of Beecher's Cascades on the r., and a long granite ridge (forest-clad)
running N. towards the Fabyan House, including Mt. Andalusite and Mt.
Rosebrook. Far away over Field is Mt. Carr; and the round tops of Mts.
Bond and Guyot are over the gap between Field and Tom, with the
peaks of Flume and Liberty beyond; and the high ridge of Moosilauke
closes the vista. To the r. of Bond is the S. Twin Mt., over which are
the sharp spire of Lincoln and the sierra of Lafayette. N. Twin is on
the r. of Lafayette, below, and has a slide on its side; and near its N. end
are Mt. Hale and the Sugar Loaves. The view now traverses a great area
of the Ammonoosuc Valley, from the Marshfield station and the white
Fabyan House to Bethlehem (with Mt. Agassiz near) and Littleton. Mt.
Gardner appears on the S. of Littleton, and Sugar Hill and the Landaff
highlands are also seen. In the foreground is the green forest which

sweeps up to the well-marked peak of Deception, over whose l. is Beech Hill, while Cherry Mt. commences on the r. and ascends on the N. to the peak of Owl's Head. Under the highest part of Cherry is Mt. Dartmouth, and Mt. Mitten is farther to the r. Over the r. of Cherry is Dalton Mt., with Mts. Prospect and Pleasant on the r.; and the great line of the Green Mts. of Vermont is beyond. The white village of Jefferson Hill appears at the foot of Mt. Starr King, with the blue cluster of the Pilot Range farther to the r.

The view is now narrowed on the immense peaks to the N., Jefferson, Adams, and Madison, whose formidable sides may be closely studied from this point. Goose-Eye is next seen, far away in the N. E.; and then follow Moriah, Imp, and Carter, down the Great Gulf, with several of the Androscoggin hills and Grafton peaks beyond. The view to the S. E. and S. is closed by Mt. Washington.

88. The Bridle-Path from the Crawford House.

The Crawford Path has two advantages over the other approaches to Mt. Washington, which may in part compensate for its difficulty and ruggedness. It commands near views of the great ravines that head into the heart of the range, and gives frequent and extensive outlooks from the high peaks over which it passes, so that an approximate idea of the prospect from above may be obtained, if Mt. Washington is capped with clouds. A practised pedestrian could ascend the Crawford Path and go down from Mt. Washington to the Glen House in a long day; and this is the best combination of routes in the mountains, since thereby one gets the exhilarating advantage of the march upward over the S. W. peaks, and on his descent to the Glen enjoys noble views of the northern mountains and the Great Gulf. The absence of vegetation (except small subalpine varieties) on the ridge makes travelling easy, and the only laborious parts of the route are on the ascents of Mts. Clinton and Washington. The cloud-effects from this path are finer than those on any other approach to the mountain. Some warm clothing should be taken, to be put on while the visitor is resting on the intermediate peaks.

The proprietors of the Crawford House attempted in 1875 to secure a charter to build a carriage-road over this route, but were refused by the State Legislature. The trail is plain and safe, and can be followed without a guide by any one of ordinary intelligence, *unless a fog rises.* In this case there is some peril, and the advance should be slow and cautious. Several people have been fatally lost from this path at such times. The Crawford House should have two or three guides qualified for this journey. No horses can be obtained for this route, and it must, therefore, be performed on foot.

More particular descriptions of the mountains on this line and their views are given in preceding routes, in order to avoid overloading the account of the path itself. The following are the times made by the Guide-Book party in easy marching along this path: From the Crawford House to the summit of Clinton, 1½ hrs.; Clinton to the top of Pleasant, 40 min.; Pleasant to Franklin, 30 min.; Franklin to the top of Monroe, 40 min.

The following distances are given on Prof. Walling's map of Coös County: Crawford House to the top of Mt. Clinton, 3 M.; to Pleasant, 4¾; to Franklin, 5¼; to Monroe, 6¾. (To Washington, 8½.)

The following table of altitudes is from Prof. Guyot's memoir on the Appalachian System: Height of the Crawford House above the sea, 1,920 ft.; of Mt. Clinton, 4,320; the gap between Clinton and Pleasant, 4,050; Mt. Pleasant, 4,764; gap between Pleasant and Franklin, 4,400; Mt. Franklin, 4,904; Little Monroe, 5,204; Monroe, 5,384; gap between Monroe and Washington, 5,100. [Mt. Washington, 6,293.]

" The ascent of Mt. Washington, the great point of interest, of course, is in many respects more satisfactory from this plateau than by any other route, as it gives a person really fond of mountain-scenery and romantic adventure as much experience of the kind as is agreeable, without becoming wearisome. To one unacquainted with mountain-scenery, the ascent by the bridle-path from the Crawford Notch affords more new sensations than can, perhaps, be gained elsewhere in this region in so few hours." (*Picturesque America.*)

The path leaves the E. side of the Crawford House, and soon parts from the Gibbs-Falls path, which diverges to the l. The ascent of Mt. Clinton is made in about 1½ hrs., the distance being 3 M. The route leads through the woods all the way, with occasional sections of corduroy, and, although it is kept in good repair, this part of the journey is considered the most difficult. About 1 M. from the Crawford House is the so-called *Sunset Grove* (on the l.), just beyond which (r.) is a spring of water, near the path. The trees grow dwarfed and spiky as the summit is approached, (many of them having been killed by the frosts of 1816), and fine vistas are afforded on the l. (Ammonoosuc Valley) and in front. Farther up, a guide-board is seen on the r., bearing the inscription, " **Mt. Clinton, 3 M.**," meaning that it is on Mt. Clinton and 3 M. from the Crawford House, and not, as some have very naturally supposed, that Mt. Clinton is 3 M. from the guide-board. The summit (see Route 80) is a short distance to the S. Fine views to the S. and W. are gained from this ridge, and in front are seen the massive peaks of the Presidential Range, close at hand.

The path from Mt. Clinton to Mt. Pleasant is in good condition, and winds along the crest-line of a high, bare, and ledgy ridge. There is a short copse of evergreens in the hollow N. of Clinton, which is, however, easily threaded by the trail. The course is about N. E., and Mt. Pleasant may be reached in 40 minutes, beyond several minor rocky knolls. The views on either side are of a highly interesting character and of great extent, including the wide Ammonoosuc Valley on the l. As the path nears the S. flank of Mt. Pleasant, it traverses a little green hollow in whose l. side is a spring of clear water, surrounded by spring flowers late in August, and by snow until July. A short distance beyond an indistinct path diverges to the l. by some low bushes, and begins the direct ascent of Mt. Pleasant by a series of zigzags. The main path continues to the r. and passes on the E. side of the mountain, far below the summit. The vague divergent trail was the first route of the Crawford path, and leads up the dome-like swell of Pleasant. It soon strikes the old Fabyan trail, coming from the l. and ascending to the r., which was originally well made and is still finely preserved. This broad path is followed, by a series of easy gradients, to the pile of stones on the summit of Pleasant. Tourists who do not chance to find the divergent path can ascend the mountain in about 20 minutes from the main trail, without any difficulty.

Mt. Pleasant and its view, see Route 81.

The Fabyan path passes from the plateau of Mt. Pleasant toward the N. E., soon descending a sharp-angled slope towards the dell in which *Red Pond* is situated. The main Crawford trail is rejoined on the ridge beyond this pool, where it is broad and well defined, but soon reaches a narrow ridge, with precipitous sides, where the profound chasm of *Oakes's Gulf* breaks down on the r. for 3,000 ft. In foggy or stormy weather this is one of the most dangerous points on the route. In front looms the bold S. face of Mt. Franklin, affording a short and breathless clamber, after which the path passes directly over the flat and characterless top of the mountain. The time from the summit of Pleasant to Franklin is less than 30 minutes.

Mt. Franklin, see Route 82.

Beyond Mt. Franklin the path descends but little, and soon reaches the verge of Oakes's Gulf again, winding along there for a considerable distance. The profound abyss is seen far below on the r., sweeping out between slide-striped walls to the greater valley of the Mt.-Washington River. On the l. are the sharp slopes which fall away to the ravine which opens on the Ammonoosuc. In front are the castellated peaks of Monroe, the most alpine of the crests between Adams and Chocorua, through whose gap is the hamlet-crowned cone of Washington. The path bends from E. to N. around the picturesque S. peak of Monroe, and is in a manner terraced narrowly between the cliffs and the profound ravine to the r. Noble views of mountain-architecture are gained along the crags of Monroe, to the l., and deep and shadowy gorges sweep away on the r., while the gray pyramid of Washington looms up ahead. Near the E. slope of the N. peak is a boggy pool, above which, on the tributary rills, are small basins of clear, cold water.

As the path passes out by Monroe, *Boott's Spur* is seen on the r., running out from Mt. Washington over the head of a great ravine within whose cool depths the Editor has found snow-banks in July. On the l. are the *Lakes of the Clouds*, which are easily accessible from the path. The cold waters of the larger lake bathe the sides of several sharp insulated rocks. S. of the lake is a marshy pool about 25 ft. long; and to the N. W. (and higher up) is a small rock-rimmed pond.

Mt. Monroe and its view are described in Route 83; in Route 76 are notes about *Oakes's Gulf*, which is visited hence; and Route 78 treats of *Boott's Spur*, which may be visited by diverging to the r. about ½ M. beyond. Tuckerman's Ravine, see Route 72.

The path now strikes the broad southern buttress of Mt. Washington, and ascends gradually over long ledges, with the lakes and the distant Ammonoosuc Valley on the l. It soon reaches the terrace which leads out on the r. to Boott's Spur; and from its succeeding section interesting views are gained towards the farther part of Tuckerman's Ravine. After some preliminary gradings, the steep final cone is attacked, and is slowly and breathlessly ascended by a well-marked but arduous path among masses of mica-slate rocks. The terminus is at a rude stone enclosure (which was formerly used for the saddle-horses) below the U. S. Signal Station, whence a short plank walk conducts to the railway and the *Summit House.*

The Rev. Henry Ward Beecher wrote the following beautiful description of a ride over this bridle-path: " Beginning at our very feet as little crevices or petty gorges, the valleys widened, and deepened, and stretched forth, until on either side they grew dim in the distance, and the eye disputed with itself whether it was lake or cloud that spotted the horizon with silver. The valleys articulated with this ridge as ribs with a backbone. As I rode along this jagged and broken path, except of my horse's feet there was not a single sound. There was no wind. There was nothing for it to sing through if there had been ever so much. There were no birds. There were no chirping insects. . . . . There was perfect peace. perfect stillness, universal brightness, the fulness of vision, and a wondrous glory in the heaven and over all the earth. The earth was to me as it were unpeopled. I saw neither towns nor cities, neither houses nor villages, neither smoke nor motion nor sign of life. I stopped, and imagined that I was as they who first explored this ridgy wilderness, and knew that as far as eye could reach not a white man lived. . . . . And so, for a half-hour, I rode alone, without the rustle of leaves, without hum or buzz, without that nameless mixture of pipes, small and great, which fill the woods or sing along the surface of the plains. There were no nuts to fall, no branches to snap, no squirrels to bark, no birds to fly out and flap away through the leaves. The matted moss was born and bred in silence. The stunted savins and cedars crouched down close to the earth from savage winds, as partridges crouch when hawks are in the air. The forests in the chasms and valleys below were like bushes or overgrown moss. If there were any wind down there, if they shook their leaves to its piping, and danced when it bid them, it was all the same to me. For motion or rest were alike at this distance.

" Out of these chambers of the air I remembered the world afar off, as one remembers the fairy-tales of his childhood. The cities we had trodden seem in the mind like pencil-traced pictures half rubbed out. The real New York seemed too impossible even for a dream. That Boston really lay sweltering by the seaside excited a smile of incredulity. As I rode along, I tried the effect of speech. I called out aloud. The sound fell from my lips, and ceased forever. No mountains caught it and nourished it in echoes. I called again, but there was in a second no voice, and none that echoed."

89. The Davis Bridle-Path

(16 M. long) was made by N. T. P. Davis, the proprietor of the Mt.-Crawford House, in 1845, but did not become popular, and was but little used. It has been abandoned for many years, and is difficult to trace at some points. Prof. Huntington ascended by it in 1871, since which it is doubtful if it has been traversed. He reports that the chief difficulty is in getting to the trail above the Giant's Stairs, beyond which it is more easily followed. The Editor has traversed only the first 5 M. and the last 3 M. of this route. The path up to the Giant's Stairs is described on page 138, and passes up between Mts. Crawford and Resolution, attacking the

Stairs on the l., 4-5 M. from Bemis Station.   From the head of the Stairs it passes nearly N. for 7 M., through a forest of black growth, and may be followed without much difficulty, especially as it keeps along the highest line.   No views are afforded, save at one point where the forest has been burned away; and the route is highly monotonous.   The long and massive ridge upon which this path runs from the Giant's Stairs to Boott's Spur lies between the valleys of the Mt.-Washington River and the Rocky Branch, and is 8-10 M. long, with a spur towards the Saco, N. of Sleeper Brook.   This prominent feature of the White Mts. has hitherto been without a name, and the Editor applies to it the provisional title of the **Montalban Ridge.**   About 11 M. from Bemis Station the path begins to open some pleasant views, which continue for about 2 M., when the ridge is left and the side of Boott's Spur is attacked.   The route lies (not over but) up by the side of the adjacent crags, and out over the Spur.   It is about 3 M. from this point to the Summit House, and the path lies nearly over the middle of the plain, being marked in some parts by piles of stones along the side and elsewhere by the discolorations on the ledges.   In case of fog the path should be closely followed; otherwise the visitor can be guided by the looming up of the great cone ahead, care being taken to keep well to the l. to avoid the depressions towards Tuckerman's Ravine. Noble views are afforded during all this last third of the route. and the mountains rise up on every side.   The path enters that from the Crawford House near the foot of the great cone, and then ascends rapidly over the steep rocks.

90. The Route over the Northern Peaks.

This is one of the grandest mountain walks in Eastern America, and one of the least known.   The absence of any trail and the difficulty of getting upon Madison render it impracticable to the majority of tourists; and the possibility of bewildering clouds and fog-banks settling on the peaks adds an element of danger to the journey.   It certainly should not be attempted unless by (or under the guidance of) expert woodsmen and mountaineers.   When the sky is absolutely cloudless, and the wind is from the N. W. or N., there is but little danger of fog coming down during the day.   No one should attempt the journey alone, since in case of a crippling accident to the ankles or legs no aid could be summoned from the distant hotels.

This route has been traversed in 12 hrs., but there is too much noble scenery along its course to be satisfactorily studied in so short a time.   It is therefore wiser to encamp one night on the way.   If the weather is warm and still, this can be done comfortably near Star Lake or Storm Lake, where water and fire-wood may be obtained.   But if the wind is blowing and a storm threatens, it would be unsafe to pass the night on the

10*   O ridge, and the travellers will be forced to go far down into one of the lateral ravines. The compass-bearings of the route in advance should be taken and recorded before encamping, so that the march can be resumed with confidence in the morning, even though clouds rest on the ridge. The camp-ground should be selected long before dark, in order that a sufficient store of dry wood may be accumulated to keep a large fire burning all night, with which to remove the chill from the air. The Guide-Book survey encamped near Star Lake, on a still night in August, with watchers detailed to keep up a roaring fire all night; but at dawn the tent was incrusted with ice. An abundant supply of provisions should be taken, in order to be prepared for contingencies. The descent is doubtless easier than the ascent, but loses the rare excitement of the progress from the less to the greater and the mysterious but certain attraction of Mt. Washington. The start should be made at dawn, in order to reach the Summit House in good season.

The chief obstacle to be overcome is the ascent of Mt. Madison, which has been made by four routes. The Guide-Book party rode from the Glen House to the height-of-land between the Peabody and Moose Rivers, whence a march of 4 hrs. led to the summit. The first 3 hrs. were in an ordinary forest, with a steady ascent; the last hour was along and over a line of hummocks, between which were almost impassable thickets.

The most arduous route by which Madison has ever been ascended was that taken by Raymond, who left the Glen House at 8 A. M., and reached the summit at 4 P. M., crossed Jefferson at twilight, and reached Washington at 3 A. M. He ascended the W. Branch of the Peabody to the stream which flows from between Adams and Madison, which was followed to its branch from Madison. This soon vanished under the rocks, and thence a tremendous thicket of dwarf trees was crossed by slow and painful labor.

The ridge which ascends from Copp's, on the E. flank, has been used as a route up Madison, and is for the most part straight and steady. The lower part is covered with a heavy forest, in which the walking is good; but farther up the ridge has been burnt over, and is nearly bare. The fire performed a good service by laying low the dwarf thickets, but there is an area about ¼ M. broad where the half-burnt tree-trunks are piled upon each other in perplexing confusion. It has been suggested that this belt should be flanked on the r., keeping near its r. edge.

The profile of the line of march from *Howker's* is as follows: Howker's, perhaps 1,300 ft. above the sea; Mt. Madison, 5,365; Star Lake, 4,912; Mt. Adams, 5,794; Storm Lake, 4,939; Mt. Jefferson, 5,714; gap towards Clay, 4,979; Mt. Clay, 5,553; gap towards Washington, 5,417; Gulf Tank, 5,800; Mt. Washington, 6,293.

The route from Howker's is nearly the same as that followed by the earlier explorers of the N. peaks. Howker's is about ½ M. from Wood's tavern, on the Jefferson road, and is on the old road to the Glen House. The ridge which runs thence to the summit of Madison is nearly straight and without hummocks. It is probably the best route to the summit. It will, however, be observed that while 1 M. an hour is the average time of marching, the passage from the region of small trees to the ledges, but little over ½ M. in extent, requires over 1½ hrs. The course from Howker's is about S. S. W.

Another route which has been used is begun by entering the woods ¼ M. from Wood's, on the Old Pinkham Road, just across the Moose-River Bridge, passing in for 1 M. to the Bumpus Brook, and following it for ½ M. to a dry brook-bed which is followed to the S. The way reaches the upper knoll of the hummocky ridge, and then follows the track of Gordon's path through the dwarf fir. There is no water for the last 3 M. This route is 5½ M. long, the hard climbing being between the second and fourth miles, succeeded by 1 M. of dwarf fir.

The following figures are taken from Prof. Pickering's notes of an ascent (in the summer of 1875) over Madison, Adams, Jefferson, and Clay, to the summit of Washington, in one day. The times and distances (pedometric measurement) are computed from Howker's to the Summit House (fractions are given in decimals): Starting-point to region of small trees, 1.83 hours, 1.7 M.; thence to low shrubs, 1.08 hours, .5 M.; thence to the ledges, .5 hour, .1 M.; thence to the summit of Madison, .8 hour, .8 M.; thence to Star Lake, .6 hour, .7 M.; thence to the summit of Adams, 1.83 hours, .8 M.; thence to Storm Lake, 1.15 hours, 1 M.; thence to the top of Jefferson, 2 hours, 1.4 M.; thence to the N. peak of Clay, 1.4 hours, 1.5 M.; thence to the depression between Clay and Washington, .5 hour, .5 M.; thence to the summit of Washington, .4 hour, .8 M. The entire march took about 12 hours, and the distance is 10 M.

Mt. Madison and its view, see Route 84.

The slope from Madison to the gap towards Adams is sharp and rugged, but the descent is not difficult, since the way is not encumbered by bushes. The difference of altitude is 453 ft., and the time about ⅓ hr. In the saddle between Madison and Adams there is a tiny lakelet, filled with clear, sweet water. To this the temporary name of *Star Lake* has been applied, on account of its extreme height and because it mirrors so perfectly the constellations above. The land slopes gradually away to dwarf forests on the N. W.; and on the S. E. falls sharply off into a tangled lateral ravine which opens towards the Great Gulf. Perhaps the best point on the range to make a camp during cold or threatening weather is about ½ M. down the ravine which falls slowly to the N. W. from Star Lake. There is plenty of wood there, and never-failing springs of water, in a well-sheltered locality. The adjacent gorge is the Madison Ravine, through which flows a small brook.

Starr King says of this place : " We are almost overhung by the lawless rocks of a subordinate peak of Mt. Adams, which we called *John Quincy Adams*, and back of that was the profile line of the higher crest, bulging off and sweeping down into a ravine deep below the general level of the ridge. The rocks were very jagged, and at first sight nothing could seem more harsh and chaotic. Yet the view was strangely fascinating."

In order to attack the ridge of Adams it is necessary to flank on the r. the lofty crags S. of Star Lake, and this course leads near the gate of **King's Ravine,** which is about 200 ft. above the gap (see Route 75). The top of Adams is about 700 ft. above the gate of the ravine, and ½ hr.'s march distant. The route lies between the precipitous verge of the ravine and the subordinate peak on the l., and is on good ground for walking. A short detour should be made to the head of the cliffs, on the r., whence a noble view may be obtained down the vast chasm and out to the Randolph road. The peak of Adams looms up boldly in front, and is reached only after a hard scramble over steep piles of rocks.

Mt. Adams and its view, see Route 85.

Lowe's Path. — Charles Edward Lowe lives in Randolph, 4 M. E. of the Mt.-Adams House and 8 M. W. of Gorham. He is the best guide to the great northern peaks, and is said to be cautious, intelligent, and companionable. He charges $3 a day. He has cut a new path, 8 ft. wide, from his house (*Woodvale*) to Mt. Adams, which starts out towards the N. peak, crosses a high foot-hill, and advances up the steep spur of Nowell's Ridge, which walls King's Ravine on the S. W. After going through 2 M. of tall mixed and spruce woods it crosses about ½ M. of 10-ft. trees and reaches the ledges, about 1½ M. from the crest of the main peak. There are none of the breast-high trees on this route. Fine views of King's Ravine may be obtained on the l. Just before reaching the W. peak of Adams a branch-path deflects to the r., running to Mt. Jefferson; and another diverges nearer the W. peak, running N. E. to Mt. Madison, between the N. peak and the head of King's Ravine. The constructor has promised to finish these paths by June, 1876, and to erect a small stone hut on the ridge of Adams. A cross-path has also been laid out from the peak of Adams to Jefferson, by two springs.

The foot of Jefferson's cone may be reached from Adams by going out on the ridge which runs towards the W. peak of Adams, and bending to the l. on the first high branch-ridge. The walking is somewhat rugged, but the views are fine. A nearer way, affording better walking, and sheltered from the N. winds, is found by striking down directly from the peak toward the gap, crossing a sloping plateau below the rugged ridge. This vicinity is rich in rare plants and flowers, — the bilberry, mountain blueberry, cowberry, arenaria, and cassiope. Near the S. E. side of the ridge is a cold spring, ¾ M. S. S. W. of the peak of Adams.

The distance from Adams down the long ridge to the S. W. and around to Storm Lake is about 1 M., the descent being 855 ft. Heart-stirring views are afforded on every side, and the mind is hardly diverted by the necessity for sure footing. *Storm Lake* is the wet-weather pond which lies in the hollow between Adams and Jefferson, occupying a locality which seems to be the nursery of clouds and tempests. This notch is 27 ft. higher than that in which Star Lake is situated, and is less closely environed.

"The long tramp which follows next, around the bending ridge between Jefferson and Adams, is rewarded by the glorious picture of Washington, superior to any other which the range affords. The long easterly slope is shown from its base in the Pinkham forests; the cone towers sheer out of the Great Gulf, and every rod of the bridle-path is visible, from the Ledge to the Summit House "

There are two lane-like tracks up the side of Jefferson from Storm Lake, of which that on the r. passes dangerously near the verge of a line of high cliffs. The lane on the l. is the one which was followed by the Gordon path, which passed to the E. of (and not over) the peak of Jefferson. These lines of easy travel are visible from Mt. Adams. On the r. lane is *Spaulding's Spring*, which discharges ten hogsheads of water an hour, and is the source of Israel's River, flowing down through the Ravine of the Castles. It was named for the Rev. H. G. Spaulding, an old comrade of Starr King.

From the Storm Lake to the top of Jefferson it is about 1½ M., the rise being 775 ft. But the grade is not uniform, and some severe climbing has to be done. The true peak of Jefferson is reached after crossing several apparent tops, and clambering over long lines of ledges and broken rocks.

Mt. Jefferson and its view, see Route 86.

The descent to the next notch is less difficult, much of the way being occupied by a slope which is covered with sedge. To this friendly and conspicuous area the name of the *Monticello Lawn* is here applied, in allusion to President Jefferson's home in Virginia. The gap towards Clay is less deep than either of those to the N., but contains no water in ordinary weather. A singular rocking-stone has been observed in this depression. The view towards the Great Gulf is partly closed by the long flank of Jefferson. The ascent of Mt. Clay is now begun, and is full of weariness, though unobstructed. The ensuing march over or between the three aligned peaks of Clay is also tiresome to the already wearied traveller.

Mt. Clay and its view, see Route 87.

The gap between Clay and Washington is reached in about ½ hr. from the N. peak of Clay, and the route is then laid up the long and monotonous slope of Washington itself, by way of the Gulf Tank.

" As to our satisfaction with the excursion, costing, as it did, no little toil, let me say that there is no approach to Mt. Washington, and no series of mountain views, comparable with this ascent and its surroundings on the northerly side. Your path lies among and over the largest summits of the range Between Madison and Adams you have the noblest outlines of rocky crest which the whole region can furnish. Mt. Jefferson glories in the afternoon light with the most fascinating contrasts of purple and orange hues. Mt. Washington shows himself in impressive and satisfactory majesty. You wind around the edges of every ravine that opens around the highest summits. You see the long and narrow gully that gapes between Madison and Adams; the tremendous hollow of Adams itself which we climbed; the precipitous gulf between Jefferson and Adams on the S. E.; the deep cut gorge in Jefferson on the N W., whose westerly bones of gray cliffs, breaking bare through the steep verdure, are perhaps the most picturesque of all the rock-views we beheld; the chasm between Jefferson and Clay, divided from the savage Gulf of Mexico by a spur of Jefferson that runs out towards the Glen House; and the rolling braces that prop Mt. Pleasant, and Franklin, and the tawny Monroe, — the boundaries of the ravines that you look into in riding to Mt. Washington over the Crawford path." Gentlemen who desire to explore the mountains N. of Mt. Washington should read pages 351 - 368 of *The White Hills*, wherein many details of these routes are given.

The Old Stillings Path

was used to carry up the lumber for Rosebrook's summit-house, on Mt. Washington. It is 9 M. long, from the valley to the summit, and may be traversed in a day. The lane can still be followed through the woods, though it is nearly filled with undergrowth. The forest shuts out the view, and makes the journey doubly tedious. It runs within 1 M. of the Castellated Ridge, and crosses spurs of Mts. Mitten, Bowman, and Clay. The entrance is at *Blair's Mill*, 1½ M. E. of the Mt.-Adams House, near the junction of the Valley and the Gorham roads.

91. The Mt.-Washington Carriage-Road.

This road is about 8½ M. long, from the Glen House to the Summit House, or 8 M. from the base of the mountain. Tolls are taken from foot-passengers, as well as from carriages. The road is finely built, and sweeps up the slope in long gradual lines of ascent, with an average grade of 12 ft. in 100, and a maximum grade of 16 ft. in 100. When it passes near steep descents, the lower side is protected by stone walls. The ascent is therefore perfectly safe and easy, even in cloudy weather. The road was commenced in 1855, under Engineer C. H. V. Cavis, by a chartered company, whose president was D. O. Macomber. It was opened as far as the Ledge in 1857, and was completed in 1861. The bridle-path by which the ascent was formerly made from this side was but little over 4 M. long, and the distance has been doubled in order to secure easy grades. Long sections of the bridle-path are seen from time to time, on the r., after the Ledge has been passed.

The carriage-fare from the Glen House is $ 5.00 for each passenger to ascend and return, or $ 3.00 either way. Two mountain-wagons run each way daily. (The Glen House, see page 106.)

The views given in the following description were derived from a careful reconnoissance of the road which was made by Prof. Huntington, in 1875.

The ascent from the Glen, being over the vast flanks and spurs of Mt. Washington itself, enables one to form a more definite idea of the immensity of the mountain than in any other way. The views of the Androscoggin and Peabody Valleys, and of the grand northern peaks, are also of surpassing interest.

For the first 3½ M. the road leads upward through the woods, passing over its heaviest grades in this section, and being usually shut out from views except those of the primeval forest, whose changes are observed as higher levels are approached. The *Half-Way House* is a small white building on the r., 4 M. up, 3,840 ft. above the sea and 2,208 ft. above the Glen House. This house is occupied by the men who repair and watch the road. Near this point the Entomological Club has its summer encampment, and from it as a centre explores the flanks of the mountain.

Near the fourth mile-stone the road turns sharply to the l. at a point called **The Ledge**, and formerly known as the *Cape of Good Hope*.

The View from this point is of remarkable impressiveness and breadth, including the depths of the Great Gulf below, with its vast forests dwindled into insignificance. On the l., the shoulder of Mt. Clay appears from behind the slope of Washington, with the immense mass of Jefferson next to the r., separated by a deep gorge from the sharp pyramidal apex of Adams, and both of them looming majestically across the Great Gulf. To the r. is Mt. Madison, flanked by the more distant Pine Mt., and farther to the r. are the rugged peaks of Hayes and Baldcap, hemming in the Androscoggin Valley. The high crests about the Grafton Notch are seen in the remote distance, with the sharp peak of Goose-Eye and the blue heights of the Puzzle Mt. and the Sunday-River White Cap. The Androscoggin Valley opens to the r. well into Shelburne, and is then cut off by the low and rounded summit of Mt. Winthrop, on whose r., across the Peabody Glen, is the ponderous mass of Mt. Moriah, with the low peak of Imp nearer at hand, and the noble heights of Mt. Carter and Carter Dome on the E., over and to the r. of the white-walled Glen House. The wooded slopes of Wild-Cat close the view on the r. A glimpse of the Nelson Crag is gained from a point a little way beyond the fourth mile-post.

The view from the Ledge is thus painted by Mr. King: "Nothing which the day can show will give more astonishment than the spectacle which opens after passing through the spectral forest, made up of acres of trees, leafless, peeled, and bleached, and riding out upon the Ledge. Those who thus make their first acquaintance with a mountain height will feel, in looking down into the immense hollow in which the Glen House is a dot, and off upon the vast green breastwork of Mt. Carter, that language must be stretched and intensified to answer for the new sensations awakened. . . . . We shall never forget the phrase which a friend once used, ... as he looked, for the first time, from the Ledge upon the square miles of undulating wilderness : ' See the tumultuous bombast of the landscape.' Yet the glory of the view is, after all, the four highest companion mountains of the range, Clay, Jefferson, Adams, and Madison, that show themselves in a bending line beyond the tremendous gorge at the r. of the path, ... and are visible from their roots to their summits. . . . . On the Glen path these grand forms tower so near us that it seems at first as though a strong arm might throw a stone across the Gulf and hit them. . . . . Except by climbing to the ridge through the unbroken wilderness of the N. side, there is no such view to be had, E. of the Mississippi, of mountain architecture and sublimity."

The dead and gnarled trees which strew the slopes at this elevation are very dry and hard. They are sometimes called buck's horns, and also (referring to their color) bleached bones. The forests on this line were probably killed during the cold season of 1812 – 16, through the latter of which years they were frozen all the time, 1816 being long known as "the year without a summer." The white-boughed thickets were formerly much more extensive than at present, and bore the poetic name of the *Silver Forest.*

Beyond the Ledge the road passes upward along the edge of the Great Gulf for some distance, affording noble and satisfactory views to the r. It then turns sharply to the l. and runs S. S. E., whereas its previous course had been N. W., and begins the slow ascent of another of the great shoulders of the mountain. In former times travellers on the Glen path considered the mountain as having five summits, each receding behind the other until the highest was reached.

The View from the fifth mile-post. — Down the long valleys of the Peabody and Androscoggin Mt. Hayes is seen, flanked by Baldcap on the r., over which is Goose-Eye, with its gradual W. slope. Over the E. part of the Androscoggin Valley is Mt. Ingalls, in Shelburne, with Speckled Mt. and the Bear-River White Cap beyond, at the Grafton Notch, and the Sunday-River White Cap and Puzzle Mt. to the r. Nearer at hand and E. of the Peabody Glen are the dark crests of Moriah, the low head of the Imp, and the noble peaks of Carter and the Carter Dome, flanked by the closely adjacent Mt. Wild-Cat, over which is the long and hotel-crowned ridge of Mt. Pleasant, with Sebago Lake beyond, nearly S. E. The view now passes over the mountains of Jackson and Chatham, — the round tops of Double-Head, the pointed knobs of Gemini, and the beautiful pyramid of Kiarsarge. The ridge running S. from Wild-Cat ends in the Eagle Mts., marked by farms on the W. slope, over which are Tin and Thorn Mts., with the Green Hills of Conway still higher. Down the deep-cut glens of the Ellis River the pleasant Saco Valley is seen, with the hamlets of Lower Bartlett and N. Conway, and the Cathedral and White-Horse Ledges on the r., over which is the Green Mt. in Effingham. On the W. side of the Ellis Valley is Iron Mt., marked by farms on its upper slope, with the fine peaks of Moat Mt. beyond. To the right of Iron, and over a ridge running S. from Boott's Spur, is Mt.

Langdon, with Table Mt. beyond, on the l., and Bear Mt on the r. Between and beyond these two is the splendid white peak of Chocorua, with the blue Ossipee Range in the background. The view then rests on the near Boott's Spur, forming the S. wall of Tuckerman's Ravine; and close at hand are the cliffs of Huntington's Ravine, with a section of its S. wall and the sharp ridge that forms the N. wall, running up by the Nelson Crag, on whose l. is the crest of Mt. Washington. The edge of Huntington's Ravine can be reached by a short detour from this point.

At the sixth mile-post the road has reached one of the upper terraces, and opens a prospect to the N. and E., that to the S. being hidden by the bold Nelson Crag and its connected ridge.

The View. — Between the crests of Clay and the formidable summit of Jefferson, nearly N. W., are the low mountains of Lancaster. Between Jefferson and the noble spire of Adams is a portion of the Pilot Range; and on the r. of Madison are seen the Pine and Chickwolnepy Mts., Goose-Eye, Speckled Mt., and the Bear-River White Cap, the Sunday-River White Cap, and Puzzle Mt., with other lofty peaks of Western Maine, the Androscoggin Valley being in the foreground. The Imp closes in on the r., S. of the Moriah Range.

About 6½ M. from the base is the sharp bend in the road which is sometimes called *Cape Horn*, near which is the sofa-shaped rock, with a comfortable back, which has been called *Willis's Seat*. A fine view is obtained from this point also. The seventh mile-post is near the *Chandler Ridge* (where Benjamin Chandler, of Delaware, died of exposure in 1856).

The View. — Close at hand below are the profound depths of the Great Gulf, over which tower the vast peaks of Clay, Jefferson, Adams, and Madison, so near as to be easily studied in all their powerful lines and rugged ravines. About W. N. W. is Cherry Mt., with the peak of Owl's Head on its N. end and the long ridge of Dalton Mt. over it. Mt. Clay forms the wall at the head of the Great Gulf, and over its l. and highest part are Mt. Niles (in Concord, Vt.), the Victory Hills, the Umpire and Burke Mts., the fine peaks of Hor and Annanance, near Willoughby Lake, and the Westmore Haystack, which is the highest point in this direction. Over the extreme r. of Clay are Mts. Prospect and Pleasant, near Lancaster; and through the gap between Clay and Jefferson the white village of Lancaster is seen, over the plains of Jefferson, with Mt. Burnside still beyond. Over the flank of Jefferson is a part of Mt. Starr King; and on the r. of Jefferson several of the Pilot Mts. are seen, the South Peak being prominent, with Deer Mt. in the foreground. Over the l. flank of Adams are the white Percy Peaks, with Long Mt. on the r. Between Adams and Madison more of the Pilot Mts. are seen; and on the right of Madison are the Chickwolnepy Mts., in Success, and the Hampshire Hills, near Lake Umbagog. Nearly N. W., over the Androscoggin ranges, is the peak of Goose-Eye, on whose l. are the Bear-River White Cap and Speckled Mt., and on the r. are the Sunday-River White Cap and Puzzle Mt., in Newry. Mts. Hayes, Baldcap, and Ingalls are in the foreground, and on the r. are the near and massive ranges of Moriah. Mt. Imp, the peaks of Carter and the Carter Dome, and the dark and wooded Mt. Wild-Cat, are farther to the r., across the Peabody Glen. The long ridge of Mt. Pleasant is nearly S. E.; and the twin crests of Double-Head close the view.

The road now bends inward towards the crest, and the railroad track is seen on the r., while on the l. a superb view is gained. When near the Glen stable, towards the summit, the road turns sharply upward, and from this point is a fine view downward on the Alpine Garden.

" To the S. E. the gorge of Tuckerman's Ravine opens into the mountain, and we see far down the S. wall, while the N. wall forms the S. extremity of the lawn. Not far from the N. wall of the Ravine we see the beginning of the depression in which is Raymond's Cataract. To the N. E. rises the Chandler Ridge, below which is the Nelson Crag, forming the N. wall of Huntington's Ravine, and we see down into the gorge. The view eastward is essentially that seen from the summit."

92. The Mt.-Washington Railway.

The distance from Marshfield (Ammonoosuc Station) to the summit is 2 13/16 M., the average grade being 1,300 ft. to the mile, and the maximum grade 1,980 ft. to the mile, or 13½ inches to the yard. There are 9 curves on the line, varying from 497 to 945 ft. radius. (These figures do not cover the new lower section, opened in 1876.) The ascent takes 1½ hrs., and the engine is supplied with water at each of the 4 tanks; but the descent is accomplished in less time. There are two trains daily each way, besides occasional extras. The fare for the ascent or descent is $3; for both ways, on the same train, $4. Trunks must be paid for as freight; valises are carried without charge.

Sylvester Marsh, of Littleton, invented the mechanism of the Mt.-Washington Railway, and fought his project through much opposition and ridicule to its successful consummation. In 1858 he exhibited a model of the line to the State Legislature, asking a charter to build steam-railways up Mts. Washington and Lafayette. The charter was granted, one of the legislators suggesting the satirical amendment that the gentleman should also receive permission to build a railway to the moon. The turnpike to the foot of the mountain was commenced in April, 1866, and the construction of the railway was begun in May. ¼ M. was built in 1866; ¾ M. in 1867 (to Waumbek); 1 M. in 1868 (to the top of Jacob's Ladder); and the work was completed in July, 1869, the track, stations, and rolling-stock having cost about $150,000. A similar road has since been constructed on Mt. Rigi, in Switzerland, after drawings and models from Mt. Washington.

"The indispensable peculiarity of this railway is its central cog-rail, which consists of two pieces of wrought angle-iron, 3 inches wide and ¾ of an inch thick, placed upon their edges, parallel to each other, and connected by strong iron pins 1½ inches in diameter, and 4 inches apart from centre to centre. The teeth of the driving-wheel of the engine play into the spaces between the bolts, and, as it revolves, the whole engine is made to move, resting upon the outer rails. These cog-rails cost about $2 per foot, delivered at the base of the mountain. The appliances for stopping trains are of the most perfect kind. Both friction and atmospheric brakes are employed, and their complete reliability has been proved by the severest tests. The speed of descent is entirely regulated by their means without the use of steam." The engines weigh 6½ tons, and are rated at 50 horse-power, though, on account of their gearing, they are practically of 200 horse-power. The engines first in use resembled pile-drivers, having small upright boilers. "The driving-shaft is connected with two cylinders, with a crank-shaft geared into the centre so as to reduce the speed and multiply the power. A 24-inch gear works into a 6-inch gear, and the engine makes four revolutions to one of the driver. Thus the contrivances in this mountain-engine are adapted to develop power at the expense of speed. Force may also be required at times to hold the train at rest upon a high grade. When moving, the engine always takes the down-hill end of the train. In ascending, a strong wrought-iron 'dog' works into a wheel rolling on the cog-rail, preventing the train from falling back a single inch. . . . . The contrivances for stopping the train are also ingenious and peculiar. First is the *friction brake*, consisting of an iron band extending around each wheel, tightened at will. Second is the power of reversing the driving-wheel. Next there are atmospheric brakes upon each side of the cars. Their application is so successful that a platform or passenger car may be detached from the engine and lowered by itself, being completely under the control of the brakeman. . . . . There are in all five or six ways of stopping the trains." In 1869 the axle of the driving-wheel broke while the train was on Jacob's Ladder, but the train was stopped instantly and without any further damage. In 1875 over 7,000 people went up on the railroad. The road has never paid any dividends, nor have its officers received any salaries, all the revenues having been laid out in strengthening and perfecting the line and equipments. The manager says (with commendable pride), "We have now run the road seven years, and have not as yet injured a single person, nor damaged their property to the value of a cent."

The railroad workmen frequently descend the mountain by sliding down on boards placed over the centre-rail. Experienced sliders have gone thus to the foot of the mountain in ten minutes; but the experiment is perilous for novices. It is related that a platform-car was once left standing on the track in front of the Summit House, when some mischievous fellows started it down the mountain. It did not leap the track at any of the curves, but reached Marshfield in three minutes, making a costly smash-up at that end of the line.

Ammonoosuc (or *Marshfield*) was, until 1876, the starting-point of the mountain railroad. It has a small hotel, which is frequently found convenient for pedestrians and trout-fishers. During the winter this place is sometimes made the headquarters of lumbering parties. It is near the point where the railroad crosses the Ammonoosuc River, and is 2,668 ft. above the sea. The B., C. & M. R. R. now connects with the mountain railway, 1,500 ft. from Ammonoosuc, the latter line having been extended downward. Trains do not begin running until July 1st, and after July 15th two trains ascend the mountain daily. The trains on the Mt.-Washington Branch of the B., C. & M. R. R. from Littleton, Bethlehem, and the Fabyan House make connections with the mountain-railway near Ammonoosuc. Two new engines were put on the mountain-track in 1876.

At Ammonoosuc the train begins a sharp ascent, with a grade of 1,700 ft. to the mile, passing along a wide belt which has been cleared through the forest. About 300 ft. above the station, the grade becomes more easy; and at ¾ M. from Ammonoosuc the train stops at a water-tank near *Cold Spring*. Another section of steep climbing soon ensues, and the train then reaches and stops at the water-tank at *Waumbek Junction*. At this point the bridle-path from Jefferson Hill intersected the old Fabyan path. The trees now become smaller; the line makes several short curves, and passes through a cutting in a ledge of andalusite gneiss. Waumbek is 3,910 ft. above the sea, and 1,242 ft. above Ammonoosuc. The next water-tank is at Jacob's Ladder, 5,468 ft. above the sea.

Jacob's Ladder was an ancient name applied to a section of the Fabyan path where it zigzagged upward over a steep shoulder of the mountain. It has since been transferred to a long and massive trestle-work on the railway, near that locality, where the track is sometimes 30 ft. above the rocks, and rises on its steepest grade (1,980 ft. to the mile). The limit of trees is now passed. Fine retrospective views have been enjoyed nearly all the way from Ammonoosuc, since the track is elevated so that the trees are overlooked in many places. The Ammonoosuc Valley is at first the limit of the prospect, but the horizon rapidly widens to the W. and N. W., affording a study of fascinating interest to the traveller. Range after range comes into view, and new villages spring from the distant plains. As higher points are reached, the massive ridge of the Twins is visible, with the Field-Willey chain far to the l. Beyond the Deception-Dartmouth range the long dark mass of Cherry Mt. slowly rises, with the plains of Jefferson, Mt. Starr King, and the stately Pilot

Mts., in the N. W. The sharp sierra of Lafayette then emerges in the W. S. W., with the dark needles of the Franconia Range. Toward the W. the view gradually widens over Bethlehem and Littleton, until it embraces the dim line of the Green Mts., far away toward Lake Champlain. Mts. Franklin and Pleasant start out, close at hand on the r., across the Ammonoosuc Ravine; and on the l. are the massive slopes of Mts. Clay and Jefferson. Minor peaks come into view one after another, — Mt. Hale and the Sugar Loaves, beyond the white Fabyan House; Mt. Agassiz, near Bethlehem; Dalton Mt., close to the Connecticut River; the highlands of Lancaster; and many others, near and far. In such an hour one could wish to have the eyes of Argus, to properly study and appreciate the vast landscape.

Above Jacob's Ladder the line soon passes along the slope towards Mt. Clay, whose family of peaks are within a short walk, on the l. A view is now gained of the Great Gulf, whose profound and shadowy depths sweep away towards the Glen House. A broad easterly prospect next follows, reaching far into Maine, beyond the dark Moriah-Carter range. Close at hand, over the humps of Clay, are the vast and formidable peaks of Jefferson and Adams. If the extra shawls and overcoats have not yet been donned, they will be at this point, where the E. wind is met.

The *Gulf Tank* is 5,800 ft. above the sea, and 3,132 ft. above Marsh-field. The ascent is now more easy, there being but 800 ft. of rise in over 1 M. of track, from Jacob's Ladder to the Summit House. "The farther ascent is gradual, the broad shoulder of the mountain presenting the characteristic features of arctic desolation, — a wide expanse of large angular blocks of schist and granite, severed from the now concealed ledges by the freezing agencies of centuries. Between the fragments may be seen clumps of saxifrages, sandworts, and reindeer moss, the same species of plants which enliven the barren wastes of Labrador and Greenland. As far as the upper limit of trees, boulders that have been transported by the glacial drift from more northern summits are common. They rapidly diminish in number and size from that point, and have not been seen above the fourth water-tank."

The view now widens so rapidly as to defy analysis, and may best be understood by looking over the description of the view from the summit. If a cloud is seen resting on the top of the mountain, this prospect should be dwelt upon with the greater care, since probably the visitor will get no outlook from that point. The Mt.-Washington carriage-road now appears on the l., winding easily along the gentle slope; and soon after passing the Bourne monument (on the r.), the train stops on the platform of the **Summit House.**

93. Mount Washington.

Hotel. — The Mt.-Washington Summit House (opened in 1873) accommodates 150 – 200 guests, charging $ 1.50 for each meal and the same for lodging. It is a long three-story wooden building, strongly constructed. and firmly bound down to the ledges below. The lower story is occupied by the dining-hall, parlors, and offices, and the two upper stories by bedrooms. The building is heated by steam. There is a telegraph-office and a post-office; and writing-paper and envelopes are sold here, bearing pictures of the hotel, on which missives are sent to all parts of the country.

Routes. — A vast majority of the visitors ascend either by the railway (see page 233) or the road from the Glen House (page 230). The other and less-travelled routes are by the Crawford-House Path (page 221), Tuckerman's Ravine (page 194), and the northern peaks (page 225). By the first-named route passengers can go through by rail from New York, Boston, Portland, and other eastern cities to the summit of the mountain.

Clothing. — Visitors to the summit who wish to see more than the stove in the hotel office should carry a plentiful supply of warm clothing, with shawls and overcoats. Still and sunny days are very rare here, — windy and chilly weather being the rule; and damp and penetrating clouds frequently lie upon the summit for days at a time. An Ulster coat would be one of the most valued possessions of a visitor, even in the month of August; and ladies should be provided with the heaviest shawls and skirts. These garments will be found especially comfortable by people who stop over to see the sunrise, and who get up at 4 A. M., since the hotel-people have issued a stringent edict against guests wearing the bed-blankets outdoors over their shoulders.

Sojourn. — The present prices at the hotel do not favor permanent boarders, but the tourist could hardly find a more interesting place in which to pass two or three days, if the weather is fair. He could then escape the annoyance of the queerly-assorted crowds who come up on the trains, and could study the noble mountain in the undisturbed intervals. The mastery of the view, the cloud-phenomena, the sunrise and sunset, and the interesting excursions in the vicinity, would furnish inexhaustible resources.

Mt. Washington is the loftiest peak E. of the Rocky Mts. and N. of the Carolinas, and is 6,293 ft. high. On account of this elevation, the summit forms an arctic island in the temperate zone, having the same climate as the middle of Greenland, at 70° N. latitude. This peculiarity is shown not only in the temperature but also in the vegetation which here exists. The latitude is 44° 16′ 25″ N., and the longitude is 71° 16′ 25″ W. The upper portions of the mountain are covered with rugged fragments of mica-slate, which have been broken from the parent-ledges by the action of frosts.

The mountain falls off more rapidly to the W. than to the E., where it is braced by two long spurs and cut into by profound ravines. There are several high secondary ridges towards the Glen House, running out one below the other. On the S. E. and E are the singular terraces of Bigelow's Lawn and the Alpine Garden. The upper cone pitches off much more steeply on the S. and E. than on the other sides, the N. slope being comparatively gradual.

The alpine hamlet on the peak consists of the Summit House, with the old Tip-Top and Summit Houses in the rear, the engine-house of the railway, the U. S. Signal-Service observatory (on the S.), and the Glen-House stables (below the cone, on the road). The observatory and hotel are connected by a plank-walk, which extends around the former building. The old hotels are low-walled stone buildings, which were used for the accommodation of travellers before the present hotel was erected. The men on duty at the observatory are under orders to keep it closed against the people. It is a small wooden building 36 × 24 ft. in area, in an exposed situation towards the S. edge of the cone.

In 1870 over 7,000 visitors reached the summit, and in 1875 the number was more than 10,000.

Prof. Hitchcock says: "The most interesting features of one's stay upon this summit are derived from meteorological sources, — the sunrise and sunset, shadows of the mountain upon clouds and adjacent ranges, wonderful colors, shapes, and movements of clouds, the perception of the beginning and progress of storms, hurricanes, frost-work, variation in temperature and humidity, fluctuations in the barometer, conflicts of winds and clouds, etc."

The most beautiful phenomenon of cold weather on the mountain is the frost-work, which is formed with great rapidity, towards the wind, and in attachment to the rocks, buildings, and poles. It occurs when the wind is from the N. or W. and the mountain is covered with clouds, the lower frost-points elongating until they form long and feather-like masses of exquisite delicacy and symmetry, and of the purest and most brilliant white. Among the other phenomena which have been noticed here are thunder and lightning breaking through heavy snow-storms; rainbows, with three supernumerary bows, remaining for hours; anthelia, or glories of light; large and small coronas; the spectre of the Bröcken; halos; and parhelia.

Geology. — Isaac Hill says: "Mt. Washington had been thousands of years in existence before the internal fires upheaved the Alps." Sir Charles Lyell says: "The period when the White Mts. ceased to be a group of islands, or when, by the emergence of the surrounding low land, they first became connected with the continent, is of very modern date, geologically speaking."

The absence of striæ on its rocks shows that the summit of Mt. Washington escaped the attacks of the ice-bearing ocean of the glacial epoch. The highest markings which are now to be identified are 5,200 ft. above the sea, or about 1,100 ft. below the peak. "This summit seems to have been the only part of the State that has not been subjected to glacial action." Pebbles transported by glaciers have been found within 500 ft. of the summit. (In a paper read before the American Association, in 1875, Prof. C. H Hitchcock argued that the peak itself had been subjected to glacial action, of which it shows traces now.)

Botany. — "The summit of Mt. Washington, or that portion lying above the limit of trees, agrees in its climate and other physical features very closely with those of the coast of Northern Labrador, as observed at Hopedale, in latitude 55° 35'. The seasons correspond very exactly, as the snow melts in the early summer, and ice is formed early in the autumn at about the same dates. As is well known, the alpine flora of the White Mts. is identical with that of the arctic regions, which extends far southward along the Atlantic shore of Labrador. Not only is the flora identical so that no species of plant is known to be restricted exclusively to our alpine summits, but the times of leafing, of flowering, and fruiting of plants is much the same. Such was observed in the *Rubus chamæmorus* and *Arenaria Grœnlandica*, for example. It is also the same, apparently, with the fauna. The *Chionobas semidea* flies late in July and early in August, in greatest abundance, at the same time that its representative species swarm over the bare, rocky hill-tops of the Labrador coast. Their appearance heralds the close of summer, both on the extreme summit of Mt. Washington and the exposed hills of Labrador." (*Amer. Assoc. Advance Sci. Proc.* XVI.)

" The wind-swept summits of our White Mts. are to the botanist the most interesting locality E. of the Mississippi, for there are found the lingering remnants of a flora once common, probably, to all New England, but which, since the close of the glacial epoch, has, with few exceptions, retreated to Arctic America. On the highest of these mountains, only, are found the conditions favorable to the growth of these arctic plants. Of these alpine areas, Mt. Washington and the adjacent peaks are the largest, being a treeless region at least 8 M. long by 2 M. wide at its broadest part. These alpine plants are of great hardihood, and sometimes bloom amid ice and snow. . . . . He who ascends to this altitude has a similar opportunity for botanic study as if he made a journey to the N., passing first from the noble forests, with which we are familiar, to those of stunted growth, and, finally leaving them behind altogether, at length arriving at the barren and bleak regions beneath the Artic Circle. In approaching these mountain summits, one is first struck by the appearance of the firs and spruces, which gradually become more and more dwarfish, at length rising but a few feet from the ground, the branches spreading out horizontally many feet, and becoming thickly interwoven. These present a comparatively dense upper surface, which is often firm enough to walk upon. At length these disappear wholly, and give place to the Lapland rhododendron, Labrador tea, dwarf birch, and alpine willows, all of which, after rising a few inches above the ground, spread out over the surface of the nearest rock, thereby gaining warmth, which enables them to exist in spite of tempest and cold. These in their turn give place to the Greenland sandwort, the diapensia, the cassiope, and others, with arctic rushes, sedges, and lichens, which flourish on the very summit." (*Geology of New Hampshire*, Vol. I.)

Among the plants found on the summit are *Arenaria Grœnlandica*, *Poalaxa*, *Juncus trifidus*, and *Carexrigida*. On the rocky slope of the cone are *Potentilla trifida*, the two *Lycopodia*, *Diapensia Lapponica*, *Solidago*, the dwarf *Cornus*, *Juncus filiformis*, *Carex canescens*, Peck's *Geum*, the *Nabalus*, etc.

Visitors who are interested in the botany of this region should study Prof. Tuckerman's paper on " The Vegetation of the White Mts.," in Starr King's *The White Hills*. Also several papers in the first volume of *The Geology of New Hampshire* (published in 1875), including lists of the plants and a map of the alpine and subalpine regions.

94. Historical Sketch of Mt. Washington.

The austere and majestic crest of Mt. Washington was both the Ararat and the Carmel of the most ancient Indian traditions, sanctified by centuries of reverent memories, and regarded by the inhabitants of the valleys of Pequawket and Coös and Ossipee as a sacred and stainless shrine. How ancient the first of these traditions may have been no one can estimate, nor if it be a local transferrence of the Noachian deluge, whose memory is scattered among all races of the world. But so wide-spread was the story, and so powerful its influences of awe and solemnity, that no hunter was bold enough to approach the sacred peaks, no war-party dared to traverse their shadowy defiles. These martial tribes of hardy and adventurous men lived for ages within sight of the mountains, and within a day's march of their deer-haunted glens and teeming brooks, but were restrained from visiting them by an ineffable awe which taught them to believe that such visits would be invasions of the shrine of the Great Spirit. The few who scoffed at these fables and boldly advanced into the highlands (alpestrian rationalists of mediæval America, the Tyndalls of their time) were reported never to have returned, but to have been condemned to wander forever alone among the gloomy ravines, whence their despairing shrieks were borne from time to time to the valleys on the wings of the stormy winds.

The first of these legends is quaintly recorded by Josselyn, in his *New England's Rarities Discovered* (published in 1672). "Ask them whither they go when they dye, they will tell you, pointing with their finger to Heaven, beyond the white mountains, and do hint at *Noah's Floud*, as may be conceived by a story they have received from Father to Son, time out of mind, that a great while agon their Countrey was drowned, and all the People and other Creatures in it, only one *Powaw* and his *Webb* foreseeing the Floud fled to the white mountains, carrying a hare along with them, and so escaped; after a while the *Powaw* sent the *Hare* away, who not returning, emboldened thereby they descended, and lived many years after, and had many Children, from whom the Countrie was filled again with *Indians.*"

According to Brinton's *Myths of the New World*, the tradition of the deluge was held by 28 of the aboriginal nations of N. and S. America. The mountains of safety were the Peak of Old Zuni, in Mexico, the Cerro Naztarny on the Rio Grande, Mt. Apoala, and other eminences.

The legend of the translation of Passaconaway from the summit of this mountain in a chariot of flame is alluded to on page 29.

The first ascent was made in June, 1642 : "Darby Field, an Irishman, living about Piscat (Portsmouth), being accompanied with two Indians, went to the top of the White Hill. He made his journey in 18 days. His relation, at his return, was, that it was about 160 M. from Saco; that after 40 M. travel he did, for the most part, ascend; and within 12 M. of the top was neither tree nor grass, but low savins, which they went upon the top of sometimes; but a continual ascent upon rocks, on a ridge, between two valleys, filled with snow, out of which came two branches of the Saco River, which met at the foot of the hill, where was an Indian town of some 200 inhabitants. Some of them accompanied him within 8 M. of the top, but durst go no farther, telling him that no Indian ever dared to go higher, and that he would die if he went. So they stayed there till his return, and *his* two Indians took courage by his example and went with him. They went divers times through thick clouds, for a good space; and within 4 M. of the top they had no clouds, but very cold. By the way among the rocks there was two ponds; one a blackish water and the other reddish. The top of all was plain, about 60 ft. square. On the N. side was such a precipice as they could scarcely discern the bottom. They had neither cloud nor wind on the top, and moderate heat. All the country about him seemed a level, except here and there a hill rising above the rest, and far beneath them. He saw, to the N., a great water, which he judged to be 100 M. broad, but could see no land beyond it. The sea by Saco seemed as if it had been within 20 M. He saw, also, a sea to the E, which he judged to be the gulf of Canada. He saw some great waters in parts to the W., which he judged to be the great lake Canada river came out of. He found there much Muscovy glass; they could rive out pieces 40 ft. long, and 7 or 8 broad. When he came back to the Indians, he found them drying themselves by the fire; for they had a great tempest of wind and

rain. About a month after he went again, wi'h 5 or 6 of his company.
Then they had some wind on the top, and some clouds above them, which
hid the sun. They brought some stones, which they supposed had been
diamonds; but they were most crystal." (WINTHROP's *Journal*.)

The marvellous stories told by Field stimulated other visits by the adventurous
settlers on the sea-coast, the chief of which was conducted by Thomas Gorges and
Mr. Vines, magistrates of the colony. "They went up Saco River in birch canoes,
and that way they found it 90 M. to Pegwagget, an Indian town; but by land it is
but 60. Upon Saco River they found many thousand acres of rich meadow; but
there are 10 falls, which hinder boats, etc. From the Indian town they went up
hill (for the most part) about 30 M. in wooded lands. They then went about 7 or 8
M. upon shattered rocks, without tree or grass, very steep all the way. At the top
is a plain about 3 or 4 M. over, all shattered stones; and upon that is another rock
or spire, about a mile in height, and about an acre of ground at the top. At the top
of the plain rise four great rivers; each of them so much water at the first issue
as would drive a mill; Connecticut River from two heads at the N. W. and S. W.,
which join in one about 60 M. off; Saco River on the S. E.; Amascoggin, which
runs into Casco Bay, at the N. E.; and Kennebec at the N. by E. The mountain
runs E. and W. 30 M., but the peak is above all the rest. They went and returned
in 15 days."

Another account says: "Fourscore miles (upon a direct line) to the Northwest
of *Scarborow*, a Ridge of Mountains run Northwest and Northeast an hundred
Leagues, known by the name of the *White Mountains*, upon which lieth Snow all
the year, and it is a Land-mark twenty miles off at Sea. It is rising ground from
the Sea shore to these Hills, and they are inaccessible but by the Gullies which the
dissolved Snow hath made; in these Gullies grow *Saven* Bushes, which being taken
hold of are a good help to the climbing Discoverer; upon the top of the highest
of these Mountains is a large Level or Plain of a days journey over, whereon
nothing grows but Moss; at the farther end of this Plain is another Hill called the
Sugar Loaf, to outward appearance a rude heap of massive stones piled one upon
another, and you may as you ascend step from one stone to another, as if you were
going up a pair of stairs, but winding still about the Hill till you come to the top,
which will require half a days time, and yet it is not above a Mile, where there is
also a Level of about an Acre of ground, with a pond of clear water in the midst of
it: which you may hear run down, but how it ascends is a mystery. From this
rocky Hill you may see the whole Country round about; it is far above the lower
Clouds, and from hence we beheld a Vapour (like a great Pillar) drawn up by the
Sun Beams out of a great Lake or Pond into the Air, where it was formed into a
Cloud. The Country beyond these Hills, Northward is daunting terrible, being full
of rocky Hills, as thick as Mole-hills in a Meadow, and cloathed with infinite thick
Woods." (JOSSELYN's *New England Rarities*; 1672.)

The hostilities with the Indians which broke out soon after 1670 prevented further explorations in this region, except such practical ones as
were made at point of sword by the companies of rangers. In 1728 and
1746 such parties visited the mountain. In July, 1784, the Rev. Manasseh
Cutler, of Ipswich, and six other gentlemen, visited the summit of Mt.
Washington for scientific observations, being the first party to ascend with
that intent. The summit was embanked in clouds, but they estimated it
to be nearly 10,000 ft. high. They engraved their names on a sheet of
lead under a rock, which being found 18 years later was the source of
great mystification to the villagers at Jackson.

It is not known by whom Mt. Washington was named, but the present name is
found in Dr. Cutler's MS. of 1784, and it is probable that it was given at that time
and by Cutler's party. These explorers stayed on the mountain through a stormy
night. The Lancaster party which named the N. peaks in 1820 also remained on the
peak all night.

The first house on the summit of Mt. Washington was built by E. A. Crawford, in 1821, and was a low-browed stone cabin, located near the spring. Its floor was covered with moss (for bedding), and the furniture consisted of a roll of sheet-lead (for a hotel-register), a stove, and an iron chest for blankets. Here rested the first ladies who attained the summit (a Portsmouth party, in 1821). In 1840 occurred the first ascent on horseback, when Abel Crawford (then 75 years old) was escorted up by E. A. Crawford. Ethan's stone cabin was swept away in the terrific storm of August, 1826.

In 1850 the summit was taken possession of by an eccentric genius named Nazro, who named it *Trinity Height*, and proposed to inaugurate here " the Christian or purple and royal Democracy." By virtue of his self-assumed title of the " Israel of Jerusalem," he built toll-gates on the bridle-paths, and taxed each visitor a dollar. In 1852 the Summit House was built, of massive rocks bolted together, and with 4 cables holding the roof on. Its dimensions were 24 ⅟ 64 ft.; and the Tip-top House, which was built the next year, is 28 × 84 ft. In 1854 an observatory 40 ft. high was erected on the summit, but it was unprofitable, and was torn down two years afterwards.

The first ascent in winter was made in Dec., 1858, by a Lancaster sheriff, in order to serve a legal process. His party found frost 1½ ft. thick on the windows, the walls, and the furniture, and they had barely reached the woods below when a deadly frost-cloud overspread the summit. Three more Lancastrians ascended in Feb., 1862, and remained on top for 2 days, experiencing a 36-hours' snow-storm. In the winter of 1870 – 71, 70 ascents were made; and many others have ensued since. In Jan., 1874, the daughter of Ethan A. Crawford walked up on the railway sleepers, without great difficulty. The most perilous ascent was made by Prof. Huntington, in Nov., 1873, when the thermometer stood at 17° below zero, and the wind blew 72 M. an hour.

The president of the Mt.-Washington Road Co. vainly endeavored, in 1853, to get an appropriation of $50,000 from the national government, in order to complete the road and to erect a lofty and massive observatory on the summit. In 1869 – 70 Messrs. Huntington and Clough demonstrated the possibility of wintering at a high altitude by passing January and February on Mt. Moosilauke; and the same gentlemen, with Mr. S. A. Nelson and two other companions, boldly attacked Mt. Washington the following winter, and remained on its summit from Nov. 12, 1870, to May 12, 1871. They occupied a portion of the engine-house, and the expense of the expedition ($3,500) was borne by friends in the Eastern States, the telegraph having been laid and operated by the U. S. Weather Bureau. The narration of singular vicissitudes experienced by these ice-bound observers, and the wonderful phenomena that they saw, makes the most interesting chapter in the alpine annals of Eastern America. They are detailed in the volume entitled *Mt. Washington in Winter;* also, in an abridged form, in *The Geological Survey of N. H.,* Vol. I. pp. 96 – 118. Since this party demonstrated the feasibility and value of winter observations from this point, small squads of U. S. Signal-Service men have occupied the summit continuously, and their meteorological observations, sent down by telegraph, form an important element in the calculations of the Weather Bureau at Washington.

Daniel Webster visited the summit, and addressed himself thus: " Mt. Washington, I have come a long distance, and have toiled hard to arrive at your summit, and now you give me a cold reception. I am extremely sorry that I shall not have time enough to view this grand prospect which lies before me; and nothing pre-

vents but the uncomfortable atmosphere in which you live." The statesman and
his guide (E. A. Crawford) were enveloped in a snow-squall on their descent.

In 1869 the summit was visited by President Grant and his family.

More than $ 25,000 have been spent in lawsuits about the ownership of the top
of the mountain, the contestants being a Mr. Bellows, of Exeter, and Coe & Pingree,
of Salem and Bangor. The latter finally compromised by purchasing Bellows's
claims.

Casualties. — In October, 1851, a young graduate of Oxford University, the son of
Sir George Strickland, went up the Crawford-House Path against the advice of the
guides, became bewildered, and perished by falling over a cliff. In 1855, Miss Lizzie
Bourne, of Kennebunk, ascended from the Glen House on foot, with her uncle and
cousin, but became exhausted and sat down to rest within 30 rods of the summit
(which was veiled by fog), and died there. The place is marked by a pyramid of
stones near the railway. In August, 1856, Benjamin Chandler, an elderly man
from Delaware, was lost from the Glen-House path, and died from exposure, on the
spur now known as the Chandler Ridge. His remains were found a year later.

In late October, 1855, Dr. B. L. Ball attempted to ascend Mt. Washington alone,
but became confused in the snow and clouds, lost his way, and spent three days in
painful wanderings on the mountain-side, enveloped in storm and encased in ice.
The nights were spent cowering under an umbrella and fighting off the fatal approach of sleep. After thus struggling for 60 hours, without food or sleep, he was
found still wandering along the ridges by a party of guides sent in search of him. In
1874, a young Pennsylvanian was lost from the Crawford-House Path, and nothing
has since been seen of him.

95. The View from Mt. Washington

has justly been called "an epic landscape." The English alpestrian,
Latrobe, said that it is magnificent, but gloomy. The view-line sweeps
around a circumference of nearly 1,000 M., embracing parts of five States
and the Province of Quebec. Within this vast circle are seen scores of
villages and hamlets and hundreds of mountains, with the widening valleys
of the chief rivers of New England. If the peak was 5,000 ft. higher, the
beauty of the view would be seriously impaired by the indistinctness
caused by the greater distance.

The first mood of the visitor (unless he is one of the dull and improvident souls
who herd by the hotel-stove) is of wonder and amazement at the vastness of the
prospect. Everything appears confused and chaotic for scores of leagues around,
and the undulations of the land seem scarcely more characteristic than would so
many suddenly arrested waves of a mighty sea. But the deeply innate topographical instinct of the world-encircling Anglo-American race soon asserts itself, and
from the recognizable villages or peaks a curiosity is excited as to the others on all
sides. In the hope of gratifying this feeling, the following analysis of the view has
been prepared with great care, and after thorough investigations. The Editor stayed
six days on the summit of Mt. Washington, and Prof. Huntington remained there
more than ten days, preparing notes for this description ; and the last-named gentleman also passed fifteen days there in December, 1875, making fresh observations
during the sharp clearness of the winter. Only a few who ascend the mountain
carry compasses, and the description of the view has therefore been subdivided with
reference to conspicuous peaks which are visible hence, rather than according to the
cardinal points.

An experienced mountain-traveller, who has spent many summers at
the Glen House, says that he has *always* secured clear days for his frequent visits to the summit of Mt. Washington, by waiting until the whole
sky was absolutely free from clouds, and the wind was from the W. or N.

W. There is no observatory on the summit, nor any one place whence the whole horizon may be surveyed. The observer will therefore be obliged to move about from point to point, and sometimes must remain in cold and wind-swept places. The need of warm wrappings will then be felt. Most of the transient visitors are on the mountain during the middle hours of the day, when the view is the least interesting, being deadened by excess of light, and unrelieved by shadows. The clearest days are just after the close of long rain-storms, when the air is washed clean.

In the preceding pages the views from Mts Jefferson, Clay, Monroe, Pleasant, Boott's Spur, the Ledge, and other points close to Mt Washington have been described in detail, so that people who are cloud-bound on the summit can get fine off-looks by descending to either of those points.

" Every morn I lift my head,
Gaze o'er New England underspread,
South from St. Lawrence to the Sound,
From Catskill east to the sea-bound.
. . . . . .
Oft as morning wreathes my scarf,
Fled the last plumule of the Dark,
Pants up hither the spruce clerk
From South Cove and City Wharf.
I take him up my rugged sides,
Half repentant, scant of breath,
Bead-eyes my granite chaos show,
And my midsummer snow ;
Open the daunting map beneath, —
All his country, sea and land,
Dwarfed to measure of his hand ;
His day's ride is a furlong space,
His city-tops a glimmering haze.
I plant his eyes on the sky-hoop bounding :
' See there the grim gray rounding
Of the bullet of the earth

Whereon ye sail,
Tumbling steep
In the uncontinented deep.'
He looks on that, and he turns pale.
'T is even so ; this treacherous kite,
Farm-furrowed, town-incrusted sphere,
Thoughtless of its anxious freight,
Plunges eyeless on forever ;
And he, poor parasite,
Cooped in a ship he cannot steer, —
Who is the captain he knows not,
Port or pilot trows not, —
Risk or ruin he must share.
I scowl on him with my cloud,
With my north-wind chill his blood ;
I lame him, clattering down the rocks ;
And to live he is in fear.
Then, at last, I let him down
Once more into his dapper town,
To chatter, frightened, to his clan,
And forget me, if he can.''

RALPH WALDO EMERSON.

** The View. — *From Mt. Adams to Mt. Moriah.*

Directly N. of Mt. Washington is the noble peak of Mt. Adams, about 3 M. distant over the dark and profound depths of the Great Gulf. Its upper parts are covered with loose fragments, and form a fine spire, flanked by long ridges and spurs. Mt. Madison adjoins Adams on the r., and is lower and less imposing. Between Adams and Madison is a portion of the heavily wooded Crescent Range, beyond which the view-line traverses the wilderness of Berlin, and rests on the fair Androscoggin Valley, dotted with the farms and clearings of Milan and Dummer. Over the r. shoulder of Adams, touching the remote horizon, and marked by a precipitous slope on the E., is Mt. Carmel, on the border between Maine and New Hampshire, and within 7 M. of the Anglo-Canadian frontier. To the r. of this view-line, and nearer, is the forest-bound sheet of Head Pond, over which are the Milan Hills; and before reaching Mt. Carmel the line passes over the rolling wilderness of the Magalloway Mts., with Mt. Pisgah in advance. Farther to the r., on the remote horizon, nearly over Head Pond, are several Canadian and border peaks, Saddleback, Mt. Gosford, Ben Durban, and Mt. Nicolet, with the distant speck of Mt. Megantic, near the lake of the same name. On this line, but nearer, are the wilderness highlands of Wentworth's Location and the Dartmouth-College Grant. Just on the l. of Madison the line passes over the white houses of Milan Corner and the forests of Cambridge and Erroll, beyond which is Mt. Dustan, lifting its sharp apex N. W. of Lake Umbagog, with the Diamond Peaks still farther out.

On the r. of Mt Madison, over the farms on Randolph Hill, the view rests on the white hamlet of Berlin Falls, and the great booms in the river can be distinguished. The broad valley of the Androscoggin extends thence to the N., the Chickwolnepy Range being on the r. and the low black mound of Mt. Forist on the l. of the hamlet. The broad bright mirror of Lake Umbagog is then seen far away, with the ledge-crowned peak of Mt. Aziscoös near its N. shore, over a hummock on the S. E. spur of Madison. To the r. of Umbagog the lakes of Welokenebacook (Richardson), Mollychunkemunk, and Moosetoemaguntic are seen, stretching away in a deep valley to the N. E., surrounded by lines of noble peaks, which tower out of the deep woods, and have never yet been visited or named.

To the right of the lakes is the remote and isolated crest of Mt. Bigelow, in Flag-staff Plantation, with Saddleback to the r., on the horizon, and about N. E. Farther to the r. is the blue group of Mt. Abraham, on the S. of which, and far beyond, is the sharp apex of Mt. Katahdin.

According to the best maps the air-line distance between Mt. Washington and Mt. Katahdin is 165 M., and it has therefore been doubted whether the latter is visible. No other theme in all the scenery of the White Mountains has drawn out such sharp and protracted discussions as this of the visibility of Katahdin, which is also asserted by the admirers of Mts. Kiarsarge and Osceola. During several remarkably clear days in December, 1875, Prof. Huntington, the foremost of the dissenters, recognized Katahdin from Washington, and thereafter withdrew his opposition. But there are not more than ten days in a year when it is clear enough to see for so great a distance.

The view now returns from its distant flight and rests on the Chandler Ridge, which is a part of Mt. Washington, running N. E. from the summit, and appearing like a long terrace below, with the Nelson Crag farther down, near the great cliffs of Huntington's Ravine. Over the Chandler Ridge, and across the mouth of the Great Gulf, is Pine Mt., which hides the village of Gorham; over which and N. E. of the Androscoggin is Mt. Hayes, a low and ledgy mountain, with the ravine of the Lead-Mine Brook on the E., on whose r. is the dark mass of Baldcap. Over the r. part of Mt. Hayes is the blue and pointed peak of Goose-Eye, sloping gradually to the W., nearly in line with the far-away Katahdin. On the l. of and beyond Goose-Eye (through the Mahoosuc Notch) are the Speckled Mt. and the Bear-River White Cap, enclosing the Grafton Notch; and close on its r. is the Sunday-River White Cap, over which rise the many mountains of Andover and Byron. Nearer at hand is a section of the

Peabody Glen, over which is the hardly perceptible knoll of Mt. Surprise, on the long flank of Moriah. A beautiful vista is now gained down the Androscoggin Valley, over which, between Baldcap and the nearer Moriah, is the crest of Mt. Ingalls. Further away in this direction are tall peaks of Maine, rising from the settled townships, Black and Puzzle Mts., the Rumford White Cap, and Mt. Blue, the latter of which cuts the horizon nearly over the lowly Mt. Winthrop, and just to the r. of the Nelson Crag. In this direction the graceful curves of the Androscoggin are followed through the valleys of Shelburne, and several of its islands are seen. The Grand Trunk Railway is also plainly visible, and the headlights of its locomotives appear on clear evenings. The lower course of the Peabody River is distinguishable on account of its deep wash-outs.

Mt. Moriah to Mt. Baldface.

Mt. Moriah is 5–6 M. E. N. E. of Washington, and is seen over the heading cliffs of Huntington's Ravine, being at the angle of the eastern range as it bends to the N. E. It has no distinct or well-marked peak, and over its r. are the hills of Bethel. To the r. is the indistinct peak of Mt. Imp, whence a ridge with high cliffs descends to the Peabody Glen. A little to the r. are the long white walls of the Glen House, diminished by distance to a mere dot of light among the shadowy forests. To the right of Imp and beyond is the more distant Mt. Calabo, in Maine. Toward the E., across the adjacent Pinkham Notch, are the massive and imposing Mt. Carter and the Carter Dome, which have two convex peaks, on the S. of which is a hollow, into which fall two white slides.

The mountains on the island of Mt. Desert are due E., 145 M. distant, over the Carter Range. It is claimed that the White Mts. are seen thence, and some people think that the island is visible from Mt. Washington, with a powerful glass and on a clear day. A trifle to the r., one or two degrees S. of E., the high-placed village of Paris Hill is visible, 42 M. distant, with Streaked Mt. beyond; over which, 115 M. distant, it is claimed that Mts. Megunticook and Battie, of the Camden Mts., on the shore of Penobscot Bay, are seen from Mt. Washington.

The view now enters more certain ground and falls on the dark and adjacent Mt. Wild-Cat, covered with forests and separated from the Carter Dome by the invisible Carter Notch. The distant highlands of Waterford, Norway, and Hebron are beyond. Over its highest point is Mt. Baldface, with its N. peak bare on the apex, and the S. peak showing white ledges.

Mt. Baldface to Mt. Kiarsarge.

On the r. of the white crest of Baldface are the round and wooded summits of Mts. Sable, Eastman, and Slope, over which are sections of the Upper Kezar Pond and of Long Pond, beyond Bridgton. Between the higher part of Wild-Cat and Sable is the long and partially cleared ridge of Black Mt., in Jackson. Close over these pastured slopes are the similar summits of Double-Head, massive and symmetrical, above which are the sharp twin crests of Mt. Gemini, on the ridge running N. from Kiarsarge.

To the l. of Double-Head is the long rampart of Mt. Pleasant, rising from the sandy plains beyond Fryeburg, and forming several low crests. A white hotel is on the central peak, and is readily seen in the afternoon, when the sunlight is reflected from its walls. On the left of Pleasant are the Kezar, Upper Moose, and Long Ponds; and on the r. are Pleasant Pond and the square-shaped sheet of Lovewell's Pond, renowned in early border-history for the desperate battle on its shore. To the right of Pleasant, over and on either side of Peaked Mt., in the town of Sebago, are the broad bright waters of Sebago Lake, far beyond which, and a little to the r., is the city of Portland, on Casco Bay. Further out is the ocean, which is seen for leagues on the l. and r., but is so nearly the color of the sky as to be discerned with difficulty. In clearest weather vessels can be seen off shore all through the broad bight of the sea which was formerly known as the Gulf of Maine.

Mt. Kiarsarge to Mt. Chocorua.

Kiarsarge is nearly S. E. and about 15 M. distant. It is a graceful mountain, of conical shape, and is readily distinguished by the house on its pointed summit. On this side of Kiarsarge are Tin and Thorn Mts., on the same ridge, the latter being the higher and farther to the r. In the foreground is the long valley of the Ellis River, with its upper part densely wooded, and farms appearing as it extends to the S. The valley over and beyond the Ellis is that of the Wild-Cat Brook, and between the two are the Spruce and Eagle Mts. At the end of the Ellis Valley is Thorn Hill, over which opens the rich and beautiful Saco Valley, wherein the white hamlets of Lower Bartlett, N. Conway, and Conway are seen in succession. The Green Hills appear E. of N. Conway and from behind Kiarsarge; and on their r. is the flashing mirror of Walker's Pond. Over the Green Hills are the Frost and Burnt-Meadow Mts , in Brownfield; Saddleback, in Baldwin; and the rolling highlands of Hiram, Cornish, and Limington. It is said that Prout's Neck and Old-Orchard Beach are seen in this direction, nearly over Mt. Cutler, in Hiram.

The great gorge near at hand in Mt. Washington is Tuckerman's Ravine, whose S. wall is seen, striped with light-colored slides. Over the S. wall are the high and massive crags of Boott's Spur (2 M. S. S. E.), above which is the rounded summit of Iron Mt., falling sharply on the N. to a cultivated plateau, and braced on the S. E. by a ledgy spur. Farther away in this direction are the highlands of Madison and Eaton, with the long ridge of the Green Mt. in Effingham still more remote. Over the l. flank of the latter, down by the sea-shore, is the low round swell of Mt. Agamenticus, the landmark for sailors. Over the r. of Iron Mt. is the ridge of Mt. Attitash, overlooked by the two rocky peaks of Moat Mt., uplifted above the forests. Over Moat are the distant waters of Silver Lake and Ossipee Lake, and the line over their centre crosses the bold hills about N. Wolfeborough and rests on Copple Crown and Big Moose Mts., in Brookfield, with the Frost Mts. in Farmington, near the birth-place of Henry Wilson, far beyond. Mt. Pawtuccaway, in Nottingham, is still more distant. On the r. of Moat is Mt. Langdon.

Mt. Chocorua to Mt. Carrigain.

Mt. Chocorua is S. by E., 22 M. distant, and appears on the r. of and beyond Table Mt. It is the sharpest and noblest peak in all the view from Mt. Washington, and lifts its white pyramidal ledges far into the sky, flanked by bare supporting ridges. On the r. of Langdon is the cone of Mt. Parker, beyond which is the long dark ridge of Bear Mt., with the curving and ledgy top of Mt. Paugus beyond, and on the r. of Chocorua. Still farther out in this direction is the long blue line of the Ossipee Range, on whose r. are the island-strewn waters of Lake Winnepesaukee, seen directly over Paugus and in several places to the r.

The view in this direction should be taken from the U. S. Signal Station. Below is the alpine terrace of Bigelow's Lawn, over which the forest-crowned Montalban Ridge is seen, running down to the hardly distinguish-able crest of Giant's Stairs, 8 M. distant. Far beyond, and a little to the l., is the double peak of Mt. Belknap, over the shining levels of Lake Winnepesaukee. Close to Giant's Stairs, on the r., is the flat top of Mt. Resolution, marked by ledges of reddish granite. Next to the r. is the lower peak of Mt. Crawford, which is in poor relief.

Over Resolution is the black top of the Bartlett Haystack; and over the r. of Crawford is the foreshortened ridge of Tremont, with its highest peak on the S., partly burnt over. Paugus and the blue Ossipee Mts. are l. of and beyond the Haystack. The high round dome of Passaconaway is over the main peak of Tremont, and under it is the singular white mound of Potash. On the r. of Passaconaway, and over the r. shoulder of Tremont, is the high cloven peak of Whiteface. The twin Uncanoonucs are far away on the l. of Passaconaway, near Manchester; and between Passaconaway and Whiteface is the massive Mt. Wachuset, in Massachu-setts, with Joe-English Hill nearer, in New Boston. To the r. of White-face are the distant Temple Mts and Crotched Mt., in Francestown.

On the r. of Tremont is the square-headed top of Green's Cliff, up the valley of Sawyer's River. Over Green's Cliff is a round peak of Tri-pyramid, whose two other peaks, on the r., are sharp. The view now passes down the long valley of the Mt.-Washington River, wh'ch runs about S. S. W. from Bigelow's Lawn and opens into the Saco Valley 14 M. below. Down this long trough is the spur from Sawyer's River to the Nancy Range, overlooked by the N. E. ridge of Tripyramid, with the long curving crest-line of Sandwich Dome beyond, flanked by the Sachem Peak and the Acteon range. Far away in this direction is the top of Mt. Prospect, near Plymouth, just to the l. of the Sachem Peak; and the scarce distinguishable top of Mt. Weetamoo is on the r. of the Sachem.

Mt. Monadnock lifts its faint blue curve 100 M. away, nearly over Prospect, and about S. S. W. On its r., and nearly over Mt. Wcctamoo, is the bold peak of Kearsarge, 65 M. distant, over the Ragged Mts. Much nearer, on the line to Kearsarge, and on the r. of the mouth of the Mt.-Washington River, is the Frankenstein Cliff, around which the lines of the Portland & Ogdensburg Railroad are visible. Below the Cliff is the Nancy Range, containing Mts. Nancy, Anderson, and Lowell, and the view then rests on Carrigain.

Mt. Carrigain to Mt. Lafayette.

Carrigain is a massive mountain, with a deep hollow turned to the E., and the rounded knoll of Vose's Spur on the E. It is one of the most striking and remarkable of the peaks which are seen from Mt. Washington, and is best viewed from the Signal Station. On its r., behind and close to it, is the wilderness peak of Mt. Hancock. Between and beyond Carrigain and Hancock is Osceola, with a precipitous slope to the E. and a bold secondary peak towards the Mad-River Notch. To the r. of Osceola, and far beyond, are the white rounded crests of Mt. Cardigan, with Sunapee Mt. still farther out and the dark Croydon Mt. on the r.

The view now falls on the S. W. peaks of the Presidential Range. From the Signal Station the Lakes of the Clouds are seen, 1,200 ft. below, over which are the dark and rugged battlements of Mt. Monroe, with the bridle-path winding along on the l. Beyond, and a little to the r., is the flat top of Franklin, marked by the serpentine trail of the Crawford-House Path, on whose r. is the high and rounded summit of Pleasant, with Clinton to the l. The bare ledge beyond is the crest of Jackson; and the wooded heights of Webster end the range. On the r. of Webster, across the Notch, is Mt. Willey, which is also nearly over Mt. Pleasant, and is marked by a sharp descent towards the S. and a bare exposure on the N. The low crest of Mt. Willard is over the r. flank of Clinton, and above it is the wooded ridge of Mt. Field, continuous with Willey. Over this range the view extends across Hancock's flank to the mountain-chain which runs from Osceola to the mouth of the Hancock Branch, whose peaks are surrounded by the wilderness and have no names. Beyond this range are the Loon-Pond Mts., marked by bare granite ledges. Far away over these ridges is the monotonous range of Mt. Carr, and still more distant is the dim blue peak of Ascutney, 80 M. distant.

Over Mts. Field and Willey the view enters the Pemigewasset Forest, near whose lower end is the Potash Mt., 3 M. above Pollard's, over which are the highlands of Woodstock, a part of the Mt.-Carr range, and Smart's Mt. On the r. of Mt. Field is the ravine of Beecher's Cascades, which enters S. of Mt. Tom; and over this depression are the long and lofty ridges of Mts. Bond and Guyot, rising out of the central wilderness, and cut by a ravine in the N. E. slope. Over Bond's r. shoulder is Mt. Flume, on whose r.

is the sharp peak of Mt. Liberty. Between Flume and Liberty, and directly over Beecher's Cascades, is the long crest of Moosilauke, crowned by a house. A short distance to the r. of Moosilauke is the sharp peak of Lincoln, beyond which rises the splendid serrated ridge of Lafayette, with its crest of gray rocks. Very far away between Moosilauke and Lafayette are the dim crests of the Killington and Shrewsbury Peaks, near Rutland.

It should be mentioned that one of the best topographers in New England (Prof. C. H. Hitchcock) has studied from Mt. Washington the line of the Green Mts. S. of the Killington Peaks, and by reason of previous familiarity with the Vermont and Massachusetts mountains has recognized (in succession to the l. from the Killington Peaks) Mts. Æolus and Equinox, Stratton Mt., and the remote Greylock, the chief of the Berkshire Hills. The latter is distinguished by its sharp apex, and is nearly over Mt. Carr. Farther to the l. he also recognized the dark disk of the Hoosac Mt., where the great tunnel is. Greylock is about 160 M. distant, a little S. of S. W. If these remote peaks can be identified at all by the unscientific traveller, it can only be done in the clearest of days and with a powerful telescope.

Mt. Lafayette to Cherry Mt. (Owl's Head).

In looking towards the high sierra of Lafayette the long ridge of the S. Twin Mt. is seen under it, with the N. Twin on the r., marked by a whitish slide. (This part of the view should be taken from beyond and above the old stone house back of the hotel.) On the r. of N. Twin, and lower, is Mt. Hale, flanked by the white-topped Sugar Loaves, near the Twin-Mountain House. The view now passes down the Ammonoosuc Valley, by Ammonoosuc Station, the square clearing of the Twin-River Farm, and the white walls of the Fabyan House, with the low curving summits of Mt. Agassiz and Round Hill apparently at its end, close to the clustered houses of Bethlehem. Between Mt. Tom and the Fabyan House are the bold hills called Mt. Andalusite and Mt. Rosebrook, the latter being near the Fabyan. Farther out, above Bethlehem, is the long Mt. Gardner range, over which are the Bread-Loaf Mt. and others of the Green Mts., with the lofty crest of Camel's Hump, nearly due W. Just to the l. of Camel's Hump, and still more remote, is the dim group of the Adirondack Mts., beyond Lake Champlain, in the State of New York. Mt. Marcy is the highest point in this cluster, and is about 140 M. distant.

Nearer at hand, across the head of the Ammonoosuc Valley, and 6 M. distant, is Mt. Deception, to the r. of the Fabyan House, with light deciduous trees on the E. slope and dark evergreens on the summit, meeting on a well-defined line. Over its l. flank is Beech Hill, N. of the Ammonoosuc, above which is the white village of Littleton (with Camel's Hump on its l.). Far away on the horizon, a few degrees N. of Camel's Hump, is the conspicuous peak of Mt. Mansfield, the greatest of the Green Mts. of Vermont. Before it is Mt. Elmore, with Worcester Mt. on the l. and Mt. Sterling on the r. These stately summits are nearly over Mann's Hills, N. of Littleton village. Over and to the r. of Deception, and but 2–3 M. from it, is the long dark mass of Cherry Mt., with three small peaks, of

11 *

which the bare one, on the N., is called Owl's Head. Nearly over the middle of Cherry is the long ridge of Dalton Mt. (l. of which is Mt. Mansfield), and far away in the Green-Mt. line are the mountains of Belvidere.

From Cherry Mt. (Owl's Head) to Mt. Jefferson.

The Owl's-Head peak of Cherry Mt. is a bare hummock on the N. end of the ridge, about W. N. W. by N., 8 – 10 M. distant. On its r. is seen the clear sheet of Cherry Pond, above which is Mt. Niles, over the Lunenburg Heights, in Vermont, and far beyond and over the latter are the picturesque peaks of the Lowell Mts., in the N. part of the Green-Mt. chain. To the r. of Cherry Pond, a few miles farther distant, and over Bray Hill, are the round-topped highlands of Lancaster, Mts. Prospect and Pleasant, and the Martin-Meadow Hills. The Martin-Meadow Pond is seen on their l.; and on their r., partly concealed by intervening highlands, is the fair white village of Lancaster, nestling down on its rich green meadows, near the winding Connecticut. It has been a theme of frequent arguments whether the waters of the Connecticut are visible from Mt. Washington, except during the spring inundations. However this may be, the white fog which lies over its course at morning is plainly seen, winding among the hills and over the lowlands for many leagues.

Far away on the horizon is the sharp apex of Jay Peak, in Northern Vermont, with the Lowell and Montgomery Mts. near it on the l. Much nearer, and on the same line of view, is the dark and massive Burke Mt., with Mts. Tug and Umpire and the Victory Hills on the r.

The view now rests on Mt. Clay, close at hand and joined to Washington below, over which are the white hotels of Jefferson Hill. The valley of Israel's River lies in this direction, and is the broadest valley visible from Mt. Washington. The roads and farms are plainly seen for many miles; and the Mt.-Adams House is distinguished, on the r. of Jefferson Hill. Back of the latter village rises Mt. Starr King, with no conspicuous peak, but recognized by a slide on the side of its ridge. Farther away, over Jefferson Hill, are the Cow and East-Haven Mts., in Vermont, beyond which are Mts. Hor and Annanance, at Willoughby Lake. The famous Willoughby Notch, between these two mountains, does not show to advantage from this point. Over the ridge which runs S. W. from Starr King is the well-marked peak of Owl's Head, which rises from the shore of Lake Memphremagog. Farther to the r., nearly over Starr King, and beyond Mt. Burnside and the highlands of Guildhall and Ferdinand, is Mt. Orford, the chief peak in the Eastern Townships of Canada. Between Mt. Washington and Mt. Starr King are the wooded heights of the Pliny Range; and over Starr King is Cape Horn, rising out of the Northumberland plains. The view now rests on the distant ranges near Maidstone Lake and about the Smuggler's Notch, over the l. slope of Mt. Jefferson.

Mt. Jefferson to Mt. Adams.

Mt. Jefferson is a high and massive peak, across the upper part of the Great Gulf, and about 3 M. N. by W. from Mt. Washington. Over it and to the r. is the great assemblage of the Pilot Mts., stretching across the wilderness of Kilkenny. The South Peak of the Pilots is over the E. point of Jefferson, and shows a sharp apex. Over the r. shoulder of Jefferson, and beyond the Pilot Range, are the singular round domes of the Percy Peaks, drawing the attention on account of their light color, which strongly contrasts with the surrounding forests and woody ridges. Back of the Percies are the Stratford and Sugar-Loaf Mts.; and the Long Mt. of Odell extends from the Peaks to the r. In the foreground, nearer to Jefferson, is the dark and well-wooded range of the Randolph Mts. A long reach of the Upper Ammonoosuc Valley is seen beyond, for the most part filled with forests and unbroken by clearings. The ridge which is on the r. of the Pilot Mts., and runs back of Mt. Adams on the r., is Green's Ledge, on whose l. are Hager's Peaks and Deer Mt., the latter forming a long line of heights. Back of all these ranges which are seen between Jefferson and Adams are the distant peaks towards the Connecticut Lakes, most of which are in the Dixville, Crystal, and Magalloway Ranges. The upper Monadnock, in Lemington, Vermont, is on the extreme l. of this vista. Far beyond, and low down on the horizon, are two or three peaks in the Eastern Townships of Canada.

Starr King gives the following description of this view: "The first effect of standing on the summit of Mt. Washington is a bewildering of the senses at the extent and lawlessness of the spectacle. It is as though we were looking upon a chaos. The land is tossed into a tempest. But in a few moments we become accustomed to this, and begin to feel the joy of turning round and sweeping a horizon-line that in parts is drawn outside of New England. Then we can begin to inquire into the particulars of the stupendous diorama. Northward, if the air is not thick with haze, we look beyond the Canada line. Southward, the 'parded land' stretches across the borders of Massachusetts, before it melts into the horizon. Do you see a dim blue pyramid on the far N. E., looking scarcely more substantial than gossamer, but keeping its place stubbornly, and cutting the yellowish horizon with the hue of Damascus steel? It is Katahdin, looming out of the central wilderness of Maine. Almost in the same line on the S. W., and nearly as far away, do you see another filmy angle in the base of the sky? It is Monadnoc, which would feel prouder than Mont Blanc, or the frost-sheeted Chimborazo, or the topmost spire of the Himalaya, if it could know that the genius of Mr. Emerson has made it the noblest mountain in literature. The nearer ranges of the Green Mts. are plainly visible; and behind them Camel's Hump and Mansfield tower in the direction of Lake Champlain. The silvery patch on the N., that looks at first like a small pond, is Umbagog; a little farther away, due S., a section of the mirror of Winnepesaukee glistens. Sebago flashes on the S. E., and a little nearer are the twin Lovell lakes, that lie more prominently on the map of our history than on the landscape. Next, the monotony of the scene is broken by observing the various forms of the mountains that are as thick as 'meadow mole-hills,'—the great wedge of Lafayette, the long, thin ridge of Carter, the broad-based and solid Pleasant Mt., the serrated summit of Chocorua, the beautiful cone of Pequawket [Kiarsarge], the cream-colored Percy Peaks, as near alike in size and shape as two Dromios. Then the pathways of the rivers interest us. The line of the Connecticut we can follow from its birth near Canada to the point where it is hidden by the great Franconia wall. Its water is not visible, but often, in the morning, a line of fog lies for miles over the lower lands, counterfeiting the serpentine path of its blue water that bounds two States. Two large curves of the Androscoggin we can see. Broken portions of the Saco lie like lumps of light upon the open valley to the W. of Pequawket. The sources of the Merrimac are on the farther slope of a mountain that seems to be not more than the distance of a rifle-shot. Directly under our feet lies the cold Lake of the Clouds, whose water plunges down the wild path of the Ammonoosuc. . . . . And in the sides of the mountain every wrinkle, E. or W., that is searched by the sunbeams, or cooled by shadows, is the channel of a bounty that swells one of the three great streams of New England. And lastly, we notice the various beauty of the valleys that slope off from the central range. No two of them are articulated with the mountain by the same angles and curves. Stairways of charming slope and bend lead down into their sweet and many-colored loveliness and bounty."

But the chances are that the visitor to the summit of Mt. Washington will see more of clouds than of landscape ; and he may be enclosed in a dense pall of mist during his entire sojourn on the mountain. Days of this character are far more numerous than clear and sunny days. The most favorable time for ascending is when the wind is in the N. W. or N., and there is not a shred of cloud on or over the peak. Even then, however, a sudden veering of the wind or other atmospheric changes may cause the summit to be thickly enveloped. The loss of the view over half of New England, the chill dampness of the driving mist, and other causes of discomfort will then shake the composure of the veteran traveller ; and he may perhaps like to know how and why such days were enjoyed by the poet-dreamer, Starr King : " Cloud-effects are the most surprising and fascinating pageants which the ascent of the mountain can disclose. . . . . Certainly the richest pictures that rise to us, as we write, out of a memory of more than a score of visits to Mt. Washington, are combined out of clouds. We see again the gray scud driving over the peak as we approach it, ... we remember how dispirited the visitors on the summit seem in the chilly gloom, and we see the fog filled with yellow light, then thinning away and knitting itself together in an instant, but soon blown apart by the breeze to let the color of the nearer forests, and then of the lowlands, glow through, dimly at first and confused, in another second distinct and blinding, but soon orderly and glorious, as perhaps the realities of another existence may break upon saintly eyes that emerge from the mists of death. . . . . We behold again the settling of heavy clouds over the slopes as we descend, wrapping us in blackness of darkness ; and, hastening on through furious gusts, we come to the lower fringes of the tempest, and look back and up to see it crouched over the ravines of Clay, from which vast sheets of vapor are swept by the wind, their lower edges sulphurous as they rush into the light, and now and then the whole mass whirling apart to show the dark masses of Madison and Adams towering in treble height through the gloom. And then such glimpses of the valleys ! Sinai behind and Beulah before ! . . . . We are overtaken by a rain that rages against us out of the W. ; and after it is spent, we see a rainbow arching over the long line of the Carter range, and painting its blue-black forests at each end with variegated flames. We stand also on the summit in the morning when the sky is clear, and view a wide plain of billowy mist, ... and see the neighboring summits jutting likewise above the foam, which rolls, and tosses, and plunges, and splinters into spray, as though with its milky spume it was appointed to mimic the passion of the sea and the majesty of Niagara.

" A mist lay over all the valleys. The mountains heaved their sharp ridges out of an ocean of stagnant foam. A wild bank of dingy fog lay along the eastern sky. Over this artificial horizon we saw the advent of the morning, — the wide flush of red around a third of the vast circuit, the bubbling of rosy glory over its fleecy rim, the peep of the burning disk, and the gradual mounting of the light, showing how

> 'tenderly the haughty day
> Fills his blue urn with fire.' "

Oakes describes the beautiful effect of the transient views opened by rifts in the mist. " With the suddenness of a flash a narrow but clear space opened to the S. E., and we saw for a moment, through the window of the mist, the hills and settlements of the low country, with the rich scarlet and yellow colors of the autumnal forest glowing in the sun, — a warm bright picture set in the cold contrasted frame of mist. In an instant it passed away like a glimpse of some happy country seen in a dream."

THE FRANCONIA MOUNTAINS.

96. Littleton.

Hotels. — The Oak-Hill House (70 guests) is a pleasant summer-hotel on the high hill beyond Littleton and over the village, in the vicinity of pleasant groves of trees. It has billiard and bowling rooms, a livery stable, and a free coach ; and its rates are $ 2.50 a day, or $ 8 – 12 a week. From the piazzas a noble view is obtained, extending from Mt. Adams to Moosilauke. The distances from the Oak-Hill House are as follows : To Bethlehem, 5 M. ; Franconia, 6 ; Mt. Washington R. R., 2½ ; Twin-Mountain House, 13 ; Fabyan House, 18 ; Howland Observatory, 9 ; Crawford House, 22 ; Flume, 17 ; Profile House, 11 ; Littleton station, ½ M.

Thayer's White-Mountain Hotel is an old and well-famed house on the main street of the village, accommodating 100 guests, at $ 2.50 – 3 a day. The Union House is further up the street, and accommodates 40. The summer boarding-houses of John Merrill, A. R. Burton, Mrs. Cobleigh, Mr. Eastman, and Jefferson Hosmer are also in the village.

Littleton is one of the most prosperous villages in Northern New Hampshire, having about 2,000 inhabitants, four churches (Cong., Bapt., Meth., Epis.), a weekly paper, a high school, a bank, and 30 or 40 stores. Various branches of manufacturing are also carried on (to the value of over $ 500,000 a year), the most interesting of which is near the station, where the Kilburn Brothers have a spacious building devoted to the preparation of stereoscopic views. This is said to be the largest establishment of the kind in the world, and turns out over 300,000 views yearly. The village is pleasant and lively, with well-stocked stores and neat houses ; and the main street runs nearly parallel with the Ammonoosuc. The high-school occupies an elevated position over the central part, and is recognizable from a great distance. Farther back on the heights is the Oak-Hill House, whence a very noble panoramic * view of the White and Franconia Mts. is gained.

The town of Littleton covers 36,000 acres, half of which is improved, making agriculture one of the leading interests of the people. The tumultuous reach of the Connecticut River known as the *Fifteen-Mile Falls* lies on the N. W., and bounds the town for 13 M. In 1764 this territory was named *Chiswick ;* in 1770, *Apthorp ;* and in 1784, *Littleton,* the ugly name which it still retains. The first settlers came in 1774.

New Hampshire had been regarded by geologists as of azoic formation, until 1870, when a reef of fossil corals was discovered near Littleton. In 1873, brachiopods, crinoids, and a gasteropod were found on Fitch Hill, and established the fact that these rocks belonged to the Helderberg period.

Mann's Hills are 1½ – 2 M. from Littleton, and afford noble views of the mountain-ranges to the E. and S. The visitor should leave the road and clamber to the crest of the hill on the r. There is a fine elevation beyond

Palmer Brook, 1 M. W. of Mann's Hills (reached through the pastures), whence a still broader view is gained. *Mt. Misery* is 4 - 5 M. from Littleton, by the road over Mann's Hills, and derives its name from certain poor families who once lived at its base. From several points in the vicinity fine views of the Vermont mountains are gained.

Eustis Hill is just S. of Littleton (1½ M. to the top), and is ascended easily through the fields and forests. Here also is a broad and pleasant prospect, including much of the Ammonoosuc Valley. *Gilmanton Hill* is 2 M. from the village, and derives its name from the fact that the adjacent country was first settled by men of Gilmanton. It gives one of the best prospects of the Franconia Range. The Parker (or Blueberry) Mt. lies near the village, on the S., and from the uppermost clearings on its slope a pleasant view of Littleton and the mountains is obtained.

It is about 5 M. to Waterford, Vt., whence the drive may be prolonged to the far-viewing heights of Concord. The drives to Bethlehem (5 M.) and Sugar Hill are also interesting. Mountain-wagons run frequently to the Flume House, the expense of the ride out and back being $ 2.50 for each person.

97. Lisbon.

Hotels. — McAllister's is the best of the village-inns; the other is not recommended. Summer boarding-houses are kept by M. Bowles, T. Atwood (on the hill), and E. W. Bartlett (1 M. up the intervale).

This town contains 1,848 inhabitants, and the villages of Lisbon, N. Lisbon, and Sugar Hill. Of its 29,130 acres, over 17,000 are improved, agriculture being the chief pursuit of the people. The annual products are 4,500 tons of hay, 111,000 bushels of potatoes, 26,000 bushels of oats, 3,500 bushels of wheat, 40,000 lbs. of butter, 12,000 lbs. of cheese, and 30,000 lbs. of maple-sugar. The manufactures amount to $ 280,000 a year, chiefly in lumber and mineral fertilizers; and the gold-ore which is now being mined in the vicinity and crushed at Lisbon yields $ 14 a ton. There are also copper-mines and deposits of limestone in the town.

Lisbon is the chief village and railroad station, and has 10 or 15 stores, 2 churches, a library, and several fine villas which were built during the excited era of the gold-discoveries. It is not a very pleasant place, but its environs are picturesque and there are interesting drives in the vicinity. Among these are Sugar Hill, 7 M.; Hunt's Mt., 8 M.; Parker Hill, 4½ M.; the Profile House, 15 M.; and Landaff Centre, 3 M. *Parker Hill* is a far-viewing knoll near the Lyman road, easily ascended, and commanding a rich and extensive prospect. It is near the chief hamlet of the farming town of Lyman; and the gold and copper mines are in the same vicinity.

The copper veins in the Mt.-Gardner range consist of schists charged with the sulphurets of iron and copper, averaging less than 5 per cent before concentration. They are, however, favorably situated with respect to drainage and water-power, and Prof. Hitchcock predicts that they will eventually prove remunerative and will employ a large number of workmen.

Hunt's Mt. is reached by the road from Lisbon to Monroe. When the summit of the pass is attained, the visitor should leave the road and climb up on the r. a short distance to the crest. It is one of the peaks of the Mt.-Gardner Range, and the view to the N. is masked by the higher point of *Bald Ledge*, the station of the U. S. Coast Survey in 1875.

The view from Hunt's Mt. is one of the grandest in the Connecticut Valley, including the Franconia, White, and Green Mts., the narrow valley on the W., and a broad area of Eastern Vermont, studded with hamlets. The Editor visited this peak during a remarkably hazy June day, and could not make precise observations on which to base a minute description of the view, the outlines of the mountains were so dimmed by the blue "smoke."

Bath is a pleasant and retired old village near the B., C. & M. R. R., S. W. of Lisbon. The well-famed hotel was burnt in 1872, together with part of the village, and has not been replaced. The town of Bath has 1,168 inhabitants, most of whom are engaged in farming, along the glens of the Ammonoosuc. The Mt.-Gardner Range traverses the E. part of the town, with one high pass; and the Landaff highlands close in on the E. A road runs S. E. through Swiftwater Village and up the narrow valley of the *Wild Ammonoosuc* to N. Benton, at the foot of Moosilauke.

Landaff Centre is a petty hamlet 3½ M. from Lisbon, in a farming town of 882 inhabitants, famous for its maple-sugar. Bald and Cobble Hills and Landaff Mt. are high eminences which separate the glens. There is good trout-fishing on Tunnel Brook and other branches of the Wild Ammonoosuc. There are no boarding-houses in Landaff.

98. Sugar Hill.

Boarding-Houses. — E. H. Goodenow, 40 guests; Jason Coma and Mr. Bowles, 25 each; and Hiram Noyes, 25. The first three are on the mountain-fronting slope of the hill; the fourth is in the hamlet on the W. slope. Their rates are $6-10 a week, and the buildings are but enlarged farm-houses. There is need of a first-class boarding-house at this point.

Sugar Hill is a bold slaty ridge in the E. part of the town of Lisbon, more than 1,000 ft. above the sea, and derives its name from a large grove of sugar-maples on the summit. It is crossed by the road from Lisbon to Franconia Iron-Works, and is 7 M. from Lisbon and 2½ M. from Franconia. There are two routes from Lisbon, one following the Ammonoosuc River for nearly 3 M., and then ascending the glen of Salmon-Hole Brook to the S. E.; the other passing from Lisbon directly up the valley of Mink Brook and by Mink Pond. The latter is more hilly, but commands noble views, especially of the Green Mts. in retrospect. The quiet hamlet of Sugar Hill consists of a long street on the upper W. slope of the ridge, with a small church and one or two shops. It commands a broad panoramic view of the Green Mts. of Vermont, extending for many leagues along the horizon. Passing upward from the street, beyond the maple-grove, the open crest of the hill is reached, whence a superb view is gained to the S. E., E., and N. E. One of the best view-points is from the cottage on the r., where a side-road turns to the r., and here a large spy-glass is kept. Farther down the E. slope a short walk to the l. will lead to a higher rocky knoll on which is a beacon of the U. S. Coast Survey. Fine crystals of staurolite and epidot are found here.

Most travellers will find it convenient to pass on to Littleton, and thence to Franconia Iron-Works on the Profile-House stage. It is an easy walk thence up the road to Goodenow's, where the view is fine, or to the signal station, or the crest of the road above.

* *The View* from Sugar Hill has never been described and is generally unknown, but the Editor regards it as nearly equal to that from Jefferson Hill, and as superior to any other in the mountain region, except those from the high peaks. Nearly N. E., many miles away, is Mt. Starr King, with Bray Hill and the white village of Jefferson Hill at its foot, and the long blue lines of the Pilot Range behind and on the W. Farther to the r. are the Pliny and Randolph Mts., with Mt. Agassiz in the foreground, within 6 M. About E. N. E. is the long black mass of Cherry Mt., over whose r. flank, up the valleys of the Gale and Ammonoosuc Rivers, are the noble crests of Mts. Adams and Jefferson, flanked by the sierra of Clay, which joins them to the supreme peak of Washington. The latter is clearly visible, and the trains can be seen winding up its sides. The crests of Monroe and Franklin are less easily distinguishable, but the rounded head of Pleasant is plainly recognized, nearly due E. Some-what nearer, and on the r. of Pleasant, is the curved top of Mt. Hale, whence the immense mass of the N. Twin Mt. extends for miles to the r., culminating in a long rolling ridge, nearly over Franconia Iron-Works. Farther to the r. is the pointed crest of Haystack, over a line of clearings in the foreground. The eye now rests on the vast and far-reaching mass of Lafayette, across the deep and narrow valley below, and so situated that its very roots are visible. The great ravines in its sides, the serrated crest-line, the plateau of the lakes, and the long spur of Eagle Cliff are all unfolded to view, forming the best possible picture of the chief Franconia mountain. From this point, in May and June, the snow in one of the Lafayette ravines presents the semblance of a vast white cross.

Far below the r. spur is the low hummock of Bald Mt., over and to the r. of which is the high curving ridge of Mt. Cannon, lying across the Franconia Notch. On its r. is the long, lofty, and formidable Mt. Kinsman, looming boldly over the glens of Landaff; and from the gentle S. slope of Kinsman rises the majestic peak of Moosilauke, gaining its highest point by a series of gradually ascending terraces.

Toward the S. W. and W., when not interrupted by nearer hills, there is a broad and beautiful panorama of the distant Green Mts. of Vermont, extending for many leagues along the horizon, and gradually falling away into the hills which hem in the Connecticut Valley.

Ore Hill is about 1½ M. S. E. of the hamlet on Sugar Hill, and is surrounded with roads. It is chiefly composed of gneiss, and contains rich deposits of iron ore, which were formerly worked by the N. H. Iron Co. The old shafts and adits are now sometimes visited by the curious.

99. Mount Kinsman.

The ascent of this formidable peak is rarely undertaken, so great is the labor in comparison with the reward. It is best accomplished from the Bolles farm, a short distance S. E. of E. Landaff, and about 10 M. from the Profile House. Orson Kendall is said to be the best guide in this section. There are two possible ways of ascent, the first of which is by a logging-road which ascends about 1 M. from Bolles's, and thence marching 3 M. through the woods, encountering much fallen timber. Nothing can be more arduous and wearisome than this way. A better route is to ascend the *Slide Brook* by a line of pretty cascades and basins, with steep smooth ledges and rocky débris under foot. Where the stream is met by a long and steep slide coming from the r., it should be left, and then the bare, loose rocks of the slide are ascended to a point within ¾ M. of the summit. The distance from the Bolles farm to the top is nearly 4 M. Mt. Kinsman is 4,200 ft. high, and forms a long and ill-defined ridge, running nearly N. and S. Its geological composition is of porphyritic gneiss. Moran Lake (see Route 103) is 2 – 3 M. from its top, and attempts have been made to attack it from that direction, but without success.

The Editor was forced to return when within ½ M. of the summit, on account of the approach of night, so that he can give no account either of the character of the top or the views therefrom. From other sources he learns that there are bare ledges on the crest, and that the view to the N. E. is closed by the Franconia Range, which also partly masks the mountains to the E. Moosilauke and the Benton Range are favorably observed from Kinsman, and also a long reach of the Green Mts.

100. The Franconia Mountains

derive their name from the town of Franconia, in which their chief peaks are situated. They are about W. S. W. from the White Mts., from which they are separated by the Field-Willey and Twin-Mountain Ranges. All the mountains around the Franconia Notch usually bear the name, though it is more correctly applied to the lofty range on the E. side of the Notch, and running N. from the East Branch to Gale River. The range is narrow and straight, and the peaks are sharp and lofty, though of less altitude than the White Mts. They have not been devastated by fires, and so the dense forests occupy all their lower parts and cover several of the peaks. The scenery is more beautiful than that of the White Mts., but far less majestic and impressive. This ridge is composed of dark felsite, the southerly peaks being coarsely granitic.

Q

Fredrika Bremer preferred the Franconia Mts. to the famous Swedish districts of Dalecarlia and Norsland. " The scenery here is more picturesque, more playful and fantastic, has more cheerful diversity, and the affluence of wood and the beautiful foliage in the valleys is extraordinary : you walk or drive continually between the most lovely wild hedges of hazel, elm, sumach, sugar-maple, yellow birch, fir-trees, pines, and many other trees and shrubs ; and on all sides is heard the singing and the roaring of the mountain-streams, clear as silver, through the passes of the hills." (*Homes of the New World.*)

101. The Franconia Notch

is a noble pass, 5 – 6 M. long, between the Franconia and Pemigewasset Ranges. It is a valley about ½ M. wide, for the most part filled with forests, and traversed by the marvellously clear waters of the upper Pemigewasset River. On either side of the narrow road rise lofty mountain-walls, clad with verdure, and assuming fine alpine shapes. By the immunity of the forests from the fires and slides which have so ruined the White Mts., the Franconia Notch has retained a pleasing aspect of primeval quietude and tranquil beauty, which induces many travellers to prefer it to any other locality among the mountains.

Starr King says : " The narrow district thus enclosed contains more objects of interest to the mass of travellers than any other region of equal extent within the usual compass of the White-Mt. tour. In the way of rock-sculpture and waterfalls, it is a huge museum of curiosities. There is no spot usually visited in any of the valleys where the senses are at once impressed so strongly and so pleasantly with the wildness and freshness which a stranger instinctively associates with mountain-scenery in New Hampshire. There is no other spot where the visitor is domesticated amid the most savage and startling forms in which cliffs and forest are combined. And yet there is beauty enough intermixed with the sublimity and the wildness to make the scenery permanently attractive, as well as grand and exciting. . . . . The Franconia Pass is not oppressive. Large portions of the wall opposite the Profile House are even more sheer than the Willey Mt , or Mt. Webster, in the great Notch ; but it bends in a very graceful curve ; the purple tinge of the rocks is always grateful to the eye ; and instead of the sandy desolation over and around the Willey House, the forest foliage that clambers up the sharp acclivities, fastening its roots in the crevices and resisting the torrents and the gale, relieves the sombreness of the bending battlement by its color, and softens its sublimity into grace."

" Those who would thoroughly enjoy a forenoon, and taste with eye and ear the freshness of the forest, the glancing light on a mountain-stream, the occasional rare beauty of the mosses on its banks, the colors at the bottom of its cool, still pools, the overarching grace of its trees, or the busy babble of its broken and sparkling tide, should walk from one hotel to the other (Profile to Flume), down the river which runs parallel with the road, but which is for the most part concealed from it by the forest."

Mr. W. C. Prime says : " The grandeur of evening in the Franconia Notch is beyond all words — nay, is beyond human ability to appreciate. There are higher mountains, deeper ravines, more precipitous cliffs in the world, but nowhere in my wanderings have I found such lights as the departing sun leaves on the White Hills of New Hampshire. . . . . No capacity for enjoyment is sufficient to appreciate the variety and change of the sunset and evening lights in the Franconia Notch, — and though one has seen them a thousand times, he sees them each evening with new and sober delight, sometimes rising into awe."

Harriet Martineau said : " I certainly think the Franconia Defile the noblest mountain-pass I saw in the United States."

MAP of the FRANCONIA MOUNTAINS.
For Osgood's White Mountain Guide

Fre
Dalec
fantas
foling
most
pines,
the ro
(*Hom*

is a r
Rang
forest
Pemi
tain-
immu
Whit
quiet
to an

Star
intere
usual
it is a
valley
the wi
scener
amid
And y
make
France
House
but it
ful to
House
in the
bendir

" T
ear tl
the o
botto
busy
to the
the ro

Mr.
beyon
mount
wande
of Nev
variety
though
and so
Har
mount

102. Littleton to the Profile House.

This route is traversed by two stages each way daily. Passengers who leave Boston at 8 A. M. reach the Profile House at supper-time. The stage-fare is $2.00; the distance, 11 M. Seats on the outside should be secured, if the weather is fair.

The route is to the S. for several miles, until the highlands are crossed and the stage descends into the valley of the S. Branch of the Ammonoosuc. Sugar Hill is on the r.; and at 6 M. from Littleton, the hamlet of Franconia Iron-Works is reached.

The Franconia Lowlands.

Boarding-Houses. — In and near the hamlet of Franconia Iron-Works are the houses of Simeon Spooner (35 guests), Charles Edson (25), and D K. Priest (20). A little to the N. are the Valley House (35) and the Lafayette House (50), and farther up towards Mt. Lafayette, affording a broad and noble view, is Oakes & Priest's Franconia House (40 – 50 guests). More than 200 summer-boarders are taken in this valley, the rates being $ 7 – 10 a week. Ivory Glover's is below the Profile-House farm (20 guests); and 3 – 4 M. beyond, on the road to E. Landaff, are the boarding-houses of Wm. and Horace Brooks.

This town contains 549 inhabitants, on an area of 32,938 acres, of which only 5,369 acres are improved. The chief occupations of the people are farming, lumbering, and keeping summer-boarders. The town was settled in 1774, and received its present fortunate name at some time before 1791. The hamlet of **Franconia Iron-Works** is in its N. W. corner, at the bottom of the great valley below Mt. Lafayette. It contains a Baptist church, and one or two stores. This glen is famous for protracted cold weather. The iron-works were founded about 1805, but have long been discontinued. The ore yielded 56 – 63 per cent of iron, of which the works made 500 – 600 tons yearly. Gale River and the Lafayette Brook afford some good trouting during the summer.

Beyond Franconia Iron-Works the road passes the Lafayette House, the junction of the Bethlehem road, and the Franconia House, and then begins a long and sharp ascent, with good views of Mts. Lafayette and Cannon in front. After passing the great Profile-House farm (on the r.), the country becomes wilder, and the ascent is slowly made to and through the Franconia Notch. Soon after passing the beautiful Echo Lake (on the l.), the stage sweeps around the great white **Profile House** and stops alongside its platform.

Bethlehem Station to the Profile House.

Stages run each way twice daily. The fare is $ 2.00; the distance, 10 M. The long steep hill S. of Bethlehem is first ascended, with Mt. Agassiz on the l., and at the summit a noble view of the Franconia Range is gained. Then ensues a long descent through open farms to the valley of Gale River, which is followed thence through forests to the Littleton road, S. of Franconia Iron-Works. The remainder of the route coincides with that from Littleton.

103. The Profile House

is one of the best summer-hotels in the United States. It accommodates 5–600 guests, at $ 4.50 a day, with reductions for permanent boarders. Its dining-hall is the finest in the mountain-region; and in point of cuisine it compares favorably with the Glen House. The parlor is 100×50 ft. in area, and is the scene of brilliant evening assemblies. There are post and telegraph offices, billiard-balls and bowling-alleys, bath-rooms, a barber-shop, a salesroom for pictures and knick-knacks, a livery-stable, and other conveniences. S. of the hotel are two villas, which are leased by the season. The halls and parlors are lighted with gas. The house is surrounded with verandas; and its environs are kept with rare neatness and order. This is a favorite resort for New-Yorkers and Philadelphians, who, at certain seasons, almost monopolize the accommodations.

Distances. — Profile House, to the Profile Lake, ¼ M.; Echo Lake, ¾; the top of Mt. Lafayette, 3¾; Moran Lake, 3; Walker's Falls, 3; the Island Falls, 5; the Basin, 3½; the Flume, 6; Mt. Cannon, 2; Franconia Iron-Works, 5¾; Sugar Hill, 8; Littleton, 11; Bethlehem, 10; Jefferson Hill, 33; Dalton, 24; Fabyan House, 22; top of Mt. Washington, 31; Woodstock, 13.

Routes. — Semi-daily stages from Littleton and daily stages from Bethlehem (see Route 102). Daily stages up the Pemigewasset Valley from Plymouth (see Route 114). Passengers from Boston by Route 2 reach the Profile House in 10 hrs. The best route from New York is up the Connecticut Valley, via Springfield, meeting Route 2 at Wells River. From Saratoga and Lake Champlain by the Portland & Ogdensburg R. R. (when it is completed).

The Profile House is near the N. end of the Franconia Notch, in a narrow glen between the precipitous walls of Eagle Cliff and Mt. Cannon. There are a few acres of level land in this little valley, most of which are occupied by the hotel and its outbuildings and lawns. The glen is 1,974 ft. above the sea, and there is, therefore, no hotel in the mountains (except the summit-houses) so loftily situated as the Profile House. There are many interesting excursions in the vicinity of the house, which serve to pass the time pleasantly for men of action; while quieter souls need but to rest on the comfortable verandas and exult in the rich beauty of the forests and cliffs in the vicinity.

Eagle Cliff is a great spur of Mt. Lafayette running to the W. and N. W., and partly separated from the mountain by a tangled ravine. It is 3,446 ft. above the sea, and 1,472 ft. above the Profile House, which it closely approaches. The name was given by the Rev. Dr. Thomas Hill, 33 years ago, after he had discovered an eagle's nest high up on these beetling crags; but the eagles have long since departed. The precipice has been scaled from near Echo Lake; and the path up Lafayette ascends on the S. side. The sunset colors on this mighty wall are magnificent; and the clouds play over its surface with fine effect.

" It is a charming object to study. Except in some of the great ravines of the Mt.-Washington Range, which it costs great toil to reach, there is no such exhibition of precipitous rock to be found. And how gracefully it is festooned with the climbing birches, maples, spruces, and vines! There are those to whom the sight of such a crag, sharply set at the angle of a mountain wall, is one of the most enjoyable and memorable privileges of a tour among the hills. Such will find the best points for appreciating the height and majesty of the Eagle Cliff by ascending a few hundred feet on the Cannon Mt. opposite, or by walking to the borders of Profile Lake." (STARR KING.)

Within ½ M. of the hotel is a nameless cascade, which is reached by following up the aqueduct back of the long house in rear of the hotel. The brook falls over a succession of sloping granite ledges, and affords beautiful water-scenery after rainy weather. From this point fascinating views are obtained of Echo Lake, Eagle Cliff, and the broad valley to the N. E.

* **Echo Lake** is about ¾ M. N. of the hotel, by the roadside, and is one of the reservoirs of the South Branch of the Ammonoosuc River. On the bluff over it is a small shop for the sale of *bric-à-brac*, below which is a boat-house, where neat row-boats may be hired. Here also is a small cannon, which is fired on the payment of 50 c., and wakes the echoes from the adjacent mountains. The lake is environed by Eagle Cliff, Bald Mt., and Mt. Cannon, and beautiful limited views may be obtained from its surface, especially towards Artist's Bluff, on the E. shore. The human voice is re-echoed with great distinctness here, and the salute of the cannon is answered by a whole park of artillery. The favorite hours to visit the lake are at early morning, or towards evening, when the adjacent cliffs are richly colored by the failing light.

" Franconia is more fortunate in its little tarn that is rimmed by the undisturbed wilderness, and watched by the grizzled peak of Lafayette, than in the old Stone Face from which it has gained so much celebrity. . . . . Its more sacred use is not narrowed to the bounds of the stream which it supplies in part with gentle pulse. Thousands have seen it whose hearts its springs have fed with unwasting water, and in whose memory its beautiful surface, swept by the gentle edges of the summer breeze, and burnished by the sunlight, is a sweet and perennial symbol of purity and peace."

* **Profile Lake** is a few minutes' walk S. of the hotel, by the roadside. It is a beautiful mountain-tarn, at the foot of Mt. Cannon, and nearly surrounded by primeval forests. It was formerly called *Ferrin's Pond,* and then the *Old Man's Washbowl.* Several boats are kept in the small house by the lakeside, for private and public use. The former are owned by regular frequenters of the Profile House, and some of them are models of grace and beauty. The visitor should row out over the lake, in order to get the effect of the adjacent ridges and Eagle Cliff, taking care to avoid disturbing the trout-fishers who are on the water. Near the lake and road are the houses in which the trout are bred, where many hundred young fish are kept. The tourist will doubtless be interested to inspect them and observe their tameness. The pond above Profile Lake is the source of the Pemigewasset River, which is one of the chief contributors to the Merrimac.

Moran Lake is on Mt. Cannon, about 1,000 ft. above the road, and is reached by an admirable bridle-path diverging to the r. from the road, 1½–2 M. S. of the Profile House. It is on the ridge and under one of the high bluffs of Mt. Cannon, and is surrounded with primitive forests. The lake and the adjoining territory are owned by Messrs. W. C. Prime and W. F. Bridge, of New York ; and they have stocked it with trout and erected a quaint woodland-cottage on its shore. *Tamarack Pond* was the old name for this sheet of water : it was then named in honor of a Mr. Moran ; and the present owners call it *Lonesome Lake.* One of the best views of the upper ridges of Mt. Lafayette is obtained from a boat on its waters, near the S. W. shore. The ragged gray crest-line appears above the adjacent forest, and an inexplicable idea of the deep gorge between is also gained.

Walker's Falls are 2½–3 M. S. of the Profile House, beyond the old Lafayette-House clearing, and are reached by a path which diverges from the road obliquely to the l. The lower falls are over ½ M. from the road, and consist of a series of step-like plunges over sheets of granite. They are surrounded by the primitive forest, and are very beautiful in time of high water. The next fall above is fully 50 ft. high, and is surrounded by ragged blocks of granite. The course of the stream should be ascended for about ½ M. farther (though the way is rough) to the splendid upper falls, where the water makes a clear leap of 60 ft., with granite ledges behind. One of the best view-points is the great boulder near the base. Just above, a tributary brook falls in from the r. ; and the line of cascades and sliding waters may be followed far up into the White-Cross Ravine.

104. The Profile.

** The Profile (formerly called the *Old Man of the Mountain*) is a wonderful semblance of the human face, formed by the ledges on the upper cliffs of Mt Cannon, finely relieved against the sky and amid picturesque surroundings. It is best seen from a point marked by a guideboard, on the road a short distance S. of the Profile House, and over Profile Lake. The face looks toward the S. E. This is the most remarkable phenomenon of the kind in the world, and has drawn the admiration of myriads of travellers. There is a tradition that it was worshipped by the Indians in ancient times, but this is doubtful. It was discovered in the year 1805, by Francis Whitcomb and Luke Brooks, who were working on the Notch road, and saw it while washing their hands in Profile Lake. They exclaimed, " That is Jefferson," he being then President. It was described in the American Journal of Science, in 1828, by Gen. Martin Field, with a grotesque picture attached. Hawthorne's tale of *The Great Stone Face*, and a later book called *Christus Judex*, celebrate this marvellous outline. There is a probability that it may not last for many years longer, on account of the rapid decomposition of the granite,

which crumbles under the hand. Prof. Hitchcock says: "I would advise
any persons who are anxious to see the Profile for themselves, to hasten to
the spot, for fear of disappointment." It is formed of three disconnected
ledges of granite, in different vertical lines, their aggregate height being
36–40 ft. (as measured by the State Survey in 1871); and their height
above the lake is 1,200 ft.   One rock forms the forehead, another the nose
and upper lip, and the third the massive chin.   Although the expression
as seen from the road is melancholy and severe, there are points farther
up the ridge where it becomes amiable and pleasant.   The best time to
make the visit is in the late afternoon, when the face is strongly relieved
against the bright sky.   With the morning light falling upon it, the cheeks
appear haggard and sunken.

"The expression is severe and somewhat melancholy, and although there is a
little feebleness about the mouth, on the whole, the face of the 'Old Man of the
Mountain' is set, and his countenance fixed and firm. He neither blinks at the
near flashes of the lightning beneath his nose, nor flinches from the driving snow
and sleet of the Franconia winter, which makes the very mercury of the thermometer shrink into the bulb and congeal. . . . . As you pass down the road to
the left . . . . the Old Man's countenance first changes to that of a toothless old
woman in a mob cap, and soon the lower part of the face becomes so distorted that
the profile is lost. In going to the right, the nose and face become flattened, and at
last the forehead only is seen." (OAKES's *White-Mt. Scenery.*)

"The most attractive advertisement of the Franconia Notch to the travelling
public is the rumor of the 'Great Stone Face,' that hangs upon one of its highest
cliffs. If its enclosing walls were less grand, and its water gems less lovely, travellers
would be still, perhaps, as strongly attracted to the spot, that they might see a mountain which breaks into human expression, — a piece of sculpture older than the
Sphynx, — an intimation of the human countenance, which is the crown of all
beauty, that was pushed out from the coarse strata of New England thousands of
years before Adam. . . . . The expression is really noble, with a suggestion partly
of fatigue and melancholy. He seems to be waiting for some visitor or message.
. . . . Those who can see it with a thunder-cloud behind, and the slaty scud driving
thin across it, will carry away the grandest impression which it ever makes on the
beholder's mind. But when, after an August shower, late in the afternoon, the
mists that rise from the forest below congregate around it, and, smitten with sunshine, break as they drift against its nervous outline, and hiding the mass of the
mountain which it overhangs, isolate it with a thin halo, the countenance, awful but
benignant, is as if a mighty angel were sitting among the hills, and enrobing himself
in a cloud-vesture of gold and purple." (STARR KING.)

105. Mount Cannon, or Profile,

is the long and massive ridge which enwalls the Franconia Notch on the
W., and is separated from Mt. Kinsman by a narrow and exceedingly
rough valley. Guyot reckoned its height at 3,850 ft., or 1,876 ft. above
the Profile House. It is of granite, with great areas of exposed ledges,
and the summit is covered with trees. Some parts are very precipitous,
especially towards the S. E. It is ascended by a rude path 1½ - 2 M. long,
which passes into the woods near the bowling-alley in front of the Profile
House, and leads up the course of a small rill. After the steepest ascent
has been accomplished, the path traverses a forest of dwarfed trees, and
emerges on the ledges.

" The whole mountain from which the Profile starts is one of the noblest speci-mens of majestic rock that can be seen in New Hampshire. One may tire of the craggy countenance sooner than of the sublime front and vigorous slopes of Mt. Cannon itself, — especially as it is seen, with its great patches of tawny color, in driv-ing up from the lower part of the Notch to the Profile House.''

The **Cannon Rock**, from which the mountain derives its name, is reached by clambering down the slope to the l. for a few rods. It is a ledge of granite so balanced as to present the appearance of a cannon when seen from below. Standing upon it the visitor looks down upon the Profile House and the narrow glen beneath, and across to the mighty ridges of Lafayette.

Returning to the point where the path emerges from the woods (where a flying handkerchief or some other mark should be left , the visitor can next examine the Profile by making a long and difficult descent, where there is no path. It is hard to know when the Profile ledges are reached, and great caution must be exercised in approaching them. The mountain itself is often called *Mt. Profile*, in allusion to the grandest of its phenomena.

The View from Mt. Cannon is not easily set forth, since, because there is no marked peak, and woods cover the ridge, it is difficult to get all the features in at any one point. One of its chief objects is Mt. Lafayette, which is so near, across the Notch, that it is all visible, with its great spurs and the deep White-Cross Ravine. This prospect is very noble and satisfactory, and is beautified by the purple lights on Eagle Cliff, below.

The alpine spires of Haystack, Liberty, and Flume are seen on the r., extending towards the S. and flanked by some of the dark mountains of Pemigewasset. On the S. is the broad and beautiful Pemigewasset Valley, sweeping away for many leagues to points below Plymouth, studded with farms and clearings, and hemmed in by bold mountain-ridges. The curv-ing course of the river can be traced through all this long plain and by the hamlets far down the valley. This view is the great attraction of Mt. Cannon, and is of remarkable beauty and variety.

106. Bald Mountain

is a high and rocky knoll N. of Echo Lake and over the Franconia Valley. It is a favorable view-point for persons who dislike much climbing; and is accessible to ladies. The best time to make the visit is in the closing hours of the afternoon, when the northern valleys are filled with level sunshine and the shadows are creeping up the sides of the Notch. The distance from the Profile House to the summit is about $1\frac{3}{4}$ M., by the road to the N., whence, a few paces beyond the refreshment-house $\frac{1}{2}$ M. from Echo Lake, a path diverges into the forest on the r. There was formerly a carriage-road nearly to the summit, but it has long since fallen into neglect, and the path follows its course. This track is about $\frac{3}{4}$ M. long, and is both plain and comfortable. At its end a short, sharp clamber must be made up a steep line of rocks where a stairway once stood, and over a few low ledges of granite.

The View includes on the S. E. the vast pile of Mt. Lafayette, with its tempest-torn crest sweeping around the profound ravine of the N. slope, scarred with white land-slides. On the r. is the bold spur which ends sharply in the precipitous walls of Eagle Cliff, nearly cut off by the ravine, and bristling with spiky trees. Over this ridge is the brown head of the S. peak of Lafayette. N. of the main summit are its long flanking buttresses, forest-covered and sloping away toward the foot-hills. Over this ridge is the sharp curving crest of the Haystack, sloping to the N., precipitous to the S., and covered with the white fragments of dead trees. Farther to the r., and below, is the blue and tranquil sheet of Echo Lake, with the white walls of the Profile House beyond, in the deeply cut Notch. About S. W. is the vast green mass of Mt. Cannon, gashed by two deep ravines, and pushing its foot-hills far out over the plain. To the W. the view passes over the hills and glens of Landaff to Mt. Gardner's long ridge; and more to the r., over a bright stream and girded by fields, are the heights of Ore Hill and Sugar Hill. Nearly N. W. is the village of Franconia Iron-Works, over whose l. is Streeter Pond, backed by Blueberry Mt. Around all the W. and N. W. parts of the horizon sweep the distant blue peaks of Vermont, towards Lake Champlain. The view to the N. crosses the broad valley of the Gale River, where the light green of the cleared lands is contrasted with the darker hues of the great forests. Nearly N. is the curving top of Mt. Agassiz, with Round Hill on its r., and the long dark ridge of Dalton Mt. far beyond. Still farther away are the blue hills of Lunenburg and Victory; and the rounded highlands of Lancaster are to the r., with the white Percy Peaks beyond. Toward the N. E. is Mt. Starr King, back of which sweep the long blue lines of the Pilot and Pliny Ranges. The dark mass of Cherry Mt. next appears, traversing the lowlands of Carroll; and then the view is cut off by the shapeless mountains of the Twin Range.

107. Mount Lafayette.

The Bridle-path. — The new path enters the forest diagonally to the l. from the front of the Profile House. The distance to the Notch is 1 M.; to the lakes, 2¼ M.; to the summit, 3¼ M. The toll for pedestrians is 50c. each; and the tourist who wishes to ride up can hire a sure-footed horse and a guide for $ 3.50. The track ascends rapidly along the gorge S. of Eagle Cliff, and enters the pass between the Cliff and the main ridge. There a restful reach of easy grade on this section, beyond which another breathless ascent begins, leading to the plateau of the W. spur, whereon the present route meets the old path, coming in on the r. The plateau is covered with copses of impenetrable dwarf trees, and affords an interesting off-look to the N. To the l. and below are the Eagle Lakes, to which the path soon descends. Their water is brackish and boggy, but not unwholesome. The main peak is now attacked, the path soon passing above the line of vegetation and traversing the bare rocks for 1½ M. Fine views of the curving ridges to the N. and S. are afforded, and fascinating retrospects. When near the top a spring of clear water is passed on the r., near a bold rock, around which the path turns sharply. On the crest are the low stone walls of the old summit-house.

12

Mount Lafayette, the greatest of the Franconia Mts., is 5,259 ft. high, and is near the N. extremity of the range. Although it resembles Mt. Washington in that it has massive spurs and deep ravines, it shows a marked difference in the sharpness and decision of its lines, and the thin, keen profile of the summit-ridge. From the N. and S. sides of the main peak extend singular ridges, very narrow and rock-bound. That to the N. is about ½ M. long, and terminates at a fine subordinate peak, surrounded with cliffs, and commanding an extensive view. To the S. there is a considerable descent to the ridge, which may be followed to the S. peak, a distance of ¾ M. The narrow crest-line slopes off sharply on either hand into dark gulfs below, and is traversed by a singular path-like trench, 1 – 2 ft. wide, which some people think has been made by countless generations of animals passing along the summit. Noble views of the White-Cross Ravine are afforded on the r., with its deep and shadowy recesses far below towards the roots of the mountain. At one point there is a narrow belt of dwarf spruces, which can be passed only by strenuous labor. After crossing *Mt. Lincoln* and several other rocky hummocks the traveller arrives at the S. peak, where he can look off into the wooded glen to the S., and see Mts. Haystack and Liberty beyond. It is possible to ride to the top of Lafayette early in the morning, and cross Lincoln to Liberty and back in one day. Perhaps this is the best way by which this mountain can be visited. The journey should not be undertaken without a guide.

Among the many objects seen from the lofty S. peak, nothing is so grand and imposing as the gray crest of Lafayette, in the N., looming formidably over Mt. Lincoln and the connecting ridge.

President Dwight was deeply impressed with the majesty of this mountain as seen from Bethlehem. "One, second only to the White Mts., and Moosehillock, exhibits in its great elevation, elegance of form, and amplitude, a very rare combination of beauty and grandeur. It is composed of three lofty conical summits, accompanied by four vast, bold, circular sweeps, formed with a grace, to which in objects of this nature I had hitherto been a stranger ; and which removed all doubts, in my mind, concerning the practicability of uniting the most exquisite beauty with the most splendid sublimity. To this mountain, hitherto without a name, I have taken the liberty to give the appellation of *Mount Wentworth*."

This mountain was called *Great Haystack* on Carrigain's map, in the year 1816 ; but an account of the ascent of " Mt. Lafayette " appears in Silliman's Journal, in 1827. It was probably named during the period of the distinguished Frenchman's sojourn in the United States, in 1824 – 5. Prof. Bond made numerous valuable topographical sketches from this point, while preparing his beautiful map of the White Mts.

" Lafayette is so differently related to the level country, as the Duke of Western Coös, that the view from his upper shoulders and summit has an entirely different character from that which Mt. Washington commands. In the first place, the Mt. Washington range itself is prominent in the landscape, and the sight of it with all its N. and W. braces certainly does much to make up for the large districts which it walls from vision. Of course, with the exception of this range, there is no other mountain whose head intercepts the sweep of the eye. But it is the lowlands that are the glory of the spectacle which Lafayette shows his guests. The valleys of the Connecticut and the Merrimac are spread W. and S. W. and S. With what pomp of color are their growing harvests inlaid upon the floor of New England ! " (Starr King.)

** *The View.* — About N. E., and 2–3 M. distant, is the bold and massive cone of the Haystack, with a long line of cliffs on the S. side. More to the E. is the great Twin Range, the N. Twin being N. of E. S. Twin is E., and the high crowns of Mts. Guyot and Bond, S. of E., with a sharp ridge running from the latter towards the East Branch. Over the Twin Range and several miles beyond is the Field-Willey Range, consisting of Mts. Tom, Field, and Willey, the last of which is marked by ledges near the top and by a sharp descent on the S. Above this chain is the main line of the Presidential Range, with a part of Webster over the S. flank of Willey, Clinton over the S. flank of N. Twin and over Field, the hemisphere of Pleasant over the crest of S. Twin, the level heights of Franklin just to the l. of Pleasant, and above them the bold crags of Monroe. The crowning dome of Washington then appears over the nearer Twin Range, with its house visible and also the long curves of the railroad track on its mighty shoulders. To the l. of Washington are the rugged humps of Mt. Clay, leading to the tall and symmetrical summit of Jefferson, on whose l. is the majestic spire of Adams, its lower approaches being covered by the ridges on the S. W. The great Carter-Moriah Range and other ridges to the N. E. are for the most part concealed by the Mt.-Washington system. Double-Head is seen far away and to the S. of Willey, with the Mt.-Crawford group a little to the r. and nearer, the high Giant's Stairs, the long terrace of Mt. Resolution, and the pinnacle of Mt. Crawford.

To the r. of Crawford, and nearly in line with Thorn, is the low broad top of Iron Mt. Nearly over Mt. Bond, and over the r. of Iron Mt., is the remarkably graceful upward sweep of Kiarsarge, with its hotel on the top; and to the r., still farther away, is the long ridge of Mt. Pleasant (in Maine), with another lofty hotel near its centre.

To the S. of Mt. Willey, and at about the same distance, is the Nancy Range, — the broad-headed Mt. Nancy on the l., Anderson next to the r., and the clear-cut pyramid of Lowell at the end. The N. peak of Moat Mt. is nearly over the notch on the r. of Mt. Lowell; and Tremont and Bear Mt. are on the l. of and beyond Carrigain. The imposing and massive form of Mt. Carrigain is next visible, about 10 M. S. E. across the trackless wilderness of Pemigewasset, nearly concealing Bear Mt. To the right is the many-headed ridge of Mt. Hancock, which throws off great spurs into the woodland country; and Tremont is seen through the Carrigain Notch. About 25 M. distant, directly S. E., is the stately white peak of Chocorua, on whose r. is the low and crouching mass of Paugus. The blue ridge seen far beyond is a part of the Ossipee Range. Farther around to the r. is the Mad-River Notch, in which the Greeley Ponds are situated, to the l. of which is Mount Kancamagus, over whose l. flank is Passaconaway. Nearly down the notch is Tripyramid, its peaks foreshortened by being seen at a slight angle with the trend of the ridge. Whiteface is just beyond Tripyramid, but does not show the blanched crest which, as viewed from the lake-country, has originated its name.   Osceola is about 14 M. in an air-line to the S. S. E., and has a sharp fall to the E. and a minor peak towards the Notch, with a long sierra running N. W. to the mouth of the Hancock Branch.   Over its r. flank is a part of the long dark ridge of Sandwich Dome, with Weetamoo on its r.   Near Osceola, and on its r., is the bold pinnacle of Tecumseh, from which the upheaval falls away to the S. W., broken by the projections of Green Mt. and Fisher Mt.   Just to the r. of Tecumseh are the double peaks of Mt. Belknap, on the W. shore of Lake Winnepesaukee; and others of the Alton and Gilford hills are visible, with parts of the adjacent towns.

To the S. the long and sharp ridge of Lafayette is seen, trending away from the peak on which the observer stands to the S. peak.   Farther out are the high and remarkable pyramids of Haystack and Mt. Liberty, while below and nearer is the ridge which was formerly known as Mt. William.   At a greater distance and about S. S. E. is Owl's Head, in the Pemigewasset country; and Mt. Flume is seen on the S.   Away down the Pemigewasset Valley is Mt. Prospect, near Plymouth, to the r. of which is Plymouth Mt.; and in this direction and far beyond is Mt. Kearsarge, overtopping the Ragged Mts.   Still more remote, to the r. of Kearsarge, is the low blue crest of Monadnock, on the verge of the horizon; and to the r. of this line is the white granite peak of Mt. Cardigan.

Toward the S. are considerable portions of the fair Pemigewasset Valley, broken by low wooded ridges and thickly dotted with farms and clearings. The intervales of Thornton and Campton reveal their beauties, and here and there the river is seen, flashing in the sunlight or blue in the shadows. The upper flanks of Mt. Stinson are seen far down the valley, while some-what nearer and more to the W. are the confused and dark-tinted ridges of Mt. Carr.   The Blue Ridge and Mt. Cushman are still nearer, and also on the W. of the valley, blending in appearance, with the Carr Range; and others of the so-called Pemigewasset Mts. run to the N.   Nearly S. W. is the lofty plateau of Moosilauke, which is marked by a hotel, — and just over its r. flank, beyond Ore Hill in Warren, is the symmetrical Mt. Cuba, in Orford.   Nearly in line with Moosilauke, but much nearer, across the Franconia Notch, is Mt. Pemigewasset, which is in the vicinity of the Flume House.   This view-line crosses the profound depths of the White-Cross Ravine.   From S. W. to W. the view is interrupted by Mt. Kinsman, a massive flat-topped range which lies on the farther side of the Notch and covers several miles with its shapeless heads and outstretching buttresses.   Far up on a shoulder of this range is seen the calm expanse of Moran Lake; and through a depression to the W. S. W. are the white peaks of Sugar Loaf and Black Mt. and the hills of Newbury, beyond the Connecticut.   Just back of the white village of Newbury is the diminutive swell of Mt. Pulaski, standing before the great assembly of the hills of Orange County.

There is a question as to whether any of the Adirondacks are visible from Lafayette, the distance being over 100 M., with several lines of intervening ridges. Marcy and McIntyre, the chief Adirondack peaks, are exactly W. of Lafayette, Whiteface being a trifle farther N. They would therefore be seen over Mt. Cannon, which is opposite Lafayette, across the Notch, and is marked by many red ledges.

The Profile House and the narrower part of the Notch are hidden from view by Eagle Cliff, on whose r. is a part of Echo Lake, beautifully clear and calm, with Bald Mt. just beyond. The plateau on which the Eagle Lakes are situated is seen still nearer the mountain-top. Beyond the N. flanks of Mt. Cannon is Sugar Hill, famed for its noble view; over which are the hills of Lyman and Hunt's Mt., of the Gardner Range. Over the S. flank of that chain is the far-away peak of Camel's Hump; and to the r. of Hunt's Mt., still more remote, is Mt. Mansfield, the chief of the Green Mts. The cultivated valleys of the Gale and Ammonoosuc Rivers are next seen, with a long stretch of the Littleton road, and part of Franconia village. The upper portion of Littleton is about N. W., apparently under Mann's Hills; and the surrounding township is dotted with other commanding heights. In this direction and low down on the horizon is the blue speck which has been recognized as Jay Peak; N. E. of which is another remote peak (seen only in clearest weather), which some claim to be Owl's Head, on Lake Memphremagog, in Lower Canada. About N. N. W., over Mt. Niles in Concord, and beyond the mountains of Burke, are Mt. Hor and Mt. Annanance, forming the famous Willoughby Notch, with Willoughby Lake between. The view of the Willoughby Notch from this point is altogether the finest of the distant prospects of this remarkable piece of scenery, showing the sharp cleft between the mountains and their boldly precipitous sides. Several miles nearer, and to the r., are the handsome outlines of the Victory Hills, well into Vermont.

Across the Gale-River valley in a northerly direction are the high knolls of Mt. Agassiz and Round Hill, with the ridge which hides Bethlehem from sight. Over these are the dark Dalton Mt. and the Lunenburg Heights, Mt. Burnside, and a vast area of the hills of Essex County, and the open meadows of the Upper Connecticut Valley, with Maidstone Lake sparkling on the l. More to the r., beyond lake-strewn Whitefield, are the low and rounded summits of Mts. Prospect and Pleasant, near Lancaster, over which is the massive bluff of Cape Horn, rising from the plains of Northumberland. To the r. and beyond are the twin domes of the Percy Peaks (over the blue confusion of the Pilot Mts.), everywhere recognizable on account of their startling brilliancy of color and symmetry of form. Still farther N. on this line are the Bowback and Stratford Mts., and the pronounced peak of Sugar Loaf; while more to the r. and farther N. are the mountains of Odell and the Dixville peaks, overlooked, in the remotest distance, by Mt. Carmel, which rises near the uniting frontiers of New Hampshire, Maine, and Quebec.

Much nearer, and to the r. of Mt. Agassiz, is Beech Hill, in Bethlehem, which is also called the *Sleeping Giant;* beyond which is the bold and lofty ridge of Cherry Mt. Still farther to the N. N. E. is Mt. Starr King, with the summer-village of Jefferson Hill at its foot and the great array of the Pilot Mts. to the N. and N. W. Bray Hill and Kimball Hill are also visible on the plains of Whitefield. The Randolph Mts. and Mt. Deception lie about N. E., and support the main line of the Presidential Range.

To the r. of Cherry Mt., and very far distant, are Mts. Dustan and Aziscoös, near the gleam of the waters of the Umbagog and Rangeley Lakes. Farther to the r. is Mt. Bigelow, deep in the wilderness of Western Maine; and about N. E., over Mt. Deception, are the mountains about the Grafton Notch. Goose-Eye is also seen, close to the flank of Jefferson, but far beyond, presenting a marked blue peak.

The Haystack is a picturesque peak 2 – 3 M. E. of Lafayette, with which it is connected by a high ridge. It is doubtful whether it has ever been explored, since the way thither is surpassingly difficult, leading through long unbroken thickets of dwarf spruce. On the N. W. side of this peak, 1,787 ft. above the sea, is the *Haystack Lake,* a tiny mountain-tarn which is 15 rods long, and forms the source of the Gale River.

108. The Flume House

is at the S. entrance of the Franconia Notch, 5 – 6 M. from the Profile House and 543 ft. below it. It can accommodate 100 – 150 guests, and charges $ 3.00 a day. It is owned by the same people who own the Profile House. The stages for Plymouth leave daily at 8 A. M.; for the Profile House daily at 6 P. M. The chief objects of interest in the vicinity are the Flume, 1 M.; the Pool, $\frac{1}{2}$; the Basin 1$\frac{1}{2}$; Mt. Pemigewasset, 1$\frac{1}{2}$; Profile Lake, 4$\frac{1}{2}$; Georgianna Falls, 3. Levi E. Guernsey is the best guide in this vicinity. He lives S. of the Flume House. His rates are $ 3.00 a day.

This hotel has one of the most beautiful situations among the mountains, and is surrounded by fine bits of rock and water scenery. From the S. veranda the fair valley of the Pemigewasset is seen for many leagues, extending away to Plymouth, and flooded with rich color every sunny afternoon. In front and to the N. of the hotel is the noble line of the Franconia Mts., breaking down to the deep pass on the N., and clothed with forests to their tops. These peaks form the profile called *Washington Lying in State,* with Mt. Liberty for his upturned face, the highest ledge being the nose, and the ridges running to the N. forming the body and limbs.

" The view from the Flume House itself is a perpetual refreshment, and one needs not seek by hard climbing or wandering for any increased temptation to contentment. No scenes can be more contrasted in spirit than those around the two hotels, 5 M. apart. From the Flume House the general view is cheerful and soothing.

There is no place among the mountains where the fever can be taken more gently and cunningly out of a worried or burdened brain. So soft and delicate are the general features of the outlook over the widening Pemigewasset Valley! So rich the gradation of the lights over the miles of gently sloping forest that sweep down towards Campton! So pleasant the openings here and there that show a cluster of farmhouses, and the bright beauty of cultivated meadows enclosed by the deeper green of the wilderness! . . . . Here, too, we can have more of the landscape beauty of the larger mountains than the greater nearness of the Profile House to them would allow. The three peaks of the highest Haystacks . . . . are in view, and at evening one can see the glorious purple mount the forests that hang shaggy on their sides,—extinguishing the green as completely as if the trees for miles had suddenly been clothed with leaves of amethyst,—and then chased by the shadow retreat upwards till it dyes the rocks with its harmless fire, and still upwards to the peaks, and then leaps to the clouds above." (STARR KING.)

* The Flume is ½ – ¾ M. from the hotel, and is reached by a good carriage-road which crosses the Pemigewasset River. The road stops near the Flume Brook, whose course is ascended by a foot-path leading over clean sheets of granite, broad and slightly tilted, across which the water slips " in thin, wide, even sheets of crystal colorlessness," for several hundred feet. Farther up the ledges are more rugged, and the limpid brook lingers in rock-rimmed basins. The Flume is a marvellous fissure in the side of Mt. Flume, through which dashes a brilliant little brook. It is about 700 ft. long, and is flanked by mural precipices 60 – 70 ft. high. These walls are perpendicular and parallel, and are from 10 to 20 ft. from each other. A plank-walk extends through the gorge, crossing the stream several times. Near the upper part of the Flume the walls are narrowed, and hold firmly between them a huge boulder, which at some past time has fallen from the mountain above. The walk leads under and close to this formidable suspended rock, whose hold between the cliffs seems so slight that one cannot pass it without a tremor. By ascending to the head of the gorge and turning down the r. bank above, the visitor soon reaches a rude bridge, formed by tree-trunks and thrown across from cliff to cliff. This is a good point from which to study the Flume from above, and gain an impression of its depth and grade. The morning is the best time to visit this locality.

" How wild the spot is ! Which shall we admire most, the glee of the little torrent that rushes beneath our feet ; or the regularity and smoothness of the frowning walls through which it goes foaming out into the sunshine ; or the splendor of the dripping emerald mosses that line them, or the trees that overhang their edges ; or the huge boulder, egg-shaped, that is lodged between the walls just over the bridge where we stand,—as unpleasant to look at, if the nerves are irresolute, as the sword of Damocles, and yet held by a grasp out of which it will not slip for centuries ? "
The predominant rock at the Flume (as also at the Basin and Pool) is a coarse common granite, composed of much flesh-colored orthoclase, rude crystals of transluceut smoky quartz, and black mica. This rock underlies all the valley of the Pemigewasset, and composes the main ridge of Mt. Cannon. The Flume was discovered by Mrs. Jessie Guernsey, while fishing along the brook.

* The Pool is a gloomy and profound chasm in the forest, where the Pemigewasset River falls into a deep black basin surrounded by high cliffs. It is reached by a pleasant forest-path ½ – ¾ M. long, entering the woods opposite the Flume House. It is over 100 ft. in diameter, and the water is 40 ft. deep, entering by a cascade above and whirling down through a mass of rocky fragments. This Stygian pool is surrounded by cliffs 150 ft. high, whose deep shadows add to its weirdness and gloom. A steep path winds down to the r., to the edge of the water. For many summers an eccentric rural philosopher has lived here, in a rude boat, amusing visitors with his quaint speculations and original cosmogonies.

* The Basin is about 3½ M. from the Profile House and 1½ M. from the Flume House, close by the roadside. It is a granite bowl, 60 ft. in circumference, filled with the cold and pellucid green water of the Pemigewasset to the depth of about 15 ft. On the upper side there is a beautiful white cascade, and on the lower side is a curious rock projecting into the water, and resembling a leg and foot. The Basin is a large pot-hole, formed by the attrition of stones whirled about by the current. The beauty of the scene is much impaired by the contiguity of the dusty road.

" The bottom is strewn with rocks, and is apparently level. Its sides and margin are beautifully regular and very smooth ; and the high bank of the same solid, continuous, gray granite is elegantly arched beneath, and stained with dark purple. . . . . In a hot day in summer the pure, clear, and cold waters below, foaming and sparkling in their clean granite bowl, are delightful and refreshing to the eye and imagination. The perpendicular cascade, white with foam, falls gracefully over its brink from the rocky bed of the river above, and striking the water of the basin nearly parallel to its side, gives to the whole a strong revolving motion. Branches of trees thrown into the whirlpool are carried round many times before they are discharged, being alternately drawn to the bottom, and brought again to the top. It would probably be a dangerous bathing-place for a weak swimmer, and its steep and slippery borders are not very safe for careless visitors.

" Such cavities, of smaller size, called ' pot-holes ' by geologists, are not infrequent in the vicinity of the White Mts. In a hollow in the rocky bed of a rapid stream, where the waters form a whirlpool, a few rocks are accidentally lodged, and carried round by the revolving water, gradually grinding and wearing a round, smooth cavity, into which, as it enlarges, rocks of greater size are brought by the freshets. The process goes on until a wide and often very deep excavation is formed, which in places where the river has shifted its course are sometimes seen as deep pits, like wells, in the dry ground." (OAKES's *White-Mt. Scenery.*)

The Dog Profile is a few rods S. of the Basin, and consists of a roadside boulder (on the E.) which closely resembles a dog's head. A few rods S. of this point a path diverges to the W. to the *Stone Flume,* a sunken channel which the Pemigewasset has worn in the ledges.

Cascade Brook (also called Basin Brook) enters the Pemigewasset from the N. W., just below the Basin. According to Mr. Prime, it is " the finest brook in America for scenery, as well as for small trout." It falls over 1,000 ft. in 3-4 M.; and in the lower part of its course slips downward for over 200 ft. over splendid sheets of granite, 100 ft. wide. There is a vague path from the mouth of the Basin up to the first fall, whence the course of the stream is followed closely. A guide is essential for visitors who are unskilled in woodcraft and blind trails. The **Tunnel Falls** are about ¾ M. from the Basin, beyond the cascades over the ledges, and consist of a white and massive perpendicular plunge of about 30 ft., thrown out in strong relief by a curving background of black and ragged cliffs. Beneath is a broad pool of great depth, enwalled by lofty ledges. The **Island Falls** are about ½ M. above, and are reached by working up through the thickets and moose-bushes on the r. of the stream. They consist of three plunges, of 8, 16, and 16 ft. respectively, and during high water they insulate two bold rocky knolls below. The upper falls are at right angles with each other. The stream should be crossed above, since the best view is from the rugged knoll on the W. bank.

It is about 1½ M. from the Island Falls to the confluence of the streams above, and about the same distance thence to Moran Lake (see page 261), up the brook to the r. The Island Falls are only ½ M. in a direct line from the Notch road, which can be reached by striking across the ridge on the E. This short-cut is only practicable in dry weather, when the Pemigewasset is low.

The **Georgianna Falls** may be visited from Guernsey's (and under his guidance), a many-gabled cottage 1 M. S. of the Flume House. There is a faint trail hence over a high ridge and down into the valley of the Harvard Brook. The falls are 1½ – 2 M. from the road by this route, and are about the same distance from Russell's (to the S.), by following up the stream. They were discovered and christened in 1858, although a party of students claimed to have found them before and named them *Harvard Falls.* The State Geologist has ingeniously compromised the difficulty by naming the stream *Harvard Brook.* This rivulet rises in Bog Pond, under Mt. Pemigewasset, and has dark-colored, woody water, in which are many large but ill-flavored trout. The falls have been visited very rarely for several years past, on account of their secluded position. There are several cascades below and above, and the beautiful basin called the *Mirror* is near the upper ledges. The main fall is about 80 ft. high, slanting in the upper part, and making a direct plunge below. The best view-point is about 200 ft. below, on the l. bank, where the white column of water is seen over its deep basin and recessed between jagged and dark-colored cliffs of coarse granite. Though hard to traverse, the forests on the ascent afford rich woodland-scenery, agreeably diversified by the ledges and the roaring brook. From the cliffs above the falls is obtained a beautiful view down the Pemigewasset Valley.

109. The Lower Franconia Peaks.

The bold summits which run S. from Mt. Lafayette to the East Branch are covered with forests, and their upper ridge is composed of dark felsite rock. This range may be traversed in two days, by ascending Mt. Flume and passing N. over Liberty, Haystack, the S. peak of Lafayette, and Lincoln, to Lafayette and down to the Profile House. Levi E. Guernsey is a competent guide on this route. Most of the way is in dense forests, where the walking is somewhat difficult. The ravines between the peaks are tangled, but shallow, and afford no serious impediments. (See also page 226.) The best camping-ground is on the ridge between Liberty and the Haystack. The first day's march is from the Flume House over Flume and Liberty; the second leads over the Haystack, the S. peak of the Lafayette range, Lincoln, and Lafayette, and down to the Profile House. The ridges between Flume, Liberty, and the Haystack are well wooded, but those between the Haystack and the S. peak are covered with dwarf spruce.

The peaks of Flume and Liberty are visible from points far down the Pemigewasset Valley, showing sharp dark points, in shape like narrow-based pyramids. They are generally known among the people as *The Haystacks.* The pointed character of these peaks is apparent from every direction, and renders them very conspicuous from distant mountains.

Mount Lincoln is between the main peak of Lafayette and its S. peak, and is the highest of the rocky knolls spoken of on page 226. It is about 5,100 ft. high, and is bare of trees, like all the rest of this ridge. The views to the E., W., and S. are broad and beautiful; and on the N. is the bold crest of Lafayette, within ½ M. This rocky pile was named in honor of Abraham Lincoln, the Martyr-President.

The S. peak of the Lafayette ridge is less than ½ M. S. of Mt. Lincoln. The wooded summit on the S. is known by the name of **The Haystack.** There is still another Haystack within 3 M. (see page 270), and several more in the State, wherefore Mr. Warren Upham justly remarks that "this peak should be named after some great statesman, to accord with the other names of the same range to the N." The walk from Lafayette over Lincoln to the S. peak and back takes about 2 hrs., and is comparatively easy (see page 226).

Mount Liberty is 4,500 ft. high, and is S. of Mt. Lincoln and the Haystack. Its summit is 3 – 4 M. from the Flume House, over a very rugged route, which is entered by diverging to the N. E. from the cascade near the entrance to the Flume, or by ascending the Flume Brook. A long day is needed for the exploration of this summit; and a skilful guide should be taken.

Mount Flume is said to be 4,500 ft. high, but there seems to have been much guess-work about the recorded altitudes of these lower Franconia peaks. It is ascended by the Flume Brook, which leads the visitor far up on the ridge. The journey is arduous, the choice being given of the rolling stones of the brook or the thickets at its side. The mountain itself is in the form of a half-moon, with its horns to the N. E. and S. W., and its concavity to the N. W. The summit consists of a remarkable ridge, only 2 – 3 ft. wide, and falling away rapidly on either side, dotted here and there by low spruce-trees. The views over the Franconia region, the Pemigewasset Forest, and the pleasant lowlands to the S. are very beautiful and extensive.

110. Mount Pemigewasset

is near the Flume House, on the W., and is a high spur of Mt. Kinsman. It is ascended by a good bridle-path, 1½ M. long, whose only difficulty is the steepness of its grade, leading steadily upward through the ancient forest. Where it emerges upon the summit is a line of ledges, from whose verge vast cliffs descend abruptly to the wilderness below.

" By an easy climb of ½ hr. up Mt. Pemigewasset, directly back of the hotel, — a climb not at all difficult in dry weather to ladies, — the sunset views will be far more impressive. The spurs and hollows of Lafayette and his associates will be lighted up by the splendor that pours into them from the W. It searches and reveals all the markings of the torrents ; it gilds the tautness of the rocky tendons that stretch from the summits to the valleys, and that run sometimes in hard lines and sometimes in curves full of rebellious energy, like a tough bow strung to the utmost tension ; and it pours upon the innumerable populace of trees which the mountain-sides support one wide blaze of purple, which slowly burns off upward, leaving twilight behind it, and gleaming on the barren crests, long after the valley, which stretches in view for 20 M., is dimmed with shade." (STARR KING)

* The View to the N. and W. is confined to the great ridges of Mts. Cannon and Kinsman, — the former filling the N. with its rocky bulwarks and the latter running from N. W. to S. W. with long and monotonous wooded heights. The great basin between Pemigewasset and Kinsman is filled with dense forests, unbroken by a single clearing, and furrowed by the depression of Harvard Brook. Nearly N. E., across the centre of the Franconia Notch, is the imposing ridge of Lafayette and Lincoln, lifting its brightly colored ledges over the White-Cross Ravine, and foreshortened by the line of vision. The S. peak is conspicuous at the nearer end, and falls off rapidly into the ravines above Walker's Falls. Below Lafayette are the lofty and symmetrical pyramids of Mts. Haystack and Liberty, cutting the sky-line; and to the E. is the forest-covered ridge of Mt. Flume. To the S. E. are the Coolidge Mts. and other unvisited and as yet unnamed heights about the East Branch and towards the Pemigewasset Forest. Beyond the rugged hills in this direction the view rests on parts of the fine peaks of Waterville, — probably Tecumseh and Osceola. To the S. S. W. are the high mountains on the W. of the valley, the Blue Ridge, Mt. Cushman, and the outer flanks of Mt. Carr.

The view down the Pemigewasset Valley is the chief feature from this point, and is full of grace and beauty. The hamlets and farm-lands to the S. are visible for many leagues, and it is claimed that even Plymouth may be seen on a clear day. Flashing belts of the river diversify the scene, and on either side are the high and rounded hills of Woodstock, Thornton, and Campton. This view is the same in character (though far greater in extent) as that from the Flume-House veranda, and affords a gracious contrast to the frowning and craggy mountains on the N. and E.

111. Western Pemigewasset.

This portion of the primitive wilderness may be entered from Pollard's, and a hard two-days' march will lead the traveller across to the Saco Valley, by the Thoreau Falls and Ethan's Pond. The forest may also be left by the New-Zealand Notch on the N., or the Mad-River Notch on the S. Care should be taken to secure a good guide and proper camping equipments. This great wild land is virgin soil for the fisherman and hunter, and the brooks and ponds are swarming with trout. On account of the difficulty of the journey it is and will be but rarely visited, even by the most ardent sportsmen. The eastern part of this wilderness is described in Route 53 (page 156).

From a rude map and itinerary prepared by Dura P. Pollard, the following distances in the Pemigewasset Forest are gathered : Pollard's to Loon-Pond Brook, ¼ M. ; to the Hancock Branch, 4 M. ; to the Franconia Branch, 9 M. ; to the S Branch, 13 M. ; to Ethan's Pond, 18 - 20 M. The distance from Ethan's Pond to the White-Mt. Notch is about 3 M　Young Pollard and Levi E Guernsey are the best guides for this region. Mts. Bond and Guyot and the Twin Mts. may be attacked from this point (see page 161) : and Mt. Carrigain has also been visited from Cedar Brook, on this side (see page 134).

Pollard's is a little over 1 M. from the Pemigewasset-Valley highway, and is reached by a good road which diverges to the E., N. of N. Woodstock and about ½ M. S. of Tuttle's, crossing the river and approaching the mountains. This is the last outpost of civilization on the borders of the vast forest of Pemigewasset, and beyond it no road nor trail goes. Mr. Pollard has rooms for about 20 guests, his rates being $ 1 a day. This is a good objective point for sportsmen; but the rude furniture and uncarpeted floors render it less attractive for families. The house is on a pretty intervale of the E. Branch, and has a broad view of mountains, the chief of which are Moosilauke and Kinsman in the W., the Coolidge Mts. on the N., and the nameless peaks of Pemigewasset on the E. Mr. Prime has written of this place, somewhat enthusiastically : "Nowhere in our Northern Alps is a more beautiful view than is spread out in every direction from Pollard's house."

THE PEMIGEWASSET VALLEY.

112. Plymouth.

Hotels. — The * Pemigewasset House is a large and first-class hotel, owned by the B., C. & M. R. R., and managed by Mr. Carlos M. Morse. It accommodates 300 guests, and charges $4 a day, with fair rates to permanent boarders. The building was erected in 1863, and has a front of 230 ft., with a long wing, and is four stories high. The main floor is occupied by the offices, parlors, reading-room, great dining hall, news-stand, and barber-shop. The rooms are high and airy, and the halls are lighted with gas. The house is built on a hillside, so that the main floor is on a level with the village-street, from which the veranda is separated by enclosed lawns; while on the rear there is a lower floor, in which is the railroad station. The up and down trains meet here about noon, and stay for ½ hr., while the passengers dine in the hotel. Just across the track is the rushing current of the Pemigewasset River. There is a cupola on top of the house whence a broad view may be obtained. This hotel is famous for the scrupulous neatness of its rooms and of the adjacent grounds. A quadrille band is kept here during the season.

The Plymouth House is a comfortable village-inn on the main street, accommodating 60 guests, at $2 - 2.50 a day. It has recently been modernized and enlarged.

Boarding-Houses. — W. G Hull's broad-verandahed house, on the hillside near the Pemigewasset House, accommodates 30 guests. Dr. J. W. Preston has rooms for 20 boarders; H. S. Chase, 20; and S. D. Baker, 50. The rates at these houses are $7 - 10 a week. J C. Blair's is a larger house (60 guests), out on the road to Franconia. Guests can go down from Plymouth to Weirs on the morning train, cross the whole length of Lake Winnepesaukee, and return the same day.

Stages leave on the arrival of the Boston train (a little after noon) for Campton, the Pemigewasset Valley, and the Profile House, reaching the latter point at supper-time. Distance, 29 M.; fare, $4. Conveyances also run out frequently, but irregularly, to Waterville, via Campton Village. There are large livery-stables connected with the Pemigewasset and Plymouth Houses.

Routes. — Passengers leaving Boston on the morning trains via Lowell, dine at Plymouth; and the afternoon trains reach that point at evening. The Profile-House stages reach this station in time for the down trains. Passengers from Mt. Washington, the Fabyan House, Lancaster, Littleton, and Bethlehem come hither all the way on the B., C. & M. R. R.; and Plymouth is reached from all points to the E. and W. by closely connecting trains.

Churches. — Congregationalist and Methodist. The old Episcopal church is often visited by summer guests. (See page 279.)

Distances. — To Livermore Falls, 2 M.; Mt. Prospect, 4½; around Plymouth Mt., 10; Squam Lake, 6; Centre Harbor, 12; Bridgewater, 6; Campton, 5½; Rumney, 7; Waterville, 18; Profile House, 29; Warren, 20; Newfound Lake, 9; Loon Pond, 5; W. Campton, 7.

Plymouth is one of the most beautiful villages in New Hampshire, and is situated near the confluence of the Pemigewasset River and Baker's River. It is on a terrace close to and above the Pemigewasset, and is built chiefly on two streets, the main street running nearly N. and S., and the Rumney road, running thence to the W. Near its centre is the court-house of Grafton County and the handsome modern building of the *State Normal School*. Several beautiful villas have been erected on the street opposite and S. of the Pemigewasset House; and the undulating character of the site of the village is favorable to picturesque scenery and architecture. The professional men necessary to a prosperous shire-town are found here, forming a pleasant circle of society. A weekly paper is published in the village. The population of the town is 1,409.

Room No. 9 in the Pemigewasset House possesses a mournful interest as the place where died Nathaniel Hawthorne, the author of *The Marble Faun, The Scarlet Letter,* etc. No American writer has ever surpassed Hawthorne in grace and strength. He died here on May 19, 1864.

On the previous day he had ridden over the beautiful road from Centre Harbor to Plymouth, in company with ex-President Franklin Pierce, with whom he was travelling in search of health. Mr. Pierce entered his room early the next morning and found him dead. "He lies upon his side, his position so perfectly natural and easy, his eyes closed, that it is difficult to realize, while looking upon his noble face, that this is death. He must have passed from natural slumber to that from which there is no waking without the slightest movement."

The old court-house of Grafton County stood in the S. part of the village until 1875, when it was taken down and the timbers were used in the new library-building. It was historically interesting as the place where Daniel Webster delivered his first plea before a jury. His client was one Burnham, who had murdered two of his fellow-prisoners in Haverhill jail, and the case was so plainly proven that Webster's plea was directed against capital punishment on general principles. Burnham was executed.

The chief business of the village is the manufacture of gloves, the so-called "Plymouth buck gloves" having been long and favorably known. Eleven firms are engaged in this industry, producing 130,000 pair of gloves annually. The business was founded in 1835. The town is also famous for its agricultural products, among which are large quantities of butter, maple-sugar, apples, and hay. A large country trade is carried on by about a score of merchants. The general offices of the B., C. & M. R. R. are located here, and three passenger trains usually pass the night at Plymouth. The town library contains 600 volumes.

The profitable patronage of the State and the railroad have combined with its fine natural situation to make Plymouth a rich and prosperous village. It is also estimated that more than 500 city-boarders pass a considerable portion of each summer here, besides thousands of more transient guests. The land is undulating, with good soil, especially along the broad and beautiful intervales; and of the 16,256 acres in town, over 10,000 are improved. Starr King says of these river-lands: " In scenery Plymouth is remarkable for the beauty of its meadows, through which the Pemigewasset winds, and for the grace of its elm-trees. Even the hurrying and careless visitor will have his attention arrested here and there by a faultless one, standing out alone over its private area of shadow, seemingly an ever-gushing fountain of graceful verdure." Beautiful views of the Franconia and Waterville mountains can be obtained over these meadows by driving to the S., along the old Bridgewater road. A more extended panorama is given to the visitor who climbs on to the *South Mt.*, or *Walker's Hill*, low eminences near the village.

The Judge-Livermore mansion and the ancient Episcopal church in Holderness are 1½ M. from Plymouth, by the road that turns to the r. and crosses the Pemigewasset N. of the Plymouth House. The church is in the angle of the second road turning to the l., and is a small and plain structure near a venerable graveyard.

Holderness was granted in 1765 to John Wentworth and 67 other Episcopalians, "a company of English emigrants ardently devoted to the creed and worship of the Church of England, and with glowing anticipations of the future for the colony. The founders hoped and believed that they were laying the basis of the great city of New England, the rival of Puritan Boston, and destined to throw it into the shade. The headquarters of heresy, they allowed, would have some *commercial* advantages, on account of its nearness to the ocean and its excellent harbor; but in population, refinement, dignity, and wealth they supposed that Holderness was to be the chief city of the New-England colonies. What a strange answer to their dream, that even the pretension with which the settlement was made is not noticed by history, and has scarcely wandered from the proprietors' records into any tradition." In the churchyard the Rev. Canon Balch was buried, in 1875.

Samuel Livermore came to Holderness in 1765, and lived in feudal state, owning half the township. He became chief justice of the Superior Court, and U. S. Senator. His son Arthur was afterwards chief justice of the highest State court.

The *Livermore Falls* are about 2 M. above Plymouth, and are seen from the bridge on the first road which diverges to the r. from the Franconia highway. They are in a picturesque narrowing of the Pemigewasset among massive rocks. The popular opinion that there are traces of volcanic action hereabouts is not borne out by facts. Mr. Rogers says: "It is one of the most charming rides in the world, for the 2 or 3 M. up the Pemigewasset, before you begin to ascend. 2 M. above the meeting of the little rivers you cross a picturesque bridge at 'the Falls,' a scene for the painters when the land shall become, like the Old World, the home of the fine arts."

"By Sawheganet and Livermore Falls were the best of fishing-places, and at the confluence of the Asquamchumauke [Baker's River] and Pemigewasset were the broad and beautiful intervales of the tribe. No place more fertile can be found in New England. Luxuriant grasses and wild-flowers growing with tropical exuberance, clusters of noble elms with waving branches, a dense forest, hills and wood-crowned summits on the border, and lofty mountains in the distance, often snow-capped in midsummer, made this spot a wild paradise. Ridges, where the corn was planted, ashes where the wigwam was built, mattocks made from the bones of a moose's thigh, rude pestles and knives of stone, gouges, and arrow and spear heads here found, show that this was the chief planting-place of the [Pemigewasset] tribe. Here also was frequently the royal residence, and without doubt the Indians had encamped here for centuries. . . . . Theirs was a beautiful country. No clearer and more sparkling rivers could be found in the world than the Asquamchumauke and Pemigewasset; no brighter and more smiling lakes than the Newfound and the Squam, and no more glorious mountains than Moosilauke and the Haystacks." (LITTLE's *Hist. of Warren.*)

In the year 1712 Capt. Baker, of Newbury, led a company of Massachusetts rangers up the valley, and surprised the Indian village near the site of Plymouth. The people were not aware of the approach of an enemy until a deadly volley at close range swept through the village, killing several of the inhabitants. The survivors fled forth to rally the Indian hunters, and Baker's men marched in, plundered the village of its accumulated stock of furs, and then burned the wigwams. The Pemige-wassets pursued the retreating Americans, and were repulsed in a sharp skirmish on the Bridgewater plains. At their next night-camp, each ranger used five sticks in roasting his meat for supper, so that when the eager pursuers entered the abandoned lines next morning, they were deluded into the belief that there were five times as many rangers as there really were. Alarmed at this apparently formidable force, the Indians hastily retreated.

Baker's fame was long preserved in the valley, and, as Rogers quaintly says : " We used to think as much of *Captain Baker*, I remember, as we now do of Bonaparte or the Duke of Marlborough, and *do still, for the matter of that.*"

During Queen Anne's War, Capt. Tyng led a party of rangers, on snow-shoes, at midwinter of 1703, up this valley, where he secured 5 Indian scalps, for which Massachusetts granted him £200. Other parties followed in the same trail, and made havoc with the Indian tribes. After Baker's attack, in 1712, the main body of the Pemigewassets retired to Canada, but Lovewell's rangers swept through the valley in 1724–5, and were followed by the companies of Willard, Fairbanks, and Col. Tyng, so that the close of King William's War left this region all desolate. In 1746, a strong St. Francis war-party marched down this valley, " on the run," and captured the fort at Hopkinton ; and Capt. Goffe then advanced uselessly to Warren.

This region of the Pemigewasset Valley was settled rapidly after the conquest of Canada. Plymouth was granted in 1763, and occupied in 1764 ; and a Congregational church was formed in 1765. Col. Webster was one of the pioneers, and founded his estate here when all the country N. of Boscawen was a wilderness.

Plymouth was the birthplace of N. P. Rogers, and of antislavery in New Hampshire. When the eminent English Abolitionist George Thompson was driven from Boston by a mob (in 1835), he was secreted by John G. Whittier, in the Merrimac Valley, and afterwards Thompson and Whittier visited Rogers at Plymouth, and abode with him. On their return, they were attacked by a vindictive mob at Concord, and were saved with difficulty by the skilful strategy of Col. Carrigain, who led the rabble on a false track while the liberators escaped in the darkness. In 1841 Rogers and Garrison came to Plymouth preaching the abolition of slavery, but they were refused the use of either of the churches, and so gathered the people in a great grove near the river, where Garrison spoke from half past three until sundown. This lecture " among the Holderness maples " was long remembered in the valley.

Rogers spoke thus lovingly of his native town : " Few places, of so little note, strike the eye of the traveller so pleasantly as the town of Plymouth, in Grafton County. A beautiful expanse of intervale opens on the eye like a lake among the hills and woods, and the pretty river Pemigewasset, refreshed with its recent tributary, Baker's River, from the foot of Moosilauke, and bordered along its crooked sides with rows of maple, meanders widely from upland to upland through the meadows, and realizes to the mind some of the sequestered spots in the valleys of the Swiss cantons."

113. Mount Prospect.

Mt. Prospect is in the N. part of Holderness, 4–5 M. from Plymouth, and its summit is reached by a good road. The river is crossed at Plymouth, and pleasant mountain-views are afforded in advance. Near the Livermore place and the ancient Episcopal church of Holderness, the road turns to the l. and proceeds towards Prospect. After entering the mountain-road at a gateway, the ascent is found to be easy and gradual, for about 1½ M. The view is obtained from a crowning ledge which was formerly occupied by a beacon of the U. S. Coast Survey. From this point the nearly level summit sweeps away to the N., while the S. and S. E. sides are more abrupt in their slopes. Most of the summit is clear, and is used for pasturage. The predominant rocks are porphyritic gneiss and ferruginous schist. Prospect was formerly called *North Hill.*

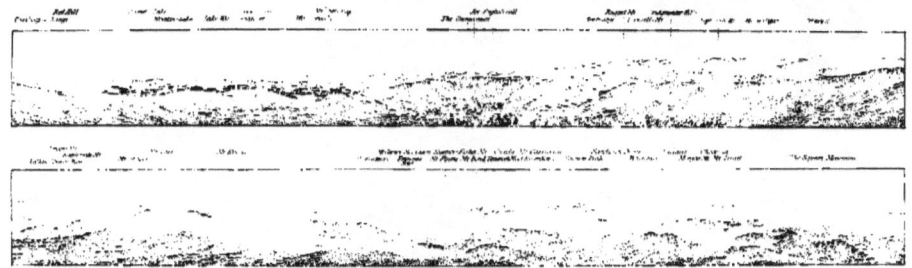

VIEW FROM MOUNT PLEASANT FLYNN

The View. — The Franconia Notch is but a trifle W. of N., and is very well marked, being seen up the long trough of the Pemigewasset Valley. On the l. of the Notch is Mt. Pemigewasset, over which is the round top of Mt. Cannon, striped with light-colored ledges. To the W. of these is the high rounded summit of Mt. Kinsman, covered with thick forests. At the end of the Notch is a portion of Eagle Cliff, coming in on the r. The Franconia Range is seen (not in profile, but) along the line of its axis, whence the more southerly peaks are not distinctly visible, being blent with the higher ridge of Lafayette. The latter assumes a sharp and detached appearance, since its narrow ridge is seen from the end. On the r. of this group is Mt. Flume, whose peak is very sharp; and over the ridge to the r. is the Franconia Haystack, N. E. of Lafayette. In this line, and close to Prospect, are the wooded lines of the Campton Mts. Over these are the Loon-Pond Mts., conspicuous by reason of their bold peaks, one of which is bare; to the r. of which is Mt. Bond, the most southerly of the Twin-Mt. Range. Nearly over Mt. Weetamoo, the bare summit near the r. of the Campton Range, is the Tecumseh group, the main peak farthest away, and the range falling away to the S. W. by Green Mt. and the light-colored ledges of Fisher and Welch Mts. To the r. of and near Tecumseh is the high crest of Osceola, with a sharp secondary peak on the r., beyond which is the well-marked Mad-River Notch, in which lie the Greeley Ponds. Through the Notch is one of the ridges of Mt. Hancock; and on the r. are the precipitous sides of Mt. Kancamagus. Over the latter is Carrigain, showing a fine outline, sloping suddenly to the l., with the Signal Ridge running to the r. Through the Sandwich Notch, some ways to the r. of Weetamoo, are the Acteon peaks, W. of Sandwich Dome, on whose l. is Mt. Washington, low down on the horizon. About N. E. is the ponderous mass of Sandwich Dome, with Sachem Peak on the l.; and on the r. is Whiteface, with white precipices on the S. Farther E. is Mt. Paugus, with long bare ledges on its front. Nearer at hand are the monotonous slopes of Morgan Mt., running off to the N. from the Squam Range, over whose N. end is Mt. Israel, a dome-like eminence, with a ridge running to the S. Over the l. of Israel is the brilliant white peak of Chocorua, clearly cut, though distant. The long Squam Range extends for a considerable distance on the S. E., near at hand, over which is the Ossipee Range, with the rocky Whittier Peak on the N. and the lofty Melvin Peak on the S. W. Over the lower extension of the Squam Range is the exquisitely beautiful Squam Lake, dotted with irregular green islands, and indented by long capes. Above the Squam Range is Red Hill, shapeless and massive, toward the high blue Ossipee Range. To the S. E. is a noble view of Lake Winnepesaukee, its blue sheet broken by hundreds of islands and promontories, and the waters of its numerous bays advancing far up into the level lowlands. Near its end, on the S. E., is the bold head of Copple Crown, flanked on the r. by the Suncook Range, and still farther N. W. by the Belknap peaks. Nearly S. is Little Squam Lake, beyond which is Lake Winnesquam; and over the extreme r. of the former are the twin peaks of the Uncanoonucs, near Manchester. About S. S. W. is the pointed crest of Kearsarge, back of the line of the Ragged Mts.; and between the Uncanoonucs and Kearsarge the high hills of New Hampton and Sanbornton are seen, nearer at hand; while far away on the horizon are the Joe-English Hill, Mt. Wachuset, in Massachusetts, and Crotched Mt., in Francestown. To the r. of Kearsarge are the highlands of Danbury and Hill, over which are Lovewell's Mt., in Washington, and Sunapee Mt., near Sunapee Lake. In the foreground are the Bridgewater Hills, and farther to the r. is Plymouth Mt. Mt. Cardigan is about S. W., and shows a round crest and minor peaks, N. of which is the broad-based Tenney Hill. To the W., up the deep-cut valley of Baker's River, is Smart's Mt., apparently precipitous on the S. On the r. of the valley are the Rattlesnake and Stinson Mts., in Rumney, the former falling off suddenly to the S., the latter making an acute peak. Beyond these are the long and curving Mt.-Carr Range and the high peak of Mt. Kinneo, in Ellsworth. Next to the r. comes Moosilauke, with its sharp S. peak and long plateau to the N. E. The valleys of Ashland and Plymouth are plainly visible below, the former being on the S. and the latter nearly W., at the mouth of Baker's River.

N. P. Rogers thus speaks of this mountain : " Old North Hill, with its bare forehead and commanding peak, which in Scotland would have been crowned with immortality in a hundred songs, standing there unhonored and unsung, a bleak hill-top, climbed now and then for prospects, but chiefly for the blueberries that grow upon its brow, or the sheep and young cattle and wild colts that pasture up its sides. . . . . If it stood where some of those renowned Scotch Bens do, and had undergone the poetic handling of their Burnses and Scotts, people would cross the ocean to see the sights from its top."

114. The Pemigewasset Valley.

This region is traversed by daily stages, which leave the Profile House early in the morning, and Plymouth early in the afternoon, after the arrival of the train from Boston. The following are the approximate distances from Plymouth to the Profile House: Plymouth to W. Campton, 7 M. ; Thornton Centre, 12; W. Thornton, 14; Woodstock, 16; N. Woodstock, 20½; Flume House, 24; Profile House, 29. The fare to the Profile House is $4. The Pemigewasset-Valley Railroad has been chartered to run up through this region, but fortunately will probably not be built.

The Pemigewasset Valley is one of the most beautiful districts among the mountains, and vies in its attractions with the valley of the Saco. Its name was given by the aboriginal Indians, and is compounded of the words *Penaqui-wadchu-coash-auke*, meaning " Crooked-mountain-pine-place." It is pronounced Pem-i-je-was-set, the accent falling on the fourth syllable. The river flows straight to the S. from the Profile Lake, descending more than 1,500 ft. before it reaches Plymouth. About 25 M. below Plymouth it unites with the outlet of Lake Winnepesaukee, and the result of this meeting of the waters is the Merrimac River. In its course through Campton and Thornton the stream is bordered by fertile intervales, and both its banks are well populated nearly up to the Flume House.

The Stage Route. — The interest of this ride is, of course, greatly heightened by going from Plymouth to the Profile House, rather than in the opposite direction, because then the Franconia Mts. are seen in front along the whole route, gradually growing nearer and more lofty, until at last the road enters their narrow defiles. Passengers should endeavor to secure seats on top of the stage, as the views from the interior are very limited.

Soon after leaving Plymouth the stage crosses Baker's River, near its confluence with the Pemigewasset, and then it keeps to the N. and N. E. near the latter stream, swinging around the base of Beech Hill (on the l.). About 4 M. from Plymouth the Campton-Waterville road is seen diverging to the r., crossing the river to the hamlet of Campton Hollow; and the mouth of Mad River is a little farther to the N. Occasional glimpses of the mountains of Rumney are obtained on the l., and on the N. E. are the peaks of Waterville. As W. Campton is approached, care should be taken to get the famous view of the Franconia Mts. from the *School-House Hill.* Beyond the petty hamlet of W. Campton the road descends to the beautiful intervales, and passes Sanborn's inn (see Routes 119 and 115). The West Branch is next crossed and Thornton is entered, with the farms of Thornton Centre across the river on the r. and the highlands of Ellsworth on the l. Beyond the inns at W. Thornton, the traveller should be on the lookout for a noble view to the N. (see Route 120).

Woodstock is now entered, and the intervales narrow rapidly, while bold and forest-covered ridges encroach on the valley on either side. Near the birch-tree, 2 M. beyond the Woodstock post-office, another grand prospect of the peaks to the N. is obtained (see Route 121). The stage now sweeps through the scattered farm-houses of N. Woodstock, and soon passes the point (on the r.) where a side-road diverges to Pollard's and the Pemigewasset Forest (see page 276). The mountains are now closing in on either side, — Flume, Liberty, and the Haystack on the r., and Pemigewasset, Cannon, and Kinsman on the l. A little way beyond Tuttle's boarding-house (which is on the l.), the Harvard Brook is crossed, and then the road passes through the narrowing glens on a long ascent. On the l. is the gabled house of Guernsey, the guide, and 1 M. beyond the Flume House is reached (see page 269). The road from this point to the Profile House rises 543 ft., and is for the most part through the dense forests of a narrow pass, with frequent glimpses of massive ridges on either side. *The Basin* (page 271) is seen on the l., about 1½ M. from the Flume House, and the slowly ascending stage passes upwards over a heavy grade. As Profile Lake is passed (on the l.), the famous Profile may be seen; and soon afterwards the weary traveller alights before the vast forest-palace of the **Profile House** (see page 260).

115. Campton Village.

Boarding-Houses. — The Black-Mountain House (O. C. Foss) accommodates 50 – 75 guests. It is finely situated at the entrance of the Mad River, and looks up at the imposing forms of Sandwich Dome (Black Mt.), Tripyramid, Welch Mt., and Tecumseh. The Hillside House (F. Chase) is W. of the village, in a secluded place towards the river, and accommodates 40 – 50 guests. The Fountain House (T. W. Mitchell) has rooms for 35 ; S. C. Willey, Chas. Cutter, and William Thornton can take 20 each ; and smaller houses are kept by F. A. Mitchell, S. D. Kinsman, Dr. W. A. Smith, and Chas. G. Webster. The general rates at these houses are $6 – 10 a week.

Distances. — Campton Village to Plymouth, 8 M ; W. Campton, 2 ; Woodstock, 10 ; Lincoln, 12 ; Profile House, 23 ; Mt. Weetamoo, 5 – 7 ; Holderness, 13 ; Mt. Prospect, 7 ; Waterville, 10 ; Sandwich, 13 ; Sandwich Dome (Black Mt.), 7 by carriage and 4 – 5 by foot-path ; Welch Mt., 5 – 6. Daily stages run between Campton Village and Plymouth, during the summer. The fare is $1.

Campton Village is on Mad River, about 2 M. from its confluence with the Pemigewasset, and is in a good position to enjoy the beauties of both these famous valleys. It is built on three streets, which form a right-angled triangle, from whose S. E. point runs the road across Mad River, and from the N. W. point the road to the W.-Campton ford diverges. The village contains a Baptist church, and three or four shops. The finest intervales are on the W.-Campton side, but the meadows above, on Mad River, are broad and verdant, and are occupied by a prosperous rural population. The view to the N. is not so good as from W. Campton, the Franconia Range being partly masked by intervening hills; but the mountains of Waterville, as seen up the Mad-River Valley, form one of the noblest vistas in the State. The brilliant ledges of Welch Mt. are on the l. and the dark and lofty mass of Sandwich Dome (Black Mt.) is on the r.; between which and closing the end of the valley is the high ridge of Tripyramid, striped from top to bottom with its great white slide.

"The general shape of Welch Mt. is extremely picturesque. Being nearly destitute of forest covering, and showing large masses of bare quartz, it presents very beautiful and striking harmonies of the grays with neutral hues of blue and white, and at sunrise and sunset exhibits proportionate increase of splendor. The Sandwich range, too, affords ample and important subjects to the dwellers in Campton for the enjoyment and study of mountain color and form. In all lights they are picturesque if not beautiful ; but there is no limit to the softness, purity, and magnificence of color with which the setting sun sometimes floods their broad and rugged sides." (STARR KING.)

The hamlets of Campton and W. Campton have for many years been favorite resorts for artists, who find here rich bits of meadow and woodland scenery. More recently they have become popular among the families who fail to find the desired summer quiet and restfulness in face of the railroads and fashionable hotels of N. Conway and Lancaster, and the other mountain-resorts. The drives in the vicinity are pleasant and of great variety, leading usually over easy roads and through peaceful farming districts. There are also several pretty rambles near the village, affording a variety of mountain and lowland views. *Campton Hollow* is a small hamlet 2½ M. S. of Campton Village, on the Beebe River, 1 M. from its union with the Pemigewasset. This stream is 12 M. long, rising between Sandwich Dome and Mt. Israel and running W. between the Campton Mts. and Morgan Mt. The Campton Mts. form a long and picturesque ridge which runs N. E. between the Beebe and Mad Rivers, the highest point being Mt. Weetamoo (see Route 116). The rural neighborhood called *New Discovery* lies to the S. of this range, near the Squam Mts., and along the Beebe River. The *Devil's Den* and the *Campton Falls* are near Campton Hollow.

Besides being a favorable centre for drives through Thornton and other adjacent towns, Campton Village is a convenient point from which to attack several of the southern mountains. Among these are Welch Mt. (Route 117), Weetamoo (Route 116), Sandwich Dome, or Black Mt. (Route 138), Morgan Mt. (Route 118), and Mt. Prospect (Route 113). The beautiful scenery about W. Campton can also be visited hence, and Waterville, Holderness, and Squam Lake are also easily accessible.

The town of Campton has 1,226 inhabitants, on 27,892 acres of land. They are nearly all engaged in farming, corn, oats, potatoes, and hay being the chief products. It is the chief maple-sugar town in Grafton County, and in 1874 produced 100,000 pounds. The forests are mainly deciduous, and 500,000 ft. of lumber are manufactured yearly. The uplands of the Campton Mts. and the hills W. of the river afford good grazing-ground. Like several others of the mountain-towns, Campton prepares large quantities of ready-made clothing for the city merchants, several thousand pairs of pantaloons being made here annually. There are three hamlets in the town, Campton Village, Campton Hollow, and W. Campton, of which the first-named is the largest.

Campton was granted in 1761, just after the Conquest of Canada, and derived its name from the fact that the proprietors erected their *camp* within its limits when they came up to survey the land. It began to be settled in 1765, and a Congregational church was established in 1774. Five Camptonians died in the Revolutionary army, and thirty laid down their lives in the late war for the Union.

116. Mount Weetamoo.

Mt. Weetamoo is a bare-topped peak of the Campton Range, in the N. E. part of that chain and S. of the Mad-River Valley. Its crest consists of highly tilted strata of White-Mountain gneiss, precipitous on the S.; and the adjacent trees have been cut away in order to afford an unbroken view. The favorable character of this peak for a view-point was discovered in 1874 by Dr. Gladwin, of Boston, and Prof. C. E. Fay, of Tufts College, under whose auspices the present path and clearings were made. To the same gentlemen the mountain is indebted for its pleasant name.

It is thought by many of the visitors to this point that the view from

Weetamoo surpasses that from Mt. Prospect, on account of the superior altitude of the former peak and its comparative vicinity to the higher mountains.

The base of Mt. Weetamoo is about 5 M. from Campton Village, and is reached by following the Waterville road for 3 M. and then turning to the r. on the Sandwich-Notch road. After crossing Mad River the road to the r. is taken and followed up among the glens of the Campton Range, finally descending to its end at the Roby farm. At this point the mountain-path begins, leaving the steep clearings about ⅓ M. back of the house and entering the woods on the r. by a high guide-board. It ascends through an unbroken forest for nearly 1 M., with a stream on the r. in a deep ravine, and is broad and well marked, being used also as a logging-road. The ascent is sometimes made on horseback. About half-way up a short path diverges to the r. to a spring, some way beyond which the main trail ascends a sharp and narrow wooded ridge by a series of natural gradients, and after following this crest-line for some distance comes out on an open ledge whence a fine view is afforded to the N. Traversing the edge of the rocky slope, the path next crosses a shallow ravine and soon ascends to the main peak, where the low trees advance from the S. to meet the inclined ledges which run W. and N.

The View. — The most conspicuous object in the field of vision is the vast mass of Sandwich Dome (Black Mt.), which rises within a few miles, toward the N. E. On its l. flank are the white and picturesque crags of the Acteon peaks, and on the r. is the conical top of Young Mt. The high crest of Whiteface appears over the r. flank of Sandwich Dome, and the noble white peak of Chocorua is seen farther away; while the hotel-crowned ridge of Mt. Pleasant is still more distant, near the dim levels of Sebago Lake. The high and partly isolated Mt. Israel is next seen, to the r. of the Dome and at about the same distance, with the Sandwich and Tamworth lowlands extending on either side, lightened by the gleam of the Silver and Ossipee Lakes, and hemmed in on the S. by the blue line of the Ossipee Range.

The view now rests on the dark and wooded Squam Range, close at hand on the S. and S. E., beyond the glens of the Beebe River, over which are seen the beauty and richness of the lake-country. Lake Winnepesaukee is overlooked for many miles, with its numerous islands and capes and the high hills of the S. shore. Copple Crown Mt. rises in the direction of Wolfeborough; and the high thickets of Rattlesnake Island are towards Alton Bay. Farther to the r. are the twin crests of Mt. Belknap, standing out in light colors. The Uncanoonucs are nearly due S., near Manchester, very distant, with the Temple Mts. and Joe-English Hill on the r., and possibly Wachuset beyond, in Massachusetts. About S. S. W., 5 M. distant, is the broad flat top of Mt. Prospect, free of forests, with the white village of Ashland beyond, towards the Bridgewater and Bristol Hills.

Farther away in this direction, over the Ragged Mts., is the high blue peak of Mt. Kearsarge.

The lower ridges of the Campton Mts. are now seen close at hand, trending away to the S. W.; and the white domes of Mt. Cardigan rise clearly beyond, with Tenney Hill on the r. A part of the Baker's-River Valley is next seen, beyond Plymouth and Loon Pond. Mt. Stinson is due W., with the high range of Mt. Carr back of it, and Smart's Mt. more distant, on the l. In the foreground the beautiful Pemigewasset Valley is visible, with the white houses of Campton Village and Thornton Street. The high wooded range beyond, and running to the N., includes Mt. Carr, the Ellsworth peak of Mt. Kinneo, Mts. Waternomee and Cushman, and ter- minates on the massive Moosilauke, which is nearly N. W.

The Mad-River Valley is in the near foreground, and Mt. Kinsman lifts its dark ridges about N. N. W. More to the r., more distant, and nearly over the adjacent wooded heights of Cone Mt., is the noble group of the Franconia Mts., with Mt. Lafayette towering over all and the ledge-faced curves of Mt. Cannon on its l.

Just across the Mad-River Valley, N. by E., is the rounded top of Welch Mt., covered with brightly colored ledges, and overlooked by the more distant white peak of Tecumseh. Mt. Bond is seen nearly over Tecum- seh. Certain of the Loon-Pond Mts. are beyond this ridge, towards the Pemigewasset Forest. About N. N. W. is the stately crest of Mt. Osceola, up the long glen of Mad River, with the Mad-River Notch on its r., flanked by the dark heights of Mt. Kancamagus. The imposing blue peak of Car- rigain is seen through the notch, and is flanked on the r. by the Nancy Range. Mt. Washington and some of the higher crests of the Presidential Range are nearly over the Nancy Mts.; and it is thought that Mt. Carter is seen on the r., still farther away. The view toward Tripyramid and other mountains in the N. E. is intercepted by the dark mass of Sandwich Dome.

117. Welch Mountain.

Welch Mt. is a lofty spur of the Tecumseh range, with which it is connected by a nearly level ridge curving about the head of a deep and wooded ravine. The greater part of its flanks is composed of granite (much of which is porphyritic in character), reaching the surface in long and slanting ledges, often of convex outlines, and affording an interesting display of Nature's architecture. In some of its aspects this mountain reminds one of the Percy Peaks, which are also faced with long granite ledges. The foliage on Welch is scanty and widely scattered, and the protrusion of such great masses of rock gives to it a peculiar and striking appearance when seen from the adjacent valleys. The summit rises like a rude dome from the long terrace below, and is attained by flanking two lines of cliffs. It is not difficult to traverse the high thin ridge to the N. W. (passing over Stone Mt.), and reach the main line of the Tecumseh range, near the base of Green Mt., whence Tecumseh may be visited. There is a spring well up on the mountain. The altitude of the peak is 3,500 ft.

Welch Mt. is visited by diverging to the N. from the Waterville road, E. of the Durgin place, about 4½ M. from Campton Village and 6½ M. from Greeley's. From this point a farm-road leads up through the fields for ½ M. to the lonely Webber house, whence a path leads to the summit in about 1 M. It passes up by the orchard back of the house, enters the maple-grove near a small apple-tree, and soon reaches the ledges, where it sometimes becomes difficult to trace. The ledges are often highly inclined, but are sufficiently rough to afford good footing, except in wet weather, when they become slippery.

The main object in visiting Welch Mt. is to see its massive sheets of granite and its symmetrical structure, since the view from the summit is of limited extent. The northern mountains are hidden by the main Tecumseh range, which sweeps off on the r. to Green Mt and the flanks of Tecumseh. On the E. also the view is intercepted by the serrated ridge of Tripyramid, whose great slide is well displayed. Nearer at hand, across the valley to the S. E., is Sandwich Dome, with its lofty crest-line and minor peaks, over whose r. flank is gained a glimpse of Lake Winnepesaukee. Farther to the r., beyond the Sandwich Notch, are the Campton Mts., a long and monotonous wooded ridge culminating in Mt. Weetamoo. The foreground on the S. and S. W. is occupied by the picturesque valley of Mad River, stretching away beyond Campton Village and toward Plymouth. Down this vista are seen the white crest of Mt. Cardigan, far away to the S. W., the high hills of Groton and Dorchester, and the dark masses of the Mt.-Carr Range, to the W. over that part of the Pemigewasset Valley which lies in the town of Thornton. Farther to the r. is the lofty plateau of Mt. Moosilauke, beyond Mts. Kinneo and Waternomee. The heavily wooded highlands near at hand, opposite and below the ledgy sides of Fisher's Mt., are known as Cone Mt.

Mt. Tecumseh, see Route 136.

118. Morgan Mountain

is the N. extension of the Squam Range, and forms the highest peak on that chain, attaining an altitude of 2,162 ft. It is easily reached from Campton Village, whence it is 6 – 7 M. distant; and the excursion is made by ladies without difficulty. It is 8 – 9 M. from W. Campton, and 9 – 10 M. from Plymouth, and is reached by driving up the Beebe-River glen from Campton Hollow, for 3 M., to the Percival farm. From this point it is about 1 M. to the summit, and most of the way is through pastures.

There is a belt of about 40 rods of woods on the route, and the remainder of the way is plain and easy.

* *The View.* — Several of the prominent peaks on the N. and N. E. are seen from Morgan, as are also the highlands surrounding the great basins on the S. and S. E. But the special charm of the view is Squam Lake, which lies outstretched below and is seen in a bird's-eye view, with all its islands and capes, its narrow straits and bays, and the delicious comminglings of land and water, forest and cove, which make Squam the fairest lake in New England. The prospect is the finest which can be obtained of Squam, surpassing even those from Red Hill and Shepard Hill. Farther away to the S. E. is the broad mirror of Lake Winnepesaukee, with its hundreds of islands and long retreating bays. This view is very lovely in the latter part of the afternoon, when the lake glitters like silver and the islands and capes are thrown into strong, dark contrast. On the E. is the long blue Ossipee Range; the low peak of Copple Crown is in the S.; and the double crest of Belknap is in the S. S. E. Farther away to the S. S. E. is the high and pointed peak of Kearsarge.

119. West Campton.

Boarding-Houses. — Sanborn's, a large and comfortable inn alongside of the stage-road from Plymouth to the Profile House, and accommodating 65 guests. D. B. Southmayd lives on the intervale near by, and takes a few boarders. J. M. Smith's (10 boarders) is on the far-viewing heights on the Ellsworth road, 2 M. from W. Campton.

The inn at W. Campton is one of the most ancient among the mountains, and was formerly known as *The Stag and Hounds.* Many years ago it was frequented in the summer-time by Durand, Gay, Gerry, Richards, Griggs, Pope, Williams, and other landscape-painters, who came out to sketch on the beautiful Campton meadows.

W. Campton is a small hamlet on the edge of a bluff over the Pemigewasset intervales, at the intersection of the Ellsworth and Franconia roads. A short distance to the N. is Sanborn's inn, pleasantly situated on the intervale, near the confluence of the Pemigewasset and the West Branch. Campton Village is reached by a road 2 M. long, which crosses the meadows and fords the Pemigewasset. Strangers should not attempt to ford alone, as the route across is not indicated, and a divergence from it would lead the carriage into deep water. The beautiful meadows between the Franconia highway and the river are the great charm of W. Campton.

"Of course the Franconia Mts. form one of the leading attractions in the landscape here, W. Campton being the southernmost point in the valley from which they can be advantageously seen. Since they are visible from the meadow as well as from the hillsides, the choice of several different combinations of middle and foregrounds is offered to artists, and to those who love complex and proportioned beauty near their summer resting-place. We have known artists to say that the marvellous middle ground of belt and copse, and meadow and river, back of which the three sharp spires of Lafayette and his associates tower, to face the heavier rocky wall which forms the western rampart of the Notch, is the most enchanting scene of the kind which this valley and that of the Saco can offer.

" Perhaps it is fair to say that the intervale alone in Campton is, in proportion to its extent, more picturesquely effective than that of N. Conway. Being finely wooded, and better united to the bordering hills, it furnishes perfectly appropriate and beautiful foregrounds to the favorite views of valleys and mountains, whose flitting moods of superior beauty and grandeur have been promoted by many painters, Mr. Durand and Mr. Gay especially, into the abiding charm of art. The windings of the river in this intervale, with the beauty, variety, and abundance of its trees, makes W. Campton rich in artist's ' bits ' of the utmost grace. Here, some elms, bordering large spaces of the smooth sward with green-domed tops, evenly poised upon their single columnar trunks, look, as an architectural friend once expressed it, like unwalled chapter-houses to the cathedral groves. There, we find a sparkling group of varied foliage which we may call voluptuous, in which the golden plumes of the ash shine, perhaps, against the brown and olive darkness of the oak, and the butternut's pale yellow spray mingles with the shimmering gray of the beech, and the dull purple and emerald of the birch and wild cherry." (STARR KING.)

S. of the hotel and just beyond the hamlet is the point called the *Starr-King View,* whence is obtained the rich prospect herewith described in the brilliant language of Mr. King himself. This point is also about 2 M. N. of Blair's boarding-house, and 6 M. from Plymouth.

" Let them study the Notch mountains of Franconia from the school-house in Campton, by the morning or evening light. They differ then from their ordinary aspects as much as rubies and sapphires from pebbles. See the early day pour down the upper slopes of the three easterly pyramids : then upon the broad forehead of the Profile Mt., kindling its gloomy brows with radiance, and melting the azure of its temples into pale violet : and falling lower, staining with rose tints the cool mists of the ravines, till the Notch seems to expand, and the dark and rigid sides of it fall away as they lighten, and recede in soft perspective of buttressed wall and flushed tower, — and then say whether, to an eye that can never be satiated with the blue of a hyacinth, the purple of a fuchsia, and the blush of a rose, the gorgeousness of the mountains is a mere exercise of rhetoric, or a fiction of the fancy. Or, towards evening of midsummer, at the same spot, see the great hills assume a deeper blue or purple ; see the burly Cannon Mt. stand, a dark abutment, at the gate of the Notch, unlighted except by its own pallor ; and, as the sun goes down, watch his last beams of crimson or orange cover with undevastating fire the pyramidal peaks of the three great Haystacks, and then decide whether language can recall or report the pomp of the spectacle, any more than the cold colors of art can exaggerate what the Creator writes there in chaste and glowing flame."

Bald Hill, or *Cook's Hill,* is about 3 M. from W. Campton, by a road which turns S. from the church 2 M. out on the Ellsworth road and sweeps around to the r., turning N. by a cemetery and soon entering pastures which overlook the Pemigewasset Valley and all the hills of Campton. Portions of the Squam and Sandwich Ranges and the chief Franconia peaks are also seen from this point. Somewhat broader views may be obtained by ascending on foot through the fields, but the crest of the ridge is covered with trees.

Ellsworth (Avery's small boarding-house) is 5 M. N. W. of W. Campton, by a road which ascends over a line of far-viewing heights, looking out on the Waterville and Franconia Mts., and passing Smith's boarding-house and the old Free-Will Baptist church. Ellsworth is a town of large area, which was granted as early as 1769 (by the name of *Trecothick*), but has only 192 inhabitants. The N. and W. parts of the town are occupied by wild mountains, of which Mt. Kinneo is a prominent object for leagues away. In the S. E part is *Ellsworth Pond*, near which is the Baptist church. There is no village, the inhabitants being engaged in farming, lumbering, and making maple-sugar. It is said that, though poor, "they enjoy the comforts of life, and are a contented, frugal, virtuous, and happy people." The scenery is not remarkable, but there is good fishing and hunting in the W. glens. (See Route 122.)

120. Thornton.

Boarding-Houses. — George Foss (25 guests) and George Jenkins (10) keep pleasant houses on the Mad-River road, 1½ - 2 M. from Campton Village. The former is known as *Brook Farm*, and is prettily situated in the valley. Henry Durgin's is farther up the valley, 4 M. from Campton. Wm. Merrill's (30 guests) is at *Thornton Centre*, on the E. side of the Pemigewasset. At *W. Thornton*, on the W. side of the Pemigewasset, are the small inns called the Union House (40 guests) and the Grafton House (25 guests), the latter of which cannot be recommended.

Routes. — The only public conveyances passing through Thornton are the daily stages between Plymouth and the Profile House, which cross the town bound S. in the forenoon, and bound N. about mid-afternoon.

The town of Thornton has 840 inhabitants, and covers an area of 28,490 acres, of which nearly half is improved, having a deep and fertile soil, prolific in corn and potatoes. The people are engaged in farming, and in the winter they get out large quantities of maple-sugar. The town was granted to the Thornton family in 1763, and was settled in 1770. A Congregational church was founded in 1780, but has since died and has been supplanted by languishing Baptist churches.

Noah Worcester. D. D., was pastor here from 1787 to 1810, his salary being $200 a year, which he eked out by farming and shoemaking. He was afterwards the founder of the Massachusetts Peace Society, and author of several theological works. The church was built in 1789, and paid for with wheat, rye, corn, and flax ; and at its dedication the following expenses were incurred : " Amount for victualling 54 persons, $9 ; for brandy and West-India rum, $5 ; for sugar, $1."
" The scenery through Thornton strongly resembles the rural districts of Scotland. It is so like it that many years ago a considerable number of Scottish emigrants, on their way perhaps to Barnet and Ryegate, settlements of their countrymen in Vermont, were induced to stop short and settle here." (N. P. ROGERS.)

There is no village in Thornton, but there are three parts of the town where the houses are less widely separated.

Thornton Street is in the Mad-River Valley, and consists of a line of farm-houses extending from near Campton Village to the Sandwich-Notch road. The scenery here is pastoral and beautiful; and the peaks of Welch, Sandwich Dome, Weetamoo, and Morgan are easily reached. There is a high and rugged road which runs S. E. from this point to Sandwich (12 M.), through the pass called the *Sandwich Notch*, which is 1,417 ft. above the sea. *Thornton Centre* is a scattered line of farm-houses (1,223 ft. above the sea) on the E. side of and parallel with the Pemigewasset River.

W. Thornton is another group of farms, on the stage-road W. of the river. It is in the vicinity of several fine view-points, and is near the bridge over the Pemigewas̄et.   On the E. side, 1½ - 2 M. from the hamlet, is a high hill, on which stands a beacon of the U. S. Coast Survey, and from this point a broad and noble prospect is obtained.

"The most striking views which the ride from Plymouth to the Flume House affords are to be found after passing the 'Grafton House,' in Thornton.   The distant Notch does not show as yet the savageness of its teeth ; but the arrangement of the principal Franconia Mts. in *half-sexagon* — so that we get a strong impression of their mass, and yet see their separate steely edges, gleaming with different lights, running down to the valley — is one of the rare pictures in New Hampshire.   What a noble combination, — those keen contours of the Haystack pyramids, and the knotted muscles of Mt. Lafayette beyond!" (STARR KING.)

"The sun, there sombred at that early hour as towards his setting, was pouring his most glorious light upon the naked Peaks, and they casting their mighty shadows far down among the inaccessible woods that darken the hollows that stretch between their bases. . . . .   Four or five of them, as distinct and shapely as so many Pyramids, some topped out with naked cliff, on which the sun lay in melancholy glory, others clothed thick all the way up with the old New-Hampshire hemlock, or the daring hacmatac.   You could see their shadows stretching many and many a mile, over 'Grant' and 'Location' ; away beyond the invading foot of Incorporation, where the timber-hunter has scarcely explored, and where the Moose browses now, I suppose, as undisturbed as he did before the settlement of the State." (N. P. ROGERS.)

The **Mill-Brook Cascades** are about 4 M. from W. Thornton (or 5 - 6 M. from Campton Village), by the road which diverges to the N. E. up the Mill-Brook valley not far from the Thornton town-house.   3 M. up this road and just beyond the white house of Mrs. Sanborn (which is at the top of a long hill), the visitor leaves the road by entering a pair of bars on the r., and walks down a path which crosses the field toward the brook. The cascade is in the edge of the woods, and may be found by following its music.   The brook here leaps over a bold ledge 42 ft. high, in a white fringe of foamy water, and dances away down a long ravine over several minor falls.   The visitor should descend to the bottom of the cascade, and view it from the broad black rocks below.   Beautiful rainbow effects may be perceived here when the sun is at the right angle.

About 2 M. beyond these falls is the end of the road, where it is intercepted by the wilderness.   The *Elkins farm* is at the foot of the Fisher Mt., and from thence the ascent of Mt. Tecumseh may be made (see Route 136).   About 2 M. up the Tecumseh ravine are two mineral springs, chalybeate and hepatic, near which a house has been built.   The road thither has been ruined by long neglect and the washing of storms.

121. Woodstock.

Boarding-Houses. — Isaac Fox has a large house (40 boarders) on the E. side of the river, about 1 M. from the hamlet of Woodstock.   At and near the neighborhood called *N. Woodstock*, on the stage-road W. of the river, are the boarding-houses of Geo. W. Russell (post-office), Curtis Parker (25 guests), Chas. Russell (20 guests), Jas. Darling, and Wm. Dearborn.   The Tuttle House (20 guests) is a little way to the N., in the town of Lincoln, and on the stage-road.

Route. — By the Plymouth-Franconia stages once each way daily.   N. Woodstock to Plymouth, 20½ M. ; to the Flume House, 4 - 5 M.

Woodstock is a forest-town, with 405 inhabitants, and covers an area of 33,350 acres, of which only 2,540 acres are improved.   It was granted in

1763, under the name of *Peeling*, and was settled in 1773. Most of the town is covered with an unbroken forest, from which large quantities of timber are cut every winter and floated down the Merrimac. The chief occupations of the people are farming and lumbering. There are three comatose churches in the town, which is not populous enough to support a chapel. In 1841 N. P. Rogers called Woodstock "the head of plough navigation," and said, "it has a noble population of Abolitionists."

Woodstock has recently been growing into favor as a place of summer-resort. It is near and in full view of the Franconia Notch, and has several tall mountains of its own. The scenery is grand and imposing on every side, and there are several famous roadside views. Trout are caught in abundance in the neighboring hill-streams; and it is but a short distance to Pollard's, at the entrance of the great wilderness. The accommodations for tourists are as yet of a humble character, though neat and comfortable. In the S. E. part of the town is the *Grafton Mineral Spring*, a sulphur-water which has been occasionally used medicinally.

In driving to the N. on the stage-road, about 2 M. N. of Woodstock, the traveller sees an old birch-tree on the r. and alongside of the road, on the edge of a slope to the N. The *view of the Franconia Mts. and Notch from this point is one of the best on the road, and has the advantages of nearness and artistic arrangement. Passengers on the stage should endeavor to catch this noble prospect.

Fox's is apparently the pleasantest public-house between Campton and the Flume House. In its vicinity are bright and verdant meadows, and the road is lined with long avenues of maple-trees. From a hill near the house a fine view of the Franconia Notch and the adjacent mountains is afforded, together with the Pemigewasset Valley. The *Gore Road* leads thence into the Thornton Gore, about 5 M. distant, ascending the valley of Eastman's Brook to the verge of the great Pemigewasset Forest. From *Wyatt Hill*, on this road, a broad view is gained over the wilderness and its many lofty peaks.

The **Agassiz Basins** are on the Moosilauke Brook, which rises on the Blue Ridge and flows into the Pemigewasset at N. Woodstock. They are reached within 2 M. of the hamlet by turning to the W. on the first divergent road N. of the highway-bridge over the Moosilauke. This by-way crosses pleasant meadows, with fine mountain-views, and then enters an ancient forest. The short path to the Basins leaves the road on the l., near a great boulder, and soon leads to the edge of the stream. At this point the water is crowded into a narrow and massive channel and plunges down perpendicularly into a chasm in the granite ledges, so deep and confined that but little of the fall can be seen, though its dull booming fills the air. Broad and ponderous granite ledges line the stream on both sides, and below and between them the water rolls through deep black basins which have been eaten into the solid rock. It is one of the finest pieces of rock-scenery in the Pemigewasset Valley, and the whole course of the brook is interesting on account of similar (though smaller) cascades and ledges. The present name was bestowed soon after Prof. Louis Agassiz visited this locality.

The *Ice Caves* are reached from the road that diverges to the W. about ¾ M. N. of Woodstock, by a path nearly 1 M. long. They are a series of dark, damp, and otherwise uninteresting crevices in a rocky hillside, wherein ice is found throughout the summer. Near this point, on the N., are the small cascades and pot-holes on *Beaver Brook* (hardly worth visiting). This road is now abandoned and in bad condition. It runs W. among the long spurs of the Blue Ridge and Mt. Cushman for 5 M., to a remote farm, whence it is 3 – 4 M. through the woods to the outer farms of Warren. This trail leads through a pass between the Blue Ridge and Mt. Cushman.

The **Blue Ridge** is in the W. part of Woodstock, and is a prolongation of the Mt.-Kinsman range. In the N. W. corner of the town, near Gordon Pond, the Wild Ammonoosuc River takes its rise, and in this direction there is a pass to Landaff, whose highest point is 1,655 ft. above the sea. The Blue Ridge is separated from Moosilauke by the ravine in which Baker's River rises; and at its S. end is a pass to Warren.

Mt. Cushman is beyond Elbow Pond, about 1 M. from the Blue Ridge, and attains a height of 3,326 ft. Farther to the S. are *Mts. Waternomee* (3,022 ft.) and *Kinneo* (3,427 ft.). All these peaks are densely wooded, and therefore are of no interest to tourists as points to visit.

THE WESTERN VALLEYS AND MOUNTAINS.

122. Rumney.

The Stinson House is a small village-inn (without a bar) near the Green. Summer-boarders are taken at the following houses, at the rate of $ 7 - 8 a week : in the village, at the homes of J. W. Pease (6 guests), G. P. French (5), and O. W. Stevens (6); near the station, C. W. Herbert (8); on farms along the intervale of Baker's River, A. D. Spalter (20), Chas. Pease (10), Mrs. J. L. Spalding (10), and C. C. Smart (12). T. G. Stevens and Josiah Quincy, respectively 1 M. and 2 M. distant, on the Plymouth road, also accommodate a few boarders. The latter is at Quincy Station, on the railroad, which is about 1½ M. from Loon Pond, and is near the abrupt hill called Hawks Mt.

Rumney is a quiet hamlet, with two churches, near the confluence of the Stinson Brook and Baker's River, between the Rattlesnake Mt. and Mt. Stinson. It is about ½ M. N. of the cluster of houses at the railroad station, 2 M. from the hamlet at Quincy Station, and 3 – 4 M. from W. Rumney. The town contains 1,164 inhabitants, most of whom are devoted to agriculture.

Mt. Stinson is a lofty ridge about 3 M. long, running N. E., and covered with alternating sections of forest and pasture-land. Its highest point is 2,707 ft. above the sea, and is covered with trees.

The peak over the village of Rumney may be ascended without much difficulty, the distance being but 3 M. The best route is by the road leading N. from the Green, crossing the Stinson Brook and passing the old coal-kilns. Just beyond Elliott's saw-mill, 1½ M. from the village, a logging-road turns off through bars, to the r., and ascends through woods and pastures upon the long W. flank of the ridge. By pressing through a narrow strip of trees far up, the visitor reaches the crest of the peak, with loftier tree-crowned heights to the N. W. The long and rugged ledges are starred with small but brilliant sheets of mica; and clusters of hardy evergreens along the narrow plateau interfere with the prospect. The white village of Plymouth is seen to the S. E., beyond which is Mt. Prospect; and the distant waters of Lake Winnepesaukee are visible on a clear day. On the E. are the bold hills of Campton, but the villages are hidden. On the S. and S. W. are strips of the Baker's-River valley and the sombre heights of the Mt.-Carr range. Rumney can be seen only by descending to the ledgy spur on the S. of the peak. The view hardly repays the toil of the ascent.

Rattlesnake Mt. is N. W. of Rumney, and may be visited by driving out for 1½ M. on the Warren road, and thence ascending on foot through the pastures for over 1 M. The chief feature of the view is the valley of Baker's River, with the villages of Wentworth, W. Rumney, Rumney, and Plymouth. On the E. is Mt. Stinson, and the high hills of Groton loom across the valley on the S.

The roads from Rumney to Ellsworth are about 6 M. long, and lead up the valley of the Stinson Brook, following the shores of Stinson's Pond for over 1 M. The road on the W. shore is the easier, and is more travelled, but the old road over the hills on the E. of the pond is the more picturesque. It leads through a region of deserted farms and abandoned houses, and gives pleasant views of the pond, embosomed between Mt. Stinson and Mt. Carr. A rude path runs N. W. from the steam saw-mill near the head of the pond to the **Glen Ponds** (or Little Ponds), a line of secluded tarns in a ravine of Mt. Carr. They are famous for trout, and are visited every year by parties of fishermen, who build camps in the vicinity. The road running S. E. from the saw-mill leads to *Ellsworth Pond*, in which pickerel have been placed; and then passes out into Thornton and Campton. There are no roads to the N. or W. from Ellsworth, since the town is enwalled on those sides by lofty and rugged mountain-ranges. Visitors should come up by the road W. of Stinson's Pond, to secure an easy gradient, and return by the hill-road, from which a succession of attractive front-views may be obtained.

The mountain, pond, and stream in Rumney are named in memory of David Stinson, a skilful hunter who came up here in 1752 with John and William Stark and Amos Eastman. They had secured £500 worth of sable, marten, mink, and beaver furs, and were about to go home, being warned by the sight of fresh moccasin-tracks. "The long shadows began to fall across the water, and the last rays of the setting sun were streaming full upon the face of craggy Rattlesnake Mt., when John Stark, who was stooping to take a steel trap from the water, was startled by a sharp hiss. Jumping up he saw the Indians; and the muzzles of half a dozen muskets, staring at him within three feet of his head, told him that escape was hopeless. That night he lay bound among his captors, and in the morning was early roused to proceed down the river, where they were to lay in ambush for the rest of the hunters. The latter had guessed the cause of Stark's absence, and at the earliest dawn packed their furs, traps, and camp equipage into their canoe and started. Eastman was upon the shore, while William Stark and Stinson guided the frail bark as it floated down in the rapid current. The Indians easily captured the former, then bade Stark hail those in the canoe, and invite them to come on shore. Stark complied so far as to tell them to pull to the opposite bank and then run for their lives, as the Indians had got him and would have them too unless they were quick in getting away. Curses and blows fell thick upon the head of the dutiful but unfrightened hunter, and then the Indians levelled their muskets to fire upon the retreating men. 'Not yet, my friends,' said the belabored Stark, as he struck up their guns at the moment of discharge. For this he got another shower of kicks and cuffs, and when a second time they attempted to fire he again endeavored to stop them, but not so successfully as before. Stinson was killed in the act of leaping upon the shore, and fell backward, his blood staining the clear water. The paddle in the hand of William Stark was shivered with bullets, but leaping from the canoe like a deer he took to the woods and escaped." John Stark was ransomed the next autumn, became a captain of rangers in the Ticonderoga campaigns, a colonel of the N. H. line at Bunker Hill, commanded the American vanguard at the battle of Trenton, defeated Burgoyne's German troops at Bennington, and then was made major-general and commander of the Northern Department.

123. Wentworth.

Hotel. — The Union House is a quaint village-inn which has come down from ancient times almost unaltered.

The hamlet of Wentworth is pleasantly situated on high ground, at the confluence of Pond Brook and Baker's River, near the railroad. It is a quiet and cosey old place, abounding in shade-trees, with an academy and a large Congregational church. Nearly 1,000 people live in the surrounding town, and are mostly engaged in farming. In the E. are the wooded heights of Mt. Carr; and on the W. are the ranges which culminate in Mt. Cuba and Smart's Mt. The town was settled before 1770, and was named in honor of the then Royal Governor of the Province. It is claimed that the people here receive $ 8,000 a year from summer-visitors.

The favorite drive is 4 – 5 M. N. to Warren, on the road W. of Baker's River, which commands pleasing views of " the river, in its ceaseless meanderings; the beautiful meadows on its banks; the uplands, gracefully sloping from the borders of the intervale to the mountain-sides; the unbroken mountain-wall on either side; and the well-cultivated farms all along the river-bottoms and on the hill and mountain sides, having neat and tasteful buildings."

Another drive is 4 – 5 M. N. W. up the Pond-Brook glens to *Baker's Ponds*, under Mt. Cuba, and near the wellnigh inaccessible *Cuba Falls*. Mt. Cuba may be ascended on this side, giving a noble view toward the Franconia and White Mts. In 1856 the dam near the ponds gave way after a two-days' storm, and part of the village of Wentworth was swept away.

In June, 1754, Capt. Powers's Rangers encamped between the Baker Ponds, while on their way toward the Upper Connecticut. " See this company of stalwart hearts, camped in storm of haile and raine and thunder, beside these exceedingly solitary ponds in the basin of the great mountains, each man eager with trusty ' Queen's arm ' to hurry further away into the wilderness, to fight what were to them veritable ' painted, red demons '; perchance to be slain, to be scalped, to be devoured by wolves, or to rot in some cold swamp — and you have the romance of Capt. Powers's expedition."

124. Warren.

Hotels. — The village-inn is known as the Moosilauke House, and is some ways S. of the station, near the railroad. The summer boarding-houses of Amos Clement (10 guests) and Russell Merrill are in the village; that of Merrill & Son (15 guests) is about 5 M. distant, at the foot of Moosilauke. Boarding-houses are also kept by Upton & Eastman, Mrs. A. Knapp, and others.

Distances. — To Ore Hill, 3½ M.; Piermont, 9½; Haverhill, 12½; Wentworth, 5.

The village of Warren is near the confluence of Black Brook and Baker's River, and is built in a straggling manner on a long street almost parallel with the railroad. It contains a Methodist church and several shops. Tourists who are interested in geology should visit the large collections of Mr. James Clement, in the old shop near the hotel. He has

13 *

for sale many hundreds of specimens of garnet, epidote, quartz crystals, brilliant ores, and other mineralogical curiosities found in this vicinity. The village itself is not very pleasant, but there are many fine bits of scenery and accessible mountains in the vicinity. There are about 900 inhabitants in the town, most of whom are engaged in farming and lumbering. There is much wild game in the forests, and the town has 50 M. of trout-streams, besides several ponds.

The local historian claims that Warren contains gold, silver, iron, copper, lead, zinc, plumbago, calc-spar, molybdenum, rutil, epidote, beryl, garnets, quartz crystals, tourmalines, and other minerals. There are valuable ore-beds on Mt. Sentinel, masses of limestone near the Summit, fine granite at Webster's Slide and Mt. Carr, and gneiss and mica slate elsewhere.

"The exact centre of Warren is the summit of *Knight Hill* [2 M. N. of the village]. Standing on the top, one is surrounded on all sides by rocky crests, and the forest hamlet appears like a huge bowl, with another bowl transparent, formed of blue sky inverted and placed over it, and resting upon the rim of mountains." *Peaked Hill* is near the village, on the S. E., and affords a comprehensive view of the Warren glens and Mt. Moosilauke. The distance to the top is about 1 M.

It is said that there are more than 100 brooks in Warren, upon which are several pretty cascades. The most picturesque of these streams is *Hurricane Brook*, which falls from the side of Mt. Carr, S. E. of the village, and is approached by the road S. of Peaked Hill. Here are found the Waternomee Falls, the Middle Cascade, and the Hurricane Falls, where the water descends over high step-like ledges for 200 ft. Next come the smaller Wolf's-Head Falls, near which is the deep pool of Diana's Bowl, by which Surveyor Leavitt camped in 1765. The *Rocky Falls* are E. of the village, on Patch Brook, between Peaked Hill and Clement Hill.

"A dozen beautiful white foamy streams come rushing down its sides, among which is . . . . that most beautiful of all streams, Hurricane Brook. On the latter are those little, white tumbling waterfalls which for so many years were almost unknown, but are now so much admired By these, it is said, in old times lived the fairies. It was here on the rich carpets of green moss they danced in the moonbeams and sang an accompaniment to the falling waters. The deep mossy-rimmed basin, set with gems, and carved in the rock high up on the mountain-side might have been their bathing-font, and in it even Robin Goodfellow and Queen Mab might have performed their ablutions. The Indians had a beautiful tradition how the fairies stole the children away, and gave them fairy bread to eat, which changed them to fairies. Then, said they, there was joy for the little folks as they revelled in the green embowering woods; and the elfin king and the fairy queen ruled long and well in the old centuries. But the period when they existed has melted into the mellow twilight of ages, and all these joyous revellers are gone forever." (LITTLE.)

Baker's River rises N. of Warren, in the ravines E. of Moosilauke. "At first a wild torrent, then a bright pebbly-bottomed stream, and lastly a deep blue river, it empties into the Pemigewasset." Its Indian name was Asquamchumauke, from *asquam-wadchu-m-auke*, meaning "Water of the Mountain-Place." The present name was given in honor of Capt. Baker, a soldier of the Indian wars (see page 279).

Mt. Sentinel is in the S. W. part of the town, and is 2,032 ft. high. It is sometimes climbed by tourists. A party of New-Yorkers, visiting Moosilauke, were deluded by the pretended trance-visions of a spiritualist, who saw 14 mines on the slopes of Mt. Sentinel. They formed a company, and worked 18 months on the ledges and in sinking a shaft, but lost all the money which they invested.

Near Marston Hill is what is called *Runaway Pond.* "Where the water burst through is plain to be seen, and on the rocks of the former beach are yet the marks scored by the tumbling waters and dashing ice. The broad acres, once the bed of the pond, are now fertile meadows. They were never fully overgrown by forest-trees. Mounds, where the Indians stored their corn ; ashes, where burned the wig-wam fires ; pieces of rude pottery, axes of stone, arrow-heads turned up by the ploughshare, and graves under the shadow of Marston Hill, tell that here once was an Indian village. By it ran the trail running to the land of the Coösucks. In front wound the Mikaseota, silent and dark, and near by the bright water of Ore-Hill Brook flashed in the rocky glen. Here the steep hills, that once sloped down to the curling waves, protected from the chill winds the Indian's maize, his pumpkins, squashes, and beans, which grew in these most fertile meadows." (LITTLE.)

One of the favorite homes of the Indians was on the meadow on the r. bank of Baker's River, below the railroad bridge. Numerous relics have been found here, and 20 rods back from the river is a high ledge called *Indian Rock*, in which are four smoothly cut bowls, at the points of the compass A pretty view of Warren and Moosilauke is gained thence. Another aboriginal settlement was near the Beech-Hill bridge, on Black Brook The old Indian trail from the Merrimac Valley to the Con-necticut Valley followed the W. bank of Baker's River as far as the point where the Mikaseota (Black Brook) comes in on the W. It then followed the latter stream (on the E. bank) to the Wachipauka Pond, whence it descended the slopes of Webster's Slide, entered the Oliverian valley, and thus reached the Connecticut River.

Ore Hill is W. of Warren, and derives its name from an old mine of silver-bearing lead near its S. end. The ore appears to have been rich, and was worked for about two years. A broad road leads N. W. from Warren by a succession of slovenly farms to the foot of Ore Hill, where the road to the mines diverges to the l. The main road ascends the long hill, and from the point where it plunges off into the deep valley beyond a pleasant view is obtained. Walking out on the plateau beyond the grove of tall trees on the r., Mt. Mist and Moosilauke are seen on the N. E., and on the N. W. is the shining plane of Tarleton Lake. On the W. are the Eastman Ponds, under the steep and forest-covered slopes of Piermont Mt. It is 3½ M. from this point to Warren, and 9 M. to the village of Piermont, on the Connecticut River.

In 1830 the mineral deposits on Ore Hill were discovered (copper, silver, and zinc), and ever since that time unavailing efforts have been made to develop the treasures. Over $ 100,000 have been expended in mining operations, and the deserted adits form a large cavern, which is frequently visited when the water will allow.

"W. of Mt. Mist, and kissing its sloping base, a crystal sheen in an emerald setting, is **Tarleton Lake**." It is soon reached by the road which descends Ore Hill to the N. W., which also passes near the Eastman Ponds. It is about 1½ M. long, and abounds in fish. On the S. W. rises the massive and forest-covered Piermont Mt., 2,167 ft. high.

Tarleton Lake is named for Col. Wm. Tarleton, who opened a tavern here in 1774, by the side of " the busiest and most travelled thoroughfare in all northern New Hampshire." There was a church here, and also a school, with a little hamlet called *Charleston*, but on the opening of the new highway to the N. this settlement decayed and it is now extinct. "The dwellers in the district by the lake are all dead, the houses and the barns have mouldered away, the spot where they stood can hardly be found, and the fields and the pastures are grown with forest trees. Even the old school-house, the church in Charleston, is gone. Nothing but the foundation remains. The burying-ground by it is overgrown ; the thistle shakes its lonely head by the tombstone, the gray moss whistles to the wind, the fox looks out of its hole by the sunken graves, and the wood-brakes and the birches wave above them."

The perfect harmony between the English settlers and the Indian tribes under Passaconaway was continued under the administration of Wonnalancet, his son and successor. But when King Philip's war broke out, the Merrimac confederation was visited by envoys both of Philip and of Massachusetts, each of whom demanded aid and warriors. Wonnalancet, bound to one of the combatants by the fellowship of race, and to the other by long friendship and religious brotherhood, declined to join either side, and held his warriors to a strict neutrality. Having been driven by the Massachusetts men successively from Pawtucket (Lowell) and Pennacook (Concord), he patiently led his people into exile, passing up the Asquamchmmauke Valley to the mountains, " where was a place of good hunting for moose, deer, bear, and other such wild beasts." Here he was joined by the Sagamore Numphow and the Christian teacher Mystic George, with the remnant of the Wachuset tribe, after a long and bitter march, wherein many of their women and children died. Numphow (who was called by Eliot " a Prince of the Bloud, a Man of a real Noble Spirit") and Mystic George also died here, and were sadly buried on the banks of Baker's River. But the valiant sagamores Monoco and Shoshamin, heroes of the long Mohawk wars, stung to revenge by the base policy of the English officers, left their pacific chief, marched into Massachusetts, and utterly destroyed Lancaster and Groton. At the end of the war (in 1676) Wonnalancet led his people back, and was soon made the victim of Major Waldron's treachery. Of the 400 Indians thus captured at Dover, in a time of peace, 100 were put to death, 200 were sold into slavery in the West Indies, and 100 were released. The liberated Pemigewassets returned to their country " firmly convinced that their pale-faced entertainers were the most honest, reliable, and pious set of cut-throats with whom they ever had the happiness to become acquainted "

During the administration of Wonnalancet, the Pemigewasset tribe was governed by Waternomee, a skilful hunter and a heroic warrior. He was with Kancamagus in his dashing attack on Dover (1689). At the outbreak of Queen Anne's War he put his tribe in array against the English, sending the women and children to the safe fastnesses of the mountains. He was killed in the battle near Squam Lake. Acteon was the last sachem of the tribe.

It is said that the name of this town was given in honor of Admiral Sir Peter Warren, who commanded the fleet in the attack on Louisbourg, and afterwards annihilated a powerful French squadron off Cape Finisterre. He married one of the New-York De Lanceys, and had rich estates in the Mohawk Valley. Other theories as to the name are that it was derived from that of an English borough ; or that the surveyors of the King's Woods had reported the town to be full of rabbits, and " all the jolly hunters say that the woods are full of moose, deer, bear, and other game, that wild ducks swim on the rivers and ponds, and that every stream is alive with the speckled trout and golden salmon." It has been claimed also that the name was given in honor of Dr. Joseph Warren (then a mere stripling), who afterwards fell at Bunker Hill.

Heber C. Kimball was a bar-tender in a Warren tavern. In later years he became a so-called " apostle " of the Mormons and " the high-priest of the order of Melchizedek " ; and on dying, left a large fortune to be divided among his 30 sons and 11 daughters. M. H Bixby, the Baptist missionary to Burmah, was also a native of this town, and was educated at the East-Parte school, with Kimball.

125. Moosilauke

is the highest elevation in New Hampshire W. of Mt. Lafayette, and is 4,811 ft. above the sea. Its name was given by the Indians from *moosi,* "bald," and *auke,* "place," with the letter *l* inserted for euphony. On Belknap's map of 1791 it was called *Mooshelock,* which has since degenerated into the absurd name of *Moosehillock.* On account of its isolation, and also because fogs or clouds rarely settle here, Moosilauke is one of the best view-points in the State, giving a noble grouping of the Franconia and White Mts., and overlooking the rich Connecticut Valley for leagues. In the adjacent forests there still remain a few deer and bears, sables and wild-cats, squirrels and hedge-hogs.

The mountain is composed of a high and pointed S. peak and a broad plateau on the N., joined by a narrow ridge and flanked by wooded foot-hills. The plateau is covered with loose stones, and has but little slope. The Prospect House stands on its S. side. The mountain is separated from the Blue Ridge by the gorge in which rises Baker's River; and from Mt. Clough, on the N. W., by a low and traversable pass.

The old bridle-path on the W. leaves the Benton road over 1 M. from the Warren-Summit station. It is rarely used. Another path is on the N. running from N. Benton, and striking into the woods above the upper saw-mill on Tunnel Brook. This is a long and tedious route, but is frequently used by the Benton and Landaff people.

On the E. side of the mountain is the deep gorge of the *Jobildunk Ravine*, in whose upper part are the woodland beauties of the **Seven Cascades**. This ravine is one of the wildest places in the State, but is difficult to traverse on account of its forests. It has, however, been explored by several parties.

N. Merrill & Son's summer boarding-house is at the foot of the mountain, 5-6 M. from Warren, on high ground which overlooks Mts. Water-nomee, Cushman, Kinneo, and Carr. The mountain carriage-road passes from this point to the Half-Way Spring, 2 M. 13 rods; the Cold Spring, 3 M. 121 rods; and the Prospect House, 4 M. 125 rods. The ascent is not difficult, and the road is frequently repaired. It leads up near the Jobildunk Ravine (on the r.) and emerges from the woods some time before reaching the crest. The ridge is met a little way N. of the S. peak, and is thence followed to the N., with broad views on either side.

" Soon we are out on the bald mountain ridge that connects the two peaks ; on either hand are wild and hideous gorges, 3,000 ft. down into the depths below. Beyond, to the W., is the bright valley of the Connecticut, garden land, with silver river ; to the E. the dark ravine of the Asquamchumauke, filled with the old primitive woods, where the trees for thousands of years, like the generations of men, have grown, ripened, and died." (Little.)

The practicability of a winter-residence on the mountains and the coincident study of atmospheric phenomena, was first demonstrated by Prof. J. H. Huntington and Mr. A. F. Clough, photographer. These gentlemen laid in a large stock of provisions on the summit of Moosilauke, and then settled down in the house there, where they remained through the months of Jan. and Feb., 1870, encountering several terrific storms and being nearly buried under vast snow-banks.

The **Prospect House** was opened in 1860, when over 1,000 people were present. It has been closed for two years, and is much dilapidated, but the proprietors promise to put it in order in 1876. It is a low and massive stone building, whence superb views are gained over the distant mountains and valleys.

The great divinity of the Pemigewassets made his home on Moosilauke, "and the Indian had a bold spirit who dared to climb the bald crest of the mountain. They heard him in the voices of the storm and the mighty torrent, and in the thunder that muttered in the dark gorges and rumbled low over the crests. They saw him in the rosy hue that kindled on the peaks in early morning, or in the sharp flash of the lightning that leaped from the murky clouds. To him they sacrificed. The first fruits of the chase, the early green maize, the golden salmon, the wild duck, the goose, and the partridge were their offerings."

There is a tradition that once the sagamore Waternomee, with a party of his men, followed the Asquamchumauke to its source, and "there they camped beside a beaver pond, where the beaver, Tummunk, had built houses. These they did not molest, but set out, just as the sun rose, to go over Moosilauke to the Quonnecticut valley. Not often did the Indians climb the mountain, and they only did it now to save time and distance. It was a hard ascent for their moccasined feet, over the stones and through the hackmatacks, as they called the dwarf firs and spruces; but upon the bald mountain crest the way was easier, and the little birds, Psukses, were whistling and singing among the lichens and rocks. When they reached the summit, the heaven, Kesuk, was cloudless, and the view unobscured. It was a sight, the like of which they had never seen before. Great mountains, Wadchu, were piled and scattered in the wildest confusion in all the land; and silver lakes, Sipes, were sparkling; and bright rivers, Sepoes, were gleaming from the forest. As they sat upon that topmost peak, the wind was still, and they could hear the moose bellowing in the gorges below: could hear the wolf, Muquoshim, howling; and now and then the great war-eagle, Keneu, screamed and hurtled through the air. A feeling of superstitious reverence took possession of these Indians as they drank in the strange sights and wild sounds, for they believed that the peak was the home of Gitche Manito, their Great Spirit. Does the unlettered Catholic have reverence at the altar? — much more was the untutored savage filled with awe as he stood in the very dwelling-place of his God, afraid that the deity would be angry at the almost sacrilegious invasion.

"As the sun, Nepauz, was going down the western sky, a light mist collected around the eastern peaks, and above all the river valleys in the west, clouds, at first no longer than a man's hand, began to gather. Soon hanging over every valley was a shower — the heaven above them clear — the sun shining brightly upon the vapor. Quickly the wind freshened, and the great clouds, purple and gold and crimson above, black as ink below, hurried from every quarter towards the crest of Moosilauke. Then thunder, Pahtuquohan, began to bellow, and the lightning, Ukkutshaumon, leaped from cloud to cloud, and streamed blinding down to the hills beneath, while the great rain-drops and hailstones, crashing upon the infinite thick woods, sent up a roar loud as a hundred mountain torrents. 'It is Gitche Manito coming to his home angry,' muttered Waternomee, as with his companions he hurried down the mountain to the thick spruce forest, Soshsumonk, for shelter." (WM. LITTLE.)

One of the saddest of the episodes in that terrible anabasis of New England, the retreat of Rogers's Rangers from St. Francis, is connected with the gloomy defiles of Moosilauke. While the remnants of the detachments were encamped on the Newbury meadows, two soldiers started off from their famished and dying comrades to seek game in the woods on the E. They reached the summit of Moosilauke, and then one of them, Robert Pomeroy, of Derryfield, sank down and died, exhausted, starved, and frozen. Here his remains were found, many years afterwards. His comrade wandered hopelessly down the E. side of the mountain, and soon fell in deadly stupor at the foot of the Seven Cascades. The legend further relates " how the ranger seemed to be dying; and when the stars shone bright above him and the moon looked in through the trees and lighted up the white foam of the cascades, distant music coming nearer seemed to mingle with that of the water, and his quickened senses heard fairy harps joined with fairy voices, and saw fairy feet dancing in the silver spray. Elfin kings and fairy queens whirled in the mazy dance for a moment and were gone. And then came a troop of nereids, with long dishevelled hair, and eyes lustrous as the stars that shone above them, to bathe in the clear crystal fountain. For an instant they seemed to hold sweet dalliance with the sparkling water, and then floated away in the thin mist that hung over the great wood and turbaned the distant mountain. Day seemed breaking, and the bright sun looked in from over the distant hills upon a crowd of mountain genii, who chanted their matin-hymns in their wild rock-hewn temples, and then mounted upon viewless steps to offer incense on their rainbow altars, golden in the flood of rosy light, and glistening in the diamond drops of the waterfall. As a dark cloud stole across the sky, veiling the moon, the scene changed. The shrieks of the dying Indians at St. Francis, the mournful peal of the chapel bell, the retreat, the famine, the terrible feast upon human heads, the dying comrade upon the mountain-top, himself perishing by the torrent, — and then, seen for a moment, the picture of a dark form bending over him, — and the famishing ranger was unconscious." (LITTLE.) He was found and brought to life by an old trapper who had followed his trail, and he afterwards reached the settlements in safety

* * *The View.* — About N. N. W. and N. W. are the stately Franconia Mts., with their sharp and alpine outlines cutting the sky-line and their deep ravines fully exposed to view. In the N. N. W. is Mt. Kinsman, a long and thickly wooded mountain with several heads, over which numerous other peaks are seen. Peeping over its r. centre is the white crest of Mt. Cannon, which runs up on the W. of the Franconia Notch and is covered with light-colored ledges. Over the r. ridge of Kinsman is the lofty and imposing serrated ridge of Mt. Lafayette, more than 1½ M. long, but foreshortened by the angle at which it is seen. The N. peak is seen on the l., then the highest crest, and then the sharply cut S. peak, beyond which opens the profound White-Cross Ravine. Near the S. peak is the culmination of Mt. Lincoln, whence a long and sweeping flank-line runs down at a high incline, before the dark ravine to the N. From the r. of Haystack a graceful curving ridge sweeps down and then ascends to the pyramidal crest of Mt. Liberty. The next peak on the r. is Mt. Flume, which assumes a pointed form beyond the ravine in which the Flume is located. Big Coolidge Mt. is to the r. of this, and is marked with white slides; and Little Coolidge and Potash Mts. lie still farther to the S., where the lower flank of the Franconia Range falls away to the East-Branch valley. Between Moosilauke and these mountains are the wooded ridges of the Pemigewasset Range, with Mt. Pemigewasset at the r. of Kinsman and below the great curve between Lincoln and Liberty.

Directly over the Franconia Range is seen the noble assemblage of mountains which compose the Twin-Mt. and Presidential Ranges. The N. Twin is concealed by Lafayette and Lincoln, but the S. Twin is seen, running to the r. from Mt. Liberty and culminating in a well-marked peak; and Mts. Guyot and Bond come next on the r., their crests being to the l. of the summit of Mt. Flume. Over the ridge which joins the S. Twin and Mt. Bond is the low-sided and symmetrical pyramid of Mt. Adams with a part of the ridge of Mt. Madison on its l.; and the high point of Mt. Jefferson is seen on the r., over the l. of Mt. Liberty. Over the l. of Bond and the r. of Liberty towers the majestic cone of Mt. Washington, with its houses plainly outlined against the sky, and with the rock-bound plateau of Boott's Spur running to the r. The mountain-chain which runs S. W. from Washington is foreshortened by the angle of vision, and resembles a gigantic stairway. Mts. Jackson and Clinton are seen in advance; the dome of Mt. Pleasant is directly over the sharp ridge which runs S. W. from Mt. Bond; and a trifle to the l. are the fortress-like crags of Mt. Monroe.

Over the r. ridge of Mt. Flume is Mt. Willey, falling off sharply to the r. and having the presidential peaks over it and the brightly striped side of Mt. Webster on the r., showing the many-colored slides in the White-Mt. Notch. To the r. of Mts. Willey and Washington, and at the sky-line, are Mts. Carter and Wild-Cat, near the Glen House, towering over the long Montalban Ridge. Above the latter ridge, r. of Wild-Cat, over the nearer Big Coolidge Mt., and cutting the sky-line, are the N. and S. peaks of Baldface, the latter of which has great white ledges on the S. To the r. of these is Mt. Eastman. The view-line now reaches a singular group of mountains, resembling two long terraces inclining to the l. and sharply cut off on the r. The upper one is Giant's Stairs, and the lower is Mt. Nancy, over whose r. end is one of the peaks of Double-Head.

The eye next follows up the East-Branch valley, by the Pollard farm, into the great wilderness of the Pemigewasset Forest, bristling with unbroken woods and broken by long dark ridges.

The long high ridge nearly E. by N., standing across the East-Branch valley about 12 M. distant, is Mt. Hancock, which is marked by a bold precipice on the S. The fine peaks of Mt. Carrigain are seen over the top of this mountain, but all its flanks are hidden. To the r. of the S. cliff of Hancock the view includes Kiarsarge, lifting its graceful cone at the end of a long section of the Saco Valley and crowned with a hotel. To the r. and nearer is the N. extension of Kancamagus, between the head-waters of Swift River and Sawyer's River. Farther to the r. is the flat top of Green's Cliff, with the sierra of Bear Mt. over and beyond it.

The immense mass of Mt. Osceola is just S. of E., with long high ridges running to the W. and S. Just to the l., over Table Mt., and very far away, are the conspicuous N. and S. peaks of Mont Mt. (near N. Conway), distinguishable by their brilliant red ledges. To the r. of Osceola is Tri-pyramid, with its three peaks, the long white slide on the S. peak being plainly seen, and also a smaller slide on the middle peak. Over the l. shoulder of Tripyramid are the more distant N. peaks of Chocorua; and just to the r. of the middle is the crest of Passaconaway.

The Fisher range is to the r. of Osceola, with the high gray crest of Tecumseh near the centre, Green Mt. on the r., and Mt. Avalanche on the l. Green Mt. divides on the W. into the ledgy spurs of Fisher Mt. and Welch Mt., between which is a shadowy ravine. The dark crest of White-face is seen beyond Tecumseh, and the long and massive Sandwich Dome looms over Welch Mt. To the r. of the latter and opening into the Pemi-gewasset Valley in the foreground is a long section of the Mad-River Val-ley, with Campton Village in the centre, backed by the Campton Mts., among which is seen the bare top of Mt. Weetamoo. Immediately to the

r. of Sandwich Dome is the distant peak of the Green Mt. in Effingham, on whose r. is the long and diversified Ossipee Range. Across the valley, and in line with the highest Ossipee peak, is Mt. Israel in Sandwich; and to the r., over the Squam Range, is Red Hill, near Centre Harbor. Long reaches of Lake Winnepesaukee are seen in the S. E., with its shining levels in many places nearly hidden by little archipelagoes or far-projecting and fertile peninsulas. Over the l. slope of the S ridge of Moosilauke, and beyond Winnepesaukee, is Copple Crown Mt., near Wolfeborough, with long and gradually sloping sides ascending to a dull point. On the l. of this summit is Smith's Pond. The ocean is sometimes seen in this direction. W. of Winnepesaukee are the white-topped Suncook and Belknap Ranges, beyond Gilford; and nearer at hand, in line with the Belknaps, is the round top of Mt. Prospect, in Holderness.

About S. S. E., above the New-Hampton hills, is a portion of Lake Winnesquam, and the Bridgewater Hills lie more to the r. Over the S. ridge of Moosilauke, and near at hand, are the wooded heights of Mt. Cushman, with Mt. Kinneo to the r., and Mt. Waternomee nearer still. Through the gap between the two former summits is a portion of Stinson Pond, with Mt. Stinson and Plymouth Mt. beyond. Far away over the r. flank of Stinson are the twin peaks of the Uncanoonucs, many leagues distant down the populous Merrimac Valley, on whose r. is Joe-English Hill, in New Boston; and Wachuset is still more distant. Nearly due S. is Mt. Carr, over which are the Ragged Mts., with Crotched Mt. in Francestown yet more remote. The shapely peak of the true Kearsarge is almost S., rising boldly over the lesser ranges, with Pack Monadnock far off on the r. S. of the latter, and even more distant, is the flattened curve of Monadnock, with Lovewell's Mt. and Sunapee Mt. on the r., and the high nearer peaks of Mt. Cardigan on the l.

The pleasant valley of Baker's River stretches away to the S., and part of the hamlet of Wentworth may be seen at its apparent end, with Croydon Mt. far beyond. The view is interrupted for a short arc by the high and near S. peak of Moosilauke, from beyond whose r. springs the bold ridge of Smart's Mt., with the sharp crest of Ascutney beyond. Just to the l. of the S. peak is Moose Mt., at Hanover, over which is a remote blue dot on the horizon, which some claim to be Greylock, in Western Massachusetts. The "Plan of Directions" of the U. S. Coast Survey locates Greylock and Saddle Mt. between Ascutney and Cuba; but this line would apparently fall to the W. of Greylock. Mt. Cuba is due S. W., and its shapeless top and long flanks are visible, far beyond which are Mt. Æolus and Mt. Equinox, on r. and l., down in S. Vermont. At the foot of Cuba the long Baker's Ponds are seen, and to the r. is the bright Tarleton Lake, at the base of Piermont Mt. Over the latter are the Peaked Mt. (or Piermont Pinnacle) and Sunday Mt., in Orford, and Fairlee Lake is over the r. of Tarleton Lake. To the l. of Piermont Mt.,

T

very far away (S. 60° W.), are the Killington and Shrewsbury Peaks, near Rutland.

Toward the W., and below, is Blueberry Mt., of which Owl's Head is a lower spur; and the Vermont villages of Bradford and Corinth are seen beyond. The view now embraces the Oliverian valley, with Webster's Slide on the l. E. Haverhill is a small near hamlet on the W., beyond which are the umbrageous streets of Haverhill Corner and the farms of S. Newbury. Above and beyond the Green Mts. sweep away for many leagues, and form a broken horizon-line. Through a deep and distant gap to the r. of E. Haverhill looms the blue crest of Mt. Marcy, the chief of the Adirondack Mts., over the plains which lie W. of Lake Champlain. Portions also of Mts. McIntyre and Wallface may be identified when the sun is in the right position.

The foreground is occupied by a broad and beautiful expanse of the meadows about the Connecticut River, with the fruitful levels of the Great Ox-Bow, the blue sinuosities of the river, and the idyllic hamlet of Newbury, on the terrace beyond, marked by its large white hotel and backed by a fine hill-country. Close at hand below are the bare-topped mountains of the Benton Range, — Blueberry, with a plateau-summit; Sugar Loaf, with its almost inaccessible white turret; and Black Mt., with ascending rocky stairways. The latter is nearly N. W. of Moosilauke, and is separated from it by Mt. Clough. The hamlet of N. Haverhill is over Sugar Loaf and on the r. of Newbury; and the prosperous adjacent villages of Wells River and Woodsville are to the r. of Black Mt.

Beyond the Connecticut Valley, many leagues away toward Lake Champlain, are the stately peaks of the Green Mts. About W. N. W., to the r. of and far beyond Newbury, is the lofty apex of Camel's Hump, over the mountains of Moretown and Duxbury, with the Fayston peaks on the S. To the N. W., out on the horizon, is the long and many-headed ridge of Mt. Mansfield, beyond Worcester Mt. and the highlands of Washington County, with the Elmore and Sterling Mts. on the r. To the r. of and beyond the white crags of Black Mt. is the long ridge of Mt. Gardner, running N. E. from Wells River and Woodsville; and more distant in this direction are the Belvidere Mts., in Northern Vermont. Farther out, over N. Benton, are the Lowell Mts., making a dim broken line on the horizon, overlooked by the remote Jay Peak, which bends over on the N. Still more remote, on the r., are the Canadian peaks of Owl's Head and Mt. Orford, on the shores of Lake Memphremagog.

The view next dwells on the adjacent highlands and hamlets of Landaff and Lisbon, far over which is the singular cleft of the Willoughby Notch, where Willoughby Lake lies (invisible) between Mts. Hor and Annanance, with the sharp Westmore Haystack on the r., and the Burke and East-Haven ranges. Over the Littleton hills are Mts. Niles and Tug and Cow

Mt. Above E. Landaff is the dark ridge of Dalton Mt., with the Bow-back and Sugar-Loaf Mts. of Stratford still farther out. Somewhat nearer are Mts. Prospect and Pleasant, rounded summits near Lancaster; and the white Percy Peaks appear over the highlands of Bethlehem, flanked on the r. by the great cluster of the Pilot Mts. The view now rests on the massive flanks and curving crest of Mt. Kinsman, near at hand across a wilderness-valley. Mt. Starr King is over the r. of Kins-man, with Aziscoös and other lofty peaks along the Maine border beyond.

"Far to the E. a vast expanse of forest stretches away over hills thickly covered with hemlock and spruce, to the purple islands of Lake Winnepesaukee, which is distinctly visible. Westward, the prospect is bounded by the rolling ranges of the Green Mts. Southward, hills rise over hills, far as the eye can see; and to the N., the Franconia Mts. and the more majestic peaks of the White Hills rise and lose themselves in the clouds. The broad valley of the Connecticut is the charm of this landscape. Its numerous villages, its hundreds of farms and orchards, and all the tributary streams that swell the river, may be traced in their devious windings by the naked eye." (BARSTOW'S *Hist. of N. H.*)

"Across the Connecticut River to the S. W. is Ascutney, and beyond it, farther down, is Saddle Mt., Greylock, and Berkshire Hills, in Massachusetts. Then wheeling around towards the N. are Killington Peaks, sharp and needle-like, shooting up above the neighboring hills; farther N. and directly W., is Camel's Hump, unmistakable in its appearance; then Mt. Mansfield, towering above the thousand other summits of the Green Mts. Above and beyond them, in the farthest distance, are counted nine sharp peaks of the Adirondacks, in New York, Mt. Marcy higher than all the rest. To-morrow morning at sunrise you will see the fog floating up from Lake Champlain, this side of them. In the N. W. is Jay Peak, on Canada line, and to the r. of it you see a hundred summits rising from the table-lands of Canada.

"Nearest and to the N E. is Mt. Kinsman, the Profile Mt; and above and over them Mt. Lafayette, its sides scarred and jagged where a hundred torrents pour down in spring, its peaks splintered by lightning. S. of this, and near by, are the Haystacks. Over and beyond the latter are the Twins, more than 5,000 ft. high; and just to the r. of them Mt. Washington, dome shaped and higher than all the rest. Around this monarch of mountains, as if attendant upon him, are Mts. Adams and Jefferson, sharp peaks on the l., and Mt. Moriah, the Imp, Mts. Madison and Monroe, Mt. Webster, the Willey-Notch precipice, Double-Head, and a hundred other great mountains standing to the r. and front.

"A little to the S. is Carrigain, 4,800 ft. high, black and sombre, most attractive and most dreaded, not a white spot nor a scar upon it; covered with dark woods like a black pall, symmetrical and beautiful, the eye turns away to return to it again and again. Mt. Pequawket in Conway, its neighbor, always seems gray in the hazy distance, Chocorua rises farther S., and Welch Mt., Osceola, Whiteface, Ossipee, Agamenticus, on the sea-coast, Mt. Prospect and Red Hill fill up the circle. This view to the N. and E. is the most magnificent mountain-view to be had on this side of the continent. The most indifferent observer cannot look upon it without feeling its grandeur and sublimity."

126. Mount Carr and Webster's Slide.

Mt. Carr is an extensive ridge which occupies portions of four townships, and attains a height of 3,506 ft. It is densely wooded even to the summit, and is therefore but rarely visited by tourists. "A hundred different kinds of rock are found upon it, and some most interesting minerals, among which are tourmaline or schorl, garnets, quartz crystals of a lovely hue, amethyst, beautiful as the summer rose, and last but not least are scattered all over it small particles of pure virgin gold."

The mountain derives its name from a man named Carr, who, a century ago, attempted to cross it from Ellsworth to Warren. He was caught in a terrific storm and laid out in the woods for two nights, sustaining life by eating frogs that he caught by the Glen Ponds. On reaching Warren and telling of his desperate adventure, the people bestowed his name on the mountain where he had so nearly found a grave.

" The grandest fires we have ever seen were the fires on Warren's mountains. Webster's Slide has blazed like a volcano. Owl's Head has burned for months, lighting up the heavens at night; Moosilauke has been wrapped in sheets of flame, completely enveloping its twin peaks, and Mt. Carr, twice within the memory of the present generation, has flashed from base to summit. It was in the summer of 1854 that the fire roared on Mt. Carr. Then a million trees burned to the wind. Then a sound came like the rushing of the tempest; like the mighty voice of the ocean. Its roaring was heard 6 M. away, and one could see to read fine print at midnight." (WM. LITTLE.)

Webster's-Slide Mt. (2,210 ft. high) derives its name from a tradition that a hunter named Webster was pursuing a deer there, when his dog and the deer suddenly reached a slanting cliff on one slope of the ridge, and were killed during the ensuing tumble. The mountain may be ascended from the Warren farm, 6 M. from Warren, on the Haverhill road. There is no reason why this journey should be made, as the ascent is arduous and the top is covered with dense woods which shut out the view. (The Editor wasted a broiling June afternoon in learning this fact.)

Mt. Mist (2,243 ft. high) is S. of Webster's Slide, and is also covered with trees. Near these ridges is *Meader Pond*, which is often visited by fishermen. It is 5 – 6 M. from Warren.

" High up in the N. E. corner of Warren is situated a pretty little sheet of water. The Indians called it *Wachipauka*, but the later generations of our mountain hamlet delight to term it Meader Pond. It is yet right in the heart of the woods, and from its E. shore springs a handsome forest-covered cape. On the N., Webster's Slide shoots sharp up 1,000 ft., its top crowned with silvery birch and waving pine ; the crannies of its rocks radiant with the blueberry, harebell, lichen, and other mountain flowers. On a warm summer day the water reflecting the rich foliage of the yet undisturbed forest is ruffled only by the great speckled trout jumping, or the wild duck swimming ; but when the autumn winds come the blue water curling smiles upon the mountain-face and laughs at the bald head of Moosilauke, looking in from the distance over the great wood." (LITTLE.)

On the shore of Wachipauka was the favorite camping-ground of the chieftain Waternomee. In 1709 Baker's Rangers encamped here, while on their way to destroy the Pemigewasset towns; and in 1754 the St. Francis warriors halted here, with their captives from Franklin. In 1758, Roger's Rangers encamped by the lake while on their way to the Lake-George campaigns. In 1767 the pioneers built a comfortable permanent camp here, as a station on the long trail from Concord to Haverhill.

When the town of Warren was granted, the royal governor, Wentworth, exercised his usual prerogative of reserving for himself 500 acres of land within its boundaries. But he was so ill-informed as to the territory that he marked as his share a tract which included Wachipauka Pond and the worthless cliffs of Webster's Slide.

127. Owl's Head and Blueberry Mt.

The Benton Range.

In the W. part of the town of Benton, and running nearly N. and S., is the chain of peaks which includes Owl's Head, Blueberry Mt., Hogsback Mt., Sugar Loaf, and Black Mt. Though not remarkable for altitude or mass, these summits are otherwise picturesque and interesting, and may be visited without great labor. The same town also contains the famous Moosilauke, another Black Mt. (now called Mt. Clough), and a part of the Blue Ridge. There are no accommodations for tourists here, and people who wish to explore the Benton Range must start out from Warren, Haverhill, or Newbury. The hotels at the latter points are better than that at Warren, and the difference in distance is small. Benton has but 375 inhabitants, and is famous for its quartz crystals and other minerals and ores.

Owl's Head is a spur of Blueberry Mt. to the S. W., and is faced by a fine precipice, several hundred ft. high, of purple and other dark-hued rocks. Thousands of bushels of blueberries are gathered yearly on this ridge. The ascent is made from the highway, near Warren Summit, and is steep, but short. A vague path conducts through the lower thickets, and along the face of the ridge which looks off on the cliffs. Large crystals of epidote are found about the cliff.

Chase Whitcher, a boy of 20 years, from Salisbury, Mass., settled in Warren in 1773, and devoted himself to hunting "At another time he was chasing a wild buck, which ran down on the rocky crest of Owl's-Head Mt. Whitcher heard the baying of his old bloodhound in the distance, at regular intervals, each time coming nearer and nearer, and cocking his rifle got behind a rock, thinking to shoot the stag as he passed. He did not have to wait long. The deer burst out of the thin woods 50 rods away, too far off for a shot, and bounded towards the edge of the precipice. He whistled on the old dog, following closely behind, whose three wild yells rang out regularly upon the clear mountain air, but could not make him bear. Neither deer nor hound heeded where they were going, and when they reached the brink of the mountain, in the excitement of the moment the hunter held his breath, as he saw the buck unable to stop, and the great black hound, intent on his prey, both leap far out over the edge of the precipice, then falling swift as lightning, disappear in the abyss a hundred fathoms down."

Blueberry Mountain is the name given to the fine peak N. of and above Owl's Head. It may be easily ascended from Owl's Head in less than an hour, although a quicker route for tourists who do not care to visit the latter summit is to go up the N.-Benton road to a point about 7 M. from Warren, and then strike up the E. flank. For about 1 M. from the summit the mountain is free from trees and is covered with alternate bands of carpet-like moss and granite ledges moderately inclined. The work of ascent and exploration is thus rendered easy and pleasant. There is but a slight depression between Owl's Head and Blueberry Mt., the former being a bold spur of the latter rather than a detached mountain. On the highest point of Blueberry Mt. is a signal-beacon of the U. S. Coast Survey (2,800 ft. above the sea).

The View from Blueberry Mt. is of great extent and beauty. On the E. is the immense ridge of Moosilauke, beyond the narrow ravine through which the N.-Benton road is carried, the hotel-building standing out in full relief. To the S. E. the wooded peak of Mt. Cushman (in Woodstock) is seen rising over the flank of Moosilauke; and the shaggy sides of Mt. Carr are farther down. The yellow spire of Warren village is almost due S., beyond which are the widening levels of the Baker's-River valley. Many leagues distant, over the hills at the foot of the valley (E. of S.), is the bold blue pyramid of Kearsarge. It is claimed (perhaps unreasonably) that Monadnock is also visible. W. of S. and near at hand is the sharp pitch-off of Webster's-Slide Mt., with Meader Pond set in the forest at its foot. Over its shoulder, and apparently connected with it, is Mt. Mist, beyond which are the great forests on Piermont Mt., with Eastman's Pond at their base. Mt. Cuba is the bold flat-topped ridge farther away, in the town of Orford; and the slender blue spire of Ascutney is many leagues beyond. The Oliverian Valley runs out to the W., soon merging into the rich and broad intervales of the Connecticut River; and within a few miles on the W. and N. W. are the white villages of E. Haverhill, Haverhill Corner, N. Haverhill, Newbury, and Bradford. In the distant W. is the vast line of the Green Mts. of Vermont, among which are seen the sharp peak of Mt. Mansfield (N. W.), the elephantine ridge of Camel's Hump, and two bold pyramids in the remote S. W., which may be Killington Peak and Shrewsbury Peak, near Rutland. On the N. is the Hogsback Mt., a long height bristling with dead trees, beyond which, and within 3 – 4 M., are seen the conical crags of Sugar Loaf and the symmetrical top of Black Mt., whose upper ledges resemble the walls of a fortress.

128. Sugar Loaf.

Sugar Loaf is to the W. slope of the mountains what Chocorua is on the E. and the Percy Peaks are on the N., — a sharp and conspicuous peak of light-colored rock, alpine in appearance, and easily recognized from a great distance. Although its height is but 2,565 ft., it will probably be a favorite point of attack when alpine exercise becomes popular in New England, on account of the fascination of its defiant cliffs, the exciting perils of the ascent, and the beautiful view from the summit.

Sugar Loaf may be reached from Newbury, Haverhill, or Warren by driving to the lime-kilns (13, 9, or 12 M.) in the glen at the foot of Black Mt. It is about 2½ M. to the summit, and the ascent is made on the S. W. side. Visitors should pass up the old road to the steam saw-mill, and thence diagonally upward across the pasture beyond and through a narrow strip of trees into the broad steep pasture on the S. W. slope. From the N. E. corner of this clearing the woods are entered and a line of bold ledges is soon reached and painfully passed. Thence the way (no path) lies through a fine belt of pine-trees, without undergrowth, which reaches to the foot of a long and formidable precipice, 80–100 ft. high. This barrier may be passed, at a point where a vast mass of rock at the top overhangs the base, by a narrow shelf which slopes upward diagonally along the face of the cliff, and is furnished with iron pins sunk in the rock at the most difficult points. A single false step on this path would hurl the incautious climber to instant destruction. It might be well for visiting parties to bring a long rope, which, being securely fastened at the top by the guide, would render the ascent more easy. From this point the way leads upward, to the r., by a chaotic grouping of evergreen thickets, fallen trees, and rugged ledges to the summit, which consists of three narrow-based piles of white rock, within a stone's-throw of each other. The cliffs fall away steeply on every side, and across the ravine to the N. is the vast pile of Black Mt., more than 1,000 ft. higher than Sugar Loaf.

The View from Sugar Loaf is limited on the N. by Black Mt., and on the E. by Mt. Kinsman, the Blue Ridge, and Moosilauke. Toward the S. the adjacent mountains of the Benton Range are overlooked, and portions of the glens and highlands of Warren are seen. The view to the W. is, however, the most beautiful, including many leagues of the rich Connecticut Valley, with its prosperous hamlets and fertile meadows. The villages of Haverhill are seen, and on the other shore are the white houses of Newbury. Farther away are the massive lines of the Green Mts., towards Lake Champlain, with many tall blue peaks cutting the horizon.

129. Black Mountain.

Black Mt. is in the N. W. part of Benton, and is the most northerly peak of the Benton Range. It reaches an altitude of 3,571 ft. From several of the Connecticut-Valley hamlets this fine peak is seen, boldly relieved against the sky, and forming the most conspicuous element in the landscape. It is separated from Sugar Loaf by the ravine of the North Branch, and descends on the N. to the Wild Ammonoosuc. The craggy character of the mountain and its fine views over the Connecticut Valley render it an interesting point for the tourist.

Black Mt. may be visited from Newbury, Haverhill Corner, or Warren, the distances to its base being 13 M., 9 M., and 12 M. respectively. The guides to this mountain are Parker Metcalf, of N. Haverhill Levi C. Whitcher and James Clement, of Warren. Carriages are left at the lime-kilns, S. W. of the peak and about 2½ M. from it. The route of ascent leads for a long distance through a steep pasture on the S. W. flank, and enters a low pine forest a few rods N. of the crags which hang over the ravine toward Sugar Loaf. A narrow but much-assisting path traverses this forest and the ledges above, leading to a group of beetling crags which seem to form the crest. On gaining this point, however, another pile of white rocks is seen beyond and far above, and as the weary but enthusiastic climber reaches this second needle, still another looms up ahead. From the latter the real peak is seen, its steep and symmetrical white walls resembling the Great Tower of Heidelberg Castle. The crest consists of a thin straight ridge several rods long, falling off rapidly on all sides, and crowned with a signal-beacon of the U. S. Coast Survey.

* *The view* from Black Mt. is of great breadth and grandeur. On the S. E. is the lofty bulwark of Moosilauke, below which are Mt. Kinneo and Mt. Carr. The sharp and craggy white peak of Sugar Loaf is close at hand across the ravine to the S., beyond which is the rounded ridge of Blueberry Mt. A portion of the Baker's-River valley is seen in the S., with the cone of Kearsarge at its end. Farther W. are Piermont Mt. and the pointed summit of Peaked Mt., in Piermont town, with blue Ascutney out beyond. In the W. is the Connecticut River, winding in wide curves through its fertile meadows, and backed by the Vermont ranges, rising from the river-hills to their culmination at Mt. Mansfield. Among the villages which are visible are Bath, on the Ammonoosuc; Woodsville and Wells River, about N. W.; N. Haverhill, with a white road running straight to the mountain; Newbury, distinguishable by its large white hotel; E. Haverhill and Haverhill Corner; and the Vermont hamlet of Bradford. On the N. and N. E. are the hills of Landaff, the outlying farms of N. Benton, the black mass of Mt. Kinsman, and the clear-cut spire of Mt. Lafayette. It is claimed that the Mt.-Washington range is visible from Black Mt., but of this the Editor is uncertain, since a rain-hurling gale which prevailed at the time of his ascent obscured the distant view and prevented the use of instruments and maps.

130. Haverhill.

Hotel. — Smith's, fronting the Green, and accommodating 30 guests. Boarding-houses. — G. W. Chapman (15 guests), Mrs. Bailey (15), Perley Ayer, A. S. Clifford, J. Woodward, and others in or near the village.

Distances. — To Newbury, 3½ M.; Piermont, 4½; Bradford, 6; N. Benton, 11½; Orford, 10; Warren, 12½; Black Mt., 9; Woodsville, 9; summit of Moosilauke, 13.

Railroads. — The B., C. & M. (Route 2) station is near the village. The Passumpsic R. R. (Route 10) is across the Connecticut River, in Newbury, and semi-daily stages run to its station at S. Newbury.

Haverhill (or *Haverhill Corner*) is a quiet and beautiful old village, standing on the edge of a plateau 150 ft above the Connecticut and viewing the intervales and the mountains. It is divided into two sections, that to the N. being composed of mills and their adjuncts, along the Oliverian Brook, while the village proper is farther S. It is built around a pretty and well-kept common, oblong in shape, enclosed by a fence, and shaded by large trees. The hotel, the village shops, the Congregational church, and the Haverhill Academy (incorporated in 1794) front on the common, together with several pleasant homes. On the street running out to the

S. E. is the antiquated semi-classic court-house of Grafton County, with the grim jail beyond. It is said that but one house has been built in this village for 25 years. The society is mainly composed of the professional element, and is dignified and conservative.

The tranquillity of Haverhill, its rural beauty, and the many interesting localities in the vicinity have combined to make it a favorite summer-home for those who love quiet and restfulness. The intervales form the distinctive feature of the scenery, and are green, level, and highly fertile meadows through which the broad Connecticut winds in long and grace-ful curves. They are best seen by driving through N. Haverhill, Woods-ville, and Newbury in a long circular route around the Ox-Bow Meadows. Pleasant scenery is also opened by going S. on the Piermont road, return-ing, perhaps, through Bradford, on the Vermont side, or by passing S. E. to Warren, by Tarleton Lake and over Ore Hill. The Piermont road is very picturesque in its lower sections.

The view from certain parts of the village includes the winding river and its broad intervales for 12 M., with the gracefully curving hills which rise towards the higher ridges. A little more than 1 M. S. is *Catamount Hill*, whence the great amphitheatre is surveyed for over 20 M., including the Ox-Bow and the rich Piermont meadows. Beautiful views are also given from *Powder-House Hill*, N. of the village; from the cemetery, near the station; and from the S. part of the village.

The mountains of the Benton Range are accessible from this point by the road running through E. Haverhill and thence to the N. It is 8–9 M. to the lime-kilns at the foot of Black Mt. and Sugar Loaf.

E. Haverhill is a cluster of houses in the S. E. part of the town, on the railroad. *N. Haverhill* (Eagle Hotel) is in the W. part of the town, on the railroad and near the intervales, over which pleasant views are gained. It is built on a straight and level street nearly a mile long, and contains two old churches. Parker Metcalf (of the village inn) is a guide to the mountains of the Benton Range.

The town of Haverhill contains 2,270 inhabitants, on an area of 34,340 acres, of which over 24,000 acres are under cultivation. It is one of the best farming towns in the State, its yearly product being 7,000 tons of hay, 70.000 bushels of potatoes, 34,000 bushels of oats and barley, 20,000 bushels of corn and wheat, 36,000 pounds of wool, and 11.000 pounds of maple-sugar, — aggregating nearly $ 300,000. The intervales are broad and rich, being composed of deep loam, annually refreshed by the spring floods. Lumber and other manufactures are also carried on with profit, on the water-power of the Oliverian Brook. Soapstone, scythe-stones, limestone, and granite are found and quarried in the town.

131. Newbury.

Hotels. — The Spring House is a large and comfortable hotel on the plateau, commanding from its cupola a noble view of the meadows, the Benton Range, Moosi-lauke, and Mt. Lafayette. It is lighted with gas, and has a livery-stable. 100 guests can be accommodated here, at $7 - 12 a week each. The boarding-house at the Springs accommodates 30 – 40 guests, and charges $8 - 10 a week. Private sum-mer boarding-houses are also kept by Messrs. Farnum and Richard Dole.

14

Distances. — Newbury to the Profile House, 25 M. ; Moosilauke, 16 ; Black Mt. and Sugar Loaf, 13 ; Haverhill, 3½ ; Wells River, 5 ; Warren, 15 ; Bradford, 7¼ ; Mt. Pulaski, ¼ ; Wright's Mt., 8 ; Wells River, 5.

Routes. — The Passumpsic R. R (Route 10) has a station at the village ; the B., C. & M. R. R (Route 2) passes through Haverhill and Woodsville.

Newbury is a neat and pleasant hamlet on the high terrace over the Connecticut intervale and near the great bend of the river. It contains Congregational and Methodist churches, the Newbury Seminary (on the Green), and the Montebello Ladies' Institute (at the Springs). The most notable characteristic of the hamlet is its remarkable quietness and tranquillity. Even the railroad trains fail to break this charmed silence, since they are far below the plateau, whose riverward spur is pierced by a tunnel. This is one of the four villages of Newbury town, which contains 2,241 inhabitants (as against 2,984 in 1850).

The chief glory of Newbury is the view of the rich meadows and the mountains beyond, which is obtained from various points in the village and on the adjacent roads. This is one of the noblest panoramas in the mountain-region, and includes by far the richest foreground and a line of craggy and picturesque peaks. Mt. Gardner is on the N., and is followed to the r. by Kinsman, Lafayette, the sharp spires of Black and Sugar Loaf, Blueberry and Owl's Head, Webster's Slide and Mt. Mist, and the immense ridges of Moosilauke. This view has been highly praised in *Picturesque America.* It is best enjoyed towards evening, by walking out on the meadows towards the mountains, and watching the rich changes of color on their rugged sides. The scenery about Newbury has been called Tyrolese in character ; it is better than that, because its foreground is composed of one of the fairest of the New-England intervales, than which Nature affords no more beautiful sight.

The intervales consist of low meadows along the Connecticut River, whose annual overflow keeps them permanently rich and productive. They are worth about $ 125 an acre, and cover several miles. The largest of them is the great Ox-Bow, which contains 450 acres, and derives its name from the long curve by which the river nearly insulates it. In their beauty and richness, and in the impressiveness of their environing walls, these meadows are not inferior to the more famous ones of Conway and Campton.

" When the freshets of spring have swelled the river to a flood, it overflows the banks, and what was a valley now seems a lake ; but when mantled in summer green or covered with golden harvest, everything growing with rank luxuriance, the meadows present the appearance of a vast plantation, shaded here and there by majestic trees, and waving with the richest crops. When the frosts of autumn have given to the woods those varied hues which constitute the peculiar charm of American forest-scenery, this valley presents a picture of many miles in extent, where, in the many-colored woods, the red, yellow, and russet brown are interspersed and blended in those rich and diverse shades, which, as they are never seen in Europe, are the wonder of European travellers." (BARSTOW's *Hist. of N. H.*)

" The expansion, in which Haverhill and Newbury are built, is seen from this point with the highest advantage. As we cast our eyes up and down the river, itself an object extremely beautiful, and with its romantic meanders extensively in view ;

a chain of intervales, sometimes on one and sometimes on both sides, extending from N. to S. not less than 10 – 12 M , spread before us, like a new Eden, covered with the richest verdure, and displaying a thousand proofs of exuberant fertility. This spot was bounded on both sides by rising grounds: now sloping; now abrupt; always interesting; and overspread alternately with forests, farms, and villages. Beyond these a train of hills, throughout the whole extent, adorned with a variety of summits, and terminating at the S. W. limit in the noble bluff of Sawyer's Mt., formed an elegant transition from the inferior to the superior parts of the landscape. The birds around us sported and sang with the highest glee. A vast multitude of neat cattle, horses, and sheep were cheerfully cropping the verdure of the rich fields beneath us, wandering about them in frolic, or quietly ruminating in the shade. The farmers were gayly pursuing the various business of the field; and the children, more gayly still, were occupied at their play. Over against us, in full view, rose the handsome village of Haverhill. 3 - 4 M. N. E., ascended two rough mountains; one an obtuse cone; the other a beautiful sugar-loaf; and with their peculiar forms finely varied the landscape. Behind the village of Haverhill, at the distance of 7 – 8 M., Moosilauke, a stupendous elevation, always reminding me of that description in Milton,

> ' The mountains huge appear
> Emergent, and their broad, bare backs upheave
> Into the clouds : their tops ascend the sky ';

rose to the height of 4,500 ft., covered with snow on the summit, of a dazzling whiteness, but gray, and grisly, as the eye descended toward the base. To finish the prospect, a chain (Franconia) which I have mentioned as skirting the White Mts. on the S. W. ascended at a distance of 40 M N. W. of Newbury, clothed in the most brilliant attire of January. This contrast of summer and winter, of exquisite beauty and the most gloomy grandeur, had the appearance of enchantment; and left an impression which can never be forgotten, until every image shall fade out of memory." (Dwight's *Travels in New England.*)

The **Montebello Sulphur and Iron Springs** are just E. of Newbury, on the edge of the intervales, and are adjoined by a large modern boarding house ($ 8 – 10 a week) which is open from May 1 to Nov. 1. The bathing-establishment is provided with copper bathing-tubs, through which pass steam-coils by which the sulphur-waters are heated without expelling the gas. The baths cost 50 c. each, and 75 c. a week to the bath-attendant. There is a resident physician in the house. The springs have been used medicinally for over 70 years; and are said to resemble those of Harrowgate and Castlemain. In each pint of the sulphur-water there are 2.2 grains of lime, 1.1 of silica, .5 of carbonate of soda, .3 each of magnesia and sulphate of soda, and traces of other minerals. It is claimed that the springs are beneficial to persons afflicted with " all cutaneous affections, and scrofulous complaints; bronchial difficulties and catarrh; derangements of the liver and urinary organs; dyspepsia; chronic rheumatism; chlorotic complaints, nervous affections of females, irregularities and functional derangements of the uterine system; general debility, poverty of blood, head affections resulting from intense study or overwork; soreness and weakness of the eyes; impotency and sterility; syphilis. The iron springs are tonic in character, and enrich and replenish the blood."

The proprietor of the Springs is William Clark, A M., for many years U. S. Consul at Milan, and afterwards founder and principal of the seminary at Bebek, on the Bosphorus. The visitor will be surprised to find in this sequestered mountain-hamlet a small but invaluable collection of authentic paintings by the old masters, gathered by Mr. Clark during his long residence abroad. Among them are The

Holy Family, *Van Dyk;* * Christ in the Garden, *Correggio* (formerly kept in the *boudoir* of the Empress Marie Louise) ; Mater Dolorosa, *Guido Reni:* St. Catherine, *Perino della Vaga;* Village Festival, *Teniers* (probably); Christ Preaching in the Desert, *Rembrandt;* Madonna, *Parmegianino;* two landscapes, by *Tempesta;* and a very ancient Byzantine Madonna.

Montebello is a rocky knoll near the springs, and on the r. of the carriage-road, whence is gained one of the most fascinating prospects over the meadows and mountains. It is but a few minutes' walk from the village, on the long promontory which projects from the terrace out over the intervales. The Passumpsic Railroad intended to build a large summer-hotel here, but was prevented by local dissensions.

Mount Pulaski is an inconsiderable elevation over Newbury, whose summit is about ¾ M. from the hotel, and is reached by an easy path. The road to the W. up along Harriman's Brook is followed to a point beyond the old ruined mill, where the path diverges to the l. and passes through fields abounding in sweet-fern and harebells. Beyond a narrow strip of forest, the visitor emerges on the bare cliffs which overlook Newbury, the broad and fruitful valley, and the noble mountains beyond. Haverhill and the Ox-Bow are plainly seen; and great Moosilauke looms up boldly on the E. The serrated crest of Lafayette rises against the horizon on the N. E., flanked by others of the Franconia Mts.; and the rugged highlands of Piermont and Orford are farther to the S.

Among the favorite drives from Newbury is that which crosses the river on an adjacent bridge to Haverhill, and thence goes N. to N. Haverhill, Woodsville, Wells River, and back to Newbury on the W. shore. Another pleasant road is to Bradford, on the S.; and the adjacent highlands of Vermont furnish pretty scenery. The excursion to the summit of Mt. Washington and back may be made in one day. Newbury is also a favorable point from which to attack the mountains of the Benton Range. The roads through the glens of Piermont and Orford, on the S., lead through some fine highland-scenery.

The rich meadows between Newbury and Haverhill were formerly the home of the Coösuck Indians, and were called *Coös,* a word signifying " pine-tree " (found also in the names of Cohasset, Acushnet, and Cushnoc). There was a large village E. of the mouth of Cow-Meadow Brook, with an ancient fort on the Ox-Bow; and the aboriginal cemetery is still pointed out. After the Deerfield massacre a party of Massachusetts troops, under Lieut. Lyman, attacked this place during a thunderstorm, at dead of night, and inflicted severe losses on the villagers. During the panic which ensued among the Northern tribes after Lovewell's battle at Pequawket the Coösucks abandoned their homes and retreated into Canada.

After their terrible retreat from St. Francis, the scattered parties of Rogers's Rangers rendezvoused on the Newbury meadows as fast as they came out from the vast and gloomy forests, in whose recesses they had supported life by eating the flesh of their dead comrades. Their commander left them here, to get supplies at Charlestown, while "the famishing rangers saw him disappear around a long sweeping bend of the river, and then lay down to wait ten days, at the end of which he had promised to return. The hours went slowly by — a week passed — and those men sat in the smoke of their fires and listened to the wind sighing about their camp. As their forms grew more attenuated, their faces more haggard, and their eyes and cheeks more sunken, they would reel into the woods to gather roots and bark, coarse food to keep the last spark of life from going out." In ten days the survivors, " looking more like ill-dressed corpses than human beings," were rescued and carried down the river to the fort at Charlestown. In his victorious attack on St. Francis, Rogers had but 142 men, and on the retreat these were divided into nine detachments. Ensign Avery's men saved the remainder by hurling themselves against the pursuing forces. They were exterminated, but the time thus gained was invaluable to the rangers. Soon afterwards the companies under Lieut. Campbell and Sergeant Evans, wandering and starving, came upon the ground and won a new lease of life by devouring the bodies of Avery's men.

In 1756 Col. Atkinson formed the following plan of settlement: " We are now upon a Project (which I believe will take effect) of settling a Tract of the finest Land on the Continent, call'd by ye Indians Co-os, which lyes upon Connecticut river about 90 miles northerly from the Province line. We have already inlisted about 400 Proper men, they are to cut a road to that place, build 2 Garrisons with sufficient accommodations for the 400 or 500 men, in this parte of the country they will raise Provisions the first year for double their number & all their land under tillage be in sight of and defended by the Garrisons ; 'tis a great undertaking and a good one, for I really believe, if we do not settle it the French will for 'tis the main passage made use of by the Indians from Canada to this country, and if we can fix there, which 500 such men as are going (barring the particular frowns of Providence) mauger all the efforts of the Indians. A regular Garrison built in each of them, encompasing perhaps 15 or more acres of Land, this to be enclosed, with Log-houses at some distance from each other, and the spaces filled up with either Palisades or Square timbers, in the middle of the square something of the nature of a Cytidall where the Public Buildings & Granarys &c. will be built & to be large enough to contain all the Inhabitants, if at any time drove from the outer Enclosure which is large enough to contain their cattle, etc."

About the year 1762 parties of pioneers came up from Newbury and Haverhill, in Massachusetts, and settled here, — those on the W. of the river naming their town Newbury, and those on the E. Haverhill. During a part of the Revolutionary War Col. Hazen's cavalry regiment (destined for service in Canada) was quartered on the Ox-Bow. The founder of the town was Gen. Jacob Bailey, of Newbury, Mass. During the Revolution a detachment of British soldiers came down to capture him, but a friend went over to the field where he was ploughing and dropped a note, saying, "The Philistines be upon thee, Samson!" On returning down the long furrow, Bailey saw the note and took the hint, flying to securer regions. The American Col. Johnson had previously been captured and paroled, near Newbury.

Wells River and Woodsville.

Wells River (*Coösuck House*) is a growing and prosperous village of Newbury, at the confluence of the Connecticut, the Lower Ammonoosuc, and Wells River, and the intersection of the Passumpsic, Montpelier and Wells River, Boston, Concord and Montreal, and White Mts. Railroads. Here passengers from the W. or on the valley lines change cars in order to go to the White Mts. There is a restaurant at the station. The village is the largest of the four in Newbury, and contains a bank, a Congregational church, and a handsome school-building.

Woodsville is in Haverhill, opposite Wells River, and on the B., C. & M. R. R., whose trains here cross the Connecticut on a high bridge, connect with the Passumpsic R. R., and then return. A large new summer-hotel has recently been erected here, on a point which overlooks the Newbury meadows and the course of the winding river. Pleasant drives may be taken from this point in several directions. N. of the village is Mt. Gardner, the S. peak of the long range of that name, which was formerly ascended by a bridle-path from Woodsville. It is now visited by crossing the Ammonoosuc by the bridge on the Bath road.

132. Bradford and Piermont.

(The Trotter House is a comfortable little hotel in Bradford.) Bradford is the chief village in a town of 1,492 inhabitants, and has Congregational and Methodist churches, an academy, a newspaper, a savings-bank, and manufactories of lumber, boxes, mackerel kitts, pails, sashes and blinds, paper, marble, patent-medicines, scythe-stones, a foundry, and grist and saw mills. It is ¾ M. from the Bradford station of the Passumpsic Railroad, and 1¼ M. from Piermont. Waite's River runs through the village, and was named for Capt. Waite, of Rogers's Rangers, who killed a deer and left it at the mouth of the river, this aid saving the lives of several of the soldiers in the direful retreat (see pages 184, 302, and 316). The first terrestrial globes in America were made in this town in 1813. Bradford sent 258 men to the late civil war, of whom 21 were killed.

Wright's Mt. (or *Virgin Mt.*) is 5 – 6 M. N. W. of Bradford, and 1,500 ft. above the river. It is faced on the W. and S. with high cliffs of argillaceous slate, and presents a bold and imposing appearance. A carriage-road formerly led to the top, by which the people of the adjacent towns ascended to merry-makings on every Fourth of July. The view from the summit is broad and beautiful. The mountain is named after a religious fanatic who imagined himself a prophet, and retired to a cavern on its side to spend forty days in fasting and prayer, but waxed hungry before the first week had passed, and returned to the fleshpots of the lowlands.

A pleasant road leads from Bradford to Piermont, crossing the ancient covered bridge where dwelt Emily R. Page, whose father was the toll-gatherer. She was the author of the poems " Be not Weary," " Only waiting till the Shadows," etc., and of " The Old Bridge," which describes the Piermont bridge and begins with the stanza:

> " Bowered at either arching entrance
>   By a wilderness of leaves :
> Clustering o'er the slant old gables,
>   And the brown and mossy eaves,
> Is the dear old bridge, which often,
>   Often in the olden time,
> Echoed to our infant footfalls,
>   And our voices' ringing chime.''

Piermont is a town of 792 inhabitants, occupying 23,000 acres, of which over 16,000 are improved. Its chief elevations are Piermont Mt. and Peaked Mt., or Piermont Pinnacle; and in the E. are the Eastman and Tarleton Ponds. Piermont village is on the large mill-stream called Eastman's Brook, near its confluence with the Connecticut, and contains two churches. Pleasant and picturesque roads lead through the glens to Haverhill, 4½ M.; Orford, 5½ M.; Wentworth, 13 M.; and Warren.

133. Orford and Fairlee.

Orford is ranked among the mountain-villages on account of its vicinity to Mt. Cuba and several other bold summits. It is one of the most beautiful hamlets in New Hampshire, and is just across the river from the Fairlee-and-Orford station of the Passumpsic R. R. It is built along a straight and level street, 1 M. long, called the Mall, which is bordered for a considerable distance by double lines of fine old trees, with promenades between. 30 – 40 rods E. of this pleasant avenue is the terrace of an old river-bank, on which is the Orford Academy and its boarding-house, with a line of fine old mansions. The Congregational and Universalist churches are on the Common, in the upper part of the village. A beautiful view of the Connecticut Valley is opened from the *Seven Pines* on the hill E. of, and 10 minutes' walk from, the mall.

Among the people whose visits have caused pride to rise in the Orfordians were Elihu Burritt, W. R. Alger, James T. Fields, Miss Robbins (the flower-painter), and Christine Nilsson. Abbott wrote considerable portions of his historical works here. Washington Irving said of Orford, during his visit there: "In all my travels in this country and in Europe, I have seen no village more beautiful than this. It is a charming place. Nature has done her utmost here."

Mount Cuba is reached by driving out through Orfordville (or along the Archertown road) to Finney's, 7 M. from Orford, whence a broad cart-road leads to the summit in 2 M. The drive may be prolonged by Baker's Ponds to Wentworth (see page 297), 13 M. from Orford; or may diverge, through picturesque glens, to Indian Pond and Soapstone Mt. Mt. Cuba is 2,927 ft. high; and derives its name from that of a hunting-dog, who was killed while fighting with a bear near its summit. *Mt. Cube* is a local corruption of this name. On account of its isolated position and considerable height, it commands a broad and noble view, by reason of which it was for one season the headquarters of the Geodetic Survey of the State. Its bases and all the country around are occupied by deserted farms and mouldering houses.

* *The view* includes nearly all the Franconia and Pemigewasset Mts., which mask the White Mts. Moosilauke is conspicuous, under the thin and lofty crest of Lafayette. The nearer range is Mt. Carr, with some of the Sandwich Mts. far away on the r. Mt. Prospect, near Plymouth, is about E. S. E., with the Ossipee Range beyond, and a glimpse of Lake Winnepesaukee on the r. Then Mt. Belknap is seen; and about S. S. E., over the bold highlands of Dorchester, are the high domes of Mt. Cardigan. Farther away Mt. Wachuset is seen, in Massachusetts.

A beautiful prospect is afforded up and down the Connecticut Valley for many leagues, with the encircling highlands of the towns adjacent. In the W. is the long line of the Green Mts. of Vermont, extending along the horizon for a great distance, with many of its high blue peaks recognizable.

Sunday Mt. derives its name from a legend that a man of Orford, many years ago, wandered away into the woods one Sunday morning, when he should have gone to church, and as he was rambling on this mountain he was caught and torn in pieces by a company of bears. This story was long current in the country-side, and was told by anxious mothers to irreligious sons. In the N. part of Orford is a high elevation which is known as *Gravestone Mt.*, because a quarry of gravestones was formerly worked on its side.

Mt. Fairlee is just across the river from Orford, and is faced on the E. by the Yosemite Cliffs, 200 ft. high. Its summit is easily reached, and commands a charming view of the Connecticut Valley for 16 M., with the included villages and the environing mountains. Back of this eminence and 1½ M. from Orford is *Fairlee Pond*, the pretty sheet of water where Morey ran his steamboat. The *Cave Falls* are near its shores, falling through a long rock-gorge, and forming beautiful water-scenery. *Fairlee Lake* is several miles distant, among the highlands. There are several pretty falls on Dayton's Brook, N. E. of Orford.

It is claimed that the first American steamboat was built and run at Orford, in the year 1792, by Capt. Samuel Morey, a citizen of the town and a skilful mechanician. The boat was a rude dug-out, and the engine was of primitive construction. The first trial-trip took place on Sunday, during church-time, the sensitive inventor hoping thereby to escape ridicule in case of failure. He succeeded in stemming the rapid current of the Connecticut, and in advancing slowly up the river. Soon afterwards he went to New York, where he imparted the particulars and showed a model of his discovery to Messrs. Fulton and Livingston, and Mr. Fulton visited Morey at Orford, and saw the operation of his boat, long before the first steamboat was launched on the Hudson River. The New-Yorkers afterwards turned a cold shoulder on the ingenious rustic, and made a glorious application of his hard-won theories. Capt. Morey afterwards built a larger steamboat, the *Aunt Sally*, on Fairlee Pond, where it was used as a pleasure boat until it was sunken, about 1820. In 1874, the N. H. Antiquarian Society made earnest efforts to find and raise this boat, but she had disappeared under the muddy bottom of the lake.

The Orford Hotel was recently destroyed by fire, and it is uncertain whether it will be rebuilt. Summer boarders can, however, find other accommodations in the village.

THE WATERVILLE AND SANDWICH MOUNTAINS.

134. Waterville.

Merrill Greeley's summer boarding-house accommodates about 50 boarders, at § 8 – 10 a week. It is reached by way of Plymouth (18 M.) and Campton (12 M.); and Mr. Greeley sends a stage to the former point when notified of the approach of his guests.

Waterville is a large town territorially, but it has only 32 inhabitants, most of whom live in the glen near the centre of the township. It was granted in 1819, and incorporated in 1829; and is covered with great forests of maple and birch, pine and hemlock, in which an abundance of game is found. The numerous brooks are also prolific in trout. The town is filled with formidable mountains and long wooded ridges, which are divided by the deep valley of Mad River.

Greeley's boarding-house is in one of the most charming glens in the mountain-region, and is frequented by families of the better class from Boston and the cities of the coast. The view on all sides is noble, including Mts. Osceola and Tecumseh, Sandwich Dome, and the bold crest of the Noon Peak. The glen seems to be enwalled with these massive highlands, and various rich combinations of scenery may be obtained by ascending Snow's Mt. and other hills in the vicinity. There are also several striking mountain-views from the road up the Mad-River Valley, which is the only road in the township. A bridle-path 15 M. long was once built from Greeley's through the Mad-River Notch and by Mt. Carrigain to the Saco Valley near Upper Bartlett, ascending the course of Mad River, crossing the high water-shed, and descending Sawyer's River. By this route people have ridden on horseback from Greeley's to the top of Mt. Washington in one day. It has long since fallen into disuse, and can be followed only by skilful woodsmen. Mr. Drake occasionally guides travellers through to the White-Mt. Notch by this path, but the journey is long and arduous.

The Cascades are about 1¼ M. from Greeley's, and are reached by a pleasant path which runs off at a right oblique from the back of the house across a hill-pasture, in a line marked by red pennons. About half-way up is a resting-place which commands the N. peak of Osceola. At the forks of the paths that to the r. is taken (that to the l. conducts to the

14 *   U

Great Slide on Tripyramid), and leads up the side of Cascade Brook to the first falls, which are of about 25 ft. aggregate height. A short distance above is a fine fall of about 30 ft. perpendicular height, above which are several other white and graceful leaps. The great charms of the Cascades are in their beautiful forest-environs, the marvellous transparency of the water, and the unique shapes assumed by the columns of falling water. The brook itself is not large, except during seasons of protracted rains.

The **Greeley Ponds** are 4 – 5 M. from Greeley's, and are reached by a plain and pleasant path leading through the forest. They are situated in a deep and narrow notch between the E. slope of Mt. Osceola and Mt. Kancamagus (3,500 ft. high) on the E. Fishermen still visit these waters in quest of trout, but better sport may be found on the larger tributaries of Mad River. The path is prolonged over the meadow at the head of the first pond, and soon reaches the second pond, which is somewhat larger and much deeper.

The Flume is about 3½ M. from Greeley's, and may be visited by following the path to the ponds for about 2 M., to a point where a guide-board points the way into a trail diverging to the r. This ascends the general course of the stream to the N. E., up to the Flume. The water here flows down for many rods between two walls of rock of about 20 ft. altitude, but well detached. One wall is perpendicular, the other falls off at a slight angle. It will scarcely pay to visit this po'nt unless one has plenty of time at his disposal. *Mt. Kancamagus* is a bold wooded ridge which may be ascended by way of the Flume Brook, but nothing is to be gained by the journey.

The **Noon Peak** is the apparently high green spur of Sandwich Dome which is thrust forward toward the Waterville glen. It was so named by an early settler, because the sun used to stand over it at mid-day. The peak is rarely ascended, being far lower than the main ridge of Sandwich Dome. To the S. E. of Noon Peak, up the ravine, is a bright cascade which bands the mountain with a white stripe after long rains. *Mt. Tecumseh* is sometimes ascended from Greeley's, by an arduous clamber up the bed of a brook, though the Welch-Mt. route is preferable. Sandwich Dome and Welch Mt. are also ascended from a point about 6 M. down the valley-road. A rude trail leads from Greeley's to Flat-Mt. Pond in 4 – 5 M., making a long and arduous ascent, and connecting with a path to Sandwich.

135. Mount Osceola

is in the upper part of Waterville, on the edge of the Pemigewasset Forest, and is 4,400 ft. high. It is flanked on the E. by a bold secondary peak, overhanging the Greeley Ponds; and a line of picturesque summits trends off in the opposite direction towards the East Branch. The view from the summit is equal to that from Mt. Carrigain or Mt. Willey for its wide sweep over the Pemigewasset Forest and the surrounding peaks.

The path from Greeley's to the top of Osceola is about 4½ M. long, and is comparatively smooth and plain. It is practicable for ladies, many of whom have enjoyed the view from the summit. The track diverges from the Greeley-Pond trail just beyond the saw-mill, turning to the l. at a guide-board. It then traverses a nearly level forest-tract for a considerable distance, crossing one of the tributaries of Mad River, whence the ascent is varied by occasional stretches of level walking. At a point about 3 M. up is a spring of cold water, and there is another by the side of the path within about ¼ M. of the summit. On the culminating ridge most of the trees have been burned away, leaving a bristling thicket of white stubs, on the farther side of which are high ledges of granite, surmounted by a signal-beacon of the U. S. Coast Survey. This point overlooks the Waterville glen, and is the best station for a general view, whose dimensions may be enlarged by following the ridge along its edge to the r. and l. The view to the S. W. is obtained to best advantage from a point on the other side of the woods from the beacon. Edward Everett Hale says: "I regard the panorama of mountains from this summit as the finest key to the mountains which I know."

** *The View.* — Toward the S. E. is the bold ridge of Tripyramid, with its three peaks rising to nearly equal altitudes; and on its side is the long and sinuous white line of the Great Slide. At the foot is the old clearing of Beckytown, long since deserted, and the dark and wooded crest of Whiteface is seen peering over the S. slope of Tripyramid. More distant is the long Ossipee Range, with the farms and clearings of Tamworth well advanced on its slope; and on the l. Portland and the ocean can be seen, on a clear day. Farther to the r., and at the foot of the low Snow's Mt., are the white buildings of Greeley's boarding-house, over which, and to the r. of Ossipee, are the long levels of Lake Winnepesaukee, with Moultonborough and Tuftonborough Bays opening into the wide and pleasant farm-lands. Long Island is also seen, standing blackly in contrast with the shining waters; and at the foot of the lake is Copple Crown Mt., near Wolfeborough.

About S. of Osceola, down the Mad-River Valley, is the vast mass of Sandwich Dome, with the low green promontory of Noon Peak thrust to the N. and the white ledges of the Sachem Peak on the W. Beyond the Dome opens a vast expanse of the Winnepesaukee-Merrimac valley, with Meredith, Lake Village, and other hamlets studded on the apparent plain, and Lakes Winnesquam, Waukawan, and numerous ponds glistening in the light. The hills of Bristol and Alexandria are in the distance, also the Bridgewater Hills and the Squam Mts. The high blue peak more distant is Kearsarge, to the r. of and below which are the Ragged Mts. The rounded ridge of Mt. Prospect is l. of Kearsarge, and the double crest of Cardigan is to the r. of the Alexandria hills. On the line toward Prospect, but much nearer, across the Mad-River Valley, are the Campton Mts., among which Mt. Weetamoo is chief. About S. S. W. is the fair and populous valley of the Pemigewasset, running down by the contrasting dark forests of Plymouth Mt. The dim blue dot on the S. W. verge of the horizon is supposed to be Mt. Ascutney, in Vermont. The ponderous mass of Mt. Tecumseh is close at hand in the S. W., toward the Mad-River Valley. Farther up, and nearly in line, one over the other, are Mt. Carr, in Warren; Smart's Mt., in Lyme and Dorchester; and Moose Mt., near Dartmouth College, in Hanover, with Mt. Kinneo farther to the r. W. by N. is Moosilauke, massive and broad-based, with many of the stately Green Mts. of Vermont stretching along the remote horizon. It is thought that both Camel's Hump and the Killington Peaks may be seen on a clear day, — the latter being far to the S. W. Beyond the nearer hills about the Hancock Branch is the Blue Ridge, near Moosilauke, to the N. of which is the high black mass of Mt. Kinsman, about N. W. Closely adjoining the latter are the white and fire-swept flanks of Mt. Cannon, over the Profile House. On the E. side of the Franconia Notch the line begins with the Big Coolidge and Potash Mts. and is prolonged by Mt. Flume to the pyramidal tops of Mt. Liberty and Haystack. Farther N is the long, thin, and conspicuous ridge of Lafayette, with its two well-defined peaks. In the foreground is the vast and unbroken wilderness of the East Branch (or true Merrimac River) and the valley of the Hancock Branch. The Owl's-Head range rises in the N. N. W., and over it is the symmetrical pyramid of the Haystack. Beyond these, on the N., are the formidable masses of the Twin-Mt. system, Mts. Guyot and Bond, and the unnamed summits adjacent.

The New-Zealand Notch opens on the r. of the Twins, and through its long vista is seen the Pilot Range, beyond Lancaster. In the foreground are the long and crouching slopes of Mt. Hancock, with a well-rounded centre; over whose W. flank is the distant Cherry Mt., while the Willey Range is over its r. centre. Next comes the great Presidential Range, — Jefferson, Clay, Adams, and Washington, with the ravine of the Mt.-Washington River sharply defined, and the long plateau of Boott's Spur running to the r. Approaching more nearly from the main summit are the characteristic peaks of Monroe, Franklin, Pleasant, Clinton, and Webster. The stripes on the latter are not visible. On the r. of this cluster, and much nearer, is Mt. Carrigain, advancing by three great terraces from E. to W., and sustained on the l. by Hancock. Above these and apparently continuous with the Presidential Range, are the blue forms of Mts. Moriah, Imp, Wild-Cat, and Carter. Under and to the r. of Carter are the red crests of Giant's Stairs, Mt. Crawford, and Mt. Resolution, and the darker Mt. Langdon. Toward the N. E., near the Maine border, are the peaks called Royce, Baldface (with a white S. slope), Eastman, and Slope. The twin mamelons of Double-Head are then seen, over Iron Mt., with Black Mt. to the l. and the pasture-bordered Thorn Mt. (in Jackson) on the r.

The prospect now comes nearer, to the Swift-River ranges, of which the white-topped Tremont comes first, over a secondary peak of Osceola. The sharp cone of Haystack is next seen, over which is the yet sharper apex of Kiarsarge, near N. Conway. The second ridge, long, monotonous, and dark-hued, is Bear Mt., over which is the white-crowned Moat Mt., with a crest-line 2 – 3 M. long. Over the S. peak of Moat is the long blue Mt. Pleasant, with a hotel just S. of the centre. In the foreground is the deep valley of the Swift River, with the clearings and houses of the Albany Intervale. On the r. is the low knoll of Potash Mt., conspicuous by its whiteness, over which the long and noble rocky ridge of Chocorua stretches along the sky-line.

It has been claimed by alpestrians most familiar with this peak that the Mt.-Osceola view commands Mt. Katahdin, in Maine; Mansfield, in Vermont; Greylock, in Massachusetts; and other distant and renowned mountains.

136. The Mount-Tecumseh Range.

There is an interesting line of mountains, 5 – 6 M. long, N. of the Mad-River Valley, and included in the townships of Thornton and Waterville. The range runs from S. W. to N. E., and consists of Fisher Mt., Green Mt., and Mt. Tecumseh, with Hogback and Spring Mts. to the N. and closely connected, and the Welch Mt. on the S., joined to the centre of the main ridge by a narrow curving crest-line about 8 M. long. The chief points of interest about the range are the peculiar architecture of the peaks, the cascades and mineral springs, and the broad views over the picturesque mountains of the Pemigewasset Forest, and of Waterville and Sandwich.

The route from the S. leads up Welch Mt. (see page 287), and thence around a ridge nearly 2 M. long, on which rises the rocky pile of Stone Mt. On the l. is a deep semi-circular ravine, over which are Fisher Mt and Green Mt. Bending around the E. side of the latter, a patch of jungle is crossed, and the traveller soon enters the dwarf forests on the slope of Tecumseh.

The mountain is sometimes ascended from Waterville by following up the course of a brook. This is by far the shortest of the routes to the summit, but the ascent is sharp and the forest is tangled.

The best point for attacking the range on the W. is from the Elkins farm, which is reached by the road up the Mill-Brook valley, diverging from the main road E. of the Pemigewasset, about ⅓ M. S. of the Thornton town-house, and near a cemetery. The road is very hilly, and passes near the Mill-Brook Cascades (see page 292). The Elkins farm is about 5 M. from W. Thornton and 7 M. from Campton Village, from either of which the ascent and return may be made in a long summer day. The visitor should drive up and leave his carriage at the farm to return with at evening. George Elkins is a quiet, tireless, and trusty young man who may usually be found at the farm, and is familiar with the adjacent mountains, over which he sometimes guides parties.

Mt. Tecumseh may be reached from the Elkins farm by either of three routes: — (a) over Fisher and Green Mts; (b) up the ravine to the mineral springs, and thence over Spring Mt.; (c) to the head of the ravine and directly up the slope of Tecumseh. The first and last ways were used by the Editor, and the second was reconnoitred. He recommends that the ascent be made by Spring Mt., and the descent by Green and Fisher Mts. (It should be remembered that Tecumseh is known to a few people by the name of *Kingsley's Peak*, a title which was recently conferred in honor of himself by a gentleman who imagined that he was its discoverer.)

The Fisher-Mt. route is entered directly from the Elkins farm by crossing long upland pastures and traversing a belt of tangled forest. Then the tourist attacks the bare white ledges of *Fisher Mt.*, whose summit is reached after an hour's breathless clambering. Pleasant views are opened in the S. and W., and in advance is the white crest of Tecumseh. A short descent and another clamber lead to a second summit, whence the *Hogback* is seen on the l., a massive pile of white rocks fringed with trees. An hour's walk along the ridge conducts to *Green Mt.*, a tall and symmetrical peak whence the view includes Moosilauke and the Pemigewasset Valley on the W., Winnepesaukee and the Sandwich Range on the S., over the Mad-River Valley, and the chief Franconia peaks on the N. Welch Mt. is seen on the S. W. below, and is joined to Green Mt. by a curving ridge about 3 M. long, clear on the crest and easy to be traversed. The climber can now inspect the main peak of the range, and mark out his route for the ascent. On the W. slope of Green Mt. are vast beds of rocky débris, among which brilliant quartz crystals may be found Their colors are white, red, yellow, and smoky. The visitor would do well to carry a hammer and chisel, in order to detach some of these beautiful crystals from the rocks.

The route from Green Mt. leads for a considerable distance over beds of broken rocks, then heads around a forest-filled ravine, and ascends the flank of the main peak. After moving up a ledgy incline, the tourist enters a forest, which must be followed along the very crest of the flank, slowly rising through a perplexing jungle of low evergreens and fallen timber. In some places the easiest way to advance is on hands and knees, so dense and spiky is the upper growth. In 1½ hrs. from Green Mt. the foot of the peak is reached, and 15 minutes of wary clambering places the tourist on the crest of the sharp pyramid of white rocks, alongside the beacon of the U. S. Coast Survey.

A pleasant forest-path, occupying the remains of the washed-out road to the spring-house, leads up the ravine of Mill Brook to the mineral springs, which are 2 M. above the Elkins farm. The lower spring is near the spring-house, and contains a perceptible amount of iron and sulphur. It has been used to advantage in cases of cutaneous diseases. The upper spring is several rods farther up the stream (on the same side), and contains iron in large amount, insomuch that the bed of the outflowing rill is stained of a bright yellowish-red color. On the l. are the sharp slopes of Mt. Avalanche, which are striped from summit to base with the white tracks of slides. Behind and above the spring-house the visitor enters upon the ascent of *Spring Mt.*, one of the lower summits of the range. After nearly an hour of arduous toil the summit is reached, with the crests of Green Mt. and Tecumseh in advance, on the r. and l. Spring Mt. is capped by a remarkable ledge, whose sides are cut with masonic precision; and on the N. E. is a long and beautiful cascade and clear fall, where a little rill plunges down the sharp slope. The tourist now descends a short distance, and then enters the shaggy forests on the flank of Tecumseh, conducting to the white pyramid of rock on which the signal stands.

The third route of ascent from the W. is by way of the Mill-Brook ravine, but it cannot be recommended. Beyond the spring-house, 2 M from the Elkins farm, the general course of the brook is followed for 1½-2 M. through a pathless jungle. The easier route is on the Mt.-Avalanche side, where many clear spaces have been made by the slides. From the head of the ravine the ascent is commenced directly up the steep slope of the main peak and through a dense forest, where the way may easily be lost. The worst feature of this climb is the insecurity of the ground, since at every few steps the foot plunges through masses of leaves and light soil into deep crevices among the rocks and roots. Bruises are inevitable, and a sprained ankle is among the possibilities.

Mount Tecumseh is over 4,000 ft. high, and lies close to Osceola, on the S. W., being separated from it by a deep and tangled ravine. Its summit is composed of a steep and lofty pile of white rocks, inaccessible on two sides, rising to a pronounced peak, and forming one of the most interesting sections of natural architecture in the mountain-region.

* The View. — Nearly N. W. across the great trough of the Pemigewasset Valley, 12 M. distant, is the long and lofty ridge of Mt. Kinsman, flanked on the r. by the ledgy sides of Mt. Cannon, and with Mt. Pemigewasset in the foreground. Black Mt. is nearer, towards the Potash and Coolidge Mts. The eye then rests on the Franconia Mts., about N. N. W., standing with the end of their line towards Tecumseh. The lower peaks of Flume,

Lincoln, and Liberty are thus somewhat indistinct, but the sharp, thin crest of Lafayette towers over their line in satisfactory majesty, with the pyramidal point of the Haystack on the r. In the foreground are the high hummocks on the ridge N. W. of Osceola, hiding part of the Pemige-wasset Forest. Nearly N. is the massive ridge which includes Mts. Bond, Guyot, and the Twin Mts., on whose r. are the distant blue lines of the Pilot Mts., with Mt. Starr King in advance.

The lofty peak of Osceola is 2 – 3 M. from Tecumseh, nearly N. E., and beyond it are the ponderous mountains of Hancock and Carrigain, the former on the l., the latter rising in long step-like terraces. Over the flank of Hancock are Mts. Field and Willey, at the White-Mt. Notch; and to the r. of these rise the loftier summits of the Presidential Range, the southwestern peaks being seen along the line of their axis. Mt. Washington can usually be recognized by its being capped with clouds, and on its l. are the lower humps of Mt. Clay and the sharp crest of Jefferson. Osceola hides the Crawford and Carter groups, and on the r. Iron Mt. and Double-Head appear, far out beyond the southward valley, with Baldface farther still. More to the r., about E. N. E., is the pyramidal peak of Kiarsarge, with a house on top, on whose r. are the ridges and crests S. of the Saco, in Bartlett. The red ledges of Moat Mt. are near Kiarsarge, with two well-marked peaks on a long ridge, Bear Mt. being nearer at hand. The summits in this direction are seen over Mt. Kancamagus, which is opposite and near to Osceola, and has no definite peak on its wooded top.

Due E., 6 – 8 M. distant, is Tripyramid, showing its four crests, and easily distinguished by the great white slide on its flank. On the r. and beyond is the stately and imposing white peak of Chocorua, flanked by rocky ridges and rising over the more lowly Paugus. Nearer, to the r. of Tripyramid, and within 10 M., are the high tops of Passaconaway and Whiteface; and the white houses in the Waterville glen are distinctly seen below. Beyond Whiteface, over Flat Mt., glimpses are gained of the Ossipee Mts. and the farming-lands of the lake-towns, with a small section of Lake Winnepesaukee. The prospect is now limited by the immense mass of Sandwich Dome, within 4 – 5 M. across the Mad-River Valley, with the Noon Peak near its centre, and the Sachem Peak lifting its weird white head on the r. Far beyond is the double peak of Mt. Belknap, in the lake-country. To the r. of the Dome are the Campton Mts., with the bare ledges of Mt. Weetamoo conspicuous, beyond which is the flat top of Prospect, near Plymouth. Kearsarge is about S. S. W., across the lower Pemigewasset Valley, over the Bridgewater Hills; and farther to the r. are the bold swells of Mt. Cardigan. Over the township of Thornton and Mt. Stinson is the blue peak of Ascutney, far away in Vermont, nearly S. W. About W. are Mts. Kinneo and Carr. across the Pemige-wasset Valley, the former being in the foreground, about 10 M. distant; and a little to the N. is the high plateau of Moosilauke, marked by a

house. The Blue Ridge is in front of and below Moosilauke, and extends
to the r. until it overlaps Mt. Kinsman. In the remote W. the Green Mts.
of Vermont are seen, stretching along the horizon for many leagues, the
peaks of Camel's Hump and Mansfield being the most conspicuous in
their line.

137. Tripyramid and the Great Slide.

* **The Great Slide** on Tripyramid is considered by many visitors the
most remarkable object among the curiosities of Waterville. It commences about 2 M. from Greeley's, at the ancient clearing known as
Beckytown, and is reached by taking the path to the Cascades and diverging to the l. at the forks (about 1¼ M. from the house). Beckytown is 280
ft. above Greeley's, and from this point the Slide may be ascended for 2
M., gaining a farther altitude of 1,015 ft. This section of the devastated
valley is followed by a small stream, and the fringes are encumbered with
the high-piled remnants of the ruined forests. It is here over 1,000 ft.
wide, where the débris spread over the meadows, and it decreases in the
ascent to 125 – 300 feet wide. The upper half-mile narrows gradually from
640 to 30 ft., and has an angle of about 34°.

At 2 M. above the foot, the Slide turns at right angles with the brook, coming
down directly off the sharp slope of Tripyramid, very broad, heaped with bare white
rocks and their disintegrated fragments, and breaking down to the bed-rock in frequent places. This section is about ½ M. long, and rises 1,100 ft., giving for the
whole Slide a length of 2½ M., and an altitude, from base to top, of 2,115 ft. The
brook section is comparatively easy of ascent, but the upper half-mile is so extremely steep as to enforce slow advance on the part of the climber, whose footing, moreover, is often very insecure. Beautiful views of Lake Winnepesaukee and the mountains to the S. and W. are afforded in retrospect. The great natural convulsion
which caused this catastrophe took place during the remarkable rains of the year
1869, but was unattended with any loss of life or property. The Slide contains many
points of interest to geologists and other scientific men. The rocks on the lower
half are labradorite and ossipyte, and the white boulders above are granitic.

Tripyramid is usually ascended by way of the Slide, its S. peak being
but 200 ft. higher than the top of the landslip. None of the other peaks
are visited, and the tourist finds his way from the Slide to the S. peak by
a straggling line of blazed trees leading off to the r. oblique from the head
of the ridge. It is about 4½ M. from Greeley's, and the march will require
5 – 6 hrs. There is no view from the peak itself, except for such as may
be pleased to clamber up a tree and hold on. The other peaks of Tripyramid are also densely wooded with tangled and bristling evergreens.
On the S. side of the S. peak is a projecting ledge which commands a fine
view over the lake-valleys.

* *The View.* — The first prominent object on the l. is the high and
densely wooded bluff of Passaconaway, towering beyond two deep ravines and an intervening ridge. On its l. is the sharp and ragged crest-
line of Chocorua, to the l. of and far beyond which is the blue ridge of Mt.
Pleasant, in Denmark. Near Pleasant is the lower knob of Mt. Tom,

near Fryeburg, with a section of the Saco Valley; and Sebago Lake glitters on the r. On the right of Passaconaway, and about 2 M. distant, in an air-line, are the dark forests which clothe the N. W. slopes of White-face; and through the opening between Passaconaway and Whiteface, the high blue hills of Madison and Eaton stand out with good effect. Just to the l. is the Green Mt. of Effingham, beyond Ossipee Lake.

On the r. of Whiteface the long and picturesque Ossipee Range stands out in bold relief across the apparently rich meadows of Sandwich. The Bearcamp Pond is in line with the ridge, and below. On the S. is the broad and highly diversified surface of Lake Winnepesaukee, with Moul-tonborough Neck projecting out towards Long Island and the level plains of Tuftonborough. On the farther shore are the double peaks of Mt. Bel-knap, to whose l. is a flotilla of islets running down toward the knoll of Rattlesnake Island. Nearer at hand is Red Hill, hiding Centre Harbor with its double ridge; and farther to the r. is Mt. Israel. Over the r. flank of Red Hill is the mirror-like flash of Lake Winnesquam, and the hill itself is girded by a belt of farms. To the S. W. is another high ridge, over whose sloping saddle is a glimpse of Squam Lake. More to the W. and in the foreground is the low plateau of Flat Mt., flanked by deep ravines. Farther away on the horizon is the noble peak of Kearsarge, over the Ragged Mts.

The great mass of Sandwich Dome is close at hand in the W., on whose r. flank is the Sachem Peak. On the r. and over the Sachem Peak is the high-notched and desolate Mt. Cardigan. Plymouth Mt. and Mt. Prospect are down the Mad-River Valley, on whose r. side is the white-striped Welch Mt., running into the Fisher-Mt. ridge. Smart's Mt. is in the dis-tant W., and Mt. Carr is farther to the r. Over Fisher Mt. is the remote blue disk of Mt. Cuba, near the Connecticut River.

The views which may be obtained from the branches of some of the trees on the S. or central peaks of Tripyramid include also the Franconia and Twin-Mt. ranges, Carrigain and Hancock, the Presidential Range, and the ridges of Bartlett and Jackson. The central peak is the highest in the series, and is but ½ M. from the Slide. It should be cleared, an opera-tion which would cost but little trouble, since it is remarkably pointed; and a path should be made to it along the crest of the ridge. The excur-sion from Greeley's to Tripyramid and back will occupy a long and work-ful day. In August, 1875, the Editor, with the party engaged in the Guide-Book survey, crossed from Greeley's to Sandwich in two days, passing over the crests of Tripyramid and Whiteface, and traversing the intervening ravines. As ascertained by barometric observations, the second ravine between the two summits is 1,300 ft. below Tripyramid, and the ridge beyond may be followed in its bend to the S. and E. until the cone of Whiteface is encountered, 325 ft. below the summit.

138. Sandwich Dome.

This immense mountain lies in the W. part of the towns of Sandwich and Waterville, and is separated from the Campton Range by the Sandwich Notch, and from Whiteface by the low plateau of Flat Mt. It is 4,000 ft. high. It is flanked by the subordinate ridge of the Noon Peak on the N. and by the sharp white spires of the Acteon Range on the N. W. The long upper ridges are bare of trees and swell into minor peaks, separated by thickets and stony levels. On account of the relation of Sandwich Dome to the White and Franconia Mts. and the lake-country, it commands one of the grandest and most fascinating panoramas in New England.

This mountain is known to most of the country people in the adjacent towns as *Black Mountain*. But this name has been applied to so many summits in New Hampshire that it has lost all value as a distinctive appellation. Recognizing this fact, Prof. Guyot bestowed upon this Black Mt. the name of *Sandwich Dome*, which has been accepted by the State Geological Survey and is gratefully adopted here. From the lake-towns it presents the appearance of a flattened dome, and the prefix of "Sandwich" is given because it is the chief and most conspicuous peak in that town; while Waterville, in which half of it lies, has several higher summits.

The old bridle-path from Sandwich has been neglected of late years, and is partly overgrown. It was at one time an easy and favorite route.

The path to the summit of Sandwich Dome leaves the Waterville road at the Dolloff farm, which is beyond the Mad-River bridge, about 5½ M. from Campton Village and 6½ M. from Greeley's. It is over 3 M. long, and requires 2½-3½ hrs. of active climbing. Much labor appears to have been bestowed upon the path in past years, but it has recently been neglected, and now requires a thorough overhauling. A little judicious labor would go far towards making out of this blind trail a comparatively easy way of access to one of the noblest view-points in New England. Guide-boards should mark out the way from Dolloff's to the ruined bridge, and should show where the path is resumed after crossing the stream in many places far up in the ravine. But the most important improvement would be to clear a path through the dense growth of stunted spruce and fir-trees surrounding the final peak — a dense thicket, which is quite impassable for ladies and nearly so for gentlemen.

Passing up the road from Dolloff's to and through the gate on the r. a few rods distant, a field-road is entered and followed towards the woods. Beyond a second gate the road grows more vague, and the traveller soon reaches a brook on the r., coming out of the woods. Without crossing, he should keep up along a path near the water until he reaches a dilapidated log-bridge, about 1 M. from Dolloff's. On crossing this, the mountain-path is entered on the l., and thenceforward ascends through the woods for a great distance, being well defined and comfortable, save for its upwardness. Glimpses of the lofty white cliffs of the Acteon Range are gained on the l., across a wide ravine, but the main peaks are hidden by the trees in advance. As the head of the ravine is approached, the path crosses the brook several times, and the climber must be careful or he will be unable to find it on the other side. The last ¾ M. is very steep, and leads over fallen fragments of rock and loose stones, débris and sharp bluffs. If the path can be followed it will be of material aid, since this narrow track has been cleared of rolling stones and angular rocks. This part of the route is above the woods, and commands broad retrospective views. At last it approaches a sharp white peak, and ten to one the unwarned traveller would strike directly thitherward as for the ultimate crest. The actual summit is to the N. E., and is much higher than the white peak, but is surrounded by a dense and formidable thicket of dwarf evergreens, through which no adequate path can be found. Visitors sometimes remain on the lower peak, weary with their long march and daunted by the jungle ahead; but they should call out their reserve of nerve-strength for a final pull of 15 minutes up through the spruces, and then be rewarded richly. The crest overlooks all the trees, and on its highest ledge is the beacon which marks it as a station of the U. S. Coast Survey. From this point the massive proportions of Sandwich Dome may be studied, — with the white-crowned Sachem

Peak flanking it on the W., the long ridge on the N., which rises into a pointed nubble on the W. and to Noon Peak on the r., and a high wooded ridge on the S. E. toward Upper Sandwich.

Professor Huntington and the Editor ascended Sandwich Dome from Dolloff's, and went down on the S. side, taking a compass-line across the great ravine to Mt. Israel. The woods were not difficult to traverse, though there were several belts of burnt ground, and after over two hours of rapid descent they reached a remote farm whence a path led out around Young Mt. to the Guinea-Hill road and Sandwich Centre. The descent on this side cannot be recommended, as a slight error in calculation would necessitate a very long march through trackless forests.

** *The View.* — The brilliant white spire of Mt. Chocorua is a little N. of E., and is flanked by its high N. peaks. Beyond and partly hidden by Chocorua is the long and uniform ridge of Mt. Pleasant, with its summit-hotel, and somewhat nearer are two or three of the Madison and Conway ponds. A little to the r., but many leagues away, are the Frost and Burnt-Meadow Mts., of Brownfield, beyond the hills of Madison and Eaton. Due E., over 40 M. distant, the waters of Sebago Lake may be seen on a very clear day, and Silver Lake (with Madison village on its N. shore) is in nearly the same line, though not half so far away. Over Great-Hill Pond, which is to the r. of and nearer than Chocorua, is the spire of Tamworth, beyond which are Gline Mt. and the more remote highlands of Baldwin and Cornish, with Mt. Prospect in Freedom to the r. To the r. of White Pond is S. Tamworth village, near the Ossipee Range, with part of Ossipee Lake beyond and the Green Mt. in Effingham on the sky-line, on whose l. is the city of Portland, at the seaside. The S. E. is occupied by the Ossipee Range, with the pastured flanks of the Whittier Peak to the l. of and beyond the islandless Bearcamp Pond, over which is the sky-attaining Melvin Peak, on the S. W. of the range. The view next includes the hamlets to the S., Centre Sandwich near by and to the l. of Mt. Israel, with the island-studded Red-Hill Pond on the l., beyond which, on the r., is Sandwich Corner, with Moultonborough still farther off.

Lake Winnepesaukee is about S. S. E., and forms the richest and most beautiful element of the view from the Dome. Moultonborough Neck projects its thousands of well-tilled acres between Moultonborough Bay and the Centre-Harbor bay, and Long Island is nearly joined to its end. Numerous islands are seen in the bay, and on its E. shore is Melvin Village, over which is the hamlet of Mackerel Corner, on the highlands; and still farther out on this line, over Smith's Pond (at Wolfeborough), is the symmetrical peak of Copple Crown, flanked on the l. by the low cliff of Tumble Down Dick. Under Copple Crown is Tuftonborough Neck, and on its r. are the high hills of E. Alton, over which the ocean off Portsmouth may be seen on very clear days. Over Long Island is the ridgy Rattlesnake Island, before the entrance to Alton Bay, with the Alton hills near the shore. More to the N. W. is Diamond Island, and a picturesque archipelago is seen farther up the lake. Across the deep wooded ravine on the S. of the Dome is Mt. Israel, about 4 M. distant, over which is the long, low, and shapeless mass of Red Hill, above whose central depression are the sister peaks of Mt. Belknap, with Mt. Retreat on the l.   To the r. of Belknap is Long Bay, running to the S. of Round Bay, and with Lake Winnesquam on the r., at whose foot is the large village of Laconia.   The Northwest Bay of Winnepesaukee is nearer, with a part of Meredith Village, on whose r. is Waukawan Lake.

Squam Lake is on the S. and all its regalia of islands is seen, with long curving promontories and deeply recessed bays.   Far away over its tranquil waters, and over the r. of Waukawan Lake, are the twin Uncanoonuc peaks, near Manchester; while equally distant, over the r. of the long island in Squam, is Joe-English Hill, in New Boston.   Over the r. part of Squam is White-Oak Pond, with an apparent island, over whose middle and very far away are the mountains in Temple and Wilton, and still more remote, over a depression in these ridges, is the crest of Mt. Wachuset, in Massachusetts.   The high hills of Sanbornton and New Hampton are nearer, and stand in confused array.   At the r. end of Squam is a small portion of Little Squam Lake, and between the Dome and these waters are the low and wooded ridges of the Squam Range, flanked on the r. by Morgan Mt.   Far down on the horizon, over the S. end of this range, is the flattened dome of Monadnock, on whose l., and almost equally distant, is Crotched Mt., in Francestown.   A trifle to the r. of Monadnock, but much nearer, is Mt. Kearsarge, with a level ridge running from the l. to a sharp pyramidal peak.   In front and to the r. are the Ragged Mts., with the high hills of Warner on their l., far beyond Ashland village; and over their r. is Lovewell's Mt., in Washington, touching the sky-line.   Near at hand, and more to the r., is the flat top of Mt. Prospect, directly over which is the long ridge of Sunapee Mt., clear-cut against the sky.   The dark forests of Plymouth Mt., just across the Pemigewasset Valley, are in like manner overlooked by Wantastiquet Mt., which is opposite Bellows Falls. The ledgy and level top of Mt. Weetamoo is seen below, on the S. W., with the white houses of Plymouth on the l. and Campton on the r.; and just to the r., due S. W., are the three graceful domes of Mt. Cardigan. Far beyond Cardigan, and below the serried ridges of the hill-country of Southern Vermont, it is thought that the Massachusetts peak of Greylock may be seen.   On the r. flank of Cardigan falls the ridge of Croydon Mt., which is crested with white ledges.

Ascutney lifts its stately blue peak on the r. of Croydon, and beyond it are the billowy Vermont hills; while far down on the horizon Mt. Equinox lifts its sharp point on the l. of Ascutney and Mt. Æolus shows a rounded crest on the r.   The last two mountains are near the Battenkill River, which is a tributary of the Hudson.   The view next extends across the Pemigewasset Valley and up the course of Baker's River, the first notable peak being Mt. Stinson, in Rumney.   Over its l. flank is the long and low Moose Mt., near Dartmouth College, and on the r. are remote ranges in

Vermont. A little more distant and on the r. of Stinson, is Smart's Mt., marked by a high round black swell with equal slopes. To the r. and very far away is the blue lance-point of Killington Peak, flanked by others of the mountains near Rutland. Just to the r. of Smart's Mt., due W. across the valley, is the Mt.-Carr Range, a long and shapeless wooded ridge reaching the sky, flanked by Mt. Kinneo and the rounded head of Mt. Cushman. Toward the r. is Ellsworth Pond, over which is the great mountain in Ellsworth township. In the foreground is the long valley of Mad River, running down to Campton Village and apparently prolonged to Mt. Stinson. By standing on the heap of stones at the beacon the blanched cliffs of the Sachem Peak may be seen below, over which are the rock-banded sides of Welch Mt., with a long purplish perpendicu'ar stripe. Still beyond are the cliffs on Fisher Mt., and further up is Mt. Waterno-mee, on whose r. is the imposing plateau of Moosilauke, high uplifted over all. Toward the N. W. and close at hand below is the semi-detached ridge of Sandwich Dome from which the so-called Jennings Peak rises on the l. and the Noon Peak on the r. Over the former is Green Mt., whence long ascending ridges mount up on the r. to the white rocks which form the crest of Tecumseh.

Over the ridge which runs to the r. from Moosilauke is the slender peak of Black Mt. in Benton, far beyond which is the Vermont summit of Camel's Hump, with its pointed top falling off suddenly to the r. It is believed that Mt. Mansfield may also be seen, a short distance to the r. of Camel's Hump. On the l. of Tecumseh, several miles away up the Pemigewasset Valley, is Mt. Kinsman, with an acute point on the r.; and on the r. of Tecumseh, beyond the Black Mt. of Thornton Gore, are the light-colored ledges on Mt. Cannon, near the Profile House. Over the great ridges S. of the Hancock Branch is the wonderfully sharp spire of Mt. Liberty, with Mt. Flume on the r. and nearer. Haystack is keenly marked against the sky; and the thin serrated crest-line of Lafayette rises beyond, at the N. end of the Franconia Range. In the foreground, over Noon Peak, are the houses and clearings about Greeley's, in Water-ville, above which towers the immense mountain of Osceola, with ledgy sides, a slide near the r. centre, and a bold peak on the r. Between two of its swelling crests is seen a point of the Twin-Mt. Range; and to the r. of the E. peak, over and beyond the gap, is the double-headed ridge of Mts. Guyot and Bond. To the r. of the deep notch N. of Greeley's is Mt. Kancamagus, over which are the clustered tops of Mt. Hancock, with the majestic Mt. Carrigain on the r., falling in terraces to the r. from its high hemispherical crest. To the l. and beyond is a portion of the Field-Willey Range; while to the r. are the bright stripes on Mt. Webster, over Mts. Lowell (marked by a white slide), Anderson, and Nancy. Beyond and over these is the Presidential Range, — Clinton, against the sky; Pleasant, under the l. of Jefferson; Franklin and Monroe, against Washington; and

Washington seen at the end of the great ravine of the Mt.-Washington River, with Boott's Spur running to the r. and falling off into the lofty Mont-alban Ridge. On the l. of Washington is the sierra of Mt. Clay, with the symmetrical top of Mt. Jefferson on its l.

Across the Waterville glen to the N. E. is the great mass of Tripyramid, abounding in peaks and brilliantly decorated by the great white slide near its centre. Between the N. peaks is Mt. Carter's crest, blue in the distance; and over the gap r. of the slide is a part of Mt. Moriah. Farther to the r. are the upper parts of Baldface, Mt. Eastman, and Double-Head, with a yet more distant peak in Maine. The view-line next reaches the Conway Kiarsarge, Mt. Pequawket, with its top marked by a hotel and its r. flank falling into the N. slope of Passaconaway, a bold black mountain not far distant. A little N. of E. and 5 – 6 M. distant, is the whole W. side of Whiteface, with high white cliffs on the r., and the blue hills of Hebron far away beyond. To the r. is Mt. Paugus, bearing long lines of bare ledges and flanked by the green cone of Mt. Wonnalancet; and yet farther to the r., at the E. end of the Sandwich Range, is the pinnacle of Chocorua.

The Acteon Range

consists of three white peaks below and N. W. of Sandwich Dome, overlooking the Mad-River Valley. They lie nearly in line with each other, and are favorably seen from the middle reaches of the path up the Dome. The peak nearest to the W. has been ascended in 1¼ hrs. by turning in from the road at Dolloff's, and bearing to the l. from the Dome path, directly towards the range. The peaks consist of vast piles of white rock, affording good open ground and but little encumbered with bushes. Plenty of blueberries are, however, found here in their season.

The journey along the ridge of the peaks takes about 1 hr., crossing the middle hummock and traversing a small hollow beyond, which is filled with dwarf trees. The third peak is one of the finest in the mountains, and has been called " Chocorua in miniature." It is a needle of white rock, accessible only on the W. and E. sides, and girt with overhanging cliffs above. It is inaccessible on the N. or S. The architecture of these three white crests is very interesting on account of its decided character and graceful outlines. They command good views over the Mad-River and Pemigewasset Valleys, and the stately mountains of Waterville.

Prof. C. E. Fay has explored these peaks, and named them collectively the *Acteon Range*. (Acteon was the last chief of the Pemigewasset tribe.) The chief peak (formerly called *Bald Knob*) he names the *Sachem Peak*.

Jennings Peak is N. of Sandwich Dome, and is reached from the Sachem Peak by descending through a ravine in which dwarf trees are found. It is a very steep and craggy summit, somewhat resembling that of Tecumseh, and commanding a broad view, which includes Mt. Washington.

139. Mounts Whiteface and Passaconaway.

Whiteface is 4,007 ft. high, and is one of the chief summits of the Sand-wich Range. Its name is derived from the color of the cliffs on the S. side of the peak, which were stripped by a great land-slide, in October, 1820. This peculiar aspect of the crest is visible from many points in the lake-country, and serves to identify the mountain. On the other sides the peak is covered with dense woods. Whiteface is connected with Sand-wich Dome by the plateau of Flat Mt., and with Passaconaway by a high ridge. The view from the summit is of exquisite beauty, especially on the side towards the lakes.

The summit of the mountain is marked by a bold and picturesque pile of white rocks, on which a signal of the U. S. Coast Survey formerly stood. The W. and N. W. parts of the crest are covered with trees, through which a path passes to a projecting ledge on the N. W., whence that sec-tion of the horizon is visible. On the N. E. part of the peak is a spring of clear water, reached by a well-trodden path. The remains of a rude camp are found near the signal, and parties occasionally pass the night there, finding an abundance of wood and water close at hand, and enjoy-ing the sunset and sunrise views.

The path up Whiteface is 4 M. long, and is considered as somewhat difficult. It leaves the McCrillis farm (see Route 161) in the N. E. part of Sandwich township, where also a guide can be procured for the ascent ($ 2.00 a day). The path goes through the forest for a long distance, and then begins the ascent of the slopes on the S. side of the mountain. It is not a good path to follow, since it often becomes very obscure and might be easily lost. The general course is on the W. edge of the great ravine which is cut into the S. E. slope, and during the last mile it passes over long reaches of ledges and annoying rocks. The ascent may be made in 3 - 4 hrs.

* * *The View* from the summit of Whiteface is one of the most beautiful in the State, its chief features being the mountain-girded and fertile plains of Tamworth and Sandwich and the outspread lakes of Winnepesaukee and Squam. N. N. E. from the signal-ledge is the peak of Mt. Washing-ton. with Monroe, Franklin, and Pleasant to the l., and the high point of Jefferson between Pleasant and Monroe. Below this cluster and to the r. are the red ledges along the slopes of Mt. Crawford, Giant's Stairs, and Mt. Resolution, on whose r. is the knoll of Mt. Parker. Still nearer, and in the same line of direction, is the elongated ridge of Tremont, marked by three low nubbles, and flanked by the Bartlett Haystack on the r. Back on the horizon, and adjoining the Presidential Range on the r. are the high blue ridges of Moriah, Wild-Cat, and Carter Dome, below which are Iron Mt. and Black Mt., in Jackson, over the nearer Bear Mt.

Passaconaway is the high and ponderous peak just across the ravine, 1½ M. N. E. of Whiteface, and its wooded and nubbly ridge runs out to the r. Over this line of heights is the bold and rock-crowned crest of Moat Mt., with the sharp cone of Kiarsarge looming beyond. To the r.

are the Green Hills of Conway, turning blue in the distance, and the bright speck of Lovewell's Pond is next seen. Returning to the nearer ridge, the red and ledgy flanks of Mt. Paugus are overlooked by the superb ridge and formidable peaks of Chocorua, beyond which is the distant blue mass of Mt. Pleasant, with a white hotel on its middle crest. It is probable that Sebago Lake can be seen in clear weather. Nearly E. are Walker's Pond (near Conway), Chocorua Lake, and Silver Lake (or Six-Mile Pond), in Madison. Beyond the lake are Lyman Mt., and Gline Mt.; and the broad waters of Ossipee Lake brighten in the S. E., overlooked by the Green Mt. in Effingham. Over the r. flank of the latter is seen the distant glitter of Province Pond, in Wakefield. The long line of the Ossipee Mts. fills the S. S. E. with its blue upheavals, and the foreground is occupied by the towns of Tamworth and Sandwich, dotted with small hamlets. This vast apparent plain is composed of a rich mosaic of dark-green woods and light-green clearings, studded with farm-buildings and banded by white highways. The open country is diversified by numerous bright sheets of water. — Great-Hill, White, and Elliott Ponds, toward Ossipee Lake; the Bearcamp and Red-Hill Ponds on the S., and the apparently isolated ramifications of Lake Winnepesaukee, far up in Moultonborough. Just to the l. of and beyond the black Melvin Peak, of the Ossipee Mts., is the low point of Copple Crown, S. of Wolfeborough.

To the r. of the Ossipee Range are the long and shining levels of Lake Winnepesaukee, seen from Moultonborough Neck to Merrymeeting Bay, with all its fleets of islands and floating bits of forest. Long Island is nearly in line with the high double peak of Mt. Belknap, in Gilford, and farther S. is the knoll of Rattlesnake Island. Nearly due S. is the little archipelago off Meredith Bay, clustering about and below Bear Island. The long double ridge S. S. W. of Whiteface, over Centre Sandwich and Red-Hill Pond, is the far-viewing Red Hill, over which is a section of Meredith Bay. On the r. of Red Hill is Squam Lake, the most beautiful element in the landscape, with its bright bosom dotted with green islands and broken into placid coves by far-projecting capes. Over the r. end of Squam is the remote blue cone of Kearsarge.

In the nearer S. W. is Flat Mt., with its light-green plateau, on whose farther terrace is the Flat-Mt. Pond. On the l. of this point is Young Mt., round and smooth, to whose r. is the sloping flank of Mt. Israel. Farther away down the valley are Mt. Prospect and Plymouth Mt., with Mt. Cardigan still beyond. The view next includes the lofty and massive Sandwich Dome, 4 M. distant, with the Sachem Peak on its r., beyond which are the light-colored ledges of Welch Mt. (over the Mad-River Valley), running N. E. to the flank of Green Mt., whence the ridge ascends to the peak of Tecumseh. Directly over Tecumseh is the long blue swell of Moosilauke, supported on the S. by Mt. Carr.

From a rock about 50 ft. from the peak (by a path through the woods) the remainder of the circuit may be viewed. First on the l. is Tecumseh, next to which comes the symmetrical peak of Osceola, about N. W. Across the nearer ravine, in the direction of N. N. W., is the S. point of Tripyramid, whose remaining summits extend up to the high ridge. On its l are the distant ranges of Franconia, and on the r. is the great wilderness watch-tower of Mt. Carrigain. The black ridge of Hancock and the slide-scarred sides of Mt. Lowell are also visible in this direction; and Mt. Willey is due N., over the nearer Mt. Anderson.

Passaconaway is the loftiest of the Sandwich Mts., exceeding either Whiteface or Chocorua, and is so nearly isolated by ravines on the E. and W. that it presents a remarkably massive and commanding appearance. It reaches the altitude of 4,200 ft. When seen from distant points, it resembles a symmetrical dome of a dark color. It is wooded clear over the summit, and is therefore of but little interest to tourists. The easiest route to its crest is from Shackford's, in Albany, by ascending the ravine of Down's Brook, the distance being 3 M. It can also be visited by following the long ridge which leads up to it from Whiteface. The distance is but $1\frac{1}{2}$ M., but the ridge bristles with stunted trees. Another route is by way of the *Birch Intervale,* where the corners of Waterville, Albany, Tamworth, and Sandwich meet.

140. Mount Chocorua

is probably the most picturesque and beautiful of the mountains of New England. It is seen from the border-towns of Maine and New Hampshire for many leagues, and from the W. borders of the lake-country, always presenting to view the same blanched peaks, sharp in outline and unique in form. The destruction of the forests on the upper part of the serrated ridge adds greatly to this striking appearance. Its various aspects to the æsthetic observer may be seen from the following adjectives which Starr King applies, in different places, to this peak; defiant, jagged, gaunt and grisly, tired, haggard, rocky, desolate, craggy-peaked, ghost-like, crouching, proud, gallant, steel-hooded, rugged, torn, lonely, proud-peaked, solemn, haughty. The height of Chocorua is 3,540 ft. The main peak is surrounded by steep cliffs, in many places impassable, and has a long pile of rocks on its crest, under which shelter may be gained from N. and W. winds. To the N. is a subordinate peak of fine proportions, separated from the main crest by a small hollow. The ravine of the Chocorua Brook cuts deeply into the mountain from the S. E.; and the gorge in which are the Champney Falls enters in the opposite direction. The mountain is locally famous for vast quantities of blueberries, in their season; and is then visited by hundreds of people from the adjacent towns.

The brilliant peaks of Chocorua are formed of a crystalline labradorite called Chocorua granite, which was erupted during the great cataclysms of the Labrador period. It stood upon the site of one of the islands of porphyritic gneiss which constituted the first dry land in New Hampshire, emerging from the ocean at the dawn of geological history. Sharp as is the present peak, it is but the dwarfed remnant of the colossal spire which stood here before the glacial currents swept and ground it away. "This action shows why we have no pinnacles of rock, such as abound in the Alps. The Swiss glaciers have plowed around these pinnacles and left them standing; but the American continental Drift was of such vast proportions that the needles disappeared as though they were pebbles in the path of the ordinary river of ice."

The most popular path is from the Hammond farm, which is reached by a by-way ⅜ M. long, diverging from the Conway road near Tibbetts's Mills, 2 M. from the Chocorua-Lake House, and 14 M. from N. Conway. This route is generally plain and easy, and is sometimes ascended by ladies. It leads upward for over an hour through the woods, and reaches the first high ridge in about 1½ hrs. from the farm-house. Ascending thence over several shoulders of the mountain, where vast quantities of blueberries grow in their season, the walking is found to be good, and the rapidly opening views on either side are full of interest. The final peak can only be reached by flanking it and ascending cautiously on the upper side.

There is another path which leaves the carriage-road at J. Piper's, about 4 M. N. of the Chocorua-Lake House and 12-13 M. from N. Conway. The distance is 2½ - 3 M., and the ascent requires fully 3 hours. The trail ascends the great ravine of the Chocorua Brook, and is easy to traverse as far as the old hut, where people sometimes spend the night. Beyond this point it is steep, and passes a broad area of fallen timber. The final pull up the summit is very steep and arduous; and near this section of the path is a refreshing spring.

There is also a path on the S. W. side of Chocorua, which is much used by the people of Tamworth and Sandwich. It enters from the Derrell farm, about 5 M. from Tamworth Centre, and is 3 M. long, striking the main ridge about ¾ M. from the peak. This route is somewhat less definite than that from the S. E. side, but it is said to be easier in its grades. There is a tradition in the adjacent country that the tomb of Chocorua was on the line of this path, near the foot of the mountain, — but no one knows where it can now be found.

The peak can also be visited by advancing from the Swift-River country up the ravine to the Champney Falls, but this route, though short, is very arduous, leading through areas of fallen timber. Whoever ascends this way will endorse Dr. Jackson's statement: "Those who wish for a laborious mountain excursion can ascend Chocorua Mt. from Albany."

"How rich and sonorous that word Chocorua is! . . . . Does not its rhythm suggest the wilderness and loneliness of the great hills? To our ears it always brings with it the sigh of the winds through mountain pines. . . . . No mountain of New Hampshire has interested our best artists more. It is everything that a New Hampshire mountain should be. It bears the name of an Indian chief. It is invested with traditional and poetic interest. In form it is massive and symmetrical. The forests of its lower slopes are crowned with rock that is sculptured into a peak with lines full of haughty energy, in whose gorges huge shadows are entrapped, and whose cliffs blaze with morning gold. And it has the fortune to be set in connection with lovely water-scenery, — with Squam, and Winnepesaukee, and the little lake directly at its base. Its pinnacle, too, that looks so sheer and defiant, is a challenge to adventurous pedestrians among the mountains, which is accepted now and then by parties every summer.

"With the exception of Mt Adams, of the Mt.-Washington range, there is no peak so sharp as Chocorua. And there is no other summit from which the precipices are so sheer, and sweep down with such cycloidal curves. One must stand on the edge of the Grand Gulf, a thousand feet below the summit of Mt. Washington, to see ravine-lines so full of force, and spires of rock so sharp and fearful. It is so related to the plains on one side, and the mountain gorges on the other, that no grander watch-tower, except Mt. Washington, can be scaled to study and enjoy cloud-scenery.

"On one side of its jagged peak a charming lowland prospect stretches E. and S. of the Sandwich range, indented by the emerald shores of Winnepesaukee, which lies in queenly beauty upon the soft, far-stretching landscape. Pass around a huge rock to the other side of the steep pyramid, and you have turned to another chapter in the book of nature. Nothing but mountains running in long parallels, or bending, ridge on ridge, are visible, here blazing in sunlight, there gloomy with shadows, and all related to the towering mass of the imperial Washington." (STARR KING.)

** *The View.* — On the W., below and adjoining Chocorua, are the ledges on Mt. Paugus, whose top is nearly level, and has no peak. Over its r. side is the dark and prominent Passaconaway, falling off sharply on the r.; and over its long S. flank, across the upper clearings of Sandwich, is Mt. Israel, rising behind the low cone of Young Mt., Mt. Wonnalancet being in the foreground, S. of Paugus. On the r. of Israel, and much higher, is the dark mass of Sandwich Dome. Whiteface is nearly W., on the l. of and adjoining Passaconaway. On the r. of and beyond Passaconaway is the long and many-headed ridge of Tripyramid, beyond which are the sharp peaks of Tecumseh and Osceola, the latter being seen on the l. of the white mound of Potash, which is below in the Swift-River Valley. Much farther away in this direction (W. by N.) is the high plateau of Moosilauke, over the Blue Ridge. About N. W., toward Mt Hancock, is the square-topped mass of Green's Cliff; and the high spires of the Franconia Range rise on the distant horizon, with the gray sierra of Lafayette most conspicuous. On the r. of Hancock is the imposing pile of Mt. Carrigain, looming up boldly out of the Pemigewasset Forest; and on its E. side is the sharply cut and profound gorge of the Carrigain Notch, through which a part of the Twin-Mt. range is seen. Close on the r. of the Carrigain Notch is the remarkably pointed peak of Mt. Lowell, flanked on the r. by Mts. Anderson and Nancy, on the same ridge. Under this range is Tremont, with its highest point between Anderson and Nancy; and Mt. Hale appears over Anderson. On the r. of Tremont, and near it, is the sharp crest of the Bartlett Haystack; and between and far beyond Tremont and Haystack are Mts. Willey and Field. The purple cliffs of Mt. Willard are over the crest of Haystack, in the White-Mt. Notch, through which a part of Mt. Deception is seen.

About N. N. W., 6 M. distant across the Swift-River Valley, is the long ridge of Bear Mt , covered with woods, and on the r. of Haystack. Between Haystack and Bear are seen the richly colored stripes on the side of Mt. Webster. Farther to the r., over Bear, is Mt. Clinton, below which is the red crest of Crawford, with Resolution and Giant's Stairs on its r. Mt. Pleasant is over the r. of Bear, showing a round and dome-like crest, beyond and above which are Franklin and Monroe, due N. of Chocorua. The houses on Mt. Washington are about N., between Mts. Parker and Langdon, beyond the Saco, and Bear and Table Mts., N. of the Swift River. Table is the mountain on the r. of Bear, in the same ridge, and Iron Mt. is over its flank. Above Iron is the deep cleft of the Pinkham Notch, through which Mt. Madison is seen. On the r. of and adjoining Table is the long and imposing ridge of Moat Mt., over whose N. peak are the crests of Thorn Mt. and Double-Head, with Baldface lifting its white ledges beyond. The pyramid of Kiarsarge rises above the S. peak of Moat, and is marked by a house; and the rocky mounds of the Eagle Ledge and the Albany Haystack are across the Swift River, toward the S.

peak. To the r. of and S. of Kiarsarge are Blackcap, Middle Mt., and others of the Green Hills of Conway, with the clearings of N. Fryeburg and Lovell visible through their gaps.

The character of the view now changes from a tumultuously upheaved land of mountains to populous plains, dotted with hamlets and ponds, and diversified here and there by low ridges. The white Conway road runs N. along the base of Chocorua, curving away from its formidable rocky flanks, and lined with farms. The beautiful meadows of the Saco emerge from behind Moat Mt., and pass away to the E. in graceful bends. The fair village of Fryeburg is about 15 M. E. N. E., on the l. of and beyond which are the bright waters of Kezar, Upper Kezar, Upper Moose, and Long Ponds. Lovewell's Pond is close to Fryeburg, on the r. Nearer at hand is the bright hamlet of Conway Corner, at the confluence of the Swift and Saco Rivers. Farther out in this direction is Mt. Pleasant, a long and rolling ridge which uplifts a white hotel near its centre. On the r. of Conway, due E., is the broad mirror of Walker's Pond, over which are the Frost and Burnt-Meadow Mts., in Brownfield. Farther to the r., over Cragged Mt. and the hills of Hiram and Sebago, is the broad gleam of Sebago Lake. To the E. S. E. the view passes over the Gline and Lyman Mts., and across two counties of lowland Maine, to the city of Portland, at the gates of the sea. On a clear day a wide extent of the ocean can be seen in this direction, and extending away to the r. Farther to the r., over the adjacent Whitton Pond, are the distant hills of Cornish and Limington; and nearly S. E., over the hamlet of Madison, is Mt. Prospect, in Freedom. About 1 M. from Madison, and 6 M. from Chocorua, is the broad oval of Silver Lake, with the formless ridge of the Green Mt. in Effingham over it. The ampler sheet of Ossipee Lake is to the r. of and beyond Silver Lake, and on its r., far out on the horizon, over the hills of N. Wolfeborough, is the crest of Copple Crown.

Chocorua Lake is close to the base of the mountain, on the S., with its gracefully curving sandy beaches, bordered with trees; and the white Chocorua-Lake House is on the hill beyond, towards the hamlet of Tamworth Iron-Works, with its tall-spired church. In the plain beyond are the hamlets of Tamworth Centre, S. Tamworth, and W. Ossipee, and the White and Elliott Ponds. Then comes the long Ossipee Range, filling the horizon from S. to S. S. W., with the ledgy sides of the Whittier Peak, below S. Tamworth. The twin Belknap peaks peer over the Ossipee Mts. and are clearly seen. On the r. of the range are portions of Moultonborough Bay, Lake Winnepesaukee, and Northwest Bay, studded with islets and divided by peninsulas. The Bearcamp and Red-Hill Ponds are next seen, with the hamlet of Sandwich Lower Corner, beyond which rises the double swell of Red Hill. About S. W., over the white village of Centre Sandwich, is the exquisite beauty of Squam Lake, with its blue bosom dotted with wooded islands. The sharp crest of Kearsarge is over its l. part; the Bridgewater Hills are over the centre; and Mt. **Prospect,** near Plymouth, is farther to the r.

Chocorua (pronounced *Choc-cór-oo-a*) was named after an Indian chieftain who was killed near its summit by white men. The legend was thus narrated to the Editor by a venerable man of Tamworth, who had written it down forty years ago as he received it from his ancestors: When the Pequawket Indians retreated to Canada, after Lovewell's battle, Chocorua refused to leave the ancient home of his people and the graves of his forefathers. He remained behind, and was friendly to the incoming white settlers, and especially with one Campbell, who lived near what is now Tamworth. He had a son, in whom all his hopes and love were centred. On one occasion he was obliged to go to Canada to consult with his people at St. Francis, and, wishing to spare his son the labors of the long journey, he left him with Campbell until his return. The boy was welcomed to the hut of the pioneer, and tenderly cared for. One day, however, he found a small bottle of poison, which had been prepared for a mischievous fox, and, with the unsuspecting curiosity of the Indians, he drank a portion of it. Chocorua returned only to find his boy dead and buried. The improbable story of his fatality failed to satisfy the heart-broken chief, and his spirit demanded vengeance. Campbell went home from the fields one day, and saw the dead and mangled bodies of his wife and children on the floor of the hut. He tracked Chocorua and found him on the crest of the mountain, and shot him down, while the dying Indian invoked curses on the white men.

This legend has been enshrined in American literature in another form. On the restoration of the Stuart dynasty, one Cornelius Campbell, who had been an active and liberal partisan of Cromwell, fled to America, and entered this remote wilderness with his beautiful and high-born wife. Chocorua was a prophet of the then powerful Pequawket tribe, and his son was on intimate terms at Campbell's house. The accidental death of the boy, the murder of the family, and the fate of the chief on the mountain-crags follow as in the authentic story. Another account says that Chocorua was a blameless and inoffensive Indian, a friend of the whites, but, during one of the Massachusetts campaigns against the red men, when the Province gave a bounty of £100 for every scalp brought into Boston, a party of hunters pursued the unresisting chieftain and shot him on this mountain, in order to get the bounty-money.

While dying, Chocorua cried out, " A curse upon ye, white men! May the Great Spirit curse ye when he speaks in the clouds, and his words are fire! Chocorua had a son, and ye killed him when the sky was bright! Lightning blast your crops! Winds and fire destroy your dwellings! The Evil Spirit breathe death upon your cattle! Your graves lie in the war-path of the Indian! Panthers howl and wolves fatten over your bones! Chocorua goes to the Great Spirit, — his curse stays with the white man!" The literary account says that the settlement was afterwards wasted by pestilence, storms, and Indian attacks, and was abandoned by its people. In point of fact, however, the towns in this vicinity were never molested by the Indians. For many years it was impossible to keep cattle in Albany (the town wherein the mountain stands), for they sickened and died soon after coming there. The people laid this strange fatality to the operation of Chocorua's curse, until queenly Science found out that it was due to the presence of muriate of lime in the water which they drank.

" On the cliff's extremest brow,
  Fearless stands Chocorua now ;
  Last of all his tribe, and he
  Doomed to death of cruelty.
  O'er the broad green vales that lie
  Far beneath, he casts his eye.

" ' Lands where lived and died my sires,
  Where they built their council-fires :
  Where they roamed and knew no fear,
  Till the dread white man drew near :
  Once when swelled the war-cry round,
  Flocked a thousand to the sound ;
  But the white men came, and they
  Like the leaves have passed away.

" ' Wo to them who seek to spoil
  The red owners of the soil !
  Wo to all who on this spot
  Fell the groves, or build the cot !
  Blighted be the grass that springs !
  Blighted be all living things !
  And the pestilence extend,
  Till Chocorua's curse shall end ! '

" On his murderers turned he then,
  Eyes shall ever haunt those men ;
  Up to heaven a look he cast,
  And around — beneath — his last!
  Far down and lone, his bones are strown,
  The sky his pall, his bed of stone."

CHARLES J. FOX.

141. Albany.

Albany is a cold and rugged township of 339 inhabitants, with only about 8,000 acres of improved land. The Swift River flows through it from W. to E., and is hemmed in by high and massive mountains on either side. There are several fine trout-streams flowing from their defiles, and the Swift River has good fishing on its upper waters. Bears and deer are also found here. In 1839 a railroad route from Portland to Vermont was surveyed up this valley and thence through Waterville, and Gov. Hill vainly prophesied that it would be built by the year 1850. The *Swift-River Railroad* has recently been chartered. The town has neither doctor, lawyer, nor clergyman, and but one merchant. It was settled in 1766, and bore the name of *Burton* until 1833. In 1860 its population was much larger than it now is, but many of the people fled to Canada to escape the draft during the civil war.

At an early day this township was sold by land-speculators to New-York and Boston merchants, even Chocorua itself being platted off and disposed of. The ill-omened name of Burton was then changed to that of Albany. The Swift River is 15 M. long, and much lumber is floated down its rapid stream. The soil in the valley is naturally rich, being formed from soft, decomposing rock.

The only road into Albany is from Conway Corner, up the valley of the Swift River, which is crossed at 7½ M. from Conway. The road is very dull until it passes Allen's Mill, since it runs through deep forests. There is no public conveyance.

The *Swift-River Falls* are about 10 M. from Conway Corner, and are directly alongside the road, on the r. The river here plunges downward for a few feet through a series of boiling eddies, and is narrowed into a straight passage between regular and massive granite walls about 20 ft. high and several rods long. The stream roars down through this contracted gorge, and overflows it during high water.

About 3 M. above the falls is *Allen's Mill*, beyond which the road enters the broad basin of the *Swift-River Intervale*, with impressive mountain views on either side. On the l. are Paugus, Passaconaway, and Whiteface; and on the r. are the ridges of Bear Mt. and Tremont. In this glen is Shackford's, a large house where fishing tourists and other summer voyagers are entertained with simple and substantial fare.

The * **Champney Falls** are in one of the ravines of Chocorua, and are very pretty in seasons of high water. They are reached by diverging to the l. from the road on to a logging-road, at the first bridge E. of Allen's Mill. The distance from the main road to the falls is nearly 2 M. The falls are best seen from the large rock at their foot, and consist of a long broken plunge of the mountain-stream over ledges 60 – 70 ft. high. On the l. and in full view is another fall, on a tributary stream, exhibiting a rare grace and delicacy of outline. This has been called the *Pitcher Fall.*

" Not a dozen rods away, but almost hidden by the trees, we discover one of the most beautiful falls in New Hampshire. . . . The sunbeams fall aslant through the trees ; the eye follows the high perpendicular ledge that runs at right angles to the stream, and through the leaves of the trees we can see a small stream where it comes over the ledge, then falls down, striking the rock that projects just enough to throw the water in spray, and break, for an instant only, the continuity of the stream. In the entire fall there are three of these projections, where the water is thrown in spray, and after the last continuous fall it rests in a quiet basin, where it flows out and runs into the stream we had followed." (HUNTINGTON.)

The Potash is a singular little mountain, whose top and sides are composed of coarse white granite, rendering it a conspicuous object even when seen from a great distance. It is near the foot of Passaconaway, and about 1¼ M. from Shackford's, whence it is easily visited. The prospect to the N. and N. E. is broad and interesting, and amply repays the labor of the ascent.

" Towards Mt Carrigain the view is almost unobstructed, and there are many gentle undulations, with here and there a granite cliff standing out in bold relief, besides magnificent forests sweeping away up to the summits of the mountains; for none of the mountains to the W. have been denuded of trees. In full view, Mt. Carrigain stands in all its massive grandeur, while N. and S. there are sharp peaks and mountain-ridges. Still to the N., and yet not so far distant but that each peak and mountain-ridge stands in sharp outline, the White Mts. rise in successive culminations, until Mt. Washington — monarch of the range — seems to touch the sky." (J. H. HUNTINGTON)

Church's Falls are on Sabba-Day Brook, and have been visited and painted by F. E. Church. They are reached from Shackford's by ascending the course of the Swift River 1¼ M., to the second stream coming in on the l., near which are the remnants of an old mill. From this point the tributary should be ascended for ½ M.

Prof Huntington gives the following vivid description of the falls on Sabba-Day Brook : " The rock is a common granite, in which there is a trap-dike, and it is the disintegration of this, probably, that formed the chasm below where the steep fall now is. Above, just before we come to the falls, the stream turns to the W., and the water runs through a channel worn in the solid rock, and then, in one leap of 25 ft., it clears the perpendicular wall of rock, and falls into the basin below almost on the opposite side of the chasm. Great is the commotion produced by the direct fall of so great a body of water, and out of the basin, almost at right angles with the fall, it goes in whirls and eddies. The chasm extends perhaps 100 ft. below where the water first strikes. Its width is from 10 to 15 ft., and the height of the wall is from 50 to 60. The water has worn out the granite on either side of the trap, so that, as the clear, limpid stream flows through the chasm, the entire breadth of the dike is seen. The fall of water, the whirls and eddies of the basin, the flow of the limpid stream over the dark band of trap set in the bright, polished granite, the high, overhanging wall of rock, all combine to form a picture of beauty, which, once fixed in the mind, is a joy forever. "

Mount Paugus is the low and massive mountain between Chocorua and Passaconaway, and is chiefly composed of bare ledges, covered and flanked with blueberry-bushes. On its S. W. side is a long and singularly curving slide. There is no well-marked peak on the broad top of Paugus, and it is broken into several irregular parts by shallow ravines. The ascent may be made from Albany, or from the vicinity of the Birch Intervale, above Tamworth Village. The lower ledges are much visited by the country people during the season of berries. The name of *Paugus* was given by one of the lake-country poets, in honor of the valiant chief of the Pequawket Indians (see page 409). The mountain also bears the name of Deer, Middle, Hunchback, Frog, and Bald.   Miss Larcom also named the ragged ridges near and below Paugus the *Wahwa Hills*, and the "bright cone of perfect emerald" on the S. W. she calls **Mount Wonnalancet** (see page 25).

SONNETS.

CHOCORUA.

THE pioneer of a great company
   That wait behind him, gazing toward the east, —
   Mighty ones all, down to the nameless least, —
Though after him none dares to press, where he
With bent head listens to the minstrelsy
   Of far waves chanting to the moon, their priest.
   What phantom rises up from winds deceased?
What whiteness of the unapproachable sea?
   Hoary Chocorua guards his mystery well:
He pushes back his fellows, lest they hear
   The haunting secret he apart must tell
To his lone self, in the sky-silence clear.
A shadowy, cloud-cloaked wraith, with shoulders bowed,
He steals, conspicuous, from the mountain-crowd.

CLOUDS ON WHITEFACE.

So lovingly the clouds caress his head, —
   The mountain-monarch; he, severe and hard,
   With white face set like flint horizon-ward;
They weaving softest fleece of gold and red,
And gossamer of airiest silver thread,
   To wrap his form, wind-beaten, thunder-scarred.
They linger tenderly, and fain would stay,
Since he, earth-rooted, may not float away.
   He upward looks, but moves not; wears their hues;
Draws them unto himself; their beauty shares;
   And sometimes his own semblance seems to lose,
   His grandeur and their grace so interfuse;
And when his angels leave him unawares,
A sullen rock, his brow to heaven he bares.

               From LUCY LARCOM'S *An Idyl of Work.*

THE OSSIPEE COUNTRY.

142. Tamworth

is a town of 1,344 inhabitants, covering 28,917 acres of land, of which 15,304 are improved. It occupies the plain between the Sandwich and Ossipee Ranges, and is surrounded with noble mountain-scenery, contrasting finely with the cultivated lowlands. It is a good town for grazing and fruit-raising, and produces much maple-sugar. About 500 tourists pass portions of every summer within its borders, spending about $ 10,000. Tamworth was settled in 1771, and is named after an English town on the river Tyne.

Tamworth Iron-Works is a small hamlet with a church, a store, and a mill, on the Chocorua River, in the E. part of the town. The iron-works were founded before the Revolution, and have been closed for over 70 years. The metal was obtained from bog-iron ore from the bottom of Ossipee Lake. Here the first American machine-made nails were turned out, in 1775; and here also the first American screw-auger was made (in 1780), the maker having seen an auger on a British prize-frigate at Portsmouth. Many anchors were cast at Tamworth, and were hauled thence to Portsmouth on sledges.

The *Chocorua-Lake House* is a well-arranged summer boarding-house near Tamworth Iron-Works, on the high hill S. E. of Chocorua Lake, and commanding a noble view of Mt. Chocorua. It accommodates 40 guests, at $ 8.50 – 10 a week. The railroad station is at W. Ossipee, 6 – 7 M. distant, over a level road. Madison station is 2 – 3 M. nearer, but the road is rugged and hilly. Mt. Washington and other northern peaks are seen from a hill ½ M. S. of the hotel; and other view-points in the vicinity are Page's Hill, 2½ M. (½ M. on foot); Deer Hill, 3 M. distant on the Madison road; Around the Circle, a route 7 M. long, revealing several fine prospects; Ordination Rock, 4 – 5 M.; Elliott Pond, 4 M.; Conway, 10 M.; N. Conway, 16; and Buttermilk Hollow. Just below the hotel is an elegant villa which was recently built by a Boston gentleman; and on the shore of the lake is Cone's quaint cottage.

Chocorua Lake is over 1 M. long, and is divided into two parts, connected by a narrow strait over which a bridge has been thrown. Row-boats are kept for the use of the summer-boarders, and afford a pleasant mode of diversion and exercise. The shores are partly formed by curving sandy beaches, overhung by shadowy trees. As a fishing-ground the lake is not profitable, though a few trout are found in the distant brooks.

15 *

S. Tamworth is a small hamlet in the S. W. part of the town, on the Bearcamp River, and at the base of the Ossipee Mts. A little way to the E., on the stage-road, is an old inn. It is claimed that large deposits of coal are to be found in the Ossipee Range near this hamlet, and several companies have attempted to develop them.

Tamworth Village is near the centre of the town, on the Swift River, and has a Congregational church, the town-hall, and several shops. The quaint old inn is kept as a summer boarding-house, by the venerable Joseph Gilman; and Levi E. Remick also has a boarding-house. The mail-stage runs daily to and from W. Ossipee station. The hamlet is 10 M. from Centre Sandwich, by a road which gives glorious views of Chocorua, Paugus, Passaconaway, and Whiteface. Nearly 1 M. to the S. W., on this road, is *Ordination Rock*, a great boulder close to the road (on the r.), which supports a handsome marble obelisk. On this rock, Sept. 12, 1792, the Rev. Samuel Hidden was ordained as minister of Tamworth, where he remained for 46 years as pastor, guide, and friend.

" Mr. Hidden was ordained on a large rock (20 × 30 ft., and 15 ft. high) on which 50 men might stand. His foundation must be secure and solid; for this rock will stand till Gabriel shall divide it by the power of God. Early in the morning the people assembled around this rock, men, women, boys, and girls, together with dogs and other domestic animals. It is an entire forest about this place. The scenery is wild. On the W. is a high hill; and N. of this is a mountain called Chocorua, which touches heaven. On the S., and in all directions, are mountains, steep and rugged. . . . . The men looked happy, rugged, and fearless. Their trousers came down to about half-way between the knee and ankle: the coats were mostly short, and of nameless shapes; many wore slouched hats, and many were shoeless. The women looked ruddy, and as though they loved their husbands. Their clothing was all of domestic manufacture: every woman had a checked linen apron, and carried a clean linen handkerchief." (LAWRENCE'S *Congregational Churches.*)

On *Marston Hill*, 1½ M. N. W. of this village, occurred the famous Siege of Wolves. On the evening of Nov. 14th, 1830, messengers rode rapidly through the towns in the valley, proclaiming the startling fact that immense packs of wolves had descended from the northern mountains and were in the forest on Marston Hill. The farmers from all the country-side hastily armed themselves and hastened toward the hill, determined to prevent the enemy from advancing into the towns. A thin line of circumvallation was formed about the forest, and as fresh parties arrived during the night the lines were strengthened on all sides, amid the unearthly howling of the wolves. By the middle of the next forenoon full 600 men were bivouacked around the hill, and the command was assumed by Gen. Quimby, of Sandwich. A strong detail of riflemen was finally sent forward into the forest, and after a sharp fusillade they exterminated the invaders, many of whom, however, had escaped to the mountains about midnight. The dead wolves were transported to Tamworth Centre under escort of the victorious army, and a general jollification closed the quaint episode of the Siege of Wolves. So late as 1815 the inhabitants of Eaton, the next town on the E., were seriously annoyed by wolves, many of which they killed.

Madison is a town of 646 inhabitants, S. of Conway and E. of Tamworth. The surface is undulating, producing good crops of corn and potatoes, except in the W. part, where Silver Lake is surrounded with broad drift-plains. Lead, zinc, and silver are found here in small quantities, and have been unprofitably mined. Much ready-made clothing for the city stores is made here, also 12,000 pairs of shoes annually. The town is somewhat frequented as a summer-resort, on account of its pretty highland scenery and the distant views of the White Mts. Madison village is N. E. of Silver Lake, and has a church and three small boarding-houses. **Silver Lake** is a beautiful sheet of water in the W. part of the town, 2½ M. long and 1 M. wide, with gracefully curving beaches. Its more common name in the country about is *Six-Mile Pond*. A small steamboat will soon be put on this lake, thus affording a pleasant short excursion for the people who sojourn at N. Conway and other adjacent villages.

Eaton is a picturesque hill-town W. of Madison and S. of Conway. The only hamlet is at Robertson's Corner, in the N. W. part of the town, about 5 M. from Madison and from Conway Corner. There is a small inn here. The town has six ponds; and among the highlands deposits of iron, lead, and zinc have been discovered. There were 656 inhabitants in 1870, with a valuation of $132,000. On its W. border are Lyman Mt. and Gline Mt., and on the E. is Cragged Mt. A rustic bard named Shepherd once dwelt here, whose quaint poems, according to Mr. Whittier, "were so bad that they were very good."

143. West Ossipee.

Hotel.—The Bearcamp-River House (formerly Banks's), accommodating 75 guests, at $7 – 14 a week. This hotel is finely situated near the W.-Ossipee station and the Bearcamp River, in a pleasant and quiet neighborhood. *Daily Stages* leave for Centre Harbor, Tamworth, and Sandwich on the arrival of the up trains. A livery stable is connected with the hotel.

W. Ossipee is a station on the Eastern Railroad, 16 M. from N. Conway, near which there are a few houses. Here the romantic stream of the Bearcamp River flows close to the base of the Ossipee Mts., bordered by pleasant meadows and graceful trees. The scenery is very attractive, and includes the Ossipee Range. the Green Mt. in Effingham, and the picturesque peaks of the Sandwich Range, — the alpine-crested Chocorua on the N., the lower ledges of Mt. Paugus, the dark round top of Passaconaway, the shining cliffs of Whiteface, the forest-covered Sandwich Dome, and the conical Mt. Israel, W. by N. The sunset effects on this range are very brilliant and beautiful, and never fail to command admiration.

A short distance from the house, on the Ossipee-Lake road, a line of trees is seen on the r. hand, the farthest of which is a broad-arching and venerable maple. This stately old tree has frequently been visited and is much admired by John G. Whittier, in honor of whom it has been named the *Whittier Maple*. A little way beyond is a break in the fence on the l., and from the meadow beyond is gained a prospect of Chocorua which is a favorite view with the great New-England poet. Mr. Whittier has visited W. Ossipee for many summers, and has often tuned his harp to the music of the Bearcamp waters and the mountain-forests.

Among the favorite drives from W. Ossipee are the pleasant roads toward Centre Harbor (Route 157), to Chocorua Lake (page 344), to Madison village, to Ossipee Lake (page 347), and among the adjacent glens of the Ossipee Mts.

The **Ossipee Mountain** (so called) is the high ridge on the N. E. of the Ossipee Range (see Route 164) and over W. Ossipee. It is easily ascended from the Bearcamp-River House, by crossing the river on the Centre-Harbor road, and ascending to the l. beyond. Most of the top of the ridge is free from trees, and the view is varied according to the standpoint of the visitor. The bold, rocky spur which is first reached overlooks the Ossipee Lake and the peaks to the N. and E. The ascent from this point to the main crest is easy, and leads over broad bare ledges.

The View from the top includes the broad heaths of Ossipee, Tamworth, and Madison, in which are the shining waters of several forest-bordered ponds, with the larger sheet of Silver Lake (Six-Mile Pond) towards Madison village. The line of the Sandwich Range runs from N. W. to N., with the noble white peak of Chocorua, a little W. of N., over Chocorua Lake, and flanked by bold peaks on the W. and N. On the r. of Chocorua, and more distant, are the ledgy tops of Moat Mt., over which are the crests of Mt. Wild-Cat and the Carter Dome. The Albany Haystack is at the S. base of Moat. The white band of the railroad runs N. from W. Ossipee, over the long plains, and enters the Saco Valley, by which it passes into the mountains. On the r. of Moat is the rounded crest of Double-Head; on whose r., and beyond, are the white ledges of Baldface. Then comes the step-like spur of Mt. Bartlett, on whose r. and connected is the high and symmetrical cone of Kiarsarge, with a hotel on its apex. The Green Hills of Conway are then seen, with Blackcap as the highest; and in the foreground is the Hedgehog Hill, with other wooded eminences.

The view then passes over the Fryeburg Valley, the land of Pequawket; and about N. E. is Mt. Pleasant, with a white hotel on its middle crest. The mountains of Brownfield and Porter are nearer and more to the r.; and Mt. Prospect, in Freedom, is nearly E. About S. E. is the great mass of the Green Mt. in Effingham, 6–8 M. distant across the basin of Ossipee Lake. The view now sweeps out over the plains of lowland Maine, towards the sea-coast, covering most of the townships of York County.

The prospect to the S., S. W., and W. is shut out by high ridges of the Ossipee Range. Partial views of the border-towns on the S. S. E., and S. may be gained by advancing along the ridge and climbing on a pile of rocks at its S. end.

The town of Ossipee has 1,822 inhabitants, and occupies 55,000 acres, of which only 17,740 are improved. Within its borders are the hamlets of W. Ossipee, Ossipee Centre, Moulton's Mills, Ossipee Corner, Water Village, and Leighton's Corners. About $ 60,000 worth of lumber, bedsteads, shoes, and hose are made here every year; but the majority of the people are farmers. The N. E. district consists of dreary drift-plains, but the highlands on the W. afford pleasant scenery and good pasturage. Several hundred summer-tourists come hither every year, to enjoy the lake and mountain scenery.

Ossipee Centre (*Cottage House*) is a small hamlet ½ M. from the station of the same name on the Eastern Railroad, and on the Beech River. It has two churches and several stores. About 1 M. W. is the hamlet of *Moulton's Mills*, whence the road is prolonged to the secluded Dan-Hole Ponds, under Black Mt. Another road runs N. W. into the remote glen of Lovewell's River, among the Ossipee Mts. *Ossipee Corner* is a petty hamlet S. of the Centre, with two inns. It is the shire-town of Carroll County. The road running W. from Ossipee to Tuftonborough gives several pleasing prospects over the lake-country. Ossipee Centre is the station for Effingham, Freedom, and Green Mt. (See Routes 145 and 146.)

144. Ossipee Lake.

Ossipee Lake is about 3½ M. long, and 2 M. broad, and, with its bays and inlets, contains nearly 7,000 acres. It is destitute of islands, and is partly surrounded by a beautiful border of silvery beaches. The waters of Pine, Lovewell's, Bearcamp, and Six-Mile Rivers empty into the lake, and are discharged by the Ossipee River, leaving its S. E. corner. The shape of the lake is oval; its waters are rather shallow, and are distinguished for their remarkable transparency. The immediate vicinity of the shores is occupied by wide and gloomy heaths, but the mountain-views enjoyed from the waters are attractive, including Green Mt. on the S. E., the great Ossipee Range on the W., and Chocorua and other peaks of the Sandwich Range on the N. and N. W. The waters of the lake are stocked with pickerel and other fish. A dam has been built near the outlet to increase and regulate the supply of water for the mills at Saco and Biddeford, and has raised the surface of the lake several feet. The name Ossipee is of Indian origin, and is derived from *coös*, " pine-tree," and *sipe*, " pond."

Daniel Smith's farm lies S. of Lovewell's River, where it flows into Ossipee Lake, and between the road and the lake, 4 M. from W. Ossipee and 2 M. from Ossipee Centre. In the field N. of the house, and visible from the road, is an ancient Indian monumental mound, about 50 ft. in diameter, and 10 ft. high, from which skeletons, tomahawks, and other relics have been taken. A fine view of Chocorua is enjoyed from this point. Farther down on the alluvial meadow, half-way to the lake, and on the S. bank of Lovewell's River, are the scanty remains of Lovewell's Fort. This was a palisade of about one acre, fronting to the E. toward the lake, and so situated that a besieging force of Indians could not cut off its supply of water.

Here Lovewell (in 1725) left his surgeon and a small detachment, with the provisions of his expedition, while he advanced to the fatal battle at Pequawket. When he met the Indians one of his men fled at the first fire, and bore back the tidings to this fort that Lovewell's command had been ambushed and annihilated. The garrison evacuated the works and retreated rapidly, so that when the survivors of the expedition reached this place they failed to find supplies and aid and were forced to continue their terrible flight. Shortly after the middle of the seventeenth century English carpenters were sent up here and built a timber-fort 14 ft. high, with projecting flankarts, to serve the Ossipee Indians as a bulwark against the forays of the warlike Mohawks. This defence was destroyed in 1676, by Capt. Hawthorne, who advanced to the lakes with a strong force from Newichawannock. Troops of Massachusetts and New Hampshire occupied the ground occasionally throughout King Philip's War, and in November, 1676, they sent a detachment of picked men 20 M. to the N., but they found nothing but "frozen ponds and snowy mountains."

When Walter Bryent was surveying the Provincial line of N. H., in 1741, he traversed this country on snow-shoes, killing deer and meeting "fearful Indians about the Ossipee Ponds and Pigwacket Plain." He often viewed "the White Hills," and saw many nameless lakes from the mountains. On a tree near the Ossipee River he cut his distance-marks, and on returning found there a carving of "a sword handsomely form'd, grasp'd by a hand."

"Lake Ossipee looked like what I fancy the wildest parts of Norway to be; a dark blue expanse, slightly ruffled, with pines fringing all its ledges; and promontories, bristling with pines, jutting into it; no dwellings, and no sign of life but a pair of wild-fowl, bobbing and ducking, and a hawk perched on the tip-top of a scraggy blighted tree." (HARRIET MARTINEAU.)

145. The Green Mountain in Effingham.

Green Mt. is usually visited from the village of *Effingham Falls*, which is reached by a dreary ride of 6 M. from the Centre-Ossipee station, crossing the barren heath on the S. of Ossipee Lake. The village was formerly the seat of iron-works, and is now a poor hamlet, with one church, situated near the outlet of Ossipee Lake. (Travellers are entertained at the house of Mr. Stokes.) About 1½ M. distant, on the E. flank of Green Mt., and but a short distance in the woods, is a natural ice-house, — a deep and cavernous flume in the rock, within whose recesses snow remains until September. The summit of the mountain is reached by following the road up to the farm-houses high on the N. flank, about 1 M. distant, whence the way lies up across the fields for nearly ½ M. Guides or precise directions for this part of the journey should be obtained at the farms. The path through the woods is about 1 M. long, with a strong ascending grade, and advances close to the crest of the ridge. It has not been used much of late years, and is rather indistinct, bushes having overgrown the old roadway; but even if the trail should be lost, it is not difficult to ascend through the forest. The mountain is sometimes visited from the obscure hamlet of *Drakesville*, on the S. side, whence one can follow the remains of the road on which the materials for the hotel were transported. The ascent is steeper on the S. side, and more ledges are encountered, but the climber is not entangled in forests.

The summit of the mountains is partly covered with ledges, among which are thickets of sturdy bushes and saplings. The view-point has to be changed in order to get the entire prospect on all sides. A small hotel was once erected here, but it was burnt about the year 1866. In the last century this ridge bore the name of *Seven Mt.* The shape of the range has been likened to that of Red Hill, all its dimensions being larger; and its base is about 4 M. long. It is higher than any part of the Ossipee Range, attaining an altitude of nearly 2,500 ft.

* *The View.* — Close at hand on the W. N. W. is the broad and beautiful Ossipee Lake, unbroken by islands, and surrounded by ancient driftplains. The white hamlet of W. Ossipee is seen near its head. Over this extend the lowlands of Tamworth, at whose end, over the N. falling flanks of the Ossipee Mts., is the pointed crest of Mt. Israel, flanked on the r. by the high and massive Sandwich Dome, with its long and dark-colored sides. Just to the l. of the Dome, very far away, is the top of Moosilauke, near the Connecticut Valley. On the E. of the Dome are the blanched cliffs of Whiteface, with the dark hemisphere of Passaconaway on its r. Between these two is Tripyramid, its N. peaks being concealed. About N. W. is the broad sheet of Silver Lake, with Madison village near its N. E. shore. Mts. Wonnalancet and Paugus are above Silver Lake, the latter being the higher, and covered with bare ledges; and farther out in this direction portions of the Carrigain group are visible. Just to the r. of and beyond Silver Lake is the splendid peak of Chocorua, covered and flanked by light-colored ledges and braced by bold ridges that shut out the farther view in that direction. On the r. of and beyond Chocorua is the long foreshortened ridge of Moat Mt., also free from forests and lifting up light-colored peaks. The view-line now passes up the depression of the Saco Valley, and on the r. flank of Moat is seen the low curve of Iron Mt., directly over which, and far beyond, the crest of Mt. Washington cuts the horizon. It may be recognized by the large house upon it, as well as by its altitude and position. On the r. of Washington, and farther away, are Mts. Adams and Madison, the former being identified by its sharp apex. A little more to the r., and nearly as distant, are Mt. Wild-Cat and the Carter Dome, separated by the sharp defile of the Carter Notch, with parts of the Carter-Moriah range beyond. The lower summits of Thorn Mt. and Double-Head are to the r. and nearer.

Mt. Kiarsarge is nearly N., over the highlands of Eaton and the Green Hills of Conway, and may be recognized by the house on its top. Nearer at hand is Freedom Village, near Swasey's Pond, with Mt. Prospect on the r. More distant are the Frost and Burnt-Meadow Mts. in Brownfield, over which is a part of the long ridge of Mt. Pleasant. The rolling hills of Porter and Hiram are more to the r.; and due E., across the plains of Parsonsfield, is Trafton Mt., in Cornish, with the Baldwin Saddleback on the N. Well to the r. of the Hosac Mt., and over the heights of Limington, the city of Portland is easily seen; and the great plains of York County sweep away to the S. E. and out to the sea-line, hardly broken by the Bonnybeag Hills and Mt. Agamenticus. It is claimed that Saco and Biddeford are visible in clear weather, with other points on the coast.

Toward the S. is the round Province Pond, near at hand, with other ponds and highlands of Wakefield; and the dark distant ridges of Moose Mt. and Copple Crown are more to the r. Portions of Lake Winnepesaukee, off Wolfeborough, are seen in the S. W., with the high double peak of Belknap beyond. The view then falls on the near Ossipee Mts., whose dark masses fill most of the W. horizon.

146. Freedom Village and Mount Prospect.

Freedom Village is 2 M. from Effingham Falls, 8 M. from Ossipee Centre (daily stage), 9 M. from W. Ossipee, and 12 M. from Cornish. The best place for visitors to stop at is the white house of Mr. Towle, near the centre of the village. The village has a church, several stores, and a savings-bank; and is surrounded by a pleasant farming-country. It is near Swasey's Pond and the Great Ossipee River.

Mount Prospect is reached from Freedom by passing out 1 M. on the E. road, turning to the r. at a school-house and ascending a farm-road for 1 M., and then clambering upward over the ledges for nearly 1 M. more. The slopes and summit are rocky, though groves of small trees have grown near the top, and the view is not materially obstructed in any direction.

The View from Mt. Prospect includes Mt. Washington, and parts of Adams and Madison, which are seen through the great depressions of the Saco and Ellis valleys, about N. N. W. A little to the r., and well out on the horizon, are Mts. Wild-Cat and the Carter Dome, separated by the remarkable gorge of the Carter Notch; and Thorn Mt. is seen farther in the foreground. Farther to the r. and nearer, are the rolling Green Hills of Conway, overlooked by the hotel on the apex of Kiarsarge. Toward the N. are the high hills of Eaton, along the boundary. Farther to the r. are the Frost and Burnt-Meadow Mts., with a part of Mt. Pleasant beyond. Toward the E. the view rests on the highlands of Western Maine, towards Porter and Hiram. Farther to the r. are the Cornish and Limington hills; and the great plains of York County stretch away on the S. E. to the shore of the sea, where it is claimed that Saco and Biddeford are visible on a clear day. On the S. is Province Pond, in front of the highlands of Wakefield; and Green Mt. lifts its long and ponderous ridges just across the Ossipee Valley, in the S. W., with glimpses of Lake Winnepesaukee beyond. Nearly W., 6 M. distant, is the beautiful oval of Ossipee Lake, with the dark Ossipee Mts. extending to the r. and l. beyond for many miles. Farther to the r. is Silver Lake; and up the long Tamworth valley are the noble and imposing peaks of the Sandwich Range, with higher summits overlooking them from the N.

147. The Whittier Peak.

The Whittier Peak is one of the N. summits of the Ossipee Range, easy of access and commanding a rich and extensive view. The Editor could not find that any distinctive name had been applied to it by the people of the adjacent country-side, and he therefore suggests the above name, in honor of the noble poet of New England who has written so often of the Ossipee Range and its valleys. The mountain differs from most of its densely wooded brethren in that it is composed of a succession of highly inclined ledges, ascending so continuously that the forest cannot obtain lodgement, and only a few small trees are scattered along the slope. The crest is clear and sharp, being formed by two low ramparts of rock, between which is a tiny grassy hollow. The nearest and most arduous route of ascent is by leaving the W.-Ossipee road about 1 M. W. of S. Tamworth, crossing the meadow, traversing a belt of woods at the base, and clambering upward over the bare ledges. The easier and better routes are: (1) by driving up the road which runs S. from S Tamworth into the Ossipee glens, bearing to the r. to the S. W. side of the mountain, ascending through the pastures, to the crest of the ridge on the W., and then scaling the rocks for a few rods to the crest; or (2) by leaving the W.-Ossipee road at the first road diverging to the r., E. of Gove's Corner, and about 3 M. from Moultonborough Corner, following the side-road for about ½ M. to a yellow farm-house, whence a path leads up to the pastures on the N. W. side of the mountain, reaching the same crest as the first route and ascending thence in the same way. The time of ascent from the W.-Ossipee road to the crest should be, in either case, 1-1½ hrs.

* *The View.* — Nearly S. S. E. is the high black mass of Melvin Peak, with a densely wooded ridge running up to Black Snout, which is about S. To the r. of its flanking ridge is Red Hill, whose long heights are nearly sundered by a deep depression in the centre. Over the opening between the Ossipee Range and Red Hill, nearly S. W, is the Sanbornton Mt., with other high hills in the town of Sanbornton. Far away over the l. centre of Red Hill is the pyramidal peak of Kearsarge. Through the gap in Red Hill are the Bridgewater Hills; and between Red Hill and Squam Lake the crest of Cardigan is seen in the distance, with Plymouth Mt. on its r. Beyond and over the high-spired hamlet of Sandwich Lower Corner is a portion of the beautiful Squam Lake, with the Squam Mts. on the N. shore, on whose r. are the bold ridges of Morgan Mt., with Prospect's flat top peering over the line, due W. In the foreground is the island-dotted Red-Hill Pond, beyond which is Sandwich Centre, with its two white spires, and over the mountains above are the blue crest-lines of Mts. Carr and Kinneo. Bearcamp Pond is seen nearer at hand, marked by a single islet, over which is Mt. Israel, with pastures extending well up its slope. Far away over and a little to the r. of Bearcamp Pond is Moosilauke, crowned by a house. To the N. W., over a ledgy spur of the Whittier Peak, is the massive and pointless Sandwich Dome, on the l. of which, through the gap towards Mt. Israel, are the mountains beyond the Mad-River valley, the r. flank of Fisher's Mt., the whole of Green Mt., and the l. flank of Tecumseh. On the r. of Sandwich Dome is the long plateau of Flat Mt., with high-crowned Osceola peering over the l. and two sharp peaks of Tripyramid over the r. In the foreground is a pretty stream, meandering down the valley between brightly colored sandy shores.

The noble peak of Whiteface next appears, with its crown of glittering white cliffs, cut into by a profound ravine, on whose l. rim is the path. To the r., and connected with it by a bristling ridge, is the high pointed mass of Passaconaway, buttressed on the r. by white-headed spurs, and supported on the r. by the lower heights of Mt. Paugus, covered with bare ledges and marked by a long and gracefully curving slide. More distant, and looking over the r. shoulder of Passaconaway, are the ascending terraces of Mts. Nancy and Willey, to the r. of which, and near to Paugus, are the white crests of Tremont. On the r. flank of Paugus is the uneven ridge of Bear Mt., beyond and over which are parts of the distant Carter Range. If seen on a clear day, the Presidential Range is nearly over Mt. Paugus. The needle-like peak of Chocorua lifts its blanched point on the r. of Paugus, and shows its bold supporting flanks on the S. and E. On the r., and apparently at right angles, is the entire ridge of Moat Mt. (over N. Conway), with its clear-cut N. peak and high bare ridge. Part of Mt. Moriah appears over this mountain, and on its r. are sections of two distant peaks, with Mt. Bartlett rising up to the symmetrical pyramid of Kiarsarge, whose apex is crowned by a house.

To the r. of Kinrsarge are the Green Hills of Conway, with Blackcap pre-eminent. Nearly N. W. is the distant Mt. Sabattos, in Lovell, with the Greenwood highlands beyond; and more to the r. is the round-headed Mt. Tom, in Fryeburg, with the Waterford mountains back of it. About E. N. E., 30 M. distant, is the long ridge of Mt. Pleasant, with a hotel on its middle crest. Much nearer, and within 6 – 8 M., is Silver Lake, on whose N. shore is the hamlet of Madison, over which are the ledges of the Gline and Lyman Mts. Elliott Pond is in the line towards Silver Lake, and S. Tamworth shows its white houses close below, at the foot of the Whittier Peak. Over Silver Lake are the shaggy highlands of Eaton, with the Frost and Burnt-Meadow Mts. of Brownfield beyond. The view is now closed by the lines of wooded heights in the E. part of the Ossipee Range, extending towards W. Ossipee.

" Through Sandwich Notch the west-wind sang
   Good morrow to the cotter ;
And once again Chocorua's horn
   Of shadow pierced the water.

" Above his broad lake Ossipee
   Once more the sunshine wearing,
Stooped, tracing on that silver shield
   His grim armorial bearing.

  .   .   .   .   .

For health comes sparkling in the streams
   From cool Chocorua stealing ,
There's iron in our northern winds ;
   Our pines are trees of healing."
          WHITTIER'S *Among the Hills.*

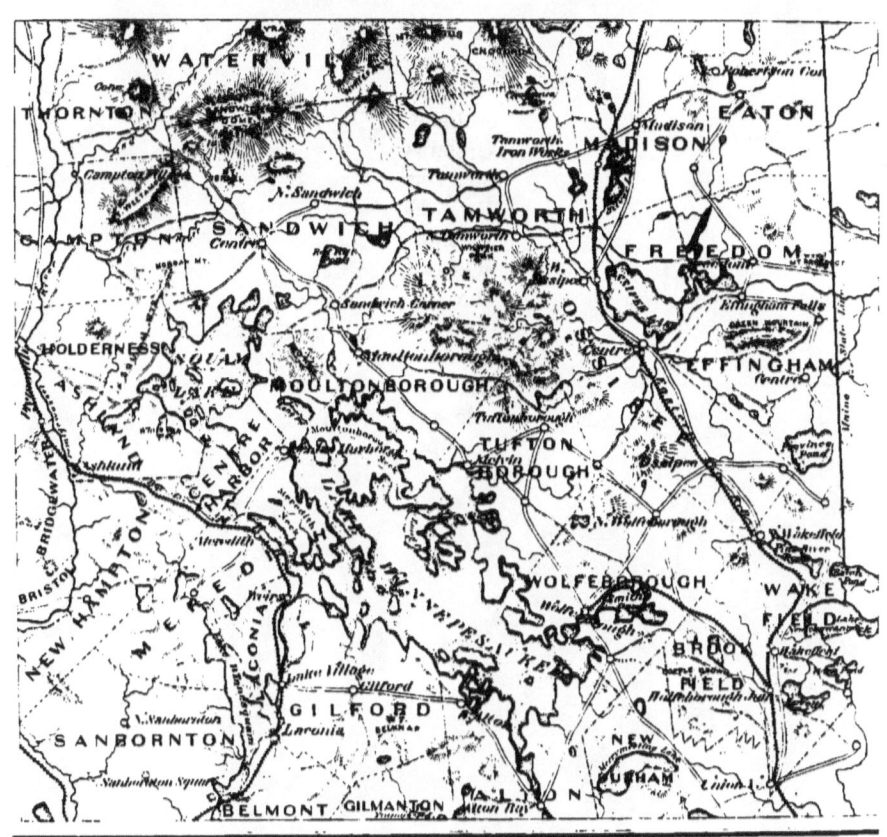

THE LAKE COUNTRY OF NEW HAMPSHIRE
FOR OSGOOD'S WHITE MOUNTAIN GUIDE.

To the
eminent.
Greenw(
Mt. Ton
E. N. E.
its mid(
whose 1
the Gli
Lake, a
of the
Eaton,
view is
Ossipe(

THE LAKE-COUNTRY OF NEW HAMPSHIRE.

148. Wolfeborough.

Hotels. — The * Pavilion accommodates 300 guests, and charges $4 a day. It is a large and first-class house, situated on high ground, whence a lawn of six acres slopes down to the lake and the boat-piers. It is not in the manufacturing part of the village. The view of Wolfeborough Bay and the Belknap peaks from this point is one of the best possible. Boats and carriages can be obtained at the hotel, which has also the usual conveniences of a large summer-resort. The Pavilion is an old and well-famed house, with an *élite* constituency of regular visitors.

The Glendon House is a new and comfortable summer-hotel at the centre of the village, near the railroad-station and steamboat-wharf. It accommodates 150 guests, at $3 a day. The Belvue House is nearly opposite the Glendon, and overlooks part of the lake. It has room for 75 guests, at $2.50 a day. The Lake House is the village inn (100 guests), and is N. W. of the station.

The Lake-View House is favorably located on the hill above the Pavilion (first street to the r.), and accommodates 50 boarders ($10-12 a week). The Belmont House (O. R. Yeaton) is in the same neighborhood, and has rooms for 25. The Glen Cottage (Levi Horn) accommodates 50, and is on the first side street to the l. beyond the Pavilion. J. L. Meader's house is 1 M. from the station, and accommodates 50 people. There are several other boarding-houses in and near the village, and many of the adjacent farmers take in summer-guests.

Distances. — Wolfeborough to Tumble Down Dick, 6 M.: to Copple Crown, 6¼: around the Short Square, 6; the Long Square, 12; around Smith's Pond, 12; to the Devil's Den, 8; to the Ossipee Falls, 14; to Alton Bay, 10; to Weirs, 16; to Centre Harbor, 20.

Railroad. — Wolfeborough is the terminus of a branch of the Eastern Railroad. The morning trains reach Boston at 2 P. M.: the morning trains from Boston arrive at 12.37 P. M. The distance to Boston is 108½ M.; the time is 4-5 hrs. Noon and evening trains run from Wolfeborough to N. Conway in about 2 hrs.

Steamboats. — The *Lady of the Lake* and the *Mt. Washington* touch at Wolfeborough several times daily. The former runs to Weirs and Centre Harbor, the latter to Centre Harbor and Alton Bay. The little steam-yacht *Nettie* is kept here for the use of tourists. Her rates are $1 an hour.

Stages run daily to Melvin Village (10 M.) The hotel-carriages frequently run to Copple Crown, the charge being $1.50 per passenger.

Wolfeborough is the largest village in Carroll County and on Lake Winnepesaukee. It is prettily situated at the foot of Wolfeborough Bay, on the outlet of Smith's Pond, whose water-power is utilized by several factories. The view from the village includes the whole extent of the bay, a part of the open lake, and the noble peaks of Mt. Belknap, beyond. A great variety of lake and mountain views may be obtained by driving out on the adjacent roads, which are well kept and easy. The lake itself is also explored with facility from this point, by excursions on the steamboats, by rowing, or by the fascinating moonlight excursions which are frequently organized. These varied attractions have made Wolfeborough an important summer-resort, where fully 10,000 visitors spend parts of their summers, resulting in an income of $ 125,000 to the residents.

The village is built on the slopes of two hills which rise about the outlet of Smith's River. It contains Congregational, Friends', and "Christian" churches, an academy called the Wolfeborough Christian Institute, a national bank and two savings-banks, a weekly newspaper, 15 – 20 stores, and several small factories.

The *Devil's Den* is 8 M. from the village, in New Durham, and is reached by a walk of ¼ M. from a neighboring farm-house. Here one can go for several rods through a dark cavern, and then climb up a ladder to the ledges above. *E. Alton* is a lofty hamlet 6 M. S. W. of Wolfeborough, whence a noble view is gained (see Route 150, *ad finem*).

The town of Wolfeborough has 1,995 inhabitants, and covers 28,680 acres of land, of which 16,694 are improved. The manufactures of the town include blankets, leather, shoes, and lumber, amounting to over $ 500,000 a year. The soil is hard, but strong, and much garden produce is raised for the use of the summer-visitors. Besides the chief village, there are the hamlets of Mill Village, Wolfeborough Centre, N. Wolfeborough, and S. Wolfeborough. The former is ½ M. from the steamboat-wharf, and contains several factories. The Centre is on the N. shore of Smith's Pond, about 3 M. from the wharf; *N. Wolfeborough* is near the foot of Batson Hill and the Whiteface and Cotton Mts., towards Ossipee Corner; and *S. Wolfeborough* is at the woollen-mills on the outlet of Rust Pond, on the Alton road. The Franklin House is a large inn at the latter point, owning the small steamboat *Morning Star.*

Smith's Pond is a pretty lake about 1 M. N. E. of Wolfeborough, and 38 ft. above the lake. It is 4 M. long, and contains several islands; the largest of which is Stamp-Act Island, nearly ¾ M. long. One of the favorite drives is around this pond, on whose E. shore are the remains of the old feudal mansion called Wentworth House. In view of the beauty of the lake and its historic memories, it would be a great improvement to change the prosaic name of Smith's Pond to the more stately and significant one of *Lake Wentworth.*

Wentworth House was 100 × 45 ft in dimensions, with five large barns. It was one of the four main stations in the surveys for Holland's map ; and here the Massachusetts youth who afterwards became Count Rumford and prime-minister of Bavaria was nobly entertained by the genial governor. After the exile of the Wentworths, the house fell into unworthy hands; and it was burnt in 1820. There is a tradition that when Wentworth was flying hence to the royal fleet at Portsmouth, he loaded his carriage here with his plate and treasure. But the weary horses were soon unable to draw the heavy load, and the alternatives were to leave Lady Wentworth or the treasure-chests. The latter were buried by night in the forest, and have often been vainly sought.

In the year 1638 a scion of the illustrious and knightly family of Wentworth passed from Yorkshire to New Hampshire, where he founded a new and noble house. One of his descendants was John Wentworth, who was born at Portsmouth and graduated at Harvard College. In 1766 he was made governor of the Province and surveyor-general of the royal forests, and he retained these high offices until 1775. At the outbreak of the Revolution he remained faithful to the King, and was driven into exile by the people of New Hampshire. He was made governor of Nova Scotia, and received the title of baronet, with the honorable privilege of bearing two keys on his coat-of-arms, emblematic of his fidelity.

President Dwight was an ardent admirer of Gov. John Wentworth, and in his honor named Lake Winnepesaukee *Lake Wentworth* and Mt. Lafayette (until then unnamed) *Mt. Wentworth.* He says: "He was a man of sound understanding, refined taste, enlarged views, and a dignified spirit. His manners, also, were elegant; and his disposition enterprising. Agriculture in this Province owed more to him than to any other man. He also originated the formation of new roads, and the improvement of old ones. . . . This gentleman was the greatest benefactor to the Province of New Hampshire, mentioned in its history."

Wolfeborough was settled in 1770, and was named in honor of Gen. Wolfe, who was killed in battle before Quebec. In 1771 the Provincial authorities commanded the construction of a road three rods wide from the Governor's House, at Wolfeborough, through Tuftonborough, Moultonborough, New Holderness, Plymouth, and Baker's River, to Dartmouth College. The ancient map of Kalm calls this section the *King's Woods.* In 1774 men were hired here to build the British barracks at Boston; and in the following year the inhabitants attacked Wentworth House.

149. Copple Crown and Tumble Down Dick.

Copple Crown is a mountain in Brookfield, 2,100 ft. high. As seen from distant points it shows a moderate peak, flanked by long and gradual slopes on either side. It affords a beautiful view of Lake Winnepesaukee and the mountains beyond, and is frequently ascended by tourists from Wolfeborough. It is claimed that 30 ponds and lakes are visible from this peak.

Copple Crown is about 6½ M. from Wolfeborough, and is reached by a pleasant road which runs to the S., with pretty lake-views on the r. 5 M. from the village an ancient church is passed on the l., just beyond which the tourist diverges on a side road to the l. It is ½ M. from this point to the farm at the foot of the mountain, which is reached by the first by-road to the r. The ascent of the mountain from the farm is about 1 M. long, of which the first half is through an open pasture, and the rest is on a pleasant and well-marked forest-path. On leaving the farm-house, the visitor ascends the open pastures on the l. flanking ridge, to a lone birch-tree, whence his course is shaped to a lone oak, up near the woods. A short distance from the oak, in the edge of the forest, the path is found. For much of the distance this route follows a nearly level ridge, broken occasionally by ledges and boulders. The labor of the ascent of Copple Crown is a mere trifle in comparison with that of certain of the minor White Mts. After advancing for about ½ M. through the forest, a divergence of the paths is reached. That which passes on leads in a few rods to a ledgy crest which overlooks the country to the S. and S. W.; and the path diverging upward to the r. conducts speedily to the main view-point on the higher peak. The young forests on the ridge have advanced so rapidly as to close out the prospect in certain directions, and greatly interfere with other lines of vision. The hotel-keepers of Wolfeborough should either have these trees cleared away, or else form a high observation-platform out of the timbers which lie on the summit. From the ledges at the end of the direct path, from the upper crest, and from a more northerly crag which is reached by a short path to the r. just beyond the timbers, a general view may be made up. The best single point for a prospect is the last-named, which looks out on Winnepesaukee and the White and Franconia Mts.

* The View. — The first feature to be noticed in the view from Copple Crown is the outstretched expanse of Lake Winnepesaukee, abounding in islands and cut by long and verdant promontories. This beautiful prospect extends to the N. W. for nearly 20 M., and is bounded by picturesque

ranges of mountains on either side. About W., and 3 M. distant, is the
rounded green ridge of Long Stack Mt., over which, and beyond the lake,
the swelling crests of the Suncook and Belknap ranges are seen, running
far to the r. Towards the lake shore is the hamlet of S. Wolfeborough,
near Rust Pond, beyond which is the long height of Rattlesnake Island;
and over the latter, 37 M. distant, are the white domes of Mt. Cardigan,
with the Bristol and New-Hampton hills nearer. Looking across the
Wolfeborough Bay and Neck, the eye traverses the broadest part of the
lake, passes Plymouth Mt. and the Bridgewater Hills, and rests on Smart's
Mt., a long ridge falling off on the S. Farther to the r., the view-line
crosses the farms on Tuftonborough Neck, Cow and Long Islands, and the
bay of Centre Harbor, and ascends the long Pemigewasset-Asquamchu-
mauke valley to Mt. Cuba, a high ridge near the Connecticut River.

In the foreground, about 5 M. distant, is the large white village of
Wolfeborough, with Smith's Pond on the r., and nearly in this line, at the
N. end of Winnepesaukee, the summer-hamlet of Centre Harbor is visible.
Over the Squam Range, back of Centre Harbor, the smooth top of Mt.
Prospect and the long dark ridges of Mt. Carr are seen; to the r. of which
is the lofty blue plateau of Moosilauke, with Kinneo's peak in advance, —
the last two being seen over the r. part of Wolfeborough. Over the little
pond r. of Wolfeborough, about 20 M. distant, is the shapeless mass of Red
Hill, with Sandwich Dome over it on the r., falling upon the l. flank of the
nearer Ossipee Range. Between Red Hill and Sandwich Dome several of
the Campton and Pemigewasset Mts. are visible. (It is claimed that the
Franconia Mts. also are visible from Copple Crown, nearly over Sandwich
Dome.) The view now rests on Smith's Pond, a beautiful island-strewn
lake in the foreground, approaching within 2½ M. of Copple Crown, with a
white and silvery beach, and containing the large Stamp-Act Island. Beyond this lake is the great dark Ossipee Range, extending for miles along
the N. To the r. of its chief peak is the blanched crest of Whiteface,
with the black dome of Passaconaway to the E. On the r. of the latter
and farther away is the vast mass of Mt. Carrigain, with the remarkable
Carrigain Notch fairly seen on the r., into which falls the precipitous side
of Mt. Lowell. The next mountain to the r. of and lower than Passaconaway is Paugus, above which is the white cap of Tremont, with part of
Mt. Willey beyond. Over the wooded cape at the r. end of Smith's Pond
are the Batson and Trask Hills, above N. Wolfeborough; and far away on
the same line is a part of the hamlet of Tamworth Iron-Works. Over the
latter rises the noble white crest of Chocorua; on whose r., and a little beyond, are the high ledges of Moat Mt. Over the latter, 50 M. away on the
horizon, a little W. of N., is the stately cone of Mt. Washington, with parts
of Adams and Madison, Monroe and Pleasant. To the r. of the ragged
peaks of Moat are the remote Mt. Wild-Cat and the Carter Dome, between
which is the sharply cut Carter Notch. Thorn Mt. is below the Carter

Dome, and on its r. is the S. peak of Double-Head. In a line towards the latter is a part of Ossipee Lake, close by which, on the E., is the great shapeless ridge of Green Mt., with pastures on its sides.

The cliff of Tumble Down Dick is just to the N., across a narrow glen, with Mt. Delight on its r. A little E. of N., and 4 – 5 M. distant, are the Whiteface and Cotton Mts., over which are the Green Hills of Conway; and still farther N. is the hotel-crowned apex of Mt. Kiarsarge, beyond N. Conway; and the white ledges of Baldface are yet more distant. Over the l. flank of Green Mt., far to the N., are Mts. Slope and Eastman, beyond which, 50 M. away, is Mt. Royce, on the edge of the Androscoggin Valley; and close by the latter is Speckled Mt., in Maine. To the r. of Green Mt. is a long array of highlands in Southwestern Maine, extending towards the Saco River.

About E., near the foot of the mountain, is Cook Pond, with the hamlet of Brookfield Corner across its r. part, beyond which is the larger village of Wakefield Corner, with a conspicuous church. Over the latter is Lake Newichiwannock, and to its r. is the historic Lovell's Pond. Mt. Teneriffe and the plains of Milton are on the S. E.; and on the S. are the high adjacent ridges of Great Moose Mt. Several villages are seen beyond, among which (it is said) are Rochester and Dover, with a broad reach of the ocean; and the Isles of Shoals are pointed out nearly over Great Falls. The long lowlands of Strafford County stretch away toward the S., beyond Great Moose Mt.

About S. W., and 4 – 5 M. distant, is the long and picturesquely irregular Merrymeeting Pond, over which, far away, are the twin Uncanoonuc peaks, with the bold Temple Mts. beyond. Near the Uncanoonucs it is claimed that Mt. Wachuset may be seen, on a clear day. Farther to the r., 65 M. distant, Monadnock is visible, with its round top low down on the horizon, and guarded on the E. by the long ridge of Pack Monadnock. Toward the S. W. the view crosses hilly Gilmanton and sweeps over leagues of the populous Merrimac Valley. About W. S. W. is Lovewell's Mt., in Washington; and glimpses of the Green Mts. may be gained beyond. Farther to the r., and nearly over the foot-hills of the Suncook and Belknap Ranges, is the high pyramid of Mt. Kearsarge.

Tumble Down Dick is a high hill 1 M. N. of Copple Crown, and 6 M. from Wolfeborough. It is covered with pasture land, and is easily reached from the road, which crosses it not far below the top. On one side there is a picturesque cliff, over which it is said a blind horse named "Dick" once fell, whence the name of the hill. There is also an Indian legend about this name. The view, though good, is far inferior to that from Copple Crown. Toward the S. is the high green-clad mass of Copple Crown, with the Long Stack Mt. on its r. The Belknap Range shows finely across a part of the lake, and Rust Pond is in the foreground (about W.),

with Rattlesnake Island over it.   Mt. Cardigan is far away beyond, over
the New-Hampton hills.   Smart's Mt. and Mt. Cuba are very distant,
nearly over the white village of Wolfeborough; Centre Harbor is seen at
the end of the long and beautiful lake, with Mt. Carr beyond to the r.,
over and far back of Squam Lake.   Over the Squam Mts. is Mt. Weetamoo, among the Campton Mts.; and the lofty crests of Kinneo and Moosilauke are still more distant, the latter being over the l. of the adjacent
Smith's Pond.   Red Hill is about N. W., over a part of Winnepesaukee;
and the dark mass of Sandwich Dome appears over it on the r., flanked
by Osceola, Whiteface, and Passaconaway.   The dark Ossipee Range
next appears, within 10 – 12 M., and covering a broad area.   On its r. is
the crest of Chocorua, the Moat Mt., Mt. Wild-Cat and the Carter Dome
(with the Carter Notch between), and the remote peaks of Double-Head,
Thorn Mt., Kiarsarge, Baldface, Royce, and the Green Hills of Conway.
The Green Mt. in Effingham is on the r. of the Ossipee Range, and several
mountains of Western Maine rise beyond.   The views to the E. and S. E.
include the ponds, highlands, and hamlets of Wakefield, Milton, and Mid-
dleton.

150. Alton Bay.

Hotel. — The Winnepesaukee (formerly Bay-View) House, looking out over the
bay and to the distant mountains on the N.   This is a large second-class house, with
a livery-stable attached.   The rates are $ 10 – 12 a week.   Row-boats and sail-boats
may be hired here.

Routes. — Alton Bay is the terminus of the Dover & Winnepesaukee branch of
the Boston & Maine R. R., and is 28 M. from Dover and 96 M. from Boston.   The
steamboat *Mt. Washington* leaves this port on the arrival of the Boston train and
goes up the lake to Wolfeborough and Centre Harbor.

Alton Bay is a collection of a few houses at the foot of the arm of Lake
Winnepesaukee which bears the same name, and at the mouth of the
Merry-Meeting River.   It is surrounded by bold hills and attractive
scenery, but has declined as a summer-resort.   There is a large camp-
meeting ground on the hill over the station.   Good fishing is found in the
vicinity, especially among the ponds in the adjacent towns.   Trouting is
good in the spring-time.   The roads are hilly, but smooth, and give a
great variety of fine prospects.   *Alton* (Cocheco House) is a small country-
village 1½ M. S. E. of the Bay.   The town has 1,769 inhabitants, most of
whom are engaged in farming.   It was settled in 1770, and named after a
town in Southamptonshire, England.   It was a favorite haunt of the
Indians, many of whose skeletons and weapons have been found.   A wharf
was built at the Bay in 1832, for the steamer *Belknap*.

Sheep Mountain is a high and rocky ridge, N. W. of Alton Bay, which
is often ascended for the sake of its broad view over Lake Winnepesaukee.
It is reached by driving out 2 M. on the road to the mountains, and then
ascending ½ M. to the l., over open ground.   By riding about ½ M farther,
visitors can avail themselves of a farm-road which leads up near the sum-

mit. The view from the highway along the N. flank is of great breadth and beauty, including the lake and the northern mountains. There is no wayside view in the lake-country that can surpass this in variety and extent.

Prospect Hill is 4 M. from Alton Bay, near S. Alton; and is reached by ascending ½ M. through the fields. It commands a very noble view of the country to the S. and S. W., and the ocean; though but little of the lake is seen.

The noble white crests of the Alton and Suncook Ranges are seen to the best advantage in driving toward Mt. Belknap. This road passes over Sheep Mt., and runs W. into Gilford, with picturesque highlands on either side. Mt. Belknap may be ascended on this side without serious difficulty, although there is no path. The visitor should turn to the r. from the Gilford road on a farm-lane, just beyond Young's Pond, 7 M. from the Bay. Thence a high and ledgy ridge is ascended, and the flank of Belknap is met beyond the intervening ravine. It is 12 M. from the Bay to the path up Mt. Belknap, and the road is very hilly.

Merry-Meeting Lake is in New Durham, 7 M. E. of Alton Bay, and is 10 M. in circumference. It is a beautiful sheet of water among the hills, S. W. of Copple Crown Mt., and affords good fishing. It received its pleasant name before 1770, when the exploring Masonian grantees met on its shores and had a great feast. *Lougee Pond* is in Gilmanton, 7 M. S. W. of Alton Bay, by the Iron-Works road. It contains large numbers of tame fish. The excursion may be made over the Sheep Mt. road, with its noble views.

Gilmanton is one of the best farming towns in the State. It adjoins Alton on the S. W. It has 1,644 inhabitants, and 23,840 acres of improved land. It was granted in 1727 to 24 persons by the name of Gilman, and many others. *Gilmanton Iron-Works* is a hamlet in the E. part, where iron was formerly made from bog-ore which was fished up with long tongs from the bottom of Lougee Pond. *Gilmanton Academy* (Washington House) is in the W. part of the town and is a pretty village with two churches. The academy dates from 1794, and has a theological department. The Suncook Range runs through this town, affording some picturesque hill-scenery.

Peaked Hill is about ½ M. E. of the Academy, and rises 450 ft. from the plain. The view from this point is of great extent and beauty, and is attained with light exertion. The chief points on the horizon, with their directions, are as follows: The Uncanoonuc Mts., S. 26° W.; the State House, at Concord, S. 31° W.; Mt. William, in Weare, S. 32° W.; Crotched Mt., in Francestown, S. 47° W.; Monadnock, S. 50° 30′ W.; Kearsarge, due W.; Mt. Ascutney, in Vermont, N. 80° 30′ W.; Mt. Cardigan, N. 51° W.; Moosilauke, in Benton, N. 19° 30′ W.; Mt. Whiteface, N. 7° E.; Mt. Washington, N. 12° E.; Mt. Belknap, N. 22° E.; Great Moose Mt., in Brookfield, N. 79° E.; Prospect Hill, in New Durham, S. 73° E.

Alton Bay to Wolfeborough.

The distance is 11 M. On reaching the top of the long hill N. E. of Alton Bay there is a good view of Alton and Gilmanton, Sheep Mt., Mt. Belknap, and the Suncook Range. These picturesque highlands are frequently seen in retrospects, while on the l. is the lake, with Red Hill

and the Sandwich Range at its head and the Ossipee Mts. on the E.
About 4 M. from Alton Bay the roads to Fort Point and Clay Point
diverge to the l. *E. Alton* is a high-placed hamlet around a large Baptist
church, commanding a rich view over the lake, with the Suncook and
Belknap Mts. on the l., followed in the distance by Prospect, Moosilauke,
Red Hill, Sandwich Dome, Flat Mt., Mt. Israel, Whiteface, and the Ossipee
Range, over whose r. flank the hotel on Kiarsarge is visible on a clear day.
The villages of W. Alton, Centre Harbor, and Wolfeborough are also seen;
and the broadest part of the lake is on the r. of the high hill of Rattlesnake
Island. The road descends through maple groves and reaches *S. Wolfe-*
borough (or Mink Brook), 3½ M. from E. Alton. Beyond this hamlet
pretty views of Copple Crown are given on the r. over Rust Pond.

151. The Weirs, Meredith, and Ashland.

The Weirs is but little more than a railroad station, where the trains
connect with the steamboat *Lady of the Lake.* Just to the N. is a large
camp-meeting ground; and to the S. E. is the outlet of Lake Winnepesaukee, over which a bridge leads into Gilford.

The *Endicott Rock* is near the Meredith shore, above the bridge over the outlet of
the lake, and is imbedded in gravel. It is 20 ft. around, and bears the carved inscription, "E. I. S. W. W. P. John Endicott, Gov." It was so inscribed in 1652 by
the officers sent by Massachusetts to find the head of the Merrimac. They were
Kentish captains, — Edward Johnson, author of *The Wonder-Working Providence
of Zion's Saviour*, and Simon Willard; aided by Jonathan Ince and John Sherman
(ancestor of the statesman of that name). The Massachusetts surveyors of 1652
were informed by the local tribe that the true Merrimac was the stream flowing
from Lake Winnepesaukee; but in 1659 Passaconaway told Major Waldron that the
name was applicable either to that stream or to the present Pemigewasset.
In the earlier days, vast numbers of alewives, salmon, and shad ascended the
Merrimac every year, the former branching off into the tributary streams, the
salmon ascending the Pemigewasset to its cold spring sources, and the shad seeking
the clear waters of Winnepesaukee. On the shallows at the outlet of the lake the
Indians laid lines of rocks joined by nets, in which they took great numbers of fish.
During the season of the annual shad-migration, the Winnepesaukee tribe encamped
about these fish-weirs, and passed many weeks in feasting and joyous hospitality.
The remnants of the old stone dams remained in place long after the occupation of
the lake-basin by the English, and the name of *The Weirs*, still attached to the steamboat and railroad stations, perpetuates their memory.
Somewhere in this valley (in 1690) occurred the mysterious battle between the
Sagamore Hope-Hood and his Saco-River warriors, and a war-party of Catholic Indians from Canada. Cotton Mather states that the chief had marched to Aquadocta
(Ahquedochtan), in the hope of enlisting the warriors of that tribe in his extermi-
nating campaign against the English settlements. Here he was met by the Canadian
invaders, who "furiously fell upon them, and in their *blind fury* slew him and a
considerable part of his company."

Meredith Village (*Elm House*, 20 guests) is pleasantly located at the
head of the Northwest Bay of Lake Winnepesaukee, and on the B., C. &
M. Railroad. It has two churches and a savings-bank, with manufactories
of hosiery and lumber. It is nearly surrounded by the rich farming town
of Meredith, which has 1,807 inhabitants and 16,586 acres of improved
land. A steamboat formerly plied between this port and the other villages

on the lake. The scenery in the vicinity is pretty, especially as seen from the hills on Meredith Neck (see Route 155, *ad finem*), and around the beautiful island-studded sheet of *Waukawan Lake*, just N. W. of the village. The distance to Centre Harbor is 5 M., over a good road. In this vicinity are the boarding-houses of G. E. Gilman (30 guests), J. P. Norris, Mrs. Smith, M. C. Pease, and the Towle house.

Ashland (*Squam-Lake House*) is a prosperous manufacturing village on Squam River, near its confluence with the Pemigewasset. It turns out annually over $600,000 worth of manufactured goods, including flannels, manilla paper, leather-board, hosiery, lumber, gloves, etc. There are two churches, a town-library, and a savings-bank. Much pleasant hill-scenery is found in the neighborhood. *Little Squam Lake* is 2 M. N. E., and the rare beauties of Squam Lake are 4 M. distant (see Route 160).

152. Laconia.

Hotels. — The Willard House, a comfortable old hotel fronted by stately trees, accommodating 75 guests; the Laconia Hotel (100 guests), a modern house on one of the main streets; and several smaller inns. The Bay-View is a first-class summer boarding-house (40 guests) near Lake Winnesquam, 1½ M. from Laconia; and the Winnesquam House is farther out, near the same lake. There are also several small summer boarding-houses in the adjacent towns of Gilford (John Follett, etc.), Tilton (Dexter House), and Belmont

Railroad. — Laconia is on the B., C. & M. R. R. (Route 2), 102 M. from Boston, and 24 M. from Plymouth. Trains run either way several times daily.

Distances. — Laconia to Lake Village, 1½ M.; to the Bay-View House, 1½ M.; to Gilford, 4; to Mt. Belknap, 17; to Alton Bay, 15 – 20; to Weirs, 6; to Gilmanton Academy, 8; to W. Alton, 12

Laconia is an important manufacturing town on the Winnepesaukee River, where it enters Lake Winnesquam. The Ranlet Car-Manufacturing Works are here, employing 200 men, and turning out annually $300,000 worth of work, from freight-cars up to palace-cars. The Belknap Mills employ 400 persons in manufacturing cotton and woollen goods; and there are several large hosiery factories, besides foundries, machine-shops, and other enterprises of a similar character. The aggregate value of the manufactures is over $1,600,000 annually. The town has five churches (Cong., Unit., Cath., Meth., F.-W. Baptist), several graded schools, a public library, a bank and two savings-banks, and a weekly newspaper (*The Laconia Democrat*). The town has about 3,000 inhabitants. The main street is lined with stores of all kinds, and presents a bright and busy appearance.

Laconia is a neat and pleasant town, showing in its public and private buildings that its citizens are well-to-do and prosperous. The factories are concentrated along the river, and the other parts of the village are quiet and attractive. The beautiful views in the vicinity and the excellency of the neighboring roads have caused this to become somewhat of a resort for summer-tourists. The streets are lighted with gas, and several of them are shaded by lines of trees.

The low hills near Laconia command a series of broad and beautiful views. The *Bay-View House*, 1½ M. S. W., overlooks Lake Winnesquam and a long line of peaks on the N. and E. From a point a little way above, on the same hill, the view includes Mts. Kearsarge, Moosilauke (up Winnesquam), Green and Tecumseh, Osceola, the black Sandwich Dome (due N.), Red Hill, Whiteface, the slide-marked Tripyramid, black-domed Passaconaway, and the beautiful peak of Chocorua. On the N. E. is the long dark Ossipee Range, with Melvin Village at its foot; and Mts. Belknap and Gunstock are much nearer. In the foreground are Lake Winnesquam, Round Bay, Long Bay, and a large part of Laconia (under the r. of the Ossipee Range).

Boyd Hill is in Gilford, about 3 M. from Laconia, and is covered with pastures. Its summit is reached from the road in about ten minutes. The view includes Mt. Wachuset, Laconia, and Lake Winnesquam, the far-distant Mt. Monadnock and the flanking ridges of Pack Monadnock, the high pyramid of Kearsarge and the Ragged Mts., Lake Village and the black Sanbornton Mt., Long Bay and the light-colored peaks of Cardigan, and the lofty ridge of Moosilauke. Over the Squam Range is the distant Mt. Kinsman; and to the r. of and far beyond the Northwest Bay are the sharp Franconia peaks. Then come Mts. Green and Tecumseh, with the long pile of Sandwich Dome, nearly N. Next is Red Hill, over which is the slide-striped Tripyramid, flanked on the r. by Whiteface and Passaconaway, over whose r. appears a part of the Presidential Range. Chocorua is over a broad part of Winnepesaukee, with the long Ossipee Range on the r. and nearer; and the bare-headed Belknap peaks are more to the r. Beautiful views over Lake Winnepesaukee and its islands are gained on the N. E. and E.

Follett Hill is not far from Boyd Hill, near Gilford, and is easily ascended. Its noble prospect includes Kearsarge, the Ragged Mts., Sanbornton Mt., Mt. Cardigan, Long Bay, Moosilauke, Meredith Village, the Squam and Morgan Mts., Red Hill, Sandwich Dome, Tecumseh, Welch Mt., Lafayette and Liberty, Passaconaway, Mts. Washington and Monroe, Paugus, the Ossipee Range, a broad area of the lake, and the Belknap peaks. This is one of the most fascinating views in the lake-country. Other fine prospects are obtained from *Liberty Hill* and *Meeting-House Hill*, in Gilford.

Lake Village (*Mt.-Belknap House*) is 1½ M. from Laconia, at the foot of Long Bay, and contains several hosiery factories, the repair-shops of the B., C. & M. Railroad, a foundry, a needle-factory, and other mechanical enterprises. It has a savings-bank, a weekly newspaper, several churches, and nearly 2,000 inhabitants. It is a port of Lake Winnepesaukee, and an excursion-steamer is kept here. The adjacent town of Gilford is one of the best farming-districts in the State, and contains the noble Belknap peaks, which are best visited from Laconia or Lake Village (see page 365).

Long Bay is an expansion of the outlet of Lake Winnepesaukee, extending from Weirs to Lake Village. It is about 4 M. long, with an average width of ½ M.; and contains 7 islands. Round Bay is another expansion, S. W. of Lake Village, covering ⅛ M.

Lake Winnesquam (" Beautiful Water ") was formerly known as *Great Bay*. It is an expansion and northward extension of the Winnepesaukee River, W. and N. W. of Laconia, and is about 6 M. long, varying in width from 2 M. to ⅛ M. It is environed by bold hills and picturesque scenery,

and is frequently navigated by boating and fishing parties.   The summer-
resorts of the Bay-View and Winnesquam Houses are on its S. E. shore, and
have stations on the railroad.

Farther to the S. is *Little Bay*, where the Provincial troops erected the
strong bulwark of Fort Atkinson in 1746.   Near this bay was a more
ancient Indian fort, with six walls.   Both of these martial monuments
have been demolished for materials to be used in building dams and other
works.

153. Mount Belknap

is a bold and bare-topped double peak in the town of Gilford.   The highest
point is 2,394 ft. above the sea, and is one of the chief stations of the U. S.
Coast Survey, whose officers call it *Mt. Gunstock*.   It is connected by a
high ridge with the second peak, which is 2,062 ft. high.   A long line of
rounded and symmetrical summits runs from Belknap to the S. E. and E.,
including the bold and picturesque highlands of Gilmanton and Alton.
This range is composed of erupted sienite.   Mt. Minor and Walnut Mt.
are S. of Belknap, and Mt. Retreat is on the N. E.

The view from Mt. Belknap is by far the finest in the lake-country, and
is one of the richest and most fascinating in New England.   Starr King
justly says that it is superior to that from Red Hill.

The path up Mt. Belknap leads in from George Morrill's farm-house, 7 M. from
Laconia and 12 M. from Alton Bay.   It is not more than 1¼ M. long, and is easily
followed.   Most of the way is through open fields, and is pointed out by guide-
boards, which Mr Morrill has thoughtfully erected   The path is sometimes steep,
but the walking is always smooth and easy.   The highest point of the plateau-crest
is marked by a beacon of the U. S. Coast Survey.   Certain points nearer the mountain, on the S. and S. E., are visible by advancing along the plateau in those
directions.

* * *The View.* — Toward the N. W. projects the second peak of Belknap,
to the r. of which, and far beyond, is Meredith Village, at the head of an
arm of Lake Winnepesaukee.   Nearly over this point, beyond the Ashland
hills, is the flat top of Mt. Prospect, and still farther to the N. W. is Mt.
Carr, ascending to a peak on the l.   Farther on, and cutting the horizon,
is the long and many-headed plateau of Moosilauke, with a hotel near its
centre, and the Blue Ridge projected against its flank and running to the
r. over the Squam Mts.   Farther to the r., over Meredith Neck, is the
beautiful Squam Lake, dotted with islands, and walled in on the N. by the
Squam Range, over which Mt. Weetamoo and the Campton Mts. are seen.
Far away in the N. are Mts. Kinsman and Cannon, the former reaching up
a well-marked peak.   Nearly over the hamlet of Centre Harbor is the long
ascending range which begins on the l. with Fisher's Mt. and Welch
Mt., rises to the crest of Green Mt., and then to the bold peak of Tecumseh, which plunges off sharply to the r.   Beyond these, and against the
sky, are the great Franconia Mts., — the sharp spire of Liberty over Green

Mt., Lincoln to the r., and the pinnacle of Lafayette just to the l. of and over Tecumseh. Above the broad arm of Winnepesaukee which extends toward Centre Harbor, and to the r. of that village, is the long dark mass of Red Hill, above which, and to the r. of Tecumseh, are the ponderous heights of Sandwich Dome, with the peak of Osceola over their l. extension. From this point the long and lowly ridge of Flat Mt. runs E. to Whiteface, and over it are seen the high crests of Mt. Hancock and the step-like ascending terraces of Carrigain. To the r. of and close to the latter (though nearer) is Tripyramid, showing a portion of its white slide and one of its sharp peaks. Below this point, and to the r. of Red Hill, is the hamlet of Sandwich Lower Corner; and the rural plains of Sandwich and Tamworth stretch far to the r. The crest of Whiteface gleams over this peaceful scene, and is flanked on the r. by the swelling dome of Passaconaway, whence step-like spurs run out to the E. Over the second of these terraces is Mt. Clinton, of the Presidential Range, and over the third is the round-topped Mt. Pleasant, with a white peak of Tremont farther in the foreground. Over the lowest part of the gap, to the r. of Passaconaway, is the ridge of Bear Mt., overlooked by the top of Mt. Crawford, which is in turn surmounted by the plateau of Mt. Franklin. Over the r. of the gap are the high crags of Monroe, and then comes the noble cone of Washington, looming over Mt. Paugus, which rises over the l. flank of the Ossipee Range. A part of Mt. Jefferson may be seen on the l. of Washington, and a part of Madison is on the r. Over the r. flank of Paugus is Table Mt., with a part of Bear Mt. Close at hand, below in Gilford, is the summit of Mt. Retreat.

The great black Ossipee Range next appears across a part of Lake Winnepesaukee, with Black Snout on the l., and the high Melvin Peak near the r. centre, over Melvin Village. The sharp white summit of Chocorua peers over the highest of the l. section of the range, to the r. of which are the upper ridges of the far-away blue Mt. Carter (back of the Glen House), and the rock-crowned Moat Mt. (over N. Conway). Between Mt. Belknap and the Ossipee Range extends a broad reach of Lake Winnepesaukee, with Bear Island on the l. and the arable plains of Welch Island on the r. Near the farther shore is Long Island, with its large boarding-houses, beyond which are the long levels of Moultonborough Neck, devoted to farming. Toward the N. E is one of the broadest reaches of the lake, bounded by Tuftonborough Neck, above which is the hamlet of Mackerel Corner. A view-line running just to the l. of the N. end of Rattlesnake Island meets Mt. Pleasant (in Maine), a long detached ridge on which a hotel may be seen. Farther to the r. are the Burnt-Meadow Mts., in Brownfield; and the Green Mt. in Effingham is somewhat nearer. To the r., more distant, is Saddleback Mt., on the shore of Sebago Lake; and the Cornish hills lie to the S.

To the r. of the high knolls of Rattlesnake Island, and over Goat Island, is the large white village of Wolfeborough, between its bay and the island-studded Smith's Pond. To the r., across the islands which front this shore, is Rust Pond, with the cliff-bound Tumble Down Dick beyond and the symmetrical peak of Copple Crown to the r., in front of Great Moose Mt., in Brookfield. Across the rolling plains of Strafford and York Counties the ocean is seen, stretching from S. S. E. to S. E., or about from Wells to Portsmouth, the best times to see it being at early morn, when the sun is reflected from the water, or at late afternoon, when the light falls on the sails of the shipping. Teneriffe Mt. in Milton is toward the S. E., and the Blue Hills are nearly in line with the more remote swell of Mt. Agamenticus; while the high bare ledges of the Suncook Range are near at hand, towards Alton Bay. Several ponds glitter in the foreground; and Gilmanton Iron-Works is about S. S. E., between Lougee Pond and the twin Suncook Ponds. Peaked Hill, in Gilmanton, lies nearly S.; and beyond, a little to the l., is Catamount Mt., in Pittsfield.

A little to the r. of this line, and nearly over Mt. Pawtuccaway, in Nottingham, the officers of the U. S. Coast Survey sighted to Thompson's Hill, which is near Gloucester, Mass.

About S. S. W., down the populous valley of the Merrimac, are the twin peaks of the Uncanoonucs, W. of Manchester. A little W., and low down on the horizon, is the blue summit of Wachuset, in Massachusetts; and farther to the W., nearly as distant, is the slowly curving Monadnock, with Pack Monadnock and the Temple Mts. to the l., and Lovewell's Mt. to the r. About W. S. W. is the high pyramid of Kearsarge, N. of which is the range of the Ragged Mts. Croydon Mt. is nearly W., and lies before Ascutney; and a line of remote peaks guards the valley of the Connecticut.

Nearly W. is the large and thriving village of Laconia, on the shore of Lake Winnesquam, beyond which is the high and near Sanbornton Mt.; and still farther to the W. are the granite crests of Mt. Cardigan. The highlands of Bristol and Bridgewater hide Newfound Lake; and beyond the head of Winnesquam is Harper Hill, in New Hampton, over which is a peak which is probably Mt. Cuba, in Orford. Lake Village is near Round Pond, which is the next widening of the Winnepesaukee River r. of Winnesquam. Plymouth Mt. is seen in the distance, and then the view-line swings over Long Bay and on to the Mt.-Carr range, towards the N. W.

The view of Lake Winnepesaukee from Mt. Belknap is one of the most beautiful and delightful prospects possible. It is best enjoyed in the afternoon, towards evening, when the light is rich and full.

154. The West Shore of Lake Winnepesaukee.

That portion of the shore-road which lies between Alton Bay and W. Alton (7 M.) is of an uninteresting character, being generally shut in by forests which prevent the obtaining of views. The road crosses the railroad near the hotel and soon diverges to the r. from the Gilmanton road, running over the high foot-hills of the Suncook Range. Occasional glimpses are given of the river-like scenery of the bay, and at 2 M. out an extensive outlook is afforded towards the northern mountains. The road then enters an almost unbroken forest of second-growth trees, through which it passes for nearly 4 M., generally within a few yards of the water, but hidden from it by a belt of dense shrubbery. There are but few breaks in this jungle, showing the narrow Alton Bay on the r. and the flanks of Mt. Major on the l.

W. Alton is a small farming neighborhood on Minge's Cove, with a store and a church. A road runs W. from this point between the mountains to Gilford, traversing a lofty and open grazing district. The shore-road continues to the N. W., and is henceforward of high interest, exhibiting an inspiring panorama of lake and mountain scenery. On the l. are the bold outer hills of the Belknap Range, and on the r. is a succession of brilliant views wherein the northern peaks are finely set forth in high relief over the broad levels of Lake Winnepesaukee.

On the r. is the high and ragged ridge of Rattlesnake Island, with Sleeper's Island farther in shore, and Diamond Island to the N., with its white hotel. The lake is here 4–5 M. broad, and beyond Diamond Island its surface is unbroken. The mountain-view includes Red Hill, near the N. shore, with the huge black mass of Sandwich Dome over it, to the r. of which is the S. peak of Tripyramid, striped by its great slide. Next comes the blanched peak of Whiteface, and then the high round crest of Passaconaway. On the N. E. shore, and extending for miles to the S., is the dark-hued Ossipee Range, with the peak called Black Snout on the N. and the Melvin Peak near the S. The appearance of these mountains remains the same as the road advances, though other peaks come into view on either side. After enjoying this prospect for over 1 M., the road ascends to higher land, whence a broader area is visible on the E. and N. There are farms all the way from W. Alton to the end of the road. After passing for about ¼ M. within hearing of the rippling water, though cut off from it by thickets, the road emerges with the open lake on the r. and in front the mountains from Moosilauke and the Squam Range to Ossipee. At some distance beyond the old town-farm and about 3 M. from W. Alton, the road skirts the lake, leaving a white-beached point on the r. rear. Near the end of this reach is an old and weather-stained house on the r., nearly opposite which is a skull-shaped knoll, whence, by a 3-minutes' climb, the tourist may gain the prettiest view on this shore.

On the l. are the Mt.-Carr and Kinneo groups, with the flat and ledgy top of Prospect next, and then the high pile of Moosilauke. To the r. of the Squam Range is the long mass of Red Hill, over whose r. is Sandwich Dome, whence Flat Mt. runs E. and is overlooked by the great slide on Tripyramid. Then Whiteface and Passaconaway come into view, and against the E. spur of the latter falls the more distant and lofty Mt. Webster, its W. side being sharply cut off where it descends into the White-Mt. Notch. To the l. and continuous are the quaint little peak of Jackson, the plateau of Clinton, the dome of Pleasant, the higher levels of Franklin, the crags of Monroe, and the lofty cone of Washington, with Jefferson seen in part behind, and on the l. The dark Ossipee Range rises on the N. E., over Welch Island; and toward the S. W. are Tumble Down Dick and Copple Crown, below Wolfeborough, with the dark knoll on Rattlesnake Island rising on the l of the latter. The quadrant between N. N. E. and E. S. E. is filled by the unbroken blue surface of Lake Winnepesaukee, here seen in its broadest part, and bounded on the N. E. by the rich farms of Tuftonborough Neck.

As the road advances up the long incline to the N. W. the view changes slowly, as Webster, Jackson, and Clinton are successively eclipsed by Passaconaway. Just beyond the white Ames farm-house on the hill-top is gained one of the richest *coups d'œil* along this route. Sandwich Dome looms over Red Hill on the l. and the Ossipee Range over Welch Island on the r. Between them are Tripyramid, Whiteface, Passaconaway, and Paugus; and between the last two are the remote but high uplifted peaks of Pleasant, Franklin, Monroe, and Washington, with Jefferson and Adams partly seen behind Washington, the former on the l. and the latter on the r. The foreground is filled with the blue mirror of the lake, unbroken for some distance, and then dotted with picturesque green islets.

At the foot of this hill are two roads, of which that to the r. is longer and easier, but the nearly disused one which ascends *Locke's Hill* gives a retrospect over the lake toward Wolfeborough, with the summits of Copple Crown, Tumble Down Dick, and Rattlesnake Island. From the top of the hill, before reaching the house one sees Prospect (near Plymouth), Moosilauke, the Squam Range, Red Hill, with Sandwich Dome over it, Flat Mt., the white gash in Tripyramid and its sharp second peak, the hamlet of Sandwich Lower Corner, Whiteface, Passaconaway, and the lowlier Paugus, marked by a sinuous slide. Between the last two are Pleasant, Franklin, Monroe, Jefferson, Washington, and part of Adams, and Ossipee is on the r. The lake view is fascinating in its breadth, variety, and richness, including scores of islands and curving sections of the E. bays. Beyond the foot of the hill the mountain features of this prospect are repeated, from near the old farm-house at the head of *Thompson's Cove*, over the fields of Locke's Island.

The road now turns S. W., away from the lake, and soon reaches a cross-road which is 13 M. from Alton Bay, 6 M from W. Alton, 3 M. from Gilford, 4 M. from Lake Village, 6 M. from Laconia, 3 M. from Weirs, and 8 M from Meredith Bridge. This reach of the shore is less interesting than the rest, as the road is practically held at 1 M. from the open lake by the intervention of farms and by the broad acres of Governor's (or Davis) Island. For nearly 1½ M. it traverses the thinly settled margin of the Gilford Intervales, crossing the Gunstock River and Meadow Brook, and approaching the bottom of a broad cove, 1 M. of inland farming country is crossed, and then the road draws near the strait inside of Governor's Island There is but little of interest here (unless the tourist wishes to drive to the r. across the bridge to the island), and in less than 1 M. farther, after views of Meredith Bay and the islands on the E, the bridge at Weirs is crossed. (Travellers who diverge to the l. at the cross-roads cross the Gilford Intervales, ascend far-viewing hills, and can quickly reach Laconia by a road which crosses Meeting-House and Follett Hills.) (See page 364.)

155. Centre Harbor.

Hotels. — The * Senter House accommodates 150 guests, at $4 a day, with reductions to permanent boarders. It is pleasantly situated above and near the lake, of which it commands a rich and beautiful view for over 20 M. The front of the house is shaded by a line of ancient elm-trees. A large livery-stable is connected with it; and a flotilla of dainty row-boats is kept on the lake below. Billiards, bowling, croquet, and other amusements are provided for the guests, and a stand for the sale of books and pictures is kept in the office.

The Moulton House adjoins the Senter House, and is a neat and comfortable hotel, accommodating 75 guests. The rates are $3 a day, or $12 to $20 a week. *Boarding-Houses.* — Almon Benson (30 guests), in the village; B. F. Kelsea (20 guests), in the village; F. and S. Wentworth, R. Fellows, R. L. Coe, Dr. Wm. Page, Mr. Weeks, also in the village; Arnold M. Graves (*Lakeside*), a short distance to the E., near the shore; D. W. Coe's, on the hill; J. B. Dow's, at the foot of Red Hill. The rates at these places vary from $6 to $10 a week.

Steamboats. — The *Lady of the Lake* and the *Mt. Washington* touch at this point 3 - 4 times daily, carrying passengers to the ports on the lake. The *Lady* belongs to the B., C. & M. R. R., and runs to Weirs and Wolfeborough. The other boat is larger and more commodious. It belongs to the Boston & Maine R. R., and runs to Wolfeborough and Alton Bay.

Stages leave Centre Harbor early every afternoon, after the arrival of the steamboats, for Moultonborough Corner, S. Tamworth, and W. Ossipee. (See route 157.) They reach W. Ossipee in time to take the evening train to N. Conway.

Distances. — Centre Harbor to Sunset Hill, 1 M.; to Centre-Harbor Hill, 1; to Squam Lake, 2; to Plymouth, 14; to Meredith, 5; to Rollins Hill, 7; to Red Hill, 6; to Moultonborough Neck, 6; to Long Island, 8; to Ossipee Falls, 10; to Centre Sandwich, 8; to S. Tamworth, 13; to W. Ossipee, 18; to Melvin Village, 12.

" Into the heart of the highlands,
    Into the north-wind free,
Through the rising and vanishing islands,
    Over the mountain sea.

" To the little hamlet lying
    White in its mountain fold,
Asleep by the lake, and dreaming
    A dream that is never told.

" And in the Red Hill's shadow,
    Your pilgrim home you make,
Where the chambers ope to sunrise,
    The mountains and the lake."

Centre Harbor is a pleasant little rural hamlet at the head of the long N. bay of Lake Winnepesaukee, and occupying decidedly the most favorable position for studying and appreciating the beauties of the lake. It contains a Congregational church, one or two stores, and a telegraph-office. The pleasantest part of the hamlet is the terrace on the road to the S., toward Meredith, where several pretty summer cottages have been erected on the ground whence the bay is overlooked. Centre Harbor is 553 ft. above the sea. It is one of the chief summer-resorts in this region, and has several hundred city boarders throughout the season. The temperature here is usually much warmer than at the villages farther up among the mountains. The scenery in the vicinity is of a high order of beauty, and is easily inspected by the aid of good roads running out in several directions. The most attractive views are those of Squam Lake (Route 160) and from Red Hill (Route 156).

Sunset Hill rises just back of the hamlet, and is easily ascended by the road. It commands a broad and noble panorama of Lake Winnepesaukee, the Ossipee Mts., Copple Crown, the Alton hills, and the double peak of Mt. Belknap. The same view is obtained from the veranda of the Senter House, but it is less favorably expanded. The hills on either side command nearly similar and thoroughly fascinating prospects. *Centre-Harbor Hill* is about 1 M. from the hotels, and gives admirable sunset views. The drive called *Around the Ring* is 4 – 5 M. long, and shows Winnepesaukee, Red Hill, and Squam Lake. Besides the pleasant drives and rambles in this vicinity, the tourist can here enjoy the varied charms of boating, visiting the adjacent islets and exploring the sequestered bays. Or, by leaving on the morning steamboat, he can traverse the lake and return by dinner-time.

Long Pond is 1 M. E. of Centre Harbor, and is a pretty sheet of water, 2 M. long, at the foot of Red Hill. It empties into the bay of Winnepesaukee between the Second and Third Moultonborough Necks, by a stream 60 rods long. *Round Pond* is connected with it on the N., and is much smaller. Light boats could be carried around the mill-dams on their outlets and rowed across the two ponds. A portage of less than ½ M. from the head of Round Pond leads to the S. E. bay of Squam Lake.

Rollins Hill is on the lower part of Meredith Neck, about 7 M. from Centre Harbor, and 1½ M. from Meredith. It is reached by diverging to the S. on a by-way, within less than 1 M. of Meredith. The hill is not high, and is covered with pastures, which are easily traversed from the adjacent farm-house. Plans are being made for a large family boarding-house on this height. * *The View* is remarkable for its breadth and beauty. Meredith Village is plainly seen at the head of the Northwest Bay, with Waukawan Lake beyond, overlooked by Plymouth Mt. Mt. Cuba is far away over the r. of Meredith, up the long valley of Baker's River; with Mt. Carr about N. W., and Moosilauke farther away, beyond Mt. Kinneo. The Squam and Morgan Mts. are next seen (nearer), with pastures on the lower flanks of their even ridges. Mt. Prospect peers over this range; and farther to the r. is Mt. Weetamoo, the chief of the Campton Mts. Green Mt. and the sharp peak of Tecumseh appear over the r. of this range; and then the ponderous black sides of Sandwich Dome are seen, with its long plateau above. About N. is the double swell of Red Hill, with pastures advancing into the central depression; and the noble peaks of Whiteface and Passaconaway rise above its l. eminence, with the avalanche-scarred Mt. Paugus over its r. On the r. of Paugus, the splendid spire of Chocorua rises into the sky, the entire mountain being visible over the lowlands of Tamworth. To the r. of this trough-like valley is the Ossipee Range, which stretches far down the E. side of the lake, whose island-strewn waters are now seen for miles. The white hamlet of Melvin is near the foot of the Melvin Peak, and Mackerel Corner is above, on the r. The Green Mt. in Effingham is over the r. flank of the Ossipee Range. Long Island is nearer, and is marked by its two large boarding-houses, and the pastures of Bear Island are still nearer. The broad bosom of Lake Winnepesaukee is then overlooked, and in the S. E. are the heights of Tumble Down Dick and Copple Crown, over Wolfeborough. Farther to the r. is the high Rattlesnake Island, with Diamond Island on the W., and the Suncook Range and the Alton hills beyond. Governor's Island is nearer, with a bridge to the mainland; and the apparently single peak of Belknap is above, on the r. The view then includes the hills and intervales of Gilford, Long Bay, and parts of Lake Village and Laconia, the Weirs, and the distant blue peak of Kearsarge. Farther to the r. are the Ragged Mts. and the three white crests of Mt. Cardigan.

The **Prospect House** is about midway between Centre Harbor and Meredith by the old hill-road, being a little over 2 M. from either village. It accommodates 65 guests, and the rates are $7 a week. The view from this point is very attractive, and includes a great portion of Lake Winnepesaukee and its islands, with the surrounding mountains. On the l. is the Ossipee Range, with Black Snout on the N. and the high Melvin Peak towards the S. Across the waters of the Centre-Harbor bay are the Moultonborough Necks, and to the r. is Long Island, with its large boarding-houses. To the S. E., at the foot of the lake, is the bold hill of Tumble Down Dick, with the symmetrical peak of Copple Crown on the r. The high knolls of Rattlesnake Island are farther up the lake, and are backed by the noble peaks of Belknap, on the r. *Gilman's Hill* is ¾ M. from the Prospect House, and overlooks Squam Lake.

The hamlet of Centre Harbor is on the line of two towns and of the counties of Belknap and Carroll. It was settled in 1765-7 by Chamberlain and Senter, who brought their families and supplies up the Merrimac River from Londonderry (Manchester). Whether its present name is a modification of *Senter Harbor*, or was given on account of its local position, cannot be certainly determined. The small town of Centre Harbor contains 515 inhabitants, most of whom are engaged in farming.

Prof. Agassiz found the glacial remains in this vicinity of great interest and scientific value. " Lateral moraines may be traced at the foot of Red Hill, a little above Long Pond ; also, along Squam Lake. Median moraines are very distinct near Centre-Harbor Hotel. Terminal moraines are also numerous near Centre Harbor, and in the neighborhood of Meredith. At the S. end of Red Hill the lateral moraines bend westward, and show their connection with the terminal moraines." (*Amer. Assoc. Advance. Sci.*, *Proc.* XIX.)

156. Red Hill

is a long ridge in Moultonborough and Sandwich, formed of two gently curving sections lying in a line nearly N. and S. Its area is elliptical, about 3 M. long and 1½ M. wide. The N. summit is 2,043 ft. high; that on the S. is 1,769 ft. high. The geological character of the ridge is of gray sienite, originating from an eruption late in the Labrador period. It is claimed that the name is derived from the berries of the *uva ursæ*, which are found here in great numbers, and from the appearance of the hill in autumn, when the foliage has changed its color from dark green to bright red. It is called *Red Mountain* on Belknap's map of 1791.

The road from Centre Harbor to the base of Red Hill is 4 M. long, and that from Centre Sandwich is about 7 M. long. A shorter route for pedestrians is by going out on the Sandwich road to the first farm-lane on the r. beyond the cemetery, and crossing the fields by retired roads. The former route diverges from the W.-Ossipee road about 1 M. from Centre Harbor, beyond Long Pond, and carriages ascend this road for 2-2½ M., turning in and stopping at a stony side-road on the r. In a shed near this diverging point are kept saddles for the use of tourists who wish to ride up the hill. The stony road ascends steeply for about ½ M. to a secluded farm-house, close around whose upper corner the mountain-path bends to the l. It is a little over 1 M. from this point to the summit, and the broad plain path leads up by comparatively easy grades, traversing second-growth woods and gaining occasional glimpses of the lake-country. This route is very easy and safe for horses, as there are no ledges and but little of stony ground. The ascent from the road should be made in less than an hour (the Editor ascended in 38 minutes, and descended in 26 minutes). When the path nears the summit, a section of open ground is crossed, from which beautiful lake-views are given Squam Lake is seen from this point to much better advantage than from the summit, nearly its whole area being visible, while the woods below the top hide a large part of it from the beacon. The summit is partly covered with ledges, and is crowned by a signal-staff of the U. S. Coast Survey.

** *The View*. — To the S. arc the broad and shining levels of Lake Winnepesaukee, stretching off into the lowlands on either side in long and sinuous bays, and flowing to the S. W. through a chain of lakelets toward the Merrimac River. Broad and fertile peninsulas reach out from the eastern towns far into the lake, with thousands of acres of valuable land, dotted with farms and clearings; and a multitude of islands breaks the surface of the water, showing every variety of form and the greatest diversity of size. On the S. W. and S. this superb sea of islands is bounded by tall and symmetrical mountains, contrasting their neutral hues with the vivid blue or silver of the far-reaching water.

Looking E. S. E., toward the middle of the dark Ossipee Range, the village of Moultonborough Corner is seen, with its little white church; and to the r., at the foot of the range, is E. Moultonborough. On the heights to the r. of the S. end of Ossipee is Tuftonborough Corner, to the r. of which and near the water's edge is Melvin Village, with the hamlet of Mackerel Corner above and beyond. Over the S. ridge of Red Hill, about S. E. from the peak, are the broad waters of Moultonborough Bay, flecked with many islets, beyond which, and on the r. of the line to Melvin, is Copple Crown Mt., with its long even slopes and decided peak. Beyond the nearer lowlands of Moultonborough Neck is Long Island, dotted with farms, and throwing its S. point far out into the lake. Over the l. of Long Island are the low plains of Cow Island and the populous Tuftonborough Neck, far beyond which are the bold hills about E. Alton. Over the r. end of Long Island is the long wooded ridge of Rattlesnake Island, beyond which is a part of Alton Bay, with the light crests of the Suncook Range to the r. Diamond Island is well out in the lake, to the r. of Rattlesnake; and more to the r. is a long line of islets running S. S. E. from Five-Mile Island to the well-cleared Welch Island. Over the r. foot of the near green ridge of Red Hill is the forest-covered Moultonborough Second Neck, beyond which are the farms on Bear Island, whence a line of smaller islets extends to the Gilford shore. To the r. of this direction, on the S. W. shore, are the graceful twin peaks of Mt. Belknap (nearly S.), with Mt. Retreat on their l. The whole extent of the bay of Centre Harbor, and all its islands, are seen in the S., and the village of that name is close below, across Long Pond. The farm-abounding Meredith Neck runs from Centre Harbor far out into the lake, its outer point coming nearly under Mt. Belknap. Over Centre Harbor, and several miles distant, is the smooth-topped Rollins Hill, with parts of Northwest Bay on either side; and also over the village, a little to the l., and beyond Weirs, is Long Bay, the outlet of Lake Winnepesaukee. At the foot of the bay are the prosperous manufacturing towns of Lake Village and Laconia, and below them extend the populous valley-towns. A little to the r. is the round hill which hides Meredith, beyond and r. of which is Crotched Mt., in Francestown, touching the sky-line. To the r. of Meredith is Lake

Waukawan, which is apparently divided into two parts by a nearer hill, over whose top, far down on the horizon, is the low round crest of Monadnock. To the l., and nearly as far away, is the blue mountain in Temple, and Pack Monadnock is seen on the r. Over the l. extremity of Squam Lake, and beyond the Ragged Mts., is the handsome peak of Kearsarge, with a long and level ridge connected with it on the l. In the foreground are the many high hills of New Hampton and Sanbornton; and to the r., beyond the Ragged Mts., is a part of Sunapee Mt. The Bridgewater and Alexandria hills next occupy the foreground; and about W. by S., down Squam Lake and Little Squam Lake, is Mt. Cardigan, with a symmetrical dome in the centre of the ridge and a minor dome on either side. Farther to the r. is the wooded Plymouth Mt.

The brilliant plain of Squam Lake now occupies the foreground, and is variegated by numerous green islets. Beyond it and curving around to the E. is the Squam Range, with a uniform and ledgy crest-line and a low elevation. Over its l. and highest peak is the sky-meeting ridge of Smart's Mt., towards the Connecticut River; and farther to the r. is Mt. Stinson, in Rumney. Next comes the Mt.-Carr Range, with a rounded crest on the S.; and then the immense mass of Mt. Moosilauke is seen, with a sharp peak on the l. and a rounded swell in the centre, overhanging Mt. Kinneo, under its l. flank. Just beyond and peering over the Squam Range, toward the r. base of Moosilauke, are the ledges on Mt. Weetamoo, the chief of the Campton Mts. Over the r. of the Squam Mts. is the remote blue peak of Mt. Kinsman, and farther to the r., over the white-topped Acteon Range, are the massive and distant Franconia Mts. To the r. of the Sachem Peak are the ascending terraces of the Fisher Range, nearly eclipsed by the huge bulk of Sandwich Dome.

A little W. of N., across a well-populated valley, are the ledgy sides of Mt. Israel, beyond which is the lofty plateau of Sandwich Dome, with its massive buttresses. To the r. of the Dome is the inconspicuous ridge of Flat Mt., over whose l. is a blue segment of Mt. Carrigain, and over the r. are the lower peaks of Tripyramid, brilliantly marked by a portion of the great slide. The beautiful valley of Sandwich fills the foreground to the N. and N. E. with its peaceful farms and pretty hamlets. The long white village of Centre Sandwich is close at hand, with its two church towers and clustering houses. Over this point are the noble peaks of Whiteface, with its marble-like cliffs and deep ravines, and Passaconaway, of about the same height but throwing out its blackish hues in vivid contrast. To the r. of the latter and below is the symmetrical green cone of Mt. Wonnalancet, on whose r. are the bare white ledges of Mt. Paugus, over N.-Sandwich village. Between Wonnalancet and Paugus, and more distant, is a portion of Bear Mt.; and part of Moat Mt. is seen on the r. of Paugus. In the foreground, to the r. of Centre Sandwich, is the island-gemmed Red-Hill Pond, over which is the superb white peak of Chocorua, with a profound ravine running from the W. to the base of its N. peaks. To the r. of the main peak is a white spur, whence a long ridge runs out to the plains of Tamworth. Over this spur, about quarter of the way from the spur to the end, and over the islands in Red-Hill Pond, is the crest of Kiarsarge, with the hotel on its summit. To the r. of Chocorua are certain of the Green Hills of Conway, above the Saco River. In the foreground is the high hamlet of Sandwich Lower Corner, near a tiny pond, and below the islandless Bearcamp Pond; and the view-line passes for leagues down the populous valleys to the N. E. A little to the r. of the Corner, and beyond, is S. Tamworth, near which is the handsome pyramid of the Whittier Peak, in the Ossipee Range. Just to the l. of this point is Mt. Pleasant, near Sebago Lake, a long ridge with four heads, on the third of which is a white hotel. On the l. of Pleasant are the mountains in Waterford, and other remote ridges in Maine. Beyond the Whittier Peak, to the r., are the remaining mountains of the Ossipee Range, stretching to the S. for miles, and culminating towards the r. in the tall Melvin Peak.

" Whoever misses the view from Red Hill, loses the most fascinating and thoroughly enjoyable view, from a moderate mountain height, that can be gained from any eminence that lies near the tourist's path. The Mt.-Washington Range is not visible, being barred from sight by the dark Sandwich chain, which in the afternoon, untouched by the light, wears a savage frown that contrasts most effectively with the placid beauty of the lake below. Here is the place to study its borders, to admire the fleet of islands that ride at anchor on its bosom, — from little shallops to grand three-deckers, — and to enjoy the exquisite lines by which its bays are enfolded, in which its coves retreat, and with which its low capes cut the azure water, and hang over it an emerald fringe." (STARR KING.)

"Red Hill, near Centre Harbor, should be ascended. The view from its summit is the most beautiful panorama which this country affords. On one side is Winnepesaukee and its still more picturesque rival, Squam, on the other Ossipee and others of less note, the whole surrounded by the lofty peaks of Kearsarge, Chocorua, Pequawket, and innumerable chains of hills which recede behind each other till lost in the horizon. The islands, which almost conceal the surface of the lake, seem to have every possible variety of form and shape, now rising to the height of several hundred feet above the water, and now seeming to float upon it, a mere tuft of evergreens, from the centre of which often projects some half-decayed, gigantic pine, whose knotty branches extend beyond the borders of the patch of soil which gave it birth." (DR. C. T. JACKSON.)

The first visit to this summit (probably) is thus described : " Having often-times travailed the country, some of the natives allwaies wth me, which hath from time to time affirmed that the lake called Winnipaseket issues into the river of Merremake, and having some Indians with me upon the north side of the said lake, upon a great mountajne, did see the s[d] lake which the Indians did affirme issues into the aforesajd river." (Peter Weare's evidence before the Mass. Legislature, 1665.)

" Red Hill is the place to behold it, and there the Indians must have stood when they gave it its name. Red Hill is near its N. extremity, and we never saw such an object in nature as Winnepesaukee seen from its top. It looks as if it had a thousand islands. They tell of 365, one for every day in the year. But there must be many more, some of them large enough for little towns, and others not bigger than a swan or a wild duck swimming on its surface of glass. . . . . Winnepesaukee, it is the very 'Smile of the Great Spirit.' And the Indians gave it that name to signify that smile. And, verily, if the propitious glance of creative Power could be left upon its inanimate works, we should think it would play there in the form of this glorious lake. . . . . The picture from Red Hill defies competition, as it transcends description. It is the perfection of earthly prospects." (N. P. ROGERS.)

Harriet Martineau said of this view: "The mountain horizon was altogether beautiful. Some had sharp peaks, some notched; the sides of some were bare, with traces of tremendous slides; others green as the spring, with wandering sun-gleams and cloud-shadows."

Isaac Hill says: "The eye never traced a more splendid prospect than the view from Red Hill. . . . On the one hand the Winnepesaukee Lake, 22 M. in length, with its bays and islands and surrounding villages and farms of party-colored fields, spreads out like a field of glass at the S. E. Loch Lomond with all its splendor and beauty presents no scenery that is not equalled in the environs of the Winnepesaukee. Its suite of hills and mountains serves as a contrast to increase its splendor.

"A more charming and delightsome view with the naked eye is not perhaps to be seen in America. . . . The view from the summit of Mt. Washington in a clear day is magnificent; the mountains in different directions seem like waves of the sea, — but the eye rests on little else besides mountains. From Red Hill the eye descries objects, — villages and isolated buildings, green fields and forests, the golden wheat and paler rye, and flowing corn, — the flocks of cattle and sheep. More beautiful than all, on the N. W., is the Squam Lake, at the S. W. foot of the Sandwich Mts., studded with islands; and the Lake Winnepesaukee nearly in its whole extent of about 30 M., with its hundreds of islands, resembling liquid silver run into a vessel of unequal surface, portions of which are too high to be covered."

President Dwight visited the summit of Red Hill in 1813, and said that the view was "worth not only the trouble of the ascent, but that of our whole journey" (on horseback from New Haven).

"Directly N. of us rose the Sandwich Mts.; a magnificent range proceeding in a N. E. direction, and terminating at the distance of 30 M. Here a succession of finely varied summits, of the boldest figures, and wrapped in a mantle of misty azure, ascended far above all parts of the visible earth. Singly they were in an eminent degree sublime; in their union they broke upon the view with grandeur inexpressible. . . . A finer object of the same nature was perhaps never seen. The lakes, which I had visited on my northern and western excursion, were all of them *undivided masses*, bordered by shores comparatively straight. This was, centrally, a vast column if I may be allowed the term, 23 M. in length, and 6-8 in breadth, shooting out with inimitable beauty a succession of arms; some of them not inferior in length to the whole breadth of the lake. These were fashioned with every elegance of figure, bordered with the most beautiful winding shores, and studded with a multitude of islands. Many of the islands in the lake are large, exquisitely fashioned, and arranged in a manner no less singular than pleasing. As they met the eye, when surveyed from this summit, they were set in groups on both sides of the great channel; and left this vast field of water unoccupied between them. . . . The points which intrude into this lake, were widely different from those which were mentioned in the description of Lake George; bold, masculine bluffs, impinging directly upon the water. These in several instances were spacious peninsulas, fitted to become rich and delightful residences of man; often elevated into handsome hills, and sloping gracefully into the lake.

"The expansion was vast and noble. Several smaller and very beautiful lakes illumined in spots the dark ground of forest by which they were encircled. Subordinate hills and intervening valleys, with houses, inclosures, and other proofs of cultivation, dispersed throughout the neighboring region, added, though in a less degree than we could have wished, a pleasing variety to the ruder scenery. As these objects receded, and vanished, the distant mountains began to ascend in misty and awful grandeur; and raised an insurmountable barrier between us and the rest of the world; while to the eye of imagination this vast array of magnificence was designed only to be the enclosure of the field of waters beneath our feet." (DWIGHT's *Travels.*)

157. Centre Harbor to West Ossipee.

The road from Centre Harbor to W. Ossipee affords some of the pleasantest views in the State, and may take rank with the famous Cherry-Mt. Road; for though the mountains visible hence are less lofty than those seen from the latter route, they are sharper and more alpine in appearance. The effect of the late afternoon light on the white peaks of Whiteface and Chocorua is very brilliant, and at the same time the shadowy ravines are brought out in strong contrast.

Soon after leaving Centre Harbor and the lake, the road approaches Red Hill, which is the most conspicuous feature of the road for several miles. Beyond Long Pond, whose clear waters are seen on the l., the S. end of the ridge is rounded, and the stage rolls on to Moultonborough Corner, 5 M. from Centre Harbor. The Ossipee Range is now boldly outlined on the r. front, with the sharp and wooded peak of Black Snout on the N. and the Melvin Peak on the S. At 2½ M. from Moultonborough Corner the stage passes the Gove's-Corner road on the r., and soon afterward the bowl-shaped summit of the Whittier Peak is seen on the r. front, nearly detached from the main Ossipee Range. The great mountains on the N. are visible from time to time, and after passing the gaunt old red farm-house (on the l.) about 8 M. from Centre Harbor, the road runs for a long distance on a terrace whence is obtained one of the finest panoramic views in this region. On the l. is an extensive belt of fertile meadows, studded with graceful elms and bounded on the N. by a low ridge, over which appear the imposing peaks of Mt. Israel, Sandwich Dome, Whiteface, Passaconaway, Paugus, and Chocorua.

They stand in the order named, and may be distinguished as follows : Israel by its dark crest and far-advanced pastures ; Sandwich Dome by its dark color and prolonged mass ; Whiteface by a pyramidal white crest, cut into by a deep ravine ; Passaconaway by a high conical dome ; Paugus by the curving white slide on its front, and by being lower than its neighbors ; and Chocorua (on the r.) by its singularly white crest and upper flanks. On the r. and near at hand is the smooth and ledgy slope of the Whittier Peak, dotted with detached trees.

A pretty glimpse of Bearcamp Pond is soon obtained on the l., out on the well-tilled plains of Sandwich. The scattered hamlet of *S. Tamworth* is passed at 12 M. from Centre Harbor, and the stage follows closely the course of the picturesque Bearcamp River, with the Ossipee Mts. on the r., until it reaches W. Ossipee (see page 347).

" The most interesting feature of the ride, however, is Chocorua, and, to those unacquainted with mountain-scenery, the first impression of this peak is very striking. Driving over the mountain-road in a hot summer afternoon, one watches the great hill-tops come up, like billows, one after another, from the sea of mountains round about, as the coach winds and twists among them. The soft afternoon light and atmosphere rest over the land, which, as the sun sinks lower, becomes streaked with pale bars of light when the sides and shoulders of the hills are developed by the failing day." (*Picturesque America.*)

158. Centre Harbor to Plymouth.

This road runs N. W. across the town of Holderness, and is a little over 12 M. long. It is very hilly in some parts, but gives beautiful views of Lakes Winnepesaukee and Squam, and the adjacent mountains. There is no public conveyance on this route, but carriages may be hired at the livery-stables in Centre Harbor or Plymouth.

After pleasant retrospects of Lake Winnepesaukee, the road reaches the highlands towards Squam Lake, of which beautiful views are afforded, with the Sandwich Range beyond. After passing White-Oak Pond (on the l.) a long upward grade leads to the upper slopes of **Shepard Hill**, from which is gained the most beautiful* prospect on the route. It may be enjoyed from the road, or (with increased breadth) from the summit of the pasture on the r. The long islands of Squam lie across the lake in parallel ranks, with lanes and squares of blue water between, and far away over this fascinating foreground, "the whole form of gallant Chocorua, with his steel-hooded head, fills the r. background to the N. (E.), towering, without any intervening obstruction, 12 – 15 M. away." To the l. of Chocorua are the crouching mass of Paugus and the bold hemisphere of Passaconaway; and Mt. Israel, the low but prominent Peaked Hill, and the Squam Range occupy the front lines to the l., back of which is the high and peakless ridge of Sandwich Dome, with two pinnacles of Tripyramid on the r. (on one of which is seen the upper part of the Great Slide). To the r. of Chocorua (and much nearer) is Red Hill, nearly eclipsing the Ossipee Range. This view was much admired by Starr King, and a picture of it is given in *The White Hills* (page 84). The best time to visit Shepard Hill is in the late afternoon, when the descending sun invests the Sandwich peaks with a wealth of color.

Beyond this point the road descends rapidly to the outlet of Squam Lake, with the calm waters of *Little Squam* on the l. The latter is followed for some distance, until the W. end of the Squam Range is turned, and the route lies more to the N., and reaches the hamlet of *Holderness.* Beyond Church Hill it passes the old Episcopal church (see page 279), and soon reaches Plymouth (page 277).

" If this road were less hilly, it would offer one of the most delightful drives among the mountains. . . . . During a large portion of the drive the two lakes — Great Squam, singularly striped with long, narrow, crinkling islands, and, like Wordsworth's river, winding in the landscape ' at its own sweet will,' — and Little Squam, unbroken by islands, fringed and shadowed by thickets of the richest foliage, that are disposed around its western shore in a long sweeping curve line which will be remembered as a delightful melody of the eye — offer themselves in various aspects that often compel us to stop and quietly drink in their beauty. . . . . There are charming reliefs of forest-path in the road, which, though uneven, is of quite civilized smoothness. And it opens at last upon a splendid surprise in the rich meadows of Holderness and Plymouth, that are studded or overlooked by tasteful country residences, and adorned with clusters and avenues of grand old elms." (STARR KING.)

159. The Drive around Squam Lake.

Distances. — Centre Harbor to Chick's Corner, 7 M.; to Squam Bridge, 15 M.; to Centre Harbor, 21 M.

This road leads through a region which is rich in quiet and pastoral beauty, and makes a pleasant day's excursion from Centre Harbor or from Plymouth. It is for the most part at some distance from the lake, and on a higher grade, and is sometimes shut off from the water-views by intervening hills; but the occasional glimpses of the lakes and of the alpine Sandwich Range are full of attractiveness and interest. The road is smooth, though very hilly, especially on the N. W. of the lake.

The road runs out from Centre Harbor to the N. E., and is the Centre-Sandwich highway, passing W. of Red Hill and overlooking Long Pond on the r. The Squam Range, Mt. Israel, and Sandwich Dome are seen in front, and on the r. is the dull mass of Red Hill. At 4-5 M. from Centre Harbor the bright waters of Squam Lake are seen, and the road soon descends towards their level, running near a thicket-fringed strand, with wooded islets off-shore. At a point where a large boulder is seen on the l. the lake-shore route diverges to the l. from the Sandwich road, and ascends to the obscure hamlet of *Chick's Corner*, with views of the Sandwich Range on the r. and front. The road turns thence towards the W., and follows the base of the Squam Range, whose long and regular ridge is faced with purple rocks. Distant views of Squam Lake are given from time to time.

Rattlesnake Mt. (or Peaked Hill) is a double-headed spur of the Squam Range which runs out nearly to the lake, with far-viewing ledges on its low and wooded crests. The road passes through the high notch between this spur and the main range, and is thus for some time shut out from open views. But on reaching the W. of the ridge the road becomes at once more hilly and more picturesque, the summit of each high rise giving views over many-islanded Squam, the dark Ossipee Range (with the Whittier Peak apparently isolated on the N.), the Red-Hill heights, and the blue mountains S. and W. of Lake Winnepesaukee, Tumble Down Dick, Copple Crown, Suncook, and the twin peaks of the Belknap Range. After 2-3 M. of hilly road, the point is reached where the road to Holderness (3 M.) and Plymouth (4 M.) diverges. The lake-drive turns to the l. and crosses the outlet of Squam, with Little Squam Lake close at hand on the l. Beyond Labarron's boarding-house the long *Shepard Hill* is ascended, whence the noblest view of the lake and Mt. Chocorua is given (see page 378).

In descending Shepard Hill, the pretty villa of Prof. Norton is seen on the r., below which is White-Oak Pond. The road runs S. E. through a populous farming country, and soon overlooks Squam Lake near Drake Island. The retrospects from the higher hills include Moosilauke and Mt. Prospect. (The return may be agreeably varied by diverging on the Meredith Road, bearing to the l., and entering Centre Harbor by the long hill on the S. W., which overlooks Lake Winnepesaukee.) After this section of the lake is passed the road traverses a series of high hills, approaching Red Hill and Long Pond on the l. In descending the last long hill to Centre Harbor, a pleasing view of Lake Winnepesaukee is obtained on the r. front.

160. Squam Lake

contains 15½ square miles, and is 510 ft. above the sea. It lies in three counties and four townships. The length is over 6 M.; the greatest breadth is 3 M. It stretches in the direction of S. W. and N. E., and has several deep bays, entering the land. It is nearly divided into two sections by promontories from the N. and S., between which are several long islands parallel with each other. There are sixteen islands in the lake, several of which are used for pasturing horses and sheep. The animals are made to swim out, behind boats, and are towed about the lake until they are very tired, before they are allowed to land. It is supposed that they thus obtain the idea that it is a prodigious distance from their islands to the mainland, and therefore they do not attempt to swim across.

There is no other lake in New England so beautiful as Squam, none in Eastern America that can surpass it in picturesqueness. The limpid purity of the waters, the singular grace of the tree-tufted islands, and the grandeur of the adjacent mountains, all combine here to make a rich and fascinating panorama. The best view-point on the adjacent roads is Shepard Hill (see page 378). The correct name of the lake is *Asquam*, and was applied by the Indians, being their word for water.

The Squam Range and Morgan Mt. are close to the lake, on the W. and N. W.; Red Hill is on the E.; and Mt. Israel and the Sandwich Range stretch along the N. E. The largest of the islands are Sturtevant (1 M. long, in the S. bay), Drake, Mile, and Great Islands; and those in the N. and N. E. parts of the lake are much smaller, being hardly more than insulated ledges. The crystalline waters contain perch and pickerel, and a few great lake-trout.

Robinson Labarron's cottage, between Squam and Little Squam, is the only place near the shore where boarders are taken. The rates are low. Andrew Smith has a large summer boarding-house about 1½ M. S. W. of the lake, on far-viewing heights. The hotels at Ashland and Centre Harbor are 4 – 6 M. from Squam.

"And then the Great Squam, connected with it on the E. by a short, narrow stream, the very queen of ponds, with its fleet of islands, surpassing in beauty all the foreign waters we have seen, in Scotland or elsewhere — the islands, covered with evergreens, which impart their hue to the mass of the lake, as it stretches 7 M. on E. from its smaller sister, towards the peerless Winnepesaukee." (N. P. ROGERS.)

President Dwight wrote of this lake (which he afterwards named *Lake Sullivan*): "This lake, notwithstanding its uncouth name, is a splendid sheet of water; finely indented by points; arched with beautiful coves, and studded with a succession of romantic islands. At its head rose the Red Mt. in its grandest attitude; and formed an appropriate background of the picture. On the eminence just mentioned we had a spacious prospect of the surrounding region, composed of valleys, hills, and mountains. Some of the mountains were very lofty. One particularly, ascending in the N. E., was distinguished by the form and sublimity of its summit (Chocorua)."

"The two largest (Squam lakes) from their romantick beauties, deserve a better name. One of them, which borders on the road to Senter Harbour, is indeed a most interesting object. Its union of wildness and beauty gives it a peculiar charm. If its good fortune had placed it in the Old World, it would not so long have remained unsung. Many a tourist would have tasked his imagination for sonorous epithets to describe its scenery, many an artist would have prepared his softest tints to paint its beauties, and many a poet would have strung his lyre to sound its praises in a name that taste and poetry might use. But, alas! its pellucid bosom, its undulating shores, its hanging woods, and all its magick beauties, are probably destined long to be veiled in obscurity deep as its own seclusion." (*Mass. Hist. Colls.*, 1815.)

"No wonder that the Indians were so strongly attached to this neighborhood, and fought so desperately before yielding the possession of it to the white intruders. The lower hills tempted them with abundance of game, and the calm water supplied them with unfailing stores of fish; while Winnepesaukee was but 6 M. distant one way, and the Pemigewasset equally near on the W. And possibly the surpassing loveliness of the landscape served as a golden thread in the cord which bound them to this peaceful dell in the centre of New England. The larger Squam Lake, not a fourth part so large as Winnepesaukee, is doubtless the most beautiful of all the small sheets of water in New England; and it has been pronounced by one gentleman, no less careful in his words than cultivated in his tastes, more charmingly embosomed in the landscape than any lake of equal size he had ever seen in Europe or America." (STARR KING.)

After the destruction of the homes of the Pemigewassets by Baker's rangers, in 1712, the tribe removed to Canada. But a few of its people were too much attached to their beautiful land of lakes and mountains to leave it, and they settled about Squam Lake, where they lingered until, one by one, they died. They were driven to the mountains when Lovewell's and Tyng's forces marched around Squam. The Canadian Indians of St. Francis frequently encamped in this vicinity when on their deadly forays against the New-England frontiers.

" I felt the cool breath of the North ;
  Between me and the sun,
O er deep, still lake and ridgy earth,
  I saw the cloud-shades run.

Before me, stretched for glistening miles,
  Lay mountain-girdled Squam :
Like green-winged birds, the leafy isles
  Upon its bosom swam."
                    JOHN G. WHITTIER.

161. Sandwich.

This is one of the lake-towns, containing 1,854 inhabitants, and covering 64,000 acres of land, of which nearly 24,000 are improved. The chief pursuit of the people is agriculture. It is estimated that fully 1,000 city-people spend their summer-vacations here, paying $ 30,000 into the hands of the residents. This is the best town in Carroll County for farming, and has many blooded horses and cattle. There is good pasturage on the hills, and the deep soil of the lowlands produces corn, hay, oats, and wheat. The rich combination of lake and mountain scenery within its borders has made Sandwich a favorite resort for summer-travellers, hundreds of whom are kept in the farm-houses, besides those in the boarding-houses.

This town was granted in 1763, and was 6 M. square, but the grantees represented that the N. and W. sides were "so loaded with inaccessible mountains and shelves of rocks as to be uninhabitable," whereupon Gov. Wentworth made it 10 M. square. It was first settled in 1768, and in 1775 it had 245 inhabitants.

Centre Sandwich (Burleigh's house receives travellers) is the chief village, and is in a picturesque glen, surrounded by an amphitheatre of mountains, in the lower part of the town. It contains Free-Will Baptist, Methodist, and Friends' churches, an academy, and several stores. Mt. Israel is near, on the N. W.; Red Hill is toward the S.; and the Ossipee Mts. are on the S. E. During the summer boarders are taken at Beede's Literary-Institute buildings. This village has been a favorite resort of Lucy Larcom, the poetess. It is one of the best points from which to visit the lovely Squam Lake, which lies within 3 M.; and the beautiful island-studded sheet of Red-Hill Pond is within 1½ M., on the E.; while the Bearcamp Pond is 4 – 5 M. to the E., towards S Tamworth. There are several points in the vicinity whence rich views of the alpine Sandwich Range are obtained; and Mts. Israel and Whiteface are ascended hence.

Many trout are caught in the brooks which flow down from the ravines of the Sandwich Range, especially about the flanks of Whiteface. Other fish are found in Squam Lake and the adjacent waters. In the autumn a favorite and exciting sport is the pursuit of red foxes, many of which are to be found among the Sandwich hills. It is 7 M. from Centre Sandwich to the path up Red Hill, 12 M. to Thornton Street, 10 M. to Tamworth, and 8 M. to Centre Harbor (via Squam Lake).

On the Guinea-Hill road, running N. W. from Centre Sandwich, about 1 M. from the village, are the summer-boarding-houses of Albert Fogg, Burleigh Hoyt, and Samuel Burleigh. The road to the Mad-River and Pemigewasset Valleys runs N. W. from Centre Sandwich through the defile between Mt. Israel and the Squam Range, and then crosses the glen in which the Beede River rises. It approaches Sandwich Dome on a hard ascending grade, and crosses the range at the lofty pass of Sandwich Notch, between the Dome and the Campton Mts. The vast pile of Sandwich Dome is conspicuous from many parts of the town, and is usually called *Black Mt.* by the residents. It cannot well be ascended on this side (see page 330).

The road from Moultonborough Corner to Centre Sandwich is 4½ M. long, and affords a pleasant drive, being mostly in a cleared country, with broad views from the hills. At 3 M. from Moultonborough is the hamlet of *Sandwich Lower Corner* (Quimby's boarding-house), situated on a high hill-top and viewing the country for miles. Beyond this point are the mansions of Paul Wentworth and the wealthy Adams family, crowning the long slope down which the road passes to Centre Sandwich. Fine views to the N. and N. E. are afforded from this hill, including the chief peaks of the Sandwich Range.

N. Sandwich is in the E. centre of the town, and is a small hamlet with a church. Near this place is the range of far-viewing heights called the *Maple Ridge*, on which are two summer-boarding-houses, kept by Messrs. Watson and Wiggins (4 M. from Centre Sandwich).

William McCrillis's farm and summer boarding-house are beautifully situated in a sequestered glen in the N. part of the town, 1,083 ft. above the sea, 4 M. from N. Sandwich, 14 M. from Centre Harbor, and 11 M. from the W.-Ossipee station. From the front the long blue Ossipee Range is seen, across a rich meadow bordered by woods; and back of the house are the great cliffs of Whiteface, which is seen and visited thence to best advantage (see page 335). McCrillis accommodates 25 guests, at $7 each per week. Young Mt., the Great-Hill Pond, and other points of interest are visited from this point; and good trouting is found in the adjacent ravines. About 2 M. S. is the petty hamlet of *Weed's Mills*, with a small church. The way to get to McCrillis's is to go on the stage from Centre Harbor to N. Sandwich, where the boarding-house carriage will be in waiting, if previously ordered.

There is a trail which leaves the end of the road, 3 M. W. of the McCrillis farm, and ascends for 1½ M. to *Flat-Mt. Pond*, whence a vague path leads in to Greeley's in 3-4 M. The pond was formerly much visited by fishing-parties, but of late years it has afforded but little sport. It lies on Flat Mt., a long and wooded plateau 2,700 ft. above the sea, extending from Whiteface to Sandwich Dome.

162. Mount Israel

is N W. of Centre Sandwich, and its base is reached in 2-2½ M. by way of the Guinea-Hill road, which is left at the forks where a road diverges to the r. The ascent from this point takes 1-1½ hrs. (being about 1½ M. long), two thirds of the way leading through pastures and the rest over highly inclined ledges. The summit is an extensive ledge, uplifted above all surrounding objects, and crowned with a beacon of the U. S. Coast Survey. It is 2,880 ft. above the sea.

* *The View.* — The northern view from Mt. Israel is shut out by the vast mass of Sandwich Dome, which closes in near at hand across a narrow wooded glen. E. of N., across the dark waters of Guinea Pond, is a spur of this mountain, over which is the blanched crest of the Sachem Peak. The lofty plateau which forms the summit of Sandwich Dome is nearly N. of and high above Mt. Israel. To the r. are the comparatively low ridges of Flat Mt., over which peer the S. and W. peaks of Tripyramid, marked by the Great Slide. Bold ridges run from Tripyramid to the high white cliffs of Whiteface, on whose r. is the massive dome of Mt. Passaconaway. The symmetrical low peak of Young Mt. lies between Israel and the S. ridges of Passaconaway, and the flattened ledges of Mt. Paugus are seen over the clearings on the r. of Young Mt. Still farther E., over the McCrillis farm, is the soaring white peak of Chocorua, girded with cliffs and as sharply cut as the Matterhorn. The view opens far away to the E. down the rural valleys of Tamworth and the Bearcamp River, forming a great trough between the Sandwich and Ossipee Ranges. Beyond N. Sandwich are the hamlets of Tamworth, with the high hills of Eaton to the E.; and farther to the S. is the white village of Centre Sandwich, near at hand, with the Red-Hill and Bearcamp Ponds beyond. S. of and far beyohd Chocorua is the rampart-ridge of Mt. Pleasant, with its summit-hotel, to the r. of which a part of Sebago Lake may be seen on a clear day, beyond Saddleback Mt. in Baldwin. Farther to the r., and below Silver Lake, is the low point of Mt. Prospect in Freedom; and still farther S. are the flanks of the Green Mt. in Effingham, nearly hidden by the Ossipee Range.

The Ossipee Range fills the S. E., and is seen throughout its entire length, including the pyramidal ledges of the Whittier Peak, and, farther to the r., the high black crown of the Melvin Peak, beyond the plains of Moultonborough. To the S. E. is a large area of Lake Winnepesaukee, making a brilliant and effective panorama of blended woods and waters, islands and bays. Copple Crown Mt. and Tumble Down Dick are on the horizon, beyond Smith's Pond and Wolfeborough Bay; and farther to the r. are the rich peninsulas of Moultonborough and Tuftonborough, Long Island, and numerous other lake-environed bits of land. Beyond Rattlesnake Island is the entrance to Alton Bay, with the white-capped Suncook Range on the W.; and broad reaches of blue water extend to the N. and N. W. In the foreground, across the narrow Sandwich valley, is the shapeless mass of Red Hill, over whose lower flank are the twin crests of Mt. Belknap, clearly outlined against the sky. Farther to the r. is Lake Winnesquam, with its long, bright waters, and the suburbs of busy Laconia.

W. of S., and close at hand below, is the fairy-like scene of Squam Lake, whose every island is overlooked, and all its retreating bays are seen as in a bird's-eye view. This is one of the most comprehensive of all the prospects over this beautiful sheet of water, and will long attract the admiring attention of the visitor. Far away over the neck which runs into the lake from the S. are the hills of Temple, and over the W. end of a long island is the flattened dome of Monadnock, very distant on the S. S. W. Near the S. W. shore of Squam is White-Oak Pond, to the r. of which is Little Squam Lake. Over the l. of the latter is the fine blue peak of Kearsarge, with the Ragged Mts. in front and on the r.; and over the r. and nearer are the broken lines of the Bridgewater Hills. Over these, to the r. of and beyond Kearsarge is Lovewell's Mt., in Washington. Beyond and to the r. of the Ragged Mts. is Sunapee Mt., near Sunapee Lake. To the l. of the ridge which runs W. S. W. from Israel is the long curving crest-line of the Squam Range, above which is Plymouth Mt., covered with dark forests. Over the r. of the ridge is the flattened top of Mt. Prospect, above and beyond which are the symmetrical triple peaks of Mt. Cardigan, on whose r. and low down on the horizon is the dim blue spire of Mt. Ascutney, in Vermont. Near at hand are the wooded flanks of Morgan Mt., under the swelling outlines of Tenney Hill, in Groton; and down the short valley of Beebe River, which runs away towards Campton, is Moose Mt., touching the sky, near Dartmouth College. On the r. of this valley is Mt. Stinson, with its central peak and similar flanks; and a little to the r., about over a nearer clearing, is the rocky crest of Mt. Weetamoo, among the Campton Mts. The next high mountain on the r. is a portion of the well-wooded Mt.-Carr Range. Above the clearings to the N. W. is the Sandwich Notch, just to the l. of and beyond which is the high peak of Mt. Kinneo, in Ellsworth. Over the Notch is Mt. Cushman, and above the hill on its r. is the imposing ridge of Moosilauke.

163. Moultonborough.

Hotels. — The Whiteface House is a small village-inn and Jaclard's is a neat summer boarding-house at Moultonborough Corner. This hamlet is traversed by the daily stages between Centre Harbor and W. Ossipee. It is 5 M. from Centre Harbor, 4½ M. from Centre Sandwich, 12 M. from Tamworth, 5 M. from the Ossipee Falls, 5 M. from the Whittier Peak, and 10 M. from Tuftonborough Corner.

The town of Moultonborough has 1,299 inhabitants, and contains 14,265 acres of improved land. It was granted to Col. Jonathan Moulton and 61 others, in 1763; and the church was established in 1773. Most of the people are farmers, wheat, corn, and potatoes being the chief products. The town includes Red Hill and a large portion of the Ossipee Range, with the rich lowlands of the Moultonborough Necks, and several islands in Lake Winnepesaukee. Near the upper centre of the town, and in close proximity, are the hamlets of Moultonborough Corner and Moultonborough Centre, the former being the larger, and containing two churches and two stores. Several score of summer-boarders sojourn in this town, and thousands pass through it.

Moultonborough Bay is the largest of the bays of Lake Winnepesaukee, having a length of 6 – 8 M. It was formerly traversed by the steamer *Red Hill*, until her boiler exploded while she lay at the wharf near Moultonborough Centre. It is now crossed only by horse-boats and occasional adventurous row-boats.

Moultonborough Neck is a long and broad peninsula which lies between Lake Winnepesaukee and Moultonborough Bay. It affords good ground for a drive, on account of its rich pastoral beauty and broad mountain-views. The objective point is *Uncle Tom's Hill,* 5 M. from Centre Harbor. The road is sandy in parts. The drive down the Neck is entered by passing out from Centre Harbor on the W.-Ossipee road for about 2 M., and diverging to the r. at a little cross-roads hamlet. The road is not level (as might be supposed), and consequently it affords views of a pleasant variety of scenery, including many of the farms of the Neck, which are famous for their richness. About 1 M. from the cross-roads, the road ascends a long slope whence is obtained a noble view of the Sandwich Range and other mountains. Across the lake are the Belknap and Suncook Ranges, and Mt. Cardigan is seen in the N. W., whose high white centre and the minor swells on r. and l. have suggested the national Capitol to imaginative tourists. Farther to the r. is Moosilauke, with Prospect and the Squam Mts. to the r. Red Hill fills the foreground in this direction, and from its r. flank springs the lofty curve of Sandwich Dome. Next to the r. is Tripyramid, with a pyramidal peak, and the upper part of its famous slide showing brightly over Flat Mt. Close to the r. is the brilliant front of Whiteface, supported on the E. by dark Passaconaway, at whose base is the green and rounded top of Wonnalancet. To the r. of and lower than Passaconaway is Paugus, marked by light-colored ledges; and on the extreme E. of the line is the superb white peak of Chocorua, clearly cut against the sky. The black Ossipee Range fills all the view to the E.; and the village of Moultonborough Corner is seen on the N., in line with Whiteface.

The view from *Uncle Tom's Hill* (at the road) includes nearly the whole of Lake Winnepesaukee and Moultonborough Bay, with their countless islands and graceful coves. This is the chief and satisfactory character of the prospect, though most of the lake boroughs are visible, and the peaks of Copple Crown, Suncook, Retreat, Belknap, and the northern ranges from Moosilauke to Passaconaway, Paugus and Chocorua being hidden by the Ossipee Mts.

Long Island is about 3 M. beyond Uncle Tom's Hill, and is reached by a bridge from the Neck road. It lies between Moultonborough Bay and Lake Winnepesaukee, and is 3 M. long and over 1 M. wide. The soil is remarkably rich, and has produced 132 bushels of corn to the acre, while the adjacent Cow Island has yielded 130 bushels. The large boarding-houses of Geo. K. Brown and B. B. Lamprey are on this island, fronting the lake. Brown's is called the Long-Island House, and is white; the other is a new brown house to the N. There are pleasant groves and beaches on the island, and conveniences for boating, bathing, fishing, hunting, and riding. The distance by road to Red Hill is 7 M.; to Centre Harbor, 8; to the Ossipee Falls, 7. The steamboats on the lake stop off the island to allow passengers to land.

The *Moultonborough Mineral Spring* is about 4 M. S. E. of Moulton-borough Corner, on the road to Melvin, and a little way N. W. of the Ossipee Falls. This water has been used locally for over half a century, and is now being exported in large quantities (at $15 a barrel). It contains iron and sulphur, and is used in cases of contagious affections, rheumatism, dyspepsia, and diseases of the lungs and kidneys.

The **Ossipee Falls** are 5 M. from Moultonborough Corner, 10 M. from Centre Harbor, 18 M. from W. Ossipee, and 14 M. from Wolfeborough. They are on a small stream which descends between the higher flanks of the Melvin Peak and a craggy ridge to the N. W. Travellers from the N. pass the white house near the Moultonborough Mineral Springs, cross a small bridge beyond, and turn to the l. up a field-road which passes through bars, in front of the divergence of the Melvin road. It is not more than ten minutes' walk from the highway to the falls, and carriages may be driven in half-way, beyond which a broad and easy footpath is followed along the bank of the stream. The falls are 35 ft. high, the water plunging at an angle of 80° over a ragged cliff of blackened rock into a deep basin of transparent clearness. This is surrounded on all sides, except at the outlet, by high and rugged walls of rock, in which deep hollows are seen. The white and resounding falls leap down from a rectangular flume in the cliff. A pretty view is gained from the ledges above the falls, and favorable glimpses of the plunging water through the abundant foliage of the glen are afforded from along the stream below. Farther up there are several smaller falls and cascades.

Going up from the brook toward the r., one reaches a grassy bluff up which a well-marked path leads. It is near a group of mountain farms, and is overlooked by a high crag on the N., whence a still broader view is gained. The bluff commands broad areas of the lake and of Moultonborough Bay, with their numerous islands. The bold peaks of Mt. Belknap are across the water, to the S. S. W. ; and the sharp blue crest of Kearsarge is farther to the r., flanked by the Rugged Mts. and other distant ranges. The Sandwich Range, Mt. Cardigan, and Moosilauke are visible from the crag to the N.

Some distance above the falls is a remarkable spring, 16 ft. in diameter, which throws up in the centre masses of crystal water and white sand to the height of two feet. On the l. of and below the falls is a small grotto, which was a place of refuge during the Indian wars.

There is a tradition that Lieut. Chamberlain (who was afterwards engaged in Lovewell's great Pequawket fight) was once hotly pursued by the Indians along the base of this mountain. When they were close upon him, he reached the chasm near the Ossipee Falls, 18 ft. wide, and cleared it at a single leap, thus securing his escape. The foremost of his pursuers essayed the same feat, but failed to reach the opposite side, fell, and was dashed to pieces on the rocks below.

164. The Ossipee Mountains

cover an irregular area of about 60 square miles, in four townships, and consist of a main range on the W., running N. and S., with long spurs to the E., cut into by Lovell's River and two tributaries of the Bearcamp. Most of the peaks are covered to their crests with heavy forests of larch, spruce, and birch. Their summits are composed of great masses of felsite, overlying the spotted granite below. Singularly, no ossipyte is to be found here. The range presents a formidable appearance when seen from the lake-towns, and appears as a long blue wall when seen from the White Mts.

"The appearance of the Ossipee Hills is peculiar for New-Hampshire mountains. Although formed of the primitive rock, which usually presents abrupt surfaces, sharp-cornered angles, rough sides, and sudden precipices, these hills are round, smooth, and capable of cultivation all over the sides, to the very summits." Dr. Jackson speaks of the great trap-dikes on the range, and highly commends the view from the peaks.

In 1672 Josselyn described "a very Princely Bird" called the *Pilhannaw*, nearly as large as an ostrich, covered with white mail, and preying upon fawns and jackals, stating that "She Ayries in the Woods upon the high Hills of *Offapy*, and is very rarely or feldome feen." This marvellous bird of Ossipee was probably the great heron, which is sometimes found in inland New Hampshire.

Black Snout is the second peak of the Ossipee Range, and is at the corners of the towns of Moultonborough, Sandwich, and Tamworth. It has received its present inelegant name from the adjacent rustics on account of its dark color, and also by reason of a supposed physical resemblance. As seen from the N. and N. W. this peak has a bold and symmetrical form, and is a tempting point for enthusiastic alpestrians, but the summit is so thickly wooded as to render the ascent useless.

After devoting a day to preliminary reconnoissances on three sides, the Editor attacked this peak (Sept. 21, 1875) on its most open flank, by walking 1 M. up the side-road which diverges from the W.-Ossipee highway at Gove's Corner, 2½ M. E. of Moultonborough Corner. By ascending through steep and weedy pastures for 1 M., the woods are met, on the E. flank of the peak. Some assistance is here gained from an old logging-road, but about ½ M. must be traversed through the thickets and a bewildering jungle of wood-choppings. From the rude clearings near the top glimpses may be gained of some of the mountains to the N. W., of the plains and hamlets on the W., and of Lake Winnepesaukee. If the summit was burned over, or otherwise cleared, it would give one of the best views in Carroll County, but at present it does not repay the labor of ascending.

The highest peak in this range is in its S. part, and is 2,361 ft. above the sea. This is also called *Black Snout* by the people of the lake-towns, but the Editor distinguishes it here by the name of the **Melvin Peak**, to avoid confusion, and because it is near Melvin Village. It is rarely visited, and is covered with woods, even over the crest.

The best points in the Ossipee Mts. for tourists to visit are the peak over W. Ossipee (Route 143) and the Whittier Peak (Route 147).

165. Tuftonborough

is an ancient and decadent town on the E. of Lake Winnepesaukee. It has 949 inhabitants (300 less than in 1820), and 12,635 acres of improved land, which is highly diversified in character and value. The name was given in honor of J. Tufton Mason, one of the original grantees. It contains the hamlets of Melvin Village, Mackerel Corner, and Tuftonborough Corner; and the neighborhoods of Canaan (in a secluded glen of the Ossipee Range, near the Dan-Hole Pond) and Tuftonborough Neck (near the

lake). A portion of the forest-covered Ossipee Range lies in this town, including the Melvin Peak, the highest point of the group (nearly 2,400 ft.).
The Peak is sometimes ascended hence, but is of slight interest to tourists,
being covered with woods. By climbing a tree the ocean may be seen.

Several arms of Lake Winnepesaukee enter the town, making a beautiful element in its scenery. Moultonborough Bay lies along its S. W.
border; and farther to the S. E. is *Tuftonborough Bay*, which is 3½ M.
long and is entered by a passage ¼ M. wide between the Tuftonborough
and Wolfeborough Necks. Near its N. end a narrow strait leads into a
beautiful inner basin, and the broad sheet of *Dishwater Pond* is close by.

Melvin Village is a quiet hamlet on the shore of Moultonborough Bay,
at the mouth of Melvin River, and near the S. end of the Ossipee Range.
It has two churches, two or three stores, and a national bank. There is a
small inn at the bridge; and several scores of summer-visitors sojourn in
the neighboring farm-houses. The steamer *Ossipee* is owned and sails
from this point, and several horse-boats belong here. The waters of the
Moultonborough Spring are exported hence. Melvin is 8 M. from Moultonborough and 10 M. from Wolfeborough. A mail-wagon (for passengers
also) runs to the latter point daily, over a road which gives occasional
views of the lake and the Belknap and Ossipee Ranges. *Mackerel Corner*
is a small hamlet on this road, 3 – 4 M. from Melvin.

The most picturesque route to Wolfeborough is by the road which lies
near the shore, and is a little longer than the mail-route. This road passes
along the heads of the 20-Mile and 19-Mile Bays, and then traverses for
1 M. the strip of land between Tuftonborough Bay and Dishwater Pond.

The hieroglyphic histories of the Ossipee Indians were found carved on the trees,
when the first settlers came up. In 1808, on the shore of the lake, N of the mouth
of Melvin River, a gigantic human skeleton was found, buried in a high tumulus.
About this was a circle of stones of different character from any found in this region.
This was the theme of Whittier's poem of 26 stanzas, "The Grave by the Lake."

" Where the Great Lake's sunny smiles
Dimple round its hundred isles,
And the mountain's granite ledge
Cleaves the water like a wedge,
Ringed about with smooth gray stones,
Rest the giant's mighty bones.

" Close beside, in shade and gleam,
Laughs and ripples Melvin stream
Melvin water, mountain born,
All fair flowers its banks adorn ;
All the woodland's voices meet,
Mingling with its murmurs sweet.

" Over lowlands, forest-grown,
Over waters, island-strown,
Over silver-sanded beach,
Leaf-locked bay and misty reach,
Melvin stream and burial-heap,
Watch and ward the mountains keep.

" Part thy blue lips, Northern lake !
Moss-grown rocks, your silence break !
Tell the tale, thou ancient tree !
Thou too, slide-worn Ossipee !
Speak, and tell us how and when
Lived and died this king of men ! "

166. Lake Winnepesaukee

lies in the counties of Belknap and Carroll, and its waters cover an area
of about 70 square miles (exclusive of the islands). Its course is S. 25°
E., with a length of 19 M.; and the greatest width is 8¼ M. A dam has
been erected at its outlet, to store up the water for the use of the manu-

facturing cities on the Merrimac River, and the lake is thus raised 6 ft., making it 502 ft. above the sea at mean tide. The water is remarkably pure and transparent, but is shallow, and nowhere attains a depth of more than 200 ft. The lake is very irregular in outline, having several long and fiord-like bays running far into the land on the S., E., and N., and flanked by broad arable peninsulas. Near the shores are high, steep, and picturesque hills, — the round-headed crests of the Suncook Range, the stately Belknap peaks, the picturesque group of the Ossipee Mts., and the pointed crest of Copple Crown.

One of the chief elements in the scenery is the great archipelago which rises in the lake, and is popularly supposed to consist of 365 islands. The actual number is 267, covering over 8 square miles, 10 of them containing over 100 acres each (three having more than 500 acres each), while 226 are less than 10 acres in area. Long, Bear, Cow, Governor's, and Rattlesnake Islands are the largest Several of the islands are inhabited, and the people communicate with the shore in summer by means of peculiar vessels called "horse-boats," and in winter by driving sleighs over the firm ice.

It is supposed that the streams which flow into this basin are altogether incompetent to create so great a mass of water, and the theory has therefore been advanced that the bottom of the lake contains many large springs. The outlet is by the Winnepesaukee River, which unites with the Pemigewasset to form the Merrimac. The largest streams which enter the lake are the Merry-Meeting and Smith's Rivers on the S., the other tributaries being short brooks or the outlets of adjacent ponds.

The Indian etymology of *Winnepesaukee* is thus explained: *Winne*, "beautiful"; *nebe*, or *nippe*, "water"; *kees*, "high"; *auke*, *ahki*, or *ohki*, "place," — wherefore *Winne-nippe-kees-auke*, or *Winnepekesauke*, or *Winnepesauke*, "The-Beautiful-Water- (in the) High-Place"; in plain English, "The Beautiful Lake of the Highlands." (The popular definition of this word in New Hampshire is "The Smile of the Great Spirit," but it has no relation whatever to its etymology.) A common way of spelling this name is *Winnipiseogee*, but this is contrary to its elemental words and to historical precedents. The name is spelt in 28 different ways in the N. H. Provincial Papers, but the terminal consonant is almost invariably either *k* or *c*. In Farmer's edition of Belknap's *History of New Hampshire* it is spelt in 18 ways, 16 of which have the *kee* ending, or its equivalent. The name *Merrimac* has the same terminal syllable, being compounded from *merru* (swift), an euphonic *m*, and *auke* (place); the meaning being therefore "Swift (or Swift-Water) Place."

"There may be lakes in Tyrol and Switzerland, which, in particular effects, exceed the charms of any in the Western world. But in that wedding of the land with the water, in which one is perpetually approaching and retreating from the other, and each transforms itself into a thousand figures for an endless dance of grace and beauty, till a countless multitude of shapes are arranged into ease and freedom, of almost musical motion, nothing can be beheld to surpass, if to match, our Winnepesaukee." (BARTOL.)

"We came upon the large Lake Winnepesaukee, which is scattered with small islands, and surrounded by broken mountain-tops, and presents splendid views of the White Mts., whose summits, Mts. Washington, Jefferson, Adams, Lafayette, and many other republican heroes, beckoned to us in Olympian majesty, in the splendor of the brightest August sun The sunset was most magnificent above that quiet, smiling lake." (FREDRIKA BREMER.)

"But if our mountains cannot challenge comparison with the Swiss and Tyrolese Alps, if we have not the glaciers with their wonderful variety and grandeur of form, we have lake and forest scenery, which, for peculiar and bewitching beauty, may be pronounced unrivalled. An immense sheet of pure and sparkling water, enclosed in an amphitheatre of mountains, from whose summits the dark foliage of a northern forest sweeps in one unbroken mass down to the very edge of the water, whose surface is dotted with innumerable islands, crested with the towering pines, whose dark mass is repeated in a crystal mirror; and first among our lakes is Winnepesaukee. The hand of man has done nothing to add to its charm; it cannot boast either the marble villas of Como, or the terraced islands of Lago Maggiore, covered with tropical foliage. Were it not for the smoke which here and there curls up from among the pines, the eye would hardly trace the presence of man, from many points of its shores and islands. From the shore the range of vision is soon stopped by the islands, which can hardly be separated from each other in the dim distance, but from the summit of any one of the numerous mountains which surround the lake, the whole extent of its surface is spread out like a map, and glitters in the sunlight, like a sheet of crystal sprinkled with emeralds." (DR. C. T. JACKSON.)

"Mildness tempered the heat; and serenity hushed the world into universal quiet. The Winnepesauke was an immense field of glass; silvered by the lustre which floated on its surface. Its borders, now in full view, now dimly retiring from the eye, were formed by those flowing lines, those masterly sweeps of nature, from which art has derived all its apprehensions of ease and grace; alternated at the same time by the intrusion of points, by turns rough and bold, or marked with the highest elegance of figure. In the centre a noble channel spread 23 M. before the eye, uninterrupted even by a bush or rock. On both sides of this avenue a train of islands arranged themselves, as if to adorn it with the finish which could be given only by their glowing verdure and graceful forms. Nor is this lake less distinguished by its suite of hills and mountains. On the N. W. ascends a remarkably beautiful eminence, called the Red Mt.; limited everywhere by circular lines, and in the proper sense elegant in its figure beyond any other mountain among the multitude which I have examined. On the S. ascends Mt. Major; a ridge of a bolder aspect and loftier height. At a still greater distance in the S. E. rises another mountain, whose name I could not learn, more obscure and misty; presenting its loftiest summit, of an exactly semicircular form, directly at the foot of the channel above mentioned, and terminating the watery vista between the islands, by which it is bordered, in a magnificent manner. On the N. E. the great Ossipee raises its long chain of summits with a bold sublimity, and proudly looks down on all the surrounding region.

"That the internal and successive beauties of the Winnepesaukee strongly resemble and nearly approach those of Lake George, I cannot entertain a doubt. That they exceed them seems scarcely credible. But the prospect from the hill at the head of Centre Harbour is much superiour to that from Fort George; a fact of which hardly anything could have convinced me, except the testimony of my own eyes. The Winnepesaukee presents a field of at least twice the extent. The islands in view are more numerous, and except one, of finer forms, and more happily arranged. The shores are not inferiour. The expansion is far more magnificent; and the grandeur of the mountains, particularly of the Great Ossipee, can scarcely be rivalled. It cannot be remarked without some surprise, that Lake George is annually visited by people from the coast of New England; and that the Winnepesaukee, notwithstanding all its accumulation of splendour and elegance, is almost as much unknown to the inhabitants of this country, as if it lay on the Eastern side of the Caspian." (DWIGHT'S *Travels in New England.*)

Starr King recommends the æsthetic exploration of the lake by boats. "This is the way to find delicious 'bits,' such as artists love for studies, of jutting rock, shaded beach, coy and curving nook, or limpid water prattling upon amethystine sand. At one point, perhaps, a group of graceful trees on one side, a grassy or tangled shore in front, and a rocky cape curving in from the other side, compose an effective foreground to a quiet bay with finely varied borders, and the double-peaked Belknap in the distance. Or what more charming than to sail slowly along and see the numerous islands and irregular shores change their positions and weave their singular combinations? Now they range themselves on either hand, and hem a vista that extends to the blue base of Copple Crown. Now an island slides its gray or purple form across, and, like a rood-screen, divides the long watery aisle into nave and choir, followed by another and another, till the perspective is confused and the vista disappears. Then in the distance, islands and shores will marshal themselves in long straight lines, fronting you as regular as the phalanxes of an army; and if the sun is low present the embattled effect more forcibly, with their vertically shadowed sides and brightly lighted tops. Or at another spot, through an opening among dark headlands, the summit of Chocorua is seen moving swiftly over lower ranges, and soon the whole mountain sweeps into view, startling you with its ghost-like pallor, and haggard crest. On a morning when the fog is clearing, is the time to be tempted towards the middle of the lake, to see the islands, whose green looks more exquisite then than in any other atmosphere, stretch away in perspectives dreamy and illusive. Two or three miles of distance seem five times as long, when measured through such genial, moist, and silvery air."

167. Historical Sketch of Lake Winnepesaukee.

" The Kingly Lyon, and the ſtrong-arm'd Bear ;
The large-limb'd Mooſes, with the tripping Deer ;
Quill-darting Porcupines, that Rackoons be
Caſtled ith' hollow of an aged Tree ;
The skipping Squirrel, Rabbet, purblind Hare,
Immured in the ſelf-ſame Caſtle are,
Left red-ey'd Ferrets, wily Foxes ſhould,
Them undermine, if Ramper'd but with Mold ;
The grim-fac'd Ounce, and rav'nous howling Wolf,
Whoſe meagre Paunch sucks like a ſwallowing Gulph ;
Black glittering Otters, and rich Coated Beaver ;
The Civet-ſcented Muſquash ſmelling ever."

OGILBY'S AMERICA.

Thus was described the fauna of the lake-country. The aboriginal inhabitants were the Winnepesaukee and Ossipee tribes of Indians, the latter of whom lived on the E. shore. The other adjacent tribes frequently visited the lake during the fishing season, encamping about the Weirs and the lower bays. The lake-country early excited the interest of the English.

In 1634 Capt. Mason wrote to Mr. Gibbins : "I have disbursed a great deale of money in your plantation and never received one penny, but hope if there were once a discovery of the lakes that I should in some reasonable time be reimbursed again " Gibbins answered : "1 perceive that you have a great mynd to the lakes, and I as great a will to assist you. If I had 2 horses and 3 men with me I would by God's helpe soon resolve you of the situation of it, but not to live there myselfe." In 1686 Mason sold the celebrated Million-acre Purchase, which included the Merrimac Valley from Souhegan to the Winnepesaukee ; and during the same year he farmed out to Hezekiah Usher and his heirs all the mines and ores of New Hampshire for 1,000 years.

"Lake Winnepesaukee, on the northern boundary, was the rendezvous for the enemy's scouting-parties, as it furnished them with fishing-ground when their game and plunder failed ; and the adjacent Mountains became their observatory or post of observation, whence, by descrying the rising smoke in the forests, they could easily learn the position of every new settler for a vast region around." (LANCASTER's *Hist. of Gilmanton.*) The favorite route of the Canadian Indians and French soldiers in their forays on the coast of New England was by Lakes Squam and Winnepesaukee, and on their shores their plunder was divided. Down this route fled some of Rogers's Rangers (see pages 184 and 316), and their remains have been found on the islands.

The unhappy captives who were taken by the Saco Indians at the destruction of Cocheco (Dover) in 1689 tracked the shores of the lake with their blood. Cotton Mather thus narrates the case of Sarah Gerrish : "The fell salvages quickly pulled her out, and made her dress for a march, but led her away with no more than one stockin upon her, a terrible march through the thick woods, and a thousand other miseries, till they came to the Norway plains. From thence they made her go to the end of Winnopisseag lake, and from thence to the Eastward, through horrid swamps, where sometimes they must scramble over huge trees fallen by storm or age for a vast way together, and sometimes they must climb up long, steep, tiresome, and almost inaccessible mountains. Her first master was one Sebundowit, a dull sort of fellow, and not such a *devil* as many of 'em were ; but he sold her to a fellow that was a more harsh and mad sort of a *dragon*, and he carried her away to Canada." She was ransomed by the Lord-Intendant of Quebec, and placed in a nunnery, whence she was delivered by exchange during Sir William Phipps's siege, and returned home after 16 months of captivity.

In 1689 Major Swayne, the commander of the Massachusetts forces on the N. border, sent a large detachment of Christian Indians, under Capt. Wiswel, to reconnoitre the lake-country. But the native soldiers deserted their colors when they reached Winnepesaukee, and affiliated with the hostile Indians there, with whom they remained many days. They gave the insurgents such information as to the English forces and movements that "the enemy then retired into the howling deserts, where there was no coming at them." Wherefore Cotton Mather wrote: "There has been little doubt that our northern Indians are originally *Scythians;* and it has become less a doubt, since it appears from later discoveries that the pretended straits of Anian are a sham ; for Asia and America it seems are there contiguous. Now, of these our *Scythians* in *America*, we have still found what Julius Cæsar does report concerning them of Asia : *Difficilius Invenire quam Interficere :* ' It is harder to *find* them than to *foil* them.' "

" A party of men were soon after sent out of Piscataqua (in 1689), under the command of Captain Wincol, who went up to Winnopiseag Ponds (upon advice of one John Church, who ran from them, that the Indians were there), where they killed one or two of the *monsters* they *hunted* for, and cut down their corn." (MATHER.)

In 1694 the French officer Villieu and the Sachem Madockawando, with a priest and 250 Indians, attacked the English settlement of Oyster River by night and destroyed five garrison-houses, killing and capturing nearly 100 people. The redhanded crusaders then retired to camps near Lake Winnepesaukee, where they divided the prisoners and plunder, some being carried to Canada and others to the Maine woods. In 1722 the Province ordered block-houses to be built near the lake, and cut a road to its shores. In 1720 three townships were laid out on the E. shore, and were surveyed by Mr. Frost in 1728. The lake was measured and mapped by Timothy Clements, in 1753, for which he received £25, New Tenor.

In the autumn of 1746, after the destruction of the French Armada, Col. Atkinson's New-Hampshire regiment was ordered into the Winnepesaukee region to form winter-quarters and to defend the frontiers against the French and Indians from Canada. They built *Fort Atkinson*, in Sanbornton, at the head of Little Bay and W. of Union Bridge. The earthworks remained for over a century, and were generally supposed to have been Indian remains; but the masonry was all carried away to build into a dam. The troops remained here for about a year, in idleness and under the lax discipline of the Provincial commanders. Much of their time was spent in hunting and fishing excursions among the mountains and out on Lake Winnepesaukee, during which the character and capabilities of this hitherto unexplored country were minutely studied. Accordingly they carried back favorable reports of that section, and after the Conquest of

Canada (in 1760) the lake townships were laid out and settled in rapid succession.

"The expedition, apparently so fruitless, had its immediate advantages, for aside from the protection afforded by it, the various scouts and fishing-expeditions explored minutely the entire basin of the Winnepesaukee, and turned the attention of emigrants and speculators to the fine lands and valuable forests in that section of the Province. And as soon as the French and Indian wars were at an end in 1760, the Winnepesaukee basin was at once granted and settled." (POTTER'S *Hist. of Manchester.*)

Steamboats have now been running on the lake for over 40 years, and several of them have been wrecked. Only three persons have been lost out of the hundreds of thousands who have crossed the lake by the steamboats.

The geological history of Lake Winnepesaukee is arranged in ten periods: the deposition of porphyritic gneiss, the formation of lake gneiss, the White-Mt. series, the metamorphosis of these three groups, the eruption of the Ossipee granites, the deposition of felsites, the eruption of sienite, the deposition of mica schists, the glacier period, and the terrace period. After the mica-schist period an enormous interval of time elapsed, and then the glaciers marked the country with moraines, pot-holes, and striated ledges. After this epoch the ocean overflowed all this region, and the slow subsidence of the waters is shown by the terraces, which are found at heights of 100, 80, 55, 30, 23, 15, and 12 ft.

168. The Voyage across Lake Winnepesaukee.

The Route of the "Lady of the Lake."

Times. — In the earlier and later parts of the season this steamboat leaves Wolfeborough at 6.45 A. M., running to Weirs. She leaves Weirs at noon, on the arrival of the train, and runs to Centre Harbor, returning to Weirs at 1 P. M., and starting thence for Wolfeborough at 5 P. M. About the middle of June she leaves Wolfeborough at 5.30 A. M. and Centre Harbor at 7.30 A. M., for Weirs, where the morning train for Boston is met. At noon she returns to Centre Harbor, starting back to Weirs at 1 P. M., and thence for Wolfeborough at 5 P. M. About the middle of July an extra trip is added to the daily route. The *Lady* then leaves Wolfeborough at 5.30 and 10 A M., and 3.15 P. M., connecting with trains for the mountains, and for New York and Boston. She leaves Centre Harbor for Weirs at 7.15 A. M. and 1 P.M.; and at 8.30 A. M., 12 M., and 2 and 5 P. M., leaves Weirs for Centre Harbor and Wolfeborough

The steamboat leaves Wolfeborough at 5.30 A. M. and Centre Harbor at 7 30, reaching Weirs at 8.25 and connecting with the down train. She then runs to Wolfeborough direct, reaching there at 10 A. M., and returning directly to Weirs, where a connection is made at noon with the express-train from Boston to the mountains. The course is then to Centre Harbor, which is reached at 1 P. M., and the boat returns immediately to Weirs, connecting there with the down express-train at 1 50. At 2 P. M. she runs to Wolfeborough direct, connecting with the Eastern R. R. trains for Boston or North Conway, and returns to Weirs direct, reaching that port at 5 P. M. and meeting the train from the south. She then passes to Centre Harbor, arriving at 6 P. M., and thence to Wolfeborough, where the trip ends, at 7.30 P. M. Hitherto the boat has called occasionally at Diamond Island, but no arrangements have been made for 1876

Fares. — From Wolfeborough to Weirs (or *vice versa*), 80 c.; to Centre Harbor, 80 c. From Weirs to Centre Harbor, 60 c. Excursionists can sail all day for $1.00, the distance traversed being 150 M.

17 *

Distances. — Wolfeborough to Weirs, 16 M. (75 to 85 minutes); Weirs to Centre Harbor, 10 M. (50 to 60 minutes); Centre Harbor to Wolfeborough, 20 M. (80 to 100 minutes).

On leaving the wharf at Wolfeborough, the steamer runs S. W. down the bay to Sewell's Point, with the stately peaks of Mt. Belknap in front, and the hamlet of S. Wolfeborough and the long mountain of Copple Crown on the l. She then runs W. to *Parker's Island*, with the Barn-Door Islands on the l. and the lowlands of Wolfeborough Neck on the r. The course is then laid for 45 minutes W. by N. across the broadest part of the lake, toward Welch Island. The views on either side are now of the most beautiful character, and are constantly diversified by the advance of the boat. The dark Ossipee Range is on the r., the rounded crests of Mt. Belknap and the Suncook Mts. are on the l., and in front are the alpine spires of the long Sandwich Range. The islands near the shores change their relative positions with kaleidoscopic rapidity, sinking one behind the other in succession. On the r. are the farms and forests of the borough necks, separated by the narrow outlet of Tuftonborough Bay. On the l. is **Rattlesnake Island,** a high hill-range emergent from the waters, covered with rocky mounds and hardy trees, and inhabited only by the deadly reptiles from which the name is derived.

" As we shoot out into the breadth of the lake, and take in the wide scene, there is no ripple on its bosom. The little islands float over liquid silver, and glide by each other silently, as in the movements of a dance, while our boat changes her heading. And all around, the mountains, swelling softly, or cutting the sky with jagged lines of steely blue, vie with the molten mirror at our feet for the privilege of holding the eye. . . . . Looking up to the broken sides of the Ossipee Mts. that are rooted in the lake, over which huge shadows loiter; or back to the twin Belknap hills, that appeal to softer sensibilities with their verdured symmetry; or, farther down, upon the charming succession of mounds that hem the shores near Wolfeborough; or northward where distant Chocorua lifts his bleached head, so tenderly touched now with gray and gold, to defy the hottest sunlight, as he has defied for ages the lightning and the storm; — does it not seem as though the passage of the Psalms is fulfilled before our eyes, — ' Out of the perfection of beauty God hath shined ' ? " (STARR KING.)

Diamond Island is N. W. of and near Rattlesnake Island, and is of small area. A second-class summer-hotel is situated here, commanding a good view to the N. and W. The *Lady* sometimes runs directly to this point from Wolfeborough, passing on the S. and W. of Rattlesnake Island and through the narrow straits off W. Alton.

Welch Island contains over 150 acres, and is used for pasturage, being comparatively level and grassy. The course thence is N. W., into narrower waters, with a small archipelago on the r. and the Gilford Intervale on the l. Locke's Island and the deep inlet of Smith's Cove are soon passed on the l., and then the boat traverses the strait between *Timber Island* (r.) and *Governor's Island* (l.). The former is hilly, with picturesquely irregular shores; and Governor's Island is a large arable tract, connected with the mainland by a bridge.

Mr. J. M. Lovett, the pilot of the *Lady*, says that " the most dangerous point on Lake Winnepesaukee for dark or thick weather is *Witch Island*, a formidable reef S. W. of Timber Island. It is well known to some of our self-made pilots, who have found themselves fast on its hidden boulders. The *Lady* struck this place 12 years ago, receiving much damage, and sinking on the beach at Governor's Island. The *Seneca* also was wrecked and lost here."

On the reach between Diamond and Welch Islands, Moosilauke, Red Hill, Sandwich Dome, and other peaks are seen. Whiteface and Passaconaway also appear, and the remoter crest of Tripyramid, striped by its white slide. While passing Welch Island, the whitish domes of Mt. Cardigan are on the l. front, flanked on the r. by Moosilauke and Mt. Carr. In the course between Timber and Locke's Islands, Mt. Paugus appears on the r. of Passaconaway; and Mt. Washington is soon seen nearly over Paugus, for a brief space of time. The running time from Welch Island to Governor's Island is 20 minutes, and thence to Weirs it is 7 minutes. As Governor's Island is rounded, a noble view of Mt. Chocorua is gained on the N. E. The hills of Meredith Neck and Stone-Dam Island now rise near at hand on the r., and the boat runs between Governor's Island and the insulated crag of Eagle Island, opening Northwest Bay on the r.

At **Weirs** (see Route 151) the trains of the B., C. & M. R. R. are met.

On leaving Weirs for Centre Harbor, the boat crosses the openings of Northwest Bay, up which Mt. Prospect and other peaks are seen. Toward the front are Red Hill, Sandwich Dome, Mt. Paugus, the sharp-spired Chocorua, and the long line of the Ossipee Mts. The course is between Eagle and Governor's Islands, and soon opens views of the noble Mt. Belknap close on the r., with Copple Crown at the foot of the lake. Leaving Stone-Dam Island on the l., and Timber and Mark Islands on the r., the *Lady* soon enters a narrow and picturesque strait between Bear Island (r.) and the hilly shores of Meredith Neck. **Bear Island** is 2–3 M. long, with singularly irregular and deeply indented shores. It is well populated, and has a wharf. After traversing the long strait the boat passes through the narrows, with Pine Island on the l., and enters the N. part of the lake. Running between Beaver Island (l.) and Three-Mile Island (r.), she passes several smaller islets, with the pretty hamlet of Centre Harbor in front.

" The most striking picture, perhaps, to be seen on the lake, is a view which is given of the Sandwich range in going from Weirs to Centre Harbor, as the steamer shoots across a little bay, after passing Bear Island, about 4 M. from the latter village. The whole chain is seen several miles away, as you look up the bay, between Red Hill on the l., and the Ossipee Mts. on the r. If there is no wind and there are shadows enough from clouds to spot the range, the beauty will seem weird and unsubstantial, — as though it might fade away the next minute. The weight seems to be taken out of the mountains." (STARR KING.)

In running from Centre Harbor to Wolfeborough, the *Lady of the Lake* leaves Three-Mile Island on the r., Five-Mile and Six-Mile on the l., Steamboat Island on the r., and runs between Parker's Island and Wolfeborough Neck. The views from the deck on this long voyage are broad and beautiful, and are about the same as those described on the next page.

169. The Voyage across Lake Winnepesaukee.

The Route of the " Mt. Washington."

The *Mt. Washington* is the largest steamboat on the lake, and can carry 1,000 passengers with safety. She makes two trips daily between Alton Bay and Centre Harbor, touching at Wolfeborough, and connecting with trains on the Boston & Maine and Eastern Railroads.

Distances. — Alton Bay to Wolfeborough, 10 M. (45 minutes); Wolfeborough to Centre Harbor, 20 M. (75 minutes).

Fares. — Alton Bay to Centre Harbor, 85 c.; Wolfeborough to Centre Harbor, 75 c.; Alton Bay to Wolfeborough, 60 c.

The *Mt. Washington* leaves her wharf at Alton Bay (page Route 150), and runs N. down the deep fiord which was anciently known as *Merry-Meeting Bay.* This river-like expanse is ¼ – ¾ M. wide and about 5 M. long, being bordered by rugged hills and forests. The scenery is highly primitive and wild, and the curves of the bay are graceful. Gerrish Point projects from the E. about 2 M. beyond Alton Bay, and *Fort Point* is about 1½ M. beyond, also on the E. The latter was at one time crowned by a Provincial fortress, intended to block up the favorite route of the hostile Indians down the Cocheco Valley. After passing this point the course is more to the N. E., and the boat passes Little Mark Island on the l. and moves out on to the lake. Her course to Wolfeborough leaves the high knolls of Rattlesnake Island on the l., and the small rocks of Ship and Moose Islands; while on the r. are the singular mounds near Clay Point and along the Alton shores, with Copple Crown beyond. On the N. are long lines of stately mountains, including the Sandwich and Ossipee Ranges, and several more distant peaks; and Mt. Belknap and the Suncook Range lift their high and pallid crests in retrospect. After leaving on the r. the cluster of islands off S. Wolfeborough, the steamer passes Sewell's Point and enters the quiet little harbor of Wolfeborough.

The course from Wolfeborough to Centre Harbor is similar to that of the *Lady of the Lake.* After passing Sewell's Point the bow is headed for Parker's Island, and the low shores of Wolfeborough Neck are passed on the r. Copple Crown is astern, and on the l. are the round heads of Mt. Belknap and the Suncook Range. Noble views open on the N., including several peaks of the Sandwich Range and the mountains towards Moosilauke. From Parker's Island the course is towards Sandy Island, with Rattlesnake Island and then the broadest part of the lake on the l., and the borough necks on the r., separated by the deep outlet of *Tuftonborough Bay.* Diamond Island soon emerges on the l. from behind Rattlesnake Island, and farther are the W.-Alton and Gilford shores, overlooked by Mts. Belknap and Retreat. The dark masses of the Ossipee Mts. appear on the r., and on the N. W. and N. are the distant peaks of Prospect, Moosilauke, Red Hill, Sandwich Dome, Tripyramid, and Whiteface. Passaconaway soon emerges from behind the Ossipee Range, showing a

symmetrical black dome. As Sandy Island is approached, the farms on
Cow Island are seen on the r., with the outlets of the far-ramifying Moul-
tonborough Bay; and on the l. are the green pastures of Welch Island.
Beyond Sandy Island the steamer runs for some distance near **Long Island**
(on the r.), on which are seen two large summer boarding-houses (see
Route 163).

When the steamboat is off Long Island, Mts. Washington, Adams, Jefferson, and Pleasant are visible for about 15 minutes, on the r. front, over
the rounded head of Mt. Paugus. The view is especially impressive in
the autumn, when the distant peaks are whitened by the early snows.
As the boat advances the black Ossipee Range is passed, on the r., and the
noble peak of Chocorua soon emerges from behind it. By the time that
Chocorua is seen Mt. Washington has been hidden by the dark dome of
Mt. Passaconaway.

" Passing by the W. declivity of the Ossipee range, looking across a low slope of
the Sandwich range and far back of them, a dazzling white spot perhaps — if it is
very early in the summer — gleams on the northern horizon. Gradually it mounts
and mounts, and then runs down again as suddenly, making us wonder, possibly,
what it can be. A minute or two more, and the unmistakable majesty of Washington is revealed. *There* he rises, 40 M. away, towering from a plateau built for his
throne, dim green in the distance, except the dome that is crowned with winter, and
the strange figures that are scrawled around his waist in snow." (STARR KING.)

On the l., beyond the Forty Islands, is the low Steamboat Island, at the
S. E. corner of which, in 8 ft. of water, is the wreck of the *Belknap*, the
first steamer on the lake. She was built at Lake Village in 1831–3, and
was driven ashore and wrecked on this island in a gale, in 1841. There
are now nine steamers on the lake. *Bear Island* lifts its rolling hills beyond
Steamboat Island: and the twin peaks of Belknap rise in noble symmetry
on the S. Where Long Island falls away to the r., a view is gained of the
deep bight of *Brawn Bay*, back of the Long Pots and Norway Point.
The course is now laid by the Six-Mile and Five-Mile Islands, between
Cook's Point and Bear Island (l.), and enters the bay of Centre Harbor.
As the steamer advances toward Centre Harbor, the Sandwich Mts. sink,
one by one, behind the high and nearer mass of Red Hill; and Moosilauke,
which has been over the village, is hidden by the nearer hills.

Centre Harbor, see Route 155.

SUMMER BY THE LAKESIDE.

NOON.

WHITE clouds, whose shadows haunt the deep,
Light mists, whose soft embraces keep
The sunshine on the hills asleep!

O isles of calm! — O dark, still wood!
And stiller skies that overbrood
Your rest with deeper quietude!

O shapes and hues, dim beckoning, through
Yon mountain gaps, my longing view
Beyond the purple and the blue,

To stiller sky and greener land,
And softer lights and airs more bland,
And skies, — the hollow of God's hand!

Transfused through you, O mountain friends!
With mine your solemn spirit blends,
And life no more hath separate ends.

I read each misty mountain sign,
I know the voice of wave and pine,
And I am yours, and ye are mine.

Life's burdens fall, its discords cease,
I lapse into the glad release
Of nature's own exceeding peace.

O, welcome calm of heart and mind!
As falls yon fir-tree's loosened rind
To leave a tenderer growth behind,

So fall the weary years away;
A child again, my head I lay
Upon the lap of this sweet day.

This western wind hath Lethean powers,
Yon noonday cloud nepenthe showers,
The lake is white with lotus-flowers!

    .   .   .   .

JOHN G. WHITTIER.

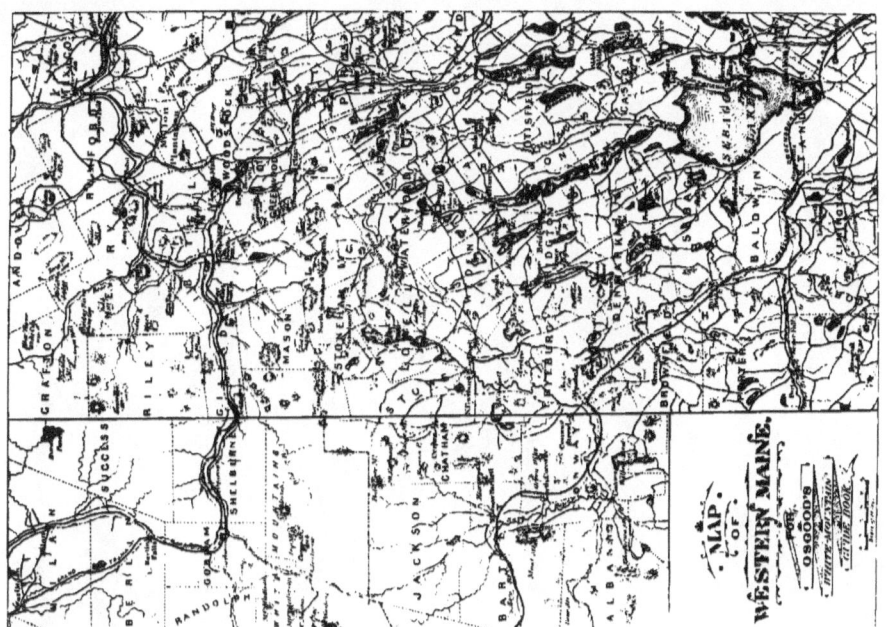

170. The Presumpscot and Lower Saco Valleys.

Portland ** (Falmouth House; Preble; United States; St. Julian Hotel)* is a busy and prosperous maritime city of nearly 35,000 inhabitants, situated on a strongly fortified harbor and in the vicinity of several popular sea-shore resorts. It is an important railroad centre, and has some imposing public buildings. From the observatory on *Munjoy Hill* a noble panorama of the White and Sandwich Mts. is enjoyed, including the peaks of Ossipee, Green, Sandwich Dome, Whiteface, Tripyramid, Passaconaway, Osceola, Tecumseh, Chocorua, Hancock, Carrigain, Tremont, Bear, Moat, Willard, Willey, Webster, Jackson, Clinton, the Green Hills, Kiarsarge, Pleasant, Franklin, Monroe, Washington, Gemini, Double-Head, Jefferson, Wild-Cat, Pleasant, Adams, Carter, Baldface, Imp, and Moriah ; and the heights of Cornish, Brownfield, etc.

This view has been drawn and published by Abner Lowell, President of the Portland White-Mt. Club. *Mt. Independence* is in the adjacent town of Falmouth, and commands a noble panoramic view of the distant White Mts., with Mts. Sabattos, Blue (in Avon), Agamenticus, and Belknap. The views from Weston's, Fort, and Town-House Hills, near Gorham village, command a long line of the White and Sandwich Mts.

Standish *(Standish House)* is 2 M. from the Sebago-Lake station of the P. & O. R. R. where stages connect with the trains. 1 M. distant, by the Bonny-Eagle road, is *Break-heart Hill,* whence are visible Sandwich Dome, Whiteface, Passaconaway, Chocorua, Hancock, Carrigain, Moat, Willey, Saddleback, Double-Head, Washington, Adams, Pleasant, and many other distant peaks. W. of Standish, and near N. Limington, is the far-viewing *Moody Mt.;* and *Wiggin* (or *Table*) *Mt.* is a lofty hill in Parsonsfield, commanding a rich view.

Baldwin is a thinly populated town on Sebago Lake, containing the low and pasture-clad eminence of **Saddleback,** which is easily reached from Steep Falls. The view thence includes the Cornish, Porter, and Ossipee hills, Green Mt., the Twin Mts. and Mt. Bond, Mt. Hancock, Sandwich Dome, Whiteface, Passaconaway, Chocorua, the Burnt-Meadow Mts., Carrigain, the Franconia Range, Moat, the Green Hills, Double-Head and Kiarsarge (under Washington), Adams and Madison, the Carter Notch,

Wild-Cat, the Carter Dome, Baldface, Pleasant, Goose-Eye, Speckled, the Sunday-River and Bear-River White-Caps, and Puzzle Mt., beyond Bethel.

Rattlesnake Mt. is across Sebago Lake, in the town of Casco, and may be visited from Naples. Its top is densely wooded, but a broad view is gained by climbing a tree, including nearly the same peaks as are seen from Saddleback.

Cornish is a small hamlet with a poor inn, 1 M. from Baldwin station. The view from *Trafton Mt.* (3 M. S.) is famous for its extent and comprehensiveness. A few miles N. W. is the *Bill-Merrill Hill*, a wooded summit, whence one of the panoramas in the *Geology of New Hampshire* was drawn.

Hiram Bridge (*Mt.-Cutler House*) is a pleasant modern village on the Saco River, amidst beautiful pastoral scenery. To the N. is the smooth hill called *Tear Cap*, which is often ascended on account of the extensive views from its ledges. 1 M. distant, on the W. bank of the Saco, is the ancient Wadsworth mansion, which was a favorite resort of Henry Wadsworth Longfellow in the days of his youth. 2 M. from the village are the * **Great Falls of the Saco**, where the river descends for 79 ft. over a long rocky slope, dashed into masses of foamy whiteness, and falling at last into a dark and quiet pool below.

Mt. Cutler is near Hiram Bridge, and is easily ascended by a path which leads in ¼ M. from the hotel to the top. The summit is wooded, but fine views are gained from ledges on the N. E. and W. The foreground is made beautiful by the graceful curves of the Saco through wide and fertile plains, and beyond are the Moat and Burnt-Meadow Mts., the Green Hills, Mts. Jackson, Pleasant, Monroe, and Washington, Kiarsarge, Adams, Madison, Gemini, Wild-Cat, the deep Carter Notch, Eastman, Imp, Baldface, Royce, Moriah, Speckled, Tom, Pleasant (with its hotel), Paris Hill, Bear and Hawk Mts. in Waterford, and many others of less interest.

Brownfield Centre (small inn) is 1½ M. from the Brownfield station, and is near the wooded heights of the *Frost* and *Burnt-Meadow Mts.* The latter are to the S., and are nearly 2,000 ft. high, commanding broad views to the W. and N.

Stages leave Brownfield station for *Denmark* and *Bridgton* on the arrival of the morning and afternoon trains from Portland. The distance is 14 M., and the fare is $ 1. Leaving Frost Mt. behind, the stage runs N. E. out of the Saco Valley, passing S. of the Boston Hills. The first village is *Denmark Corner*, which is about 5 M. from the station, and is between the Moose and Granger Ponds. As the stage passes the ancient Congregational church on the hill beyond, a fine view of the White Mts. is gained, including most of the peaks of the main range, with Kiarsarge in front of the line. The road now approaches the S. spurs of Mt. Pleasant (l.); runs along the S. end of Wood's Pond; passes through the hamlet of *Pinhook;* and thence runs N. to **Bridgton.**

171. Sebago Lake.

Steamboats. — The *Mt. Pleasant* and the *Sebago* are neat little side-wheel steamboats, with covered promenade-decks. They connect with the Portland & Ogdensburg Railroad at the Sebago-Lake station. One trip daily is made from May 1st to Nov. 1st, the steamboat leaving Harrison in the morning and connecting with the train which reaches Portland at noon ; and returning up the lakes on the arrival of the early afternoon train from Portland. During the season of summer-travel an extra trip is made daily, the steamboat leaving the railroad station on the arrival of the morning train from Portland, and getting back in time for the evening train for Portland. This arrangement enables tourists to go up from Portland, traverse the whole length of the lakes, and get back to the city in the evening. The trip takes ten hours. (Dinner at Bridgton or Harrison.)

Distances. — Sebago-Lake station to Naples, 22 M. ; to Bridgton, 30 ; N. Bridgton, 32 ; Harrison, 34. *Fares.* — Portland to Naples, $1.75 ; to Bridgton, N. Bridgton, or Harrison, $2.

> " Around Sebago's lonely lake
>   There lingers not a breeze to break
>   The mirror which its waters make.
>
> " The solemn pines along its shore,
>   The firs which hang its gray rocks o'er,
>   Are painted on its glassy shore.
>
> " The sun looks o'er with hazy eye,
>   The snowy mountain-tops which lie
>   Piled coldly up against the sky."
>
>   WHITTIER'S *Funeral Tree of the Sokokis.*

Sobago Lake is in Cumberland County, Maine, and is bounded by the towns of Standish, Sebago, Naples, Casco, and Raymond. It is 12 M. long and 9 M. wide, and receives the waters of 23 ponds. Its greatest depth is 400 ft. There are but few islands in the lake, the greater part of which is an unbroken expanse, flanked by low shores from which rise gracefully curving ridges. " The water of Sebago Lake is shown by analysis to approach as near to absolute purity as any mass of water of large volume in the world, so far as known." It is now used to supply the city of Portland, to which it is conveyed by a long aqueduct. The immediate vicinity of the lake possesses but little scenic interest, but the distant views of the mountains are beautiful, and the voyage is further diversified by the unique and interesting episode of the ascent of the Songo River.

The name *Sebago* is derived from the Indian words *sipe*, or *sebe*, " pond," and *auke*, " place." There is but little of historic interest attaching to this region since it was not settled until after the Conquest of Canada. About the year 1670 Josselyn romanced thus about it : —

" Twelve mile from *Casco-bay*, and paſſible for men and horſes, is a lake called by the *Indians Sebug*, on the brink thereof at one end is the famous Rock ſhap'd like a *Mooſe-Deere* or *Helk*, Diaphanous, and called the Mooſe-Rock. Here are found ſtones like Cryſtal, and *Lapis Specularıs* or *Muscovia* glass both white and purple.

Soon after leaving the wharf at the Sebago-Lake station the red gate-house of the Portland Water-Works is seen on the r., and the steamer runs to the N. E., by the groves of Indian Island. She then approaches *Frye's Island*, which contains 1,000 acres of woodland and is situated with

Z

relation to Raymond Cape as Sicily is to Italy. Sometimes, during the summer, the course is laid to the E. around the island. Views are given past Sloop and Squaw Islands, up *Jordan Bay*, a deep arm of the lake which runs in to **Windham.**

The latter is an ancient town which was granted in 1734, and was fortified by Massachusetts in the fifth Indian war. During an attack which was made here by the savages (in May, 1756), the celebrated chief Poland was killed. The town bore the name of *New Marblehead* for 25 years, because most of its people came from Marblehead, Mass. For six years the inhabitants were forced by fear of the hostile Indians to live in the fortress. In Windham, John Albion Andrew, the great War-Governor of Massachusetts, was born, in 1818.

Near *N. Windham* (Nemasket House) is the outlet of Sebago Lake, which discharges through the Presumpscot River into Casco Bay. About 1 M. E. of the hamlet is **Little Sebago Lake,** which is about 7 M. long and is prettily divided by *Sabbath-Day Point,* on the S. W.

The passage between Frye's Island and *Raymond Cape* is called *The Notch,* and is a little over 1 M. long, commencing about 5 M. from the Lake station. The varying shores of the island on the l. and *The Images* rocks on the r. (60 – 80 ft. high) afford pleasant bits of scenery. In the E. shore is the Cave, a square crevice in the rock 6 × 4 ft. large and 25 ft. long, approachable by boat and having an exit above on the land side. There is a local tradition to the effect that Nathaniel Hawthorne, the eminent author, was wont to row his boat into the Cave and meditate there. Hawthorne spent some of his earlier years in the little hamlet at the head of Raymond Cape, near Thomas Pond, living in the house next to the present Union Church.

When Capt. Hawthorne died his son Nathaniel was but four years old, and the widowed Mrs. Hawthorne moved with him from Salem to a sequestered and lonely farm-house in Raymond, where he lived for several years. He says: "I lived in Maine like a bird of the air, so perfect was the freedom I enjoyed. But it was there I first got my cursed habits of solitude. How well I recall the summer days, when, with my gun, I roamed at will through the woods of Maine. Everything is beautiful in youth, for all things are allowed to it then." He was in the habit of boating and skating on Sebago Lake, which was not far from the Hawthorne house. He always said that this was the happiest part of his life. The house looked out on the lake and over to Mt. Washington. "The spot must ever have had the utter loneliness of the pine forests upon the borders of our northern lakes. The deep silence and dark shadows of the old woods must have filled the imagination of a youth possessing Hawthorne's sensibility with images which later years could not dispel."

Usually the boat runs on the W. side of Frye's Island, and does not traverse the Notch. The Images and Raymond Cape are then seen on the r., and the steamer runs out into the widest part of the lake, whence fine views are enjoyed to the N. W. The main lines of the White Mts. are well defined, but are sometimes partly eclipsed by the intrusion of Saddleback Mt. (in Baldwin), and Peaked Mt. (in Sebago township). The first line of mountains consists of the sharp cone of Kiarsarge, flanked on the l. by Blackcap and the Green Hills of Conway, and on the r. by Mt. Gemini, back of and over which is one of the crests of Double-Head. Mt. Wild-Cat and the Carter Dome are in the extreme N. W., and on their l. are

Adams, Washington, Monroe, Franklin, and Pleasant. Nearer, on the l. rear, are the singular hills of Cornish; and Mt. Pleasant, with its hotel-crowned ridge, is farther to the N. On the N. E. is the tall mound of Rattlesnake Mt., in Casco. These views experience frequent relative changes as the steamer advances, and some of the more southerly mountains come into view. The Moat range, the red peaks of Chocorua, the round dome of Passaconaway, and the crests of Whiteface are grouped along the W.; and a few momentary glimpses are afforded of the remote Mt. Carrigain, far beyond and to the r. of Chocorua. The color-effects on these distant ridges toward sunset (when seen from the afternoon boats) are remarkably brilliant. The air-line distance from Sebago Lake to Mt. Washington is about 42 M.

Passing, on the r., the deep inlet of *Kettle Cove*, in the town of Casco, the hills and farms of Sebago town are seen on the l., and the boat soon crosses a shallow sandy bar and passes between the brushwood jetties at the mouth of the Songo River. The usual time for running up the lake is 1 hr. The * **Songo River** is 6 M. long, though the distance from its inlet to the outlet is but 2½ M. in a straight line. The boat makes 27 turns in traversing this singularly crooked stream, and is often within leaping distance of the banks. The passage of the "sweetly sinuous Songo" is the most interesting part of the trip, and its best feature is the perfect reflection which the forests and banks make in the sluggish and tranquil stream. The most vivid colors and the most delicate foliage are duplicated in the dark mirror of the waters with marvellous accuracy.

> " Nowhere such a devious stream,
>   Save in fancy or in dream,
>   Winding slow through bush and brake,
>   Links together lake and lake.
>
> " Walled with woods or sandy shelf,
>   Ever doubling on itself,
>   Flows the stream, so still and slow,
>   That it hardly seems to flow.

> " Never errant knight of old,
>   Lost in woodland or on wold,
>   Such a winding path pursued
>   Through the sylvan solitude.
>
> " In the mirror of its tide
>   Tangled thickets on each side
>   Hang inverted, and between
>   Floating cloud or sky serene."
>
>   LONGFELLOW'S *Songo River.*

About 5½ M. from the mouth of the Songo, the *Crooked River* is seen coming in on the r., flowing down from near Bethel. The steamer now enters a lock and is closed in by gates, and after the water has risen to the level of the river above, the upper gates are opened and she passes out. 1 M. above this point the **Bay of Naples** is entered. This pretty sheet of water is 2 M. long and about ¾ M. wide. It was until recently known as *Brandy Pond*, but received its new name on account of the township in which it is situated. When about half-way up the Bay, glimpses are gained of Mts. Carter, Adams, and Washington, on the l. oblique, over a high sand-bank. In front the vessel approaches the dark spire of Naples and the French-roofed summer-hotel on the r. Traversing a short strait and passing through a drawbridge, Long Pond is entered, and the steamer rounds in at **Naples**, a prettily situated hamlet on the S. W. shore.

Beyond the bridge is the small and sequestered summer-hotel known as the *Elm House.*

Long Pond is about 13 M. long, and from ½ M. to 1½ M. wide. The shores are low and unpicturesque, and are inhabited by a large population of farmers. The most attractive features of the journey are the distant mountain-views on the W. and N., which include the peak of Washington, with Jefferson on its r., then Mt. Carter, over which is the crest of Adams, then Eastman and Baldface, with Moriah on the r. Mt. Pleasant is close at hand in the N. W. To the N. is the symmetrical cone of Goose-Eye, with Bear-River White-Cap and Sunday-River White-Cap more to the r., all of these being very lofty and distant.

At *Bridgton Landing*, on the W. shore, stages are in waiting to carry passengers 1 M. W. to the prosperous village of **Bridgton** (*Bridgton House;* *Cumberland House*). There are several woollen-factories here, on the outlet of Highland Lake, and a large packing establishment for the Winslow green corn. The main portion of the village is pleasantly situated on high land, and is provided with four churches and numerous stores. It is much frequented in summer on account of the beauty of the surrounding lakes and hills, and the facilities for riding, boating, and fishing. Just N. of Bridgton is **Highland Lake** (formerly known as *Crotched Pond*), a lovely sheet of water 3 M. long, embosomed in hills and studded with wooded islets. 3 – 4 M. to the S. W. is *Wood's Pond*, which is nearly 1½ M. long. Near the village are the umbrageous shades of *Forest Avenue*, a short reach of road which is completely overarched by closely growing trees. The favorite drive is to *Bridgton Heights*, where the road traverses a line of highlands on the E. of Highland Lake, whence are obtained views of the mountains and ponds of the adjacent towns, — Long Pond, Highland Lake, Mts. Pleasant, Kiarsarge, Washington, and other prominent peaks of the White-Mt. range. There is a pleasant circular drive 7 M. long, leading over the Heights to N. Bridgton and returning by a road which follows Long Pond for some distance. This ride may be prolonged into Waterford, which is 9 M. from Bridgton. The Moose-Pond circuit is another drive, 8 M. long, leading to the hills which overlook the extensive and beautiful *Moose Pond*, under Mt. Pleasant, and returning by Highland Lake.

Bridgton is 10 M. from the Mt.-Pleasant House; 14 M. from Fryeburg; 9 M. from Naples; and 14 M. from Brownfield. Stages leave Bridgton daily at 7 A. M. for Brownfield ($ 1), and also for N. Bridgton, Norway, and Paris Also at 7 A. M. on Monday, Wednesday, and Friday, for Naples, Casco, Raymond, Windham, Westbrook, and Portland (returning alternate days). Stages connect with all steamers on Sebago Lake. Daily stages also run to Mt. Pleasant (fare to the Half-Way House, $ 1).

Bridgton was formerly called *Pondicherry*, on account of the numerous ponds and abundance of wild cherries found in its limits, though others say that the name is of Indian origin. In 1767 the town was named *Bridgton*, in honor of Moody Bridges of Andover, one of its proprietors. The early settlers of the town were all from Essex County, Mass.

After leaving the Bridgton Landing, the steamer runs N. 4 – 5 M. to **N. Bridgton** (*Lake Hotel*), a small but attractive hamlet on the W. shore. It has an academy and a Congregational church, situated on the tree-lined main street, and is visited by summer-tourists. During the last part of the course the steamer runs toward the bold ridges of Bear Mt. and Hawk Mt. (in Waterford), the latter being marked by precipitous sides.

Waterford Flat is a pretty hamlet 5 M. N. of N. Bridgton, on the meadows by the shore of a picturesque and sandy-beached lake, over which rises *Mt. Tire'm.* It is celebrated as the birthplace of Artemus Ward (Charles F. Browne), the great humorist, whose remains are buried in the village cemetery.

The Browne homestead is an old-fashioned house on the l. of the road which is entered by turning to the l. at the church. In after years, when his mirthful fame was spread over America and Great Britain, the genial wit retained the pleasantest memories of Waterford, and often wrote in its praise.

Seba Smith was educated at the N.-Bridgton Academy. He afterwards became the immortal "Major Jack Downing," poet, editor, humorist, and author of the quaint political satires of fifty years ago.

Thomas Stone, the author of *The Rod and the Staff*, lived near the base of Bear Mt.; and Cyrus Hamlin, the missionary to Asia, was born near by.

Bear Mountain is ascended on this side by a well-marked path, and although the old observatory has been destroyed, the clear places on the summit afford fine views of the White Mts. and their southerly peaks, Kiarsarge, Mt. Pleasant, and the blue lakes to the S. and S. W. There are several other small mountains in the town, and 6 ponds, whose tributary brooks afford fair trout-fishing. The *Albany Basins* and the beautiful little *Kezar Falls* (3 – 4 M.) are often visited from this point; and there are picturesque drives into Bridgton, Sweden, and Norway. James Walker, D. D., for 7 years President of Harvard College, was brought up near the Kezar Falls.

The Waterford House was new in 1875, and is situated at Waterford Flat, opposite Dr. Shattuck's water-cure. The Point-Grove House is nearer N. Bridgton, and the Bear-Mt. House is a small establishment near *Bear Pond.*

Stages leave N. Bridgton for Waterford on the arrival of the afternoon boat, returning early the next morning in time for her down trip. Stages also leave Waterford for S. Paris (12 M. distant, on the Grand Trunk Railway) daily, at 9 A. M.; returning from S. Paris at 4 P. M.

After touching at N. Bridgton, the steamer crosses the lake to its head at the village of **Harrison** (*Elm House*), 1½ M. distant, at the outlet of Anonymous Pond. The drives from this point are into Waterford and Norway; S. along the E. shore of Long Pond; and E. 7 – 8 M. by Bolster's Mills and Saturday Pond to E. Otisfield, on *Thompson's Pond*, which is about 7 M. long. This town and the nearly contiguous towns of Gray and Otisfield were named in honor of Harrison Gray Otis, of Boston.

The first settler on the shores of Long Pond was an old sea-captain from Ipswich, Mass., who established an inn and an Indian trading-station on the site of N. Bridgton, in 1768. Some portion of his old nautical tastes seems to have remained, and prompted him to build a two-ton sail-boat, after which he made a six-years' contract with the township, binding himself to carry freight or passengers between N. Bridgton and the foot of Sebago Lake whenever desired. After the Indian attack on Bethel the settlers at Bridgton became panic-stricken, and were about to retreat by the boat, when tidings came that the enemy had turned back toward Canada.

172. Fryeburg.

Hotel. — The Oxford House is an ancient and comfortable hotel, to which extensive additions have recently been made and furnished with modern improvements. It accommodates 100 guests, at § 7 - 12 a week. There are several summer boarding-houses in the village, accommodating about 150 visitors in the aggregate.

Routes. — Fryeburg is a station on the Portland & Ogdensburg Railroad, 49 M. from Portland, and 11 M. from N. Conway. Stages leave this point daily for Mt. Pleasant and for Lovell, Sweden, Waterford, and Norway; also for N. Fryeburg, Lovell, and Chatham (in summer).

Distances (from the Oxford House). — To Pine Hill, ¼ M.; Stark's Hill, ½ M.; Swan's Falls, 1½ M.; Jockey Cap, 1 M.; Lovewell's-Pond Battle-Ground, 2 M.; Mt. Pleasant, 7 M.; N. Conway, 10 M.

Fryeburg is a quiet and ancient village of 500 inhabitants, situated in a great bend of the Saco River, near the frontier of Maine and New Hampshire. It is one of the most beautiful places in the vicinity of the mountains, and is surrounded by a pleasant and attractive country. The main street is bordered by venerable houses and shaded by lines of large and umbrageous trees. In the E. part is the commodious brick building of the *Fryeburg Academy*, whose grounds are surrounded by a quaintly constructed stone-wall. Daniel Webster was one of the early preceptors of this academy, when he resided in this town. The present building is near the site of the ancient academy, and contains an interesting museum of Indian relics and other curiosities. The village churches are of the Congregational and Methodist persuasions.

Fryeburg has not become known as a summer-resort, and lacks accommodations for many visitors. But its rare tranquillity, the beauty of the village, and the noble lake and mountain scenery in the vicinity, give it strong claims as a home for the lover of nature. There are no manufactures here, the prosperity of the place being kept up by a steady influx of country trade from the towns on the N. and N. E. The intervales in this vicinity cover over 10,000 acres, and are famous for their richness and beauty. They are overflowed and fertilized every spring by the Saco River. On these meadows is the winter home of the large droves of cattle which graze on the mountains during the summer. There are several thousand acres of forest in the town, and it is claimed that Fryeburg has more standing timber now than it had forty years ago.

Pine Hill is about 10 minutes' walk from the hotel, and is reached by following the river-road to a low wooded hill on the r., which is ascended by a good path. From the ledges on the summit one overlooks the village, Jockey Cap, the black mound of Mt. Tom beyond, and the long and rampart-like ridge of Mt. Pleasant, with its crowning hotel. Below the latter is a blue strip of Lovewell's Pond, with its yellow eastern beach; and to the r. are the well-marked hills of Hiram and Cornish, with Mt. Cutler. Close by and to the S. are the wooded slopes of Stark's Hill, much higher than Pine Hill. In the S. W. is the long blue Ossipee Range, with parts of the Sandwich Mts. on the r.; and about W. S. W. is the high alpine crest-line of Chocorua, brilliant in color and noble in form. Beyond Chocorua is the black hemisphere of Mt. Passaconaway; and still farther W. are the four peaks of Tripyramid. Across the valley to the W., and on the right of Chocorua, the long bare ridge of Moat Mt. is favorably seen, with its handsome peaks covered with red ledges. In this direction and far to the r. extend the luxuriant meadows of the Saco, of the richest green color, banded by the sinuous river, and dotted here and there by white houses and by clumps of graceful elms and other trees. This rich and peaceful plain extends to the foot of the Green Hills, which intervene between Fryeburg and N. Conway. The symmetrical pyramid of Mt. Kiarsarge rises on the N. of these hills; and on its r. are the twin peaks of Mt. Gemini. Mt. Carter is seen on the horizon, nearly over the latter; and on the r. are Mts. Eastman and Baldface, the latter being marked by its white peaks. Nearly N. and on the N. H. border, is Mt. Royce, on whose r. is Speckled Mt.; and to the r. is the great basin of the Kezar Ponds.

Stark's Hill is across the road from Pine Hill, ½ M. from the hotel. It is much higher than Pine Hill, and commands noble views of the mountains and meadows from its upper ledges. This hill was named for Capt. Wm. Stark (brother of Gen. John Stark) of Rogers's Rangers, who led Col. Frye up on its crest to view the town, on his first visit.

Jockey Cap is a mass of very coarse granite, rising over the tree-tops of the forest near Lovewell's Pond. It is about 1 M. from Fryeburg, and is reached by the road which diverges from the Lovell highway at the academy, and passes out by the quaint stone-walls of the village cemetery. A well-travelled forest-road is thence followed to the l., and the tourist soon reaches the open space where carriages are left. Of the two paths which run from this point to the top of the rock, that to the l. is longer and more easy, and that to the r. passes up steeply along the base of the cliff.

* *The View* from Jockey Cap is one of the most beautiful that can be obtained from a slight elevation in the whole mountain-region. Towards the E. is the long dark ridge of Mt. Pleasant, with its hotel near the central summit; and the wide woodlands of Denmark are in the foreground, beyond the Saco River. Near at hand on the S. E. is Lovewell's Pond, with its broad blue sheet sweeping around two tiny islets and bordered here and there by brightly colored sandy beaches. Farther away down the valley of the Saco are Trafton Mt. and the rolling hills of Cornish

and Hiram. To the S. W., are Stark's Hill and other low elevations near the border; and the foreground on the S. W. and W. is occupied by the rich intervales of the Saco, studded with white farm-houses and bowering elms, and banded by the blue and yellow of the river and its beaches. The pleasant village of Fryeburg is close at hand below, and nearly all its buildings are visible. Over the smaller spire are the distant crests of the Ossipee Mts, near Lake Winnepesaukee; on whose r. is Mt. Israel, in Sandwich. Above the village, and many miles away, is the alpine ridge of Chocorua, with its rugged peaks of reddish rock vividly outlined; and over its r. flank are the high dome of Mt. Passaconaway and the pointed peaks of Tripyramid. On the r. of Chocorua is the entire ridge of Moat Mt., with serrated crests of red rock, overtopping the Green Hills of Conway, which commence under the N. peak of Moat and run N. to the tall and graceful cone of Kiarsarge. Above and on the r. of Kiarsarge is the summit of Washington, beyond which, just to the l. of Mt. Gemini, is the small but sharp peak of Adams. The long ridge of Mt. Carter is next seen, against which rests the white crown of Baldface. The rounded swells of Mts. Eastman and Sable are to the left of Baldface, and on its r. are the massive outlines of Mt. Royce and Speckled Mt. To the N. and N. E. are the broad farm-lands and forests of Fryeburg and Lovell, with a succession of low wooded ridges and highland clearings.

Lovewell's Pond is 1 M. from Fryeburg and is a picturesque lakelet among the woods, about 1½ M. long, and containing two islets. The five-ton yacht *Paugus* is kept here for the use of visitors, and gives the means for a pleasant sail. The scene of the battle was on the N. shore, between the outlets of two small brooks.

Beautiful mountain views are obtained from the Lovell and Conway roads, near Fryeburg. Mt. Pleasant (see page 410) is 7 M. distant, and is best visited from this point, though the roads are sandy. An excursion of a novel character may be made by drifting down the current of the Saco River, through the rich rural scenery of the intervales, with noble views of the mountains. The river here makes a long loop toward the N., so that, after descending its current for 20 M., the voyager finds himself within little more than 1 M. of his starting-point.

The aboriginal name of this region was Pequawket, derived from *pequawkis*, "crooked," and *auke*, "place," because the Saco here winds for 30 M. in an area of 6 M. square; or, according to others, from *pequawket*, the Delaware name for a pelican or white swan, a rare bird which used to frequent the shores of Lovewell's Pond, the last of which was caught in 1785. The ancient Indian village was in a bend of the river, just N. of Fryeburg, and on the E. shore. When the English began to occupy the sea-coast the Sokokis tribe retired from the mouth of the Saco River, and joined the Pequawket and Ossipee tribes. The Marquis de Vaudreuil, Governor-General of Canada, wrote to the King of France a full account of these communities, and especially of the warlike Pequawkets.

In 1703 the New-Englanders sent 360 soldiers to attack Pequawket, aided by strong forces that advanced into the Ossipee country. Col. March, of Casco, captured the town and killed or made prisoners 12 of the Indians, following up this advantage by a series of snow-shoe expeditions through the mountain-defiles. But the Pequawkets were alert and skilful, and every Indian killed or captured cost the Provinces over £1,000. In 1717 the Sachem Adeawando of Pequawket was present at the great conference at Georgetown, on the coast of Maine, when Dudley, Penhallow, Wentworth, Walton, Sewall, Quincy, and other Provincial magnates met the chiefs of the Eastern tribes in council.

The Battle of Pequawket.

Capt. John Lovewell, the son of an ensign in Cromwell's Puritan army, was an able partisan officer of the colonies. In April, 1725, he led 46 men from the Massachusetts frontier-towns by a long and arduous march into the heart of the Pequawket country. He had 34 men left when he reached Saco (now Lovewell's) Pond, and here he encamped for 36 hours, near the chief village of the enemy. On Saturday, May 6, while the rangers were assembled around the chaplain on the beach, and ere the morning devotions had been finished, a gun was heard, and an Indian was seen watching them. They left their packs near the pond, and advanced towards the intervales, but met an Indian in the forest, who shot and mortally wounded Lovewell, though his own death followed instantly. Meantime the Sachem Paugus and 80 warriors had found and counted the packs and laid an ambuscade near them, which completely entrapped the Americans on their return. The magnanimous Paugus ordered his men to fire over the heads of the invaders, and then to bind them with ropes. With horrid yells the Indians leaped forth and asked Lovewell if he would have quarter. "Only at the muzzles of your guns!" shouted the brave captain, and led his men against the enemy. They drove the Indians some rods, but were repulsed by a fierce counter-charge, in which Lovewell and 8 of his men were killed. Then the Americans retreated slowly, fighting inch by inch, to a position with the pond on their rear, Battle Creek on the r., and Rocky Point on the l. This sheltered position they maintained for eight hours against continual assaults, and at sundown the Indians retreated, under the command of Wahwa, leaving 39 killed and wounded, including Paugus, who fell late in the contest. Throughout the long day the yells of the Indians, the cheers of the Americans, and the pattering of musketry resounded through the forest, while Chaplain Frye, mortally wounded when fighting among the foremost, was often heard praying for victory. In the moonlit midnight hour the rangers retreated, leaving 15 of their number dead and dying on the field, while 10 of the 19 others were wounded. After suffering terribly on the retreat, most of the little band reached the settlements. A long and mournful ballad of 30 stanzas (like the old Scottish ballad of Chevy Chase) commemorates this forest-fight (see Farmer's *Hist. Colls.*, Vol. III. pages 64 and 94).

> "What time the noble Lovewell came
> With fifty men from Dunstable,
> The cruel Pequa'tt tribe to tame
> With arms and bloodshed terrible.
>
> .   .   .   .   .   .
>
> "Nor, Lovewell, was thy memory forgot!
> Who through the trackless wild thy heroes led,
> Death, and the dreadful torture heeding not,
> Mightst thou thy heart's blood for thy country shed,
> And serve her living, honor her when dead.
> Oh, Lovewell, Lovewell, nature's self shall die,
> And o'er her ashes be her requiem said,
> Before New Hampshire pass thy story by,
> Without a note of praise, without a pitying eye."

The villagers of Pequawket then retired to the populous Indian town of St. Francis de Sales (the ancient *Nessawakamighe*), on the St. Lawrence River, which was from the earliest times inhabited by a clan of their nation (the Abenaquis). The Concord, Kennebec, and Pemigewasset tribes also retired to the same place, and the St. Francis Indians were long the terror of the New-England border. In 1858, 387 Abenaquis remained at St. Francis (near Beçancour), engaged in continual feuds because part of their number had become Methodists.

This township was granted in 1752 to Gen. Joseph Frye, of Andover, Mass., a brave veteran of the French and Indian wars, who was ordered to divide it into 64 parts, which were to be given to 60 families, and for the support of a Protestant clergyman, a parsonage, a school-fund, and Harvard College. During the next year the Osgoods, Evanses, and other families moved into the town, after long forest-marches from the lower counties of New Hampshire. Fryeburg was soon well populated, and for many years it remained the metropolis of the mountain-region, where all the settlers came for supplies and traffic. The nearest town was Saco, on the sea-coast; and Fryeburg drew her supplies from Sanford, 60 M distant. After the Indian attack on Shelburne, in 1781, thirty men of Fryeburg armed themselves and pursued the savages far into the north country

In 1775 the Rev. William Fessenden was ordained as minister of this parish, at a salary of $200 a year, and remained for many years, exerting a beneficial influence on all the inhabitants. In 1791-2 Fryeburg Academy was founded, its first preceptor being the brilliant but intemperate Paul Langdon, son of President Langdon of Harvard College. His successor was Daniel Webster, who taught for nine months, and then entered the more congenial occupations of law and politics. In 1817-18 the canal was cut across the great bend of the Saco, to reduce the danger of inundations. Gov Enoch Lincoln lived at Fryeburg from 1811 to 1819, and wrote a long poem called *The Village*, which was partly didactic and partly "descriptive of the beautiful scenery of the fairest town on the stream of the Saco."

> " Range after range, sublimely pil'd on high,
> Yon lofty mountains prop the incumbent sky.
> Such countless tops ascend, so vast the heap,
> As if when gush'd the deluge from the deep,
> The rushing torrents wrecked the guilty world,
> And all the rocky fragments thither whirl'd."

> " Ethereal beings, so traditions tell,
> High o er the wide-spread Sachemdoms around,
> Dwelt in that topmost height s empyreal bound,
> Watch'd o er the tribes, each wise design inspir'd,
> Advis'd in council and in battle fir'd.

> " The nation's boast, in undisturb'd repose,
> Pequawkett, then thy numerous wigwams rose.
> Thy active hunters, arm'd with bow and spear,
> The stately Moose pursued and bounding Deer,
> For howling Wolves contrived the secret snare,
> Or trapp d the Sable, or waylaid the Bear."

173. Mount Pleasant, or Pleasant Mountain,

is a line of wooded heights in Denmark and Bridgton, between the Saco Valley and the Moose Ponds. It consists of several rounded crests, separated by shallow ravines; and from distant points presents the appearance of a long wall. The peak which is generally visited, and on which the hotel stands, is near the middle of the line, and is 2,018 ft. high. On account of its isolated position, the mountain is one of the best points from which to view the White Mts., and opens a broad and magnificent prospect.

Mt. Pleasant is usually visited from *Bridgton Centre*, from which it is 10 M. distant. The road goes from Bridgton to the N W., and follows the shore of *Highland Lake* for about 2 M., when it branches to the l. and passes out by the island-surrounding Beaver Pond to *Upper Moose Pond*, a long and picturesque sheet of water which is crossed by a strong causeway, resting on an island. The road next traverses the rural street of *W. Bridgton*, and soon enters the township of Fryeburg, where it bends S. around the N. W base of Mt. Pleasant. The mountain-road diverges to the l. and ascends an open ravine by steep and rugged grades, to the Half-Way station, where there are stables, a rude reception-shed, and a cold, clear spring. The route hence is 1 M. long and is very steep, insomuch that visitors usually leave their carriages here and ascend slowly on foot, passing upward through small forests.

It is stated that a new stage-line is to be started in the summer of 1876, between Fryeburg and Mt. Pleasant. This route would be much more convenient and speedy than the Bridgton way. Another proposed new route is from Brownfield.

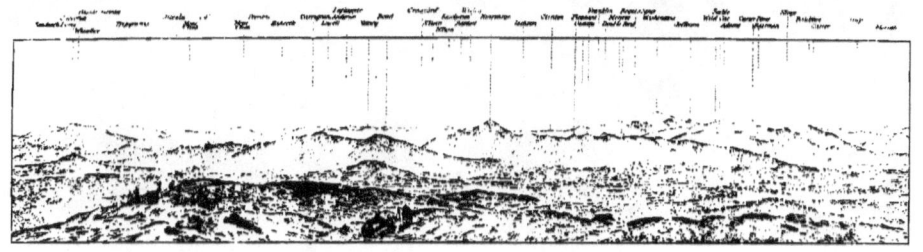

The **Mt.-Pleasant House** is the most commodious and pleasant of the hotels on the mountain-tops of New England. It is a spacious wooden building, two stories high, surrounded with broad verandas, and commanding glorious views. It has a bowling-alley, a billiard-room, and a croquet-ground. The accommodations are for 75 guests; and the rates are $2 a day, and $12 a week. The house is open from July 1 to Sept. 15.

** *The View.* — About S. W. is the long blue line of the Ossipee Range, with the craggy point of the Whittier Peak conspicuous on the r., the view-line crossing the Saco Valley and passing near the Frost and Burnt-Meadow Mts. in Brownfield and Gline Mt. in Eaton. Red Hill adjoins Ossipee on the r., and is apparently continuous with it. In the foreground is Pleasant Pond, over whose r. and near it is Lovewell's Pond, beyond which are the high hills of Madison, over the long Walker's Pond. Looking out over the S. end of Lovewell's Pond, far down on the W. horizon, Mt. Prospect is seen, beyond Squam Lake, near which the view includes the dark Plymouth Mt. and a part of the white-crested Mt. Cardigan, 55 M. distant. The Squam Mts. are next seen, nearly as far away; and the first prominent peak over the S. of the pond is Mt. Israel, on whose r. are Sandwich Dome and Flat Mt. In this direction extend the romantic lowlands of Tamworth and Sandwich, dotted with hamlets and ponds. Over the yellow beach near the centre of Lovewell's Pond, are the long ridge and double peaks of Whiteface, looking over the nearer heights of Chocorua. Close to the pond, on the W., is the rounded Stark's Hill, over which is the lofty and majestic serrated ridge of Chocorua, strewn with light-colored ledges, and cutting sharply against the dark background. Beyond Chocorua, toward the W., is the black dome of Passaconaway; the peaks of Tripyramid, resembling the teeth of a saw, are more to the r.; and the crest of Osceola is yet more distant, in the same line. Under and to the r. of the N. Tripyramid is the skull-shaped white hill called the Potash.

The white village of Fryeburg is a little S. of W., over the N. end of Lovewell's Pond, and nearly under the N. peak of Tripyramid. Beyond and about the village on the N. and W. are the beautiful intervales of the Saco, the famous ancient plain of Pequawket, covered with rich farms and striped by the sinuous bands of blue and yellow which mark the course of the river, its rapid waters coursing between brilliantly colored gravel-banks and over pebbly shoals. To the r. of Fryeburg is the scarcely distinguishable rock of Jockey Cap, near Lovewell's Pond; and farther to the r. and nearer is the bluff mound of Mt. Tom, towards Fryeburg Centre. To the r. of Stark's Hill, Conway Centre is seen, beyond which the Saco Valley enters the mountain-land. The red ledges of Moat Mt. are near its head, to the r. of Chocorua, the ridge being apparently broken by deep ravines into several sections. Over the S. peak of Moat is the massive Osceola, with Tripyramid on the l. On the opposite side of the great plain are the Green Hills of Conway, with a part of Peaked Mt. visible on the l. of Blackcap, and the low rounded top of Middle Mt. farther to the l. beyond a ravine. Over the latter is the sharp N. peak of Moat Mt., and a portion of Bear Mt. is visible beyond. Over the first ridge beyond Moat are the N. and S. peaks of Tremont. The view-line next passes up the long Saco Valley to the distant crest of Mt. Hancock, which is on the r. of and beyond Tremont and over the l. flank of the highest of the Green Hills. Next to the r., nearly equidistant, and over the top of this hill, is the immense mass of Mt. Carrigain, rising out of the Pemigewasset Forest.

The next view-line passes over Mt. Tom, the bold rocky and forest-crowned hill which rises so abruptly above the Saco plain, about W. N. W. and but a few miles distant. In this direction, and just to the r. of Carrigain, is the Nancy Range, Mt. Lowell being the sharp pyramid on the l., Mt. Anderson the middle peak, and Mt. Nancy that on the r. Over the latter, about 40 M. distant, rises the serrated ridge of Mt. Lafayette, with Mt. Lincoln on its l. and dimly seen on the remote horizon. Over the N. Green Hill, and about 35 M. distant, is the high top of Mt. Bond, on whose r. are Mt. Guyot and the Twin Mts., with the apex of Haystack peering through a depression between them.

Beyond the Saco plain and on the r. of the Green Hills of Conway is the graceful pyramid of Mt. Pequawket, the Conway Kiarsarge, on whose l. is the connected ridge of Mt. Bartlett. Over the extreme l. of the latter is the sharp little peak of Mt. Crawford, on whose r. is the slightly curving plateau of Mt. Resolution, with Mt. Parker on the l. Over the r. flank of Mt. Bartlett, and meeting the l. flank of Kiarsarge, is the long ascending terrace of Giant's Stairs, sharply cut off on the l.; and over this is the noble alpine peak of Mt. Willey, its precipitous l. flank falling nearly on the crests of Giant's Stairs and Mt. Bartlett. This cluster of mountains is nearly over a large sand-bank in the Saco, of a vivid yellow color. Over the first depression on the r. of Kiarsarge are Mts. Jackson and Clinton, in the Presidential Range, the former having a sharp point. At the foot of the range which bounds the Saco plain on the W. are the small Kimball Ponds, over which are the twin peaks of Mt. Gemini, with the round and lofty dome of Mt. Pleasant still farther away. Over the ridge which runs to the r. from Mt. Gemini are the tops of Double-Head. Mt. Franklin is on the r. of Mt. Pleasant, and is marked by a slide which falls into Oakes's Gulf; and farther to the r. is Mt. Monroe, which is over the N. peak of Double-Head. The culmination of the range is next reached in Mt. Washington, whose buildings and ravines can be discerned minutely on a clear day.

Mt. Washington is about 28 M. from Mt. Pleasant, in an air-line, and is nearly N. W. in direction. The first ridge this side of Washington is a part of the Wild-Cat range; the second is Black Mt., in Jackson; and the third is Sable Mt. Over the latter is the crest of Mt. Wild-Cat, between which and Washington is a part of Mt. Jefferson. On the r. of Mt. Wild-Cat is the deep gorge of the Carter Notch, through which the sharp apex of Mt. Adams is seen, with the secondary crags on the r. which overlook King's Ravine. A portion of Mt. Madison also is visible through the Carter Notch, on whose r. rises the lofty ridge of the Carter Dome. In the foreground is the azure and triangular Kezar Pond, beyond which the heights of Mts. Eastman and Slope arise. To the r. of the Kimball Ponds, and a little to the l. of and beyond Kezar Pond, are the hamlets of N. Fryeburg and Stow, over the latter of which rise the white caps of Mt. Baldface, nearly over Charles Pond and to the r. of Mt. Slope. The Imp Mt. is over Baldface. Over and to the r. of Kezar Pond is the long and narrow Upper Kezar Pond, which is apparently divided into several sections by intruding lines of highlands; and above its r. centre is the lofty Mt. Moriah, with Bald Mt. on the r. Over its extreme r. is Lovell Village, beyond and above which is the rounded and massive Mt. Royce, covered with dark bare ledges. Lovell Centre and Mt. Sabattos are to the r. of the N. section of Upper Kezar Pond, above which, and on the r. of Mt. Royce, are the elephantine flanks of Speckled Mt. Nearly over Sabattos, out on the horizon, are the Bear-River and Sunday-River White-Caps and the Grafton Speckled Mt. The sharp spire of Goose-Eye is over the Red-Rock and Calabo Mts., more to the l.

The view to the N. passes over the rolling highlands of Sweden, and rests on the distant peaks of Newry, filling the horizon beyond Bethel. Farther to the r. are Bear and Hawk Mts., in Waterford, near the head of Long Pond, with the peaks of Woodstock beyond. On the distant horizon are Mts. Blue and Bigelow, lifting sharp points against the sky. Much nearer is the beautiful Highland Lake, with its islands, far over which is the village of Paris Hill, with Streaked Mt. on its r. Bridgton is at the S. end of Highland Lake, over which several sections of Long Pond are seen, running from the white hamlet of Harrison S. to Naples, beyond which is the expanse of the Bay of Naples.

The view now sweeps over a broad area of the southern counties of Maine, lake-strewn, dotted with villages, and bordered by the open sea. It is claimed that the Camden Mts. are visible, a little N. of E., near Penobscot Bay; and Cape Small and other points on the coast are identified by the aid of a telescope. Nearly S. E. is the broad sheet of Sebago Lake, over which the city of Portland is visible. Farther to the r. the view passes over the near Saddleback Mt., in Baldwin, and follows the line of the lower Saco Valley to the sea. The rolling highlands of Hiram and Cornish are on the S.; and on their l. and far beyond is Mt. Agamenticus, near the ocean. Between S. S. W. and S. W. are the nearer mountains of Brownfield, beyond which are Copple Crown, the Green Mt. in Effingham, and other peaks near the border.

174. Chatham

is a mountain-town adjoining Conway on the N., and shut in on the N., W., and S. by high ranges. It has 445 inhabitants and two churches; and of its 26,000 acres of land, less than 4,000 are improved. The greater part of the town is not susceptible to cultivation, and will for ages be the resort of lumbermen in search of pine and hemlock timber, and of sportsmen after game. The road through S. Chatham is rugged and hilly; but that through N. Fryeburg and Stow is easy and level. The drive N. through Chatham gives a continuous view of the great mountain-wall on the W., composed of Mts. Kiarsarge, Slope, Eastman, Gemini, and Baldface. *Asa Chandler's* is in a rich and beautiful glen 17 M. from Fryeburg, and commands noble near views of Baldface on the W. and Mt. Royce and Speckled Mt. on the N. This is one of the most picturesque valleys in the White Mts., and is the point from which Baldface and Royce are attacked.

There are no public-houses in Chatham, but several of the farmers receive boarders. Good trout-fishing is found in Bradley's, Langdon's, and McDonald's Brooks; and pickerel are caught in the pretty Kimball Ponds, near S. Chatham. Forest-trails lead from S. Chatham to Kiarsarge Village and N. Conway; from the centre of the town by Mountain Pond to Jackson; and from the N. through Evans' Notch to Shelburne. *Mountain Pond* is high among the mountains, 2 M. from the highway, and covers about 100 acres. It is claimed that the largest speckled trout ever caught in New England have been taken here.

Province Pond is 1¾ M. from the road, and is famous for its many trout, which range from ½ pound downward.

Clay's Cascade is 1½ M. from the road, in the N. part of the town, and falls about 60 ft. It is on Carlton Brook, where many trout are found.

Mt. Eastman is about 3,000 ft. high, and is covered to its summit with dense spruce forests, whose timber is being cut and floated down the Cold and Saco Rivers. The crest of Eastman is 3 M. from the highway.

Mt. Slope has pastures well up its flanks, whence the ascent is steep and rugged. The peak is 1½ M. from the highway. The ancient and proper name of this summit is *Sloop Mt.*

Baldface is a formidable mountain in the N. W. part of Chatham, and is considered as part of the Mt.-Carter group. It has two high spur-ranges connected with it, of which the first culminates in Sable Mt., and the second is composed of Mts. Eastman and Slope. The mountain is 3,600 ft. high, and is composed mainly of a fine-grained common granite. It is remarkable for the singular whiteness of its long upper ridges, which may be recognized many leagues away on account of their unique appearance.

This mountain has been attacked from Jackson, but the route thence is long and arduous, traversing a gloomy and tangled forest. The best point from which to enter on the ascent is from Asa Chandler's, in Chatham, 17 M. from Fryeburg. It is

4 M. from Chandler's to the summit, nearly 2 M. of which is traversed by a logging-road which follows the course of the brook N. of Chandler's, crossing it several times on rude bridges. The climber should get one of the neighbors to put him well up on this road, as there are other and similar wood-tracks near the foot of the mountain, and the road itself becomes very vague on the pastures below. The course follows a ravine on the N. E. side of the peak, and when 1½–2 M. from the pastures, the visitor should turn to the l. into a foot-path which ascends the slope rapidly. In this way the ledges are soon reached, over which the summit is attained by a long scramble. The path is usually plain in August and September, during and just after the blueberry season, when it is much used.

The Guide-Book party got on the wrong trail and was forced to retire from Baldface by the approach of evening. without having reached the summit. The view thence must be very broad and noble, in view of the number of peaks from which Baldface is visible. Among the mountains in sight are Kiarsarge, Double-Head, both peaks of Moat, Chocorua, Ossipee, Bear, Passaconaway (due S. W.), Tripyramid, Sandwich Dome, Iron, Tremont, Langdon, Osceola, Moosilauke, Lowell, Carrigain, Hancock, Resolution, Giant's Stairs, the Montalban Ridge, Monroe, Washington, Wild-Cat, the Carter Dome, Adams, Madison, Carter, Moriah, Baldcap, Ingalls, and nearly all the mountains of Western Maine, from the Grafton Notch to Mt. Pleasant and the Cornish hills. Parts of Long Pond are also seen ; and Portland and the ocean are visible on a clear day.

Mount Royce is a double-peaked summit, with the Maine and New-Hampshire frontier running between its crests. It is near Baldface, on the N. E., and is rarely ascended, the slopes being terribly rugged. The summits are great rounded masses of dark rock, and are weirdly picturesque towards evening, when the shadowy ravines make strong contrasts. The W. peak is said to be 2,600 ft. high, but is probably higher. *Blanchard's Basin* is a sequestered pond on this mountain, famous for its many trout. The distance from the road to the top of Royce is 2½–3 M. *Evans' Notch* leads by the base of this mountain into Shelburne, and has been traversed by wagons and sleighs. There is no road, however. The scenery in this pass is very striking. *Speckled Mt.* is a vast mass E. of Royce, partly covered with ledges, and but rarely visited.

" The broad summit of Speckled Mt., opposite Mt. Royce, which two mountains guard the S. entrance of Evans' Notch, is glaciated both on the N. W. and N. E. flanks. Here also is a *col*, down which the ice must have moved in both directions."

Lovell.

The American House is a neat and cosey inn on the main street of the village. Its prices are $6–8 a week. Stages leave daily at 7 A. M. for Fryeburg, returning on the arrival of the 4.30 P. M. train from Portland. The distances from Lovell are : — to Kezar Falls, 6 M. ; the Albany Basins, 12 ; Mt. Pleasant, 12 ; Lovewell's Pond, 11 ; Mt. Sabattos, 3 ; N. Conway, 18 ; Fryeburg, 10.

Lovell Village is a pleasant and tranquil place on the plain S. of the Upper Kezar Pond. It is a favorable point from which to make excursions in the upper Pequawket country. The road from Fryeburg gives a succession of fine views of Mts. Kiarsarge, Gemini, Slope, Eastman, Baldface, and Royce, which form an unbroken wall along the W. Chatham may be reached either by way of Stow or by the hard hill-road through *Foxboro.* 1½ M. from the village is a large boarding-house, on the shore of the Upper Kezar Pond. There are also pleasant drives to the E., among the mountains of Sweden; and Mt. Pleasant is easily reached by a road which leads by Kezar Pond.

The *Upper Kezar Pond* is a beautiful lake in Lovell, 8 M. long and about 1 M. wide, with three receding bays. Roads border it on either side; and on the E., about midway of the pond, is the small hamlet of Lovell Centre. *Kezar*, from whom so many localities in this vicinity are named, was a famous hunter in the early days.

Mt. Sabattos is 4 M. from Lovell Village, and is easily ascended. It commands a broad prospect, the most beautiful part of which is the long expanse of the Upper Kezar Pond. Besides a rich view of Kiarsarge, Gemini, Baldface, Washington, and others of the White Mts., it commands Mt. Blue, the Camden Mts., the Edgecombe hills, Mt. Pleasant, and Mt. Independence, near Portland.

175. Paris and Norway.

Paris Hill (*Hubbard House*) is a pleasant hamlet situated on a hill 831 ft. high, whence an extensive panoramic view of the mountains is enjoyed. It contains about 60 buildings, the court-house of Oxford County and an academy. The railroad station is 2 – 3 M. distant, at the thriving village of *S. Paris*, and stages connect with the trains. **Mount Mica** is near Paris Hill, on the E., and is "the most interesting locality of rare minerals in the State of Maine." Here are found plates of mica 6 – 10 inches square; green beryls; limpid, smoky, and rose quartz; black, green, blue, and red tourmaline; feldspar, garnets, and other minerals.

Streaked Mountain is about 5 M. S. E. of Paris Hill, and is 1,756 ft. high. It commands a beautiful view, including the whole ridge of Mt. Pleasant, Kiarsarge, Moat, Chocorua, several bold peaks of the Presidential Range, Baldface, Speckled, Moriah and Carter, Goose-Eye, and many other notable summits along the border. In the foreground are Paris Hill, the Hebron Peak, the village and pond of Norway, and a wide area of pleasant and well-populated country.

Tri-weekly stages run from S Paris to Norway, Harrison, Bridgton, and Fryeburg, 33 M. (fare, $ 2).

Tudor says that Paris is "a place as little resembling its European original as a cottage does a palace. At the same time it may be said, that to the extent in which it falls short of its great prototype as to architectural beauty, does it exceed it in the beauties of nature, being surrounded by a circle of mountains of the most imposing and romantic features."

Norway (*Beal's Hotel*) is 2 M. from S. Paris, and is situated amidst delightful scenery. It adjoins the mountain-town of Waterford (page 405), and is studded with ponds and highlands. The *Great Pennessewassee Pond* is just N. W. of the village, and is a beautiful sheet of water 9 M. long. *Pike's Hill* is S. of Norway, and is 600 – 700 ft. high. It commands a noble view over Pennessewassee and out to the line of the White Mts. *Singe-Poll* is another hill which is often ascended.

Bryant's Pond (small inn) is a highland hamlet and station of the Grand Trunk Railway, near the pretty lakelet of the same name. It is in the town of Greenwood, famous for quartz crystals. Pleasant prospects are afforded from *Mt. Christopher*, over the pond; and there are several far-viewing highlands in Woodstock, the next town to the E.

From Bryant's Pond tri-weekly stages run to Milton Plantation, Rumford, and Andover (21 M.; fare, $1.50); also to Rumford, Mexico, Dixfield, and N. Jay. Another line runs through Mexico to the picturesque mountain-town of *Byron*. The high peaks of White-Cap and Glass-Face are in Rumford; and at the **Rumford Falls** the Androscoggin River falls 150 ft. in several plunges over ragged granite ledges. The third fall has a nearly perpendicular descent of over 70 ft., and its roaring is heard at a great distance. There are three taverns in Rumford. *Andover* is a favorable point from which to visit the lower Rangeley Lakes.

176. Bethel.

The chief boarding-houses at Bethel are on the village-green, and are entitled "The Elms" and the Bethel House, each of which accommodates about 70 guests, their rates being $8-14 a week. There are also the summer boarding-houses of John Russell (30 guests), Abiel Chandler (25 guests), Woodson Mason and Payson Grover (20 each), Mrs. Chapman Kimball, and Dr. Twitchell. A. W. Valentine's Spring-Grove House is on the opposite side of the river, near the chalybeate springs. The rates at these houses are $7-9 a week.

Distances. — Bethel to Paradise Hill, 1¼ M.; Sunset Rock, 1½; Songo Pond, 4; Kendall's Ferry, 5; Bryant's Pond, 9; Crystal Ledge, 12; Albany Basins, 12; Screw-Auger Falls, 15; Rumford Falls, 22; Lake Umbagog, 27; Mt. Washington, 23. *Stages*, see Route 179, *ad finem.*

Bethel is a pleasant old village on the Androscoggin River, in the State of Maine and the county of Oxford. It contains two churches, an academy, and many comfortable mansions; and the streets are over-arched with large trees. On account of its slight elevation above the intervales the place is sometimes called *Bethel Hill.* The hotels are about ¾ M. from the railway-station, and the chief roads converge in their vicinity, at the spacious Green. The chief business of the town is farming, and there are several fine estates on the Androscoggin meadows. The richness and fertility of these intervales add greatly to the beauty of the scenery in the vicinity; and are contrasted with the ruggedness of several bold mountains on the N. The summits about Bethel present an appearance far different from those of the other mountain-towns, and are not so thickly wooded. The claims of Bethel as a summer-resort are manifold, and it was called by Starr King "the N. Conway of the E. slope." The drives in the neighborhood are pleasant and diversified, and lead into several quaint and sequestered districts. One of the best of these is along the river-road to the Gilead bridge, and back on the other shore.

Paradise Hill is a gentle elevation which is crossed by the road 1¼ M. S. of Bethel. It is a favorite view-point for the White Mts. and the peaks

18*

A A

N. of the valley, towards the Grafton Notch. **Sunset Rock** is about 1½ M. distant, in an opposite direction, and is reached by ascending through a pasture a few rods from the road. Thence is obtained a pleasant view up the valley to the dominant peaks of the Presidential Range, which rise boldly beyond Mt. Moriah.

The scenery to the S. W. of Bethel, up the *Pleasant-River valley*, is of an interesting character, though the roads are rugged. In this direction is *Mt. Calabo* (or *Caribou*), from whose summit a broad view is gained. Farther to the S. is *Red-Rock Mt.*, on whose sides are long and lofty cliffs.

Bethel was granted at an early day to veterans of the French and Indian wars from Sudbury, Mass., and received the name of *Sudbury-Canada*. It was settled in 1773; and in August, 1781, it was attacked by an Indian war-party from Canada, and 3 men were killed and 3 were led into captivity. This was the last attack ever made by the Indians on the towns of New England.

The **Albany Basins** are 12 M. from Bethel, and are reached by a good road which runs S. 4 M. to *Songo Pond*, and thence down the glens of the Crooked River, through a sparsely populated and wild country. When about 3 M. from the Basins the road swings to the r. into the hill-country, traverses dense overarching forests, passes Little Papoose Pond, and enters a remote and sequestered clearing. A foot-path leads from the farthest house, in about ¼ M., to the head of the Basins. The Albany-Basin House is a small inn where fishing tourists may stop. The Basins are on a tributary of the Crooked River, in a glen 500 – 600 ft. long, and consist of a series of immense pot-holes cut in the hard talcose rock. The cavities are partly filled with fragments of rock, and cavernous hollows are seen on the sides, overhung by projecting trees which arch and shade the ravine. The roar of the stream through the holes and among the water-worn rocks is audible far away. The largest of the cavities is 40 ft. deep and 120 ft. in circumference, and several others are nearly as large. The present stream seems incompetent to their excavation, and there is a plausible theory that at some early period a large river occupied this channel, forming whirlpools in which the attrition of loose and whirling blocks of granite wore out the bed-rocks below. A subsequent subsidence of the country to the N. diverted the stream into other channels and left the pot-holes exposed. The deep pools and rock-bound recesses along the brook are frequently visited by trout-fishers.

Gilead was settled about the year 1780, and derived its name from a large Balm-of-Gilead tree near the centre of the town. There is no village or boarding-house in the town. The only habitable region is the long and narrow valley of the Androscoggin, which affords very beautiful scenery, its verdant glens being overlooked by massive mountains. The savage gorges of the Wild River debouch into the Androscoggin Valley in this town. The rugged heights which enclose the fruitful vales below are the haunts of bears and other wild beasts. *Tumble Down Dick* is a bold cliff N. of the river, near the centre of the town. **Peaked Hill** commands views of Mts. Madison, Adams, Moriah, and Washington, and the rich intervales of the Androscoggin. In 1826 the sides of this summit were swept by destructive slides, flashing like fire through the night. The terrible desolation of the *Wild-River Forest* extends S. W. from Gilead to the Pinkham Notch and Jackson. Mt. Calabo may be visited from this town; and Goose-Eye is accessible up to the glens of Lary Brook.

The fertility of the narrow valley of Gilead is remarkable. 600 bushels of potatoes to the acre and 100 bushels of corn to the acre have been raised. It is shielded by the adjacent ranges from early frosts, and a continual current of air draws through the valley. The early settlers of these remote glens were all from Massachusetts, and were distinguished for religious enthusiasm.

177. Mount Abram

is a picturesque elevation near Bethel, on the edge of the Androscoggin Valley. It is easily reached from the village, by a drive of 3¼ M.; and the ascent from the road is ¾ M. long.

* *The View.* — The foreground is occupied by the broad and verdant intervales of the Androscoggin, through which the river winds in graceful convolutions. The white village of Bethel is plainly visible, with its little suburb N. of the river. Sparrow-Hawk Mt. and other heights rise in the W., and the deep glens of Gilead and Shelburne are seen opening out from the dark mountain-land. On the r. of this valley are the peaks of Mts. Ingalls, Baldcap, and Hayes; on the l. are Moriah and the mountains of the Wild-River Forest. Over these latter rise the higher crests of the Presidential Range, — the rounded swell of Madison, the sharp spires of Adams and Jefferson, and the supreme head of Washington, capped by houses. The high Carter Range is on the l. of and below Washington, and ends in the convex slopes of the Carter Dome. At the apparent end of the Androscoggin Valley are the undulations of the Crescent and Pilot Ranges; and nearer and more to the r. is the high peak of Goose-Eye, over Black Mt. in Newry. Nearly N. W. are the mountains about the Grafton Notch, Speckled, and the Bear-River White-Cap, with the Sunday-River White-Cap nearer. More to the N. are the mountains of Newry and Andover; and the Blue Mt. in Byron is nearly over the Rumford White-Cap. Farther away are the blue crests of Mt. Bigelow and Saddleback. To the E. and S. E. extend lines of low peaks and bold highlands in Milton, Woodstock, and the adjacent towns. The mountains of Waterford are nearly S., across the forests of Albany; and Mt. Pleasant is more to the r. The house-crowned cone of Kiarsarge is next seen, with the sharp crests of Chocorua and Whiteface far beyond and the blue lines of the Ossipee Range on the l. On the line extending N. from Kiarsarge Mt. Gemini is visible, with Double-Head to the W.; and the white crests of Baldface are nearer, with Mt. Royce and Speckled Mt. on its l. In this direction are the dark heights of Mt. Calabo and the bright ledges of Red-Rock Mt. The ocean appears on the horizon to the S.

178. Goose-Eye.

Goose-Eye is the quaint name of a mountain near the centre of the town of Riley (in Maine), which attracts the attention of visitors to the more northerly White Mts. by its sharp and spire-like peak. It is about 3,200 ft. high, and consists of a long wooded ridge running nearly E. and W., with a bold projection of rocks rising into a pronounced cone. In the N. E. part of the mountain is one of the deepest and boldest ravines in the region, cutting directly down from well-marked walls. N. of Goose-Eye is the picturesque pass of the *Mahoosuc Notch.* There are three eminences on the mountain-ridge, of which that on the N. is 100 ft. the highest.

Goose-Eye is far from the haunts of civilization, and is but rarely visited by tourists. One route of attack is from Berlin Falls, whence one can drive 6 M. N. E. to the Blake farm, in the town of Success. From this point the mountain is 6 M. distant, through woods where no path exists, and the only direction is a compassline. The march from Blake's to the summit requires 6-8 hours, and the whole excursion demands 2 days, the intervening night being spent in camp on the ridge (where water may be found within ¼ M. of the peak). The whole ascent on the W. side is of a gradual character, and the woods are easily traversed, being nearly free from undergrowth or fallen timber. The geological character of the ridge is White-Mountain gneiss.

Another route is from Bethel for 10 M. up the Sunday-River valley, and then by a logging-road for 1½ M. to Blake's camp, near the foot of the mountain. Goose-Eye was successfully attacked by members of the Portland White-Mt. Club, in 1875.

The View from the summit is both extensive and picturesque, and is unique in character. The Carter-Moriah and Presidential Ranges are seen in the S. W., with parts of the remote Franconia group; and in the W. are the blue undulations of the Randolph and Pilot Mts., with the white Percy Peaks. Nearly S. are Kiarsarge, Baldface, Moat, and Pleasant; the first and last having hotels on their crests. But the characteristic feature of the prospect is the vast forest of Maine, extending away to the N. and E. for many leagues. Portions of the Androscoggin Valley are seen, a little S. of E., and to the N. is the great basin of the Umbagog country. The massive peaks of Speckled Mt. and Bear-River White-Cap, at the Grafton Notch, are near at hand; and beyond them the view sweeps away over scores of nameless peaks in the remote forests toward the Rangeley Lakes. This is one of the most favorable watch-towers from which to reconnoitre the bold peaks towards Mt. Bigelow and Abraham, and throughout the Kennebec section of the White Mts.

Puzzle Mountain is a fine rocky mass in Newry, 12 M. from Bethel (10 M. by road), and commanding a superb view.

179. The Dixville and Grafton Notches.

Colebrook (*Parsons House*) is a pretty and prosperous village in the N. part of Coös County, N. H., and is reached by daily stages from N. Stratford, 13 M. distant on the Grand Trunk Railway (see page 66). The adjoining country is fertile, and great quantities of potato-starch are made in the vicinity.

* The Dixville Notch is 10 M. S. E. of Colebrook, and is reached by a road leading up the valley of the *Mohawk River*, a pretty stream which affords good trout-fishing. The Dix House is at the mouth of the Notch, and is a new summer-hotel. The Notch is not a mountain-pass, but a deep ravine among high hills whose impending cliffs are worn and broken into strange forms of ruin and desolation. The first view is disappointing, since the pass is entered at a high level by the road which has been ascending all the way from Colebrook. No mountainous line is seen in front, and it is only after leaving the great forest and making a sharp turn to the r. and a short, steep ascent, that the high and columnar sides are seen frowning at each other across the narrow chasm.

Table Rock is on the r. of the road, and is reached by a rude stairway of stone blocks, called *Jacob's Ladder*, whose divergence from the road is marked by a guide-board near the top of the first steep rise. The rock is 56 ft. above the road, and 2,450 ft. above the sea, and is a narrow pinnacle, about 8 ft. wide at the top, with sharp and precipitous sides. It commands a broad view, including Monadnock, the Dixville and Magalloway Mts., the plains of Errol, and the upper Androscoggin Valley. Above Table Rock a short path leads to the *Ice Cave*, a profound chasm where snow and ice may be found throughout the summer. *** The Profile** is seen from a guide-board on the r. of the road, high up on the cliffs, and is preferred by some visitors to the Franconia Profile, on account of its variety of expressions. Farther on are the refreshing waters of *Clear Spring*, on the r.; and a guide-board on the l. points out *Washington's Monument* and *The Pinnacle*, remarkable rock-formations which have been exposed by clearing away the forests. Then the path to *The Flume* diverges on the l., leading to a gorge 20 ft. deep and 10 ft. wide, which has been formed by the erosion of a trap-dike. At the foot of the Notch (which is 1½ M. long) a board directs to the r. to *Huntington's Cascades*, where, from a seat high up on the cliffs, several graceful falls are visible. The *Clear-Stream Meadows* are below the E. side of the Notch, and present a scene of pastoral beauty that strongly contrasts with the desolate region behind.

"It is one of the wildest and most imposing pieces of rock and mountain scenery on the Atlantic side of our country. Totally different from, and therefore not to be compared with, any of the passes among the White Mountains, it has peculiar characteristics which are not equalled elsewhere. In general it may be said that the Notch looks as if it had been produced by a convulsion of nature, which broke the mountain ridge from underneath, throwing the strata of rocks up into the air, and letting them fall in all directions. The result is that the lines of stratification in the solid part of the hills point upward, sometimes nearly perpendicularly, and several pinnacles of rock, like the falling spires of cathedrals, stand out against the sky. . . . . The marked characteristic of all the view was the worn-out, used-up appearance of everything. The rocks were all decayed and crumbling; the mosses were brown and dry; the bushes were little, old, weazen-faced bushes; the very sky seemed brown or brassy overhead. It is a very remarkable, a wonderful piece of scenery, and taking in connection with this the various views along the road, I have no hesitation in saying that the drive from Colebrook, through Dixville Notch to Bethel, is the finest drive I have ever found in America " (W. C. PRIME.)

" The serrated cliffs of mica slate on either hand shoot into the empyrean in clear and sharply defined pinnacles and lances, to the height of 700 – 800 ft., reminding one of the turrets and minarets of Saracenic palaces. Here and there, along its walls, on some knotty spur, or in some deep fissure, cluster a few spruces and white birches, forlorn hopes of vegetation, as it were, struggling against the sliding avalanche and almost invulnerable sterility; and the bottom of the defile is encumbered with shattered rocks and the débris from the bristling crags above. The locality is indeed a second Arabia Petræa, where solitude has an abiding-place. A never-ceasing gale howls its mournful anthem among its sharp ledges, and tortured fountains, winding through secret glens, send out a gurgle that seems ominous of evil. Other sound there is none, unless it be the bark of an occasional wolf or the shout of the wayfarer, to whom the echo of his own voice in such a solitude is companionable." (BECKETT.)

The road to the S. E. runs through the woods to **Errol Dam** (*Akers House*), 13 M. distant. A steamer leaves this point semi-weekly for the *Upper Magalloway River*, and also for the *Lake House*, in Upton, at the foot of Lake Umbagog. From the Lake House semi-weekly stages run to Bethel (see page 417). The road passing from Errol Dam to Upton leads in some places over high ground, whence noble views of the White Mts. are obtained. **Lake Umbagog** covers an area of 18 square miles, and is 1,256 ft. above the sea. The length is 11 M. Tributary to it are the trout-abounding lakes of Welokenebacook, Molechunkemunk, Moosetocmaguntic, Cupsuptic, and Rangeley, covering 59 square miles.

Connecticut Lake (*Connecticut-Lake House*) is 25 M. N. E. of Colebrook, and is navigated by a small steamer. It is 5½ M. long by 2½ M. wide, and abounds in fish. 4 M. N. E. through the forest is *Second Lake*, 2½ × 2 M. in area, while still farther N. is *Third Lake*, covering 200 acres. *Fourth Lake* contains 3 acres, and is the source of the Connecticut River. It is near the Canadian border, 2,500 ft. above the sea. These lakes are in Pittsburg, a town of 200,000 acres, with but 400 inhabitants. Game abounds in the forests, and fish in the streams.

Mount Carmel is several miles N. E. of Third Lake. Dr. Jackson, then State Geologist, was the first white man who ascended this peak. In 1841 he encamped for two days near the summit, and found its altitude to be 3,615 ft. The slopes were dotted with amorphous masses of hornstone, varying in color from apple-green to black. " The view from the summit is one of surpassing interest and grandeur. Northward stretches the lofty range of hills which divide the waters flowing into the St. Lawrence from those of the Magalloway and Connecticut; and beyond these the broad prairies or table-lands of Canada. Southward are seen Umbagog Lake and the Diamond Hills, with the numerous waters in their vicinity, and far beyond them the lofty heights of the White Mts. Westward are the lakes and tributary streams of the Connecticut, and, along the horizon's verge, the Green Mts. Eastward the view is bounded by the granite peaks of Maine, Mt. Bigelow and Mt. Abraham."

The * **Grafton Notch** is traversed by the road which runs from Errol
Dam and Upton to Bethel. The short valley of the Cambridge River is
ascended from Lake Umbagog through a sparsely settled country, strewn
with forests. The scenery of the Grafton Notch is of a high order of
majesty and impressiveness, and is visited by many tourists. On the S.
W. are the ponderous heights of **Speckled Mountain**, fronted on the N. E.
by the lofty mass of the **Bear-River White-Cap**. Both of these mountains are accessible, and command broad views over the northern lake-
country. The Moose Chasm, Mother-Walker's Falls, and the Screw-
Auger Falls are among the natural curiosities here. The course of the
Bear River is followed downward, and at 3 M. beyond the Screw-Auger
Falls, the *Poplar Tavern* is passed. 6 M. below this point the road
traverses the hamlet of *S. Newry*, having given frequent views of the
Sunday-River White-Cap, Goose-Eye, Bald Mt., and Puzzle Mt. 6 M.
from S. Newry the stage enters **Bethel** (see page 417).

"The drive down the Bear-River Notch is hardly inferior in scenery to that
through Dixville Notch. All along the roadside we found streams with abundance
of small trout, and mountain and valley views which are nowhere to be surpassed."
(W. C. PRIME.)

During the summer stages leave Bethel on Mondays, Wednesdays, and Fridays, after
the arrival of the Boston train, for the Grafton Notch, Lake Umbagog, the Rangeley Lakes, and the Dixville Notch. Passengers arriving on the intermediate days
will be forwarded by the stage proprietors. The stages connect with the steamer
Diamond, which runs up Lake Umbagog to Angler's Retreat, Errol Dam, and the
Magalloway River. At Errol Dam the steamer connects with stages for Dixville
Notch and Colebrook. Stages leave Colebrook for Errol Dam on Mondays, Wednesdays, and Fridays. The *Diamond* leaves the Magalloway River for Errol Dam and
the Richardson-Lake road on Tuesdays, Thursdays, and Saturdays.

180. The Kennebec Peaks.

The Editor uses this term in speaking of the noble mountains below mentioned,
because, although they are several leagues from the Kennebec River, they are on
the W. side of its valley, and are drained by its tributaries. Many stately and formidable mountains rise between these and the White Mts. of New Hampshire, but
the accommodations for transit and sojourn in the adjacent forest-towns are so limited that but few travellers have yet reached and explored them.

Phillips (*Barden House*) is the most favorable point from which to
visit the great mountains on the N. E. It is reached by daily stages from
Farmington (20 M.), which is the terminus of a branch of the Maine Central Railroad.

Mount Blue is 4 – 5 M. from Phillips, and is 2,804 ft. high. It is a very
noble and symmetrical peak, finely pointed, and visible for many leagues.
The Saddleback and Bigelow Mts. are finely displayed in the W.; and the
Camden Mts. appear in the S. E., with a long reach of the ocean beyond
and on the r. Peaked Mt. is seen E. of Bangor; the Phipsburg peninsula, nearly S.; and a vast area of Western Maine. The White-Mt. wall
is on the S. W., and stretches along the horizon for many leagues.

The path up to the crest of Mt. Blue is ½ M. long, and is easily followed. When within ⅓ M. of the summit it emerges from the woods, and thenceforward follows a line of ledges. The peak is bare, and the visitor is not troubled by trees or shrubs.

Saddleback Mountain is 8 M. from Phillips, to the N. W., and is 3 - 4 M. beyond the hamlet of *Madrid* (Madrid House). The path to the summit is about 3 M. long, 2¼ M. being on an old logging-road, and most of the rest over bare ledges. The peak is free from trees and bushes. Saddleback is nearly 4,000 ft. high, and consists of several crests rising from a lofty ridge. The view in every direction is of vast extent, including all the chief lakes of Western Maine, the Rangeley region, the Presidential Range, the mountains along the Anglo-Canadian border, and the upper Kennebec valley.

Mt. Abraham is 10 M. from Phillips, to the E. N. E., and is 3,387 ft. high, having several prominent peaks. The path is 4 - 5 M. long.

Rangeley Lake is 20 M. from Phillips, and 8 M. farther on is the *Indian-Rock House*, a famous resort for trout-fishers. The lake is 7 M. long and 2 M. wide, 1,511 ft. above the sea; and is surrounded by bold mountains, prominent among which is Mt. Saddleback. **Moosetocmaguntic Lake** is reached by boat, or by a rude forest-road from Indian Rock, and is 10 M. long by 2 - 4 M. wide. A steamboat will soon be placed on this lake. A chain of lakes extends from Rangeley to Umbagog, embracing 80 square miles of water surface, and abounding in blue-back trout and other game-fish. Travelling in this remote wilderness is difficult, and good guides should be obtained. The usual entrances to the Rangeley region are by Farmington or Andover (from Bryant's Pond). The *Green-Vale House* is at the head of Rangeley Lake, 3 - 4 M. from the Sandy-River and Dead-River Ponds, in a good fishing locality. A steamboat now runs on the quiet waters of Rangeley Lake.

INDEX.

Index to Historical and Biographical Allusions.

Index to Quotations.

Index to Railroad, Steamboat, and Stage Lines.

Authorities Consulted in the Preparation of this Volume.

The Editor acknowledges his obligations to the officers of the Boston Athenæum, the New-England Historic-Genealogical Society, the State Library of New Hampshire, Dartmouth College, and the Massachusetts Historical Society. The latter society allowed him to consult and copy (in part) the ancient MS. of Dr. Jeremy Belknap, entitled "A Tour to the White Mountains."

The Geology of New Hampshire (Vol. I.); by C. H. Hitchcock, J. H. Huntington, etc. Concord, 1874.

The White Hills; by Thomas Starr King. Boston, 1859.

A History and Description of New England; by Coolidge and Mansfield. Boston, 1859.

New Hampshire as it is; by E. A. Charlton. Claremont, 1857.

New Hampshire Book; edited by C. J. Fox and S. Osgood. Nashua, 1842.

Gathered Sketches of New Hampshire and Vermont; edited by Francis Chase. Claremont, 1856.
Statistical Gazetteer of New Hampshire; by A. J. Fogg. Concord, 1874.
Gazetteer of the State of New Hampshire; by John Farmer and Jacob B. Moore. Concord, 1823.
Provincial and State Papers of New Hampshire (8 vols.); edited by Nathaniel Bouton. Concord, 1867-74.
Collections of New Hampshire Historical Society (8 vols.). Concord, 1824-66.
History of New Hampshire (2 vols.); by Jeremy Belknap, D. D. Boston, 1791.
History of New Hampshire; by George Barstow. Concord, 1842.
    History of Coös County; by Rev. Grant Powers. Haverhill, N. H., 1841.
    History of Gilmanton, N. H.; by Daniel Lancaster. Gilmanton, 1845.
    History of Manchester, N. H.; by C. E. Potter. Manchester, 1856.
    History of Dunstable, N. H.; by C. J. Fox. Nashua, 1846.
    History of Concord, N. H.; by Nathaniel Bouton. Concord, 1856.
    History of Dunbarton, N. H.; by Caleb Stark. Concord, 1860.
    History of Warren, N. H.; by William Little. Manchester, 1870.
Historical Relics of the White Mountains; by J. H. Spaulding. Mt. Washington, 1855.
History of the White Mountains; by B. G. Willey (revised by F. Thompson). New York, 1870.
Centennial Celebration at Orford. Manchester, 1865.
Sketches of the History of New Hampshire; by John M. Whiton. Concord, 1834.
Sketches of Oxford County. 1830.
The Willey Family. N. H. Historical Colls. III.

A Statistical View of the District of Maine; by Moses Greenleaf. Boston, 1816.
Ancient Dominions of Maine; by R. K. Sewall. Bath, 1859.
Water-Power of Maine; by Walter Wells. Augusta, 1869.
Collections of Maine Historical Society (6 vols).
History of Maine; by W. D. Williamson. 2 vols. Hallowell, 1832.
    History of Gorham, Me.; by Josiah Pierce. Portland 1862.
    History of Norway, Me.; by David Noyes. Norway, 1852.
    History of Windham, Me.; by T. L. Smith. Portland, 1873.
    History of Portland, Me.; by Wm. Willis. Portland, 1865.

The Farmer's Monthly Visitor; edited by Isaac Hill. Concord, 1839-40.
History of Bradford, Vt.; by Rev Silas McKeen, D. D. Montpelier, 1875.
Records of Narragansett No. 1 (Buxton); by Wm. F. Goodwin. Concord, 1871.
The Portland & Ogdensburg Railroad Line; by Walter Wells. Portland, 1872.
A Tour to the White Mts.; by Dr. Jeremy Belknap. MS. in possession of the Mass. Hist. Society.
The MS. Field-books of Prof. G. P. Bond, of Harvard University.
Some Account of the White Mts.; by Dr. Jacob Bigelow. In the New England Journal of Medicine and Surgery, Vol. V.
Sketches of the White Mountains. In Farmer and Moore's Historical Collections, Vol. II.
Purchas's Pilgrims 10, chap. 1. "Mavosheen."
Collections of the Massachusetts Historical Society.
Travels in New England (4 vols.); by Timothy Dwight, D. D., President of Yale College. 1821.
Sketches of Scenery and Manners in the United States; by Theodore Dwight.
Adventures in the Wilds of America; by Charles Lanman. 2 vols. Philadelphia, 1856.
Three Days on the White Mountains; by Dr. B. L. Ball.
Narrative of a Tour in North America; by Henry Tudor, Esq., Barrister at Law. 2 vols. London, 1834.
The Rambler in North America; by C. J. Latrobe. 2 vols. London, 1836.
Retrospect of Western Travel; by Harriet Martineau. 3 vols. London, 1838.
Society in America; by Harriet Martineau. 2 vols. London, 1836.
The Eastern and Western States of America; by J. S. Buckingham. 3 vols. London, 1842.
Letters from the United States; by the Hon. Miss Amelia M. Murray. New York, 1856.
North America: its Agriculture and Climate; by B. Russell. Edinburgh, 1857.

Homes in the New World; by Fredrika Bremer.

North America; by Anthony Trollope. 2 vols. London, 1862.

The Carter Notch. In Putnam's Magazine, Dec., 1858

The White Mountains (by Thomas Hill). In the Christian Examiner, Nov., 1853.

Notice of Profile Mountain (by Gen. Martin Field) In Silliman's Journal, XIV.

Notes of Excursion among the White Mountains, etc. (by James Pierce). Silliman, VIII.

Notices of Mountain Scenery. Silliman, XV.

Mount Washington in Winter. Boston, 1871.

The Notch in the White Mountains. In the Boston Book, 1841.

Sketches of New England; by John Carver.

The Works of Nathaniel Peabody Rogers.

On the Appalachian Mountain System; by Prof. Arnold Guyot. Silliman's Journal, 1861.

Prof. E. Tuckerman's articles on plants, in Silliman's Journal, Vols. XLV. and VI., New Series.

Also essays by Messrs Shepard, Hitchcock, Packard, Rogers, Field, and others, in Silliman's Journal.

First and Final Reports on the Geology of New Hampshire; by Dr. Charles T. Jackson, Concord, 1841 and 1844.

First and Second Reports on the Geology of Maine; by Dr. Charles T. Jackson. Augusta, 1837 and 1838.

Travels in the United States; by Sir Charles Lyell (Second Visit).

The Proceedings of the American Association for the Advancement of Science. Articles by Guyot (Vol. II.), Jackson (III.), Morgan (XIII.), Vose and Packard (XVI.), and Agassiz (XIX.).

Hovey's Magazine, XIII. (1847). William Oakes.

Life of Jeremy Belknap, D. D., by Miss J. Belknap.

Hawthorne; by James T. Fields. Boston.

The Poetical Works of John G. Whittier. Boston, 1874.

The Village: A Poem; by Enoch Lincoln. Portland, 1816.

An Idyl of Work; by Lucy Larcom. Boston, 1875.

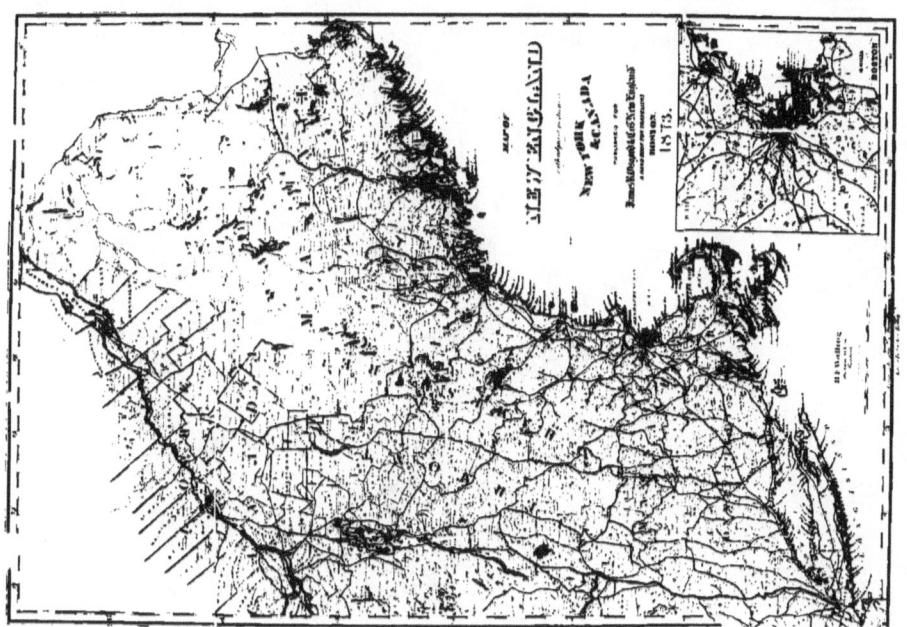

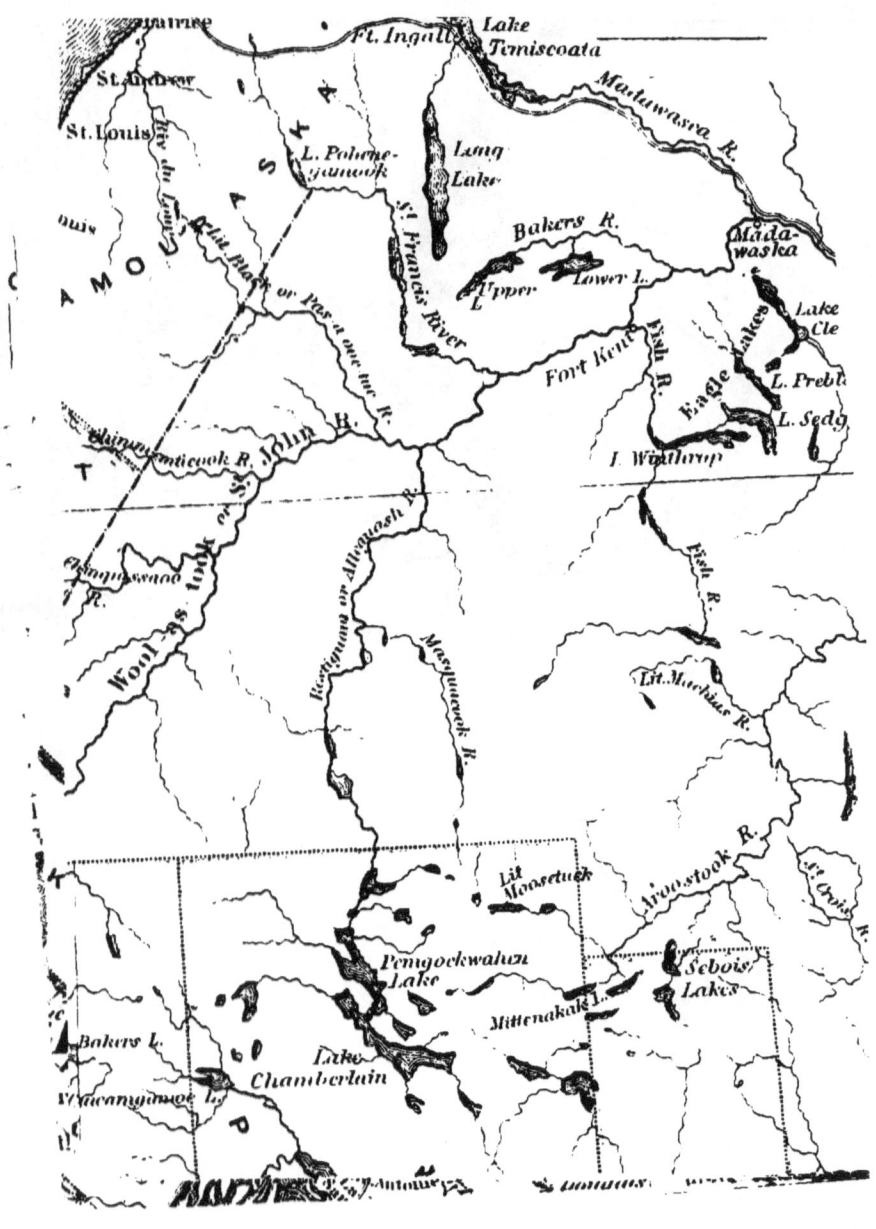

MT. PLEASANT HOUSE,

Open July 4, 1876.

This new and commodious house will be open to the public on the above date. The house is situated on the highest peak of Mt. Pleasant, 2,018 feet above the level of the sea, commanding one of the most extensive and varied land and water views to be found in New England.

Persons afflicted with hay fever find instantaneous relief here, where the pure mountain air is unequalled.

All the appointments of the

Mount Pleasant House

are first-class; the cuisine will be under the charge of a gentleman of long experience; and the public may rest assured that the tables will be fully up to the standard of our best Summer Hotels.

CHAS. E. GIBBS, Proprietor,

Bridgton, Maine.

PEMIGEWASSET HOUSE,
Plymouth, N. H.

This elegant, spacious, and well-appointed hotel, situated at Plymouth, N. H., upon the line of the Boston, Concord, and Montreal Railroad, and at the gateway to the whole mountain region, has been put in the most thorough order for the season of 1876. With every point of interest in the mountain and lake regions easily accessible, and surrounded by romantic and beautiful scenery, it offers peculiar attractions to both tourists and summer boarders. The hotel, which occupies a pleasant and healthful location near the Pemigewasset River, contains One Hundred and Fifty spacious, light, and well-ventilated rooms, en suite or single, with bath-rooms, hot and cold water, and every modern comfort. Mount Prospect, which commands a magnificent view, is only four miles distant, and a good carriage-road leads to the summit. The romantic Livermore Falls are in the neighborhood, and Lake Winnipesaukee, Squam Lake, Moosilauke, and many other points of interest, may be visited in excursions of a single day.

An excellent Quadrille Band has been engaged for the season.

A Livery Stable is connected with the house, and good teams will be in readiness at all times.

Cars leave morning and noon for Profile, Twin Mountain, and Fabyan Houses, Summit Mount Washington, and Crawford House, via Littleton, Bethlehem, and Lancaster, and Stages via Pemigewasset Valley and Franconia Notch, for the various points of interest in the mountain region.

C. M. MORSE, Manager.

BOSTON, CONCORD, MONTREAL,

AND

WHITE MOUNTAINS R.R.

The only Rail Route to Summit of Mount Washington.

NOW OPEN TO THE BASE OF MOUNT WASHINGTON AND TO NORTHUMBER-
LAND ON THE GRAND TRUNK RAILWAY.

THE SHORTEST, QUICKEST, AND BEST ROUTE

TO THE

White and Franconia Mountains, Montreal, and Quebec.

This is the only line running Day Palace Cars and Express Trains between Boston, Providence, Worcester, New London, Stonington, and the White and Franconia Mountains. This Line, passing, as it does, up the valley of the Merrimac and Connecticut Rivers, through the Cities of Lowell, Lawrence, Nashua, Manchester, and Concord, and along the borders of Lake Winnepesaukee for 30 miles, terminating at the foot of Mt. Washington, passing River, Lake, and Mountain Scenery unequalled in New England, and in the immediate·vicinity of the principal Hotels and Summer Boarding-Houses in the Northern part of the State, makes the popular travellers' route for Tourists visiting the

Lake and Mountain Scenery
OF NEW HAMPSHIRE,

For further information regarding time, connections, tickets, &c.,
see the principal R. R. Guides, or apply to

No. 5 State Street, Boston,
And the principal Offices on the Line.

☞ See Guides.

J. A. DODGE, Sup't, Plymouth, N. H.

(PROFILE HOUSE AND ECHO LAKE, FRANCONIA NOTCH, WHITE MOUNTAIN S. N. H.)

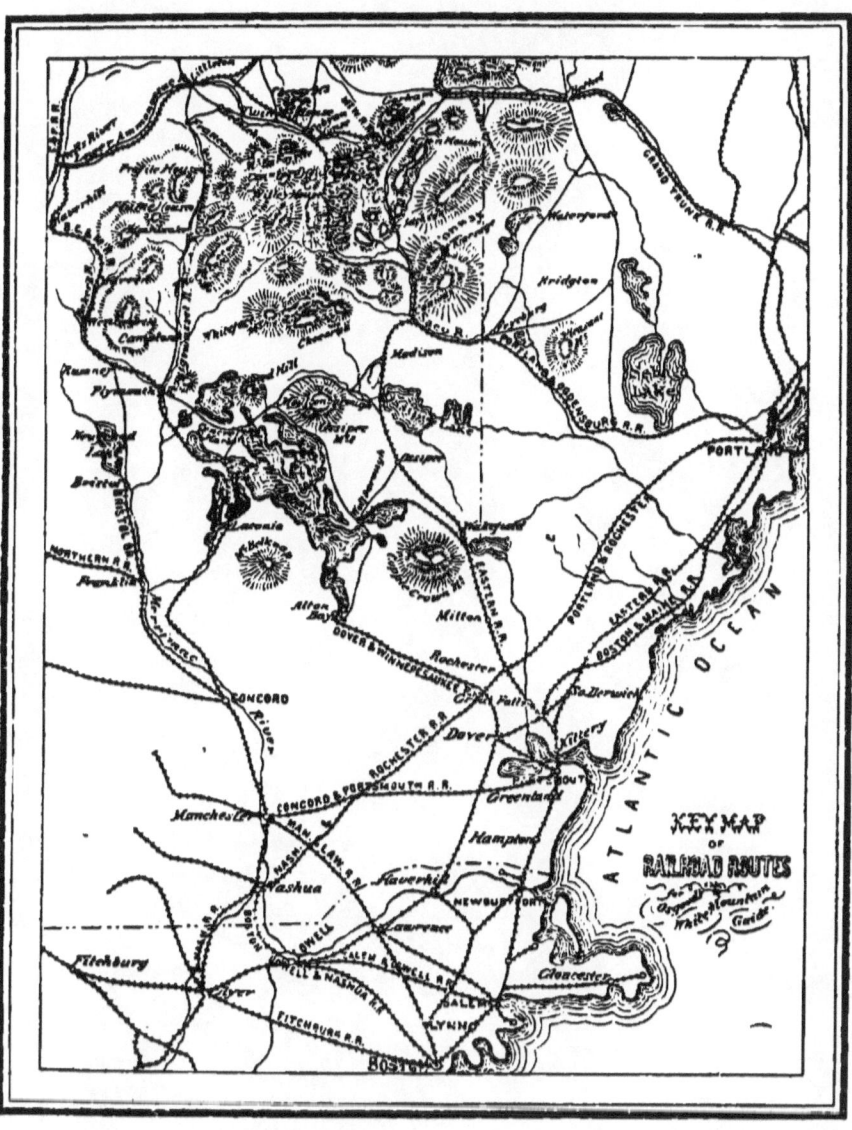

www.ingramcontent.com/pod-product-compliance
Lightning Source LLC
Chambersburg PA
CBHW032017110726
47901CB00004B/1118